D1383816

Men and Masculinity

A Text Reader

THEODORE F. COHEN
Ohio Wesleyan University

Australia • Canada • Mexico • Singapore • Spain • United Kingdom • United States

Sociology Editor: Lin Marshall
Assistant Editor: Dee Dee Zobian
Marketing Manager: Matthew Wright
Editoral Assistant: Analie Barnett
Project Editor: Jerilyn Emori
Print Buyer: Karen Hunt
Permissions Editor: Robert Kauser
Production Service: Hockett Editorial Service
Copy Editor: Sheryl Rose
Cover Designer: Stephen Rapley
Cover Image: Tony Stone
Compositor: ColorType, San Diego
Text and Cover Printer: Webcom, Ltd.

Printed in Canada
1 2 3 4 5 6 7 04 03 02 01 00

For more information, contact
Wadsworth/Thomson Learning
10 Davis Drive
Belmont, CA 94002-3098
USA
http://www.wadsworth.com

International Headquarters
Thomson Learning
International Division
290 Harbor Drive, 2nd Floor
Stamford, CT 06902-7477
USA

UK/Europe/Middle East/South Africa
Thomson Learning
Berkshire House
168-173 High Holborn
London WC1V 7AA
United Kingdom

Asia
Thomson Learning
60 Albert Street, #15-01
Albert Complex
Singapore 189969

Canada
Nelson Thomson Learning
1120 Birchmount Road
Toronto, Ontario M1K 5G4
Canada

Library of Congress Cataloging-in-Publication Data

Men and masculinity: a text reader / edited by Theodore F. Cohen.
p. cm.
Includes bibliograhical references.
ISBN 0-534-53658-1
1. Men. 2. Men—United States. 3. Masculinity. 4. Masculinity—United States. 5. Sex role—United States. I. Cohen, Theodore F.
HQ1090 .M4282 2000
305.31—dc21 00-025739

This book is printed on acid-free recycled paper.

Contents

Preface

Over much of the past four decades we have witnessed an explosion of interest in gender issues. As a result, we now recognize how important gender is in both the most personal and the most public dimensions of social life. The earliest attention to gender was directed at women's lives both because there was more serious neglect of women's lives than of men's and because of the political and economic inequalities from which women suffered. Looking closely and critically at women's experiences led to a thorough investigation of how embedded gender is in social institutions and relationships. By having ignored the significance of gender we had, in fact, also failed to examine those privileges and problems that were common to men, and defined and shaped male experience. Thus, the reanalysis of women's lives ushered in an eventual reanalysis of men's lives. Now, men were cast not as some generic category whose experiences define human experience but as a gender, subjected to role expectations and institutional constraints as real as, though often quite different from, those to which women are subjected. For the past twenty-five years we have been exploring how men are affected by masculinity.

Men and Masculinity: A Text Reader emerges out of this context. It is a product of my twenty years of experience teaching about and researching men's lives. I designed it for courses in the sociology or psychology of gender, masculinity, or men's studies, or as "the book about men" in a more women-focused gender or women's studies course. Though improved, most gender textbooks still offer too little attention to men and masculinity or gloss over inconsistencies or discontinuities in male experience. Men's ex-periences sometimes feel "tacked on," and are neither as developed nor well integrated as the books' more dominating emphases on women's experiences. Furthermore, textbooks often inadequately reflect diversity in male experience, even when emphasizing diversity in women's experiences. Thus, *Men and Masculinity: A Text Reader* could usefully supplement such texts by revealing how men's experiences fit within the more women-focused coverage of text material.

THEMES AND ORGANIZATION

Men and Masculinity: A Text Reader looks at the roles and relationships that make up men's lives and the factors that shape men's experiences within those roles and relationships. In it I have gathered recently published material from books and academic journals, personal narratives, and chapters written specifically for this volume to illustrate how men's experiences are shaped by cultural notions of masculinity and structural realities of gender. Although the emphasis is on the experience of men in the United States, I include material about men from a variety of cultural backgrounds (e.g., Japan, Mexico, Puerto Rico, Iran, and New Guinea), and about American men of varying ethnic or racial backgrounds, socioeconomic statuses, and sexual orientations. Throughout the volume, I try to portray what men experience by trying to live within or challenge notions of expected male behavior.

The selection, organization, and discussion of readings in *Men and Masculinity* are built around the following themes.

1. Men's lives are the products of the gender socialization they have encountered and the structured opportunities and constraints associated with gender.

   Three important points follow from this first theme. First, gender more than sex explains why men experience what they do. I stress the experiences men have "becoming" and later "being" men, more than the biological characteristics of males versus females to explain men's experiences. Men's lives derive their shapes from the social statuses they occupy and the particular expectations they confront. Second, gender socialization is introduced early and emphasized throughout the book. Men are subjected to a variety of messages about what it means to "be a man." These vary across time and space, but in any given period and place they define what is expected and what will be accepted of male behavior. Third, men's lives are also shaped by the structural constraints and opportunities that define their roles in various social institutions. While the readings illustrate the socialization men undergo that shapes their beliefs and guides their behavior, I also point out the structured constraints and opportunities that limit men's actions. In this way, I put socialization into its proper place, as an important but not sufficient explanation of what men experience.

2. Men's lives are diversified by culture, class, race, and sexuality. In each section of the reader, selections are included that emphasize the wide range of male experience. Each chapter introduction includes a section, "men versus men," depicting important ways in which race, class, and/or sexuality differentiate men from other men.

3. Meaningful differences separate men's experiences from women's. Careful not to exaggerate the extent of difference in men's and women's lives, the selections and introductions nonetheless depict what men experience as boys, friends, husbands, fathers, workers, athletes, criminals—because they're men, and not women. Thus, each chapter introduction contains a section, "men versus women," where important differences and similarities between men and women are highlighted.

4. Men's lives are a mixture of privilege and problems that result from being men in the circumstances that they face.

   Compared with women, men are enriched and empowered by structural inequalities of gender that reward them more extensively and offer them more opportunities to prosper. In most of the more popular gender textbooks these are the kinds of facts about men that define men's positions in society. While accurate and important, these are not the entirety of men's realities. Men's lives are also constrained and curtailed by efforts to embody conventional notions of masculinity.

Although men's rewards and costs are further differentiated depending on other structural matters (e.g., race, class, or sexuality), all men pay in ways for their various privileges. Implicit in this last statement is the suggestion that men would be better off if their ideas of acceptable roles and their visions of masculinity were modified. This is explored in the latter chapters of the book.

Organization

The book is divided into seven chapters. In Chapter 1, the readings illustrate four important elements in the conceptualization of masculinity and the sociological study of men's lives: the complex multidimensionality of gender, the cultural construction and variability of gender, the antifeminine and homophobic emphases in American masculinity, and the subcultural diversity among American men. Each of these elements resurfaces throughout subsequent sections and chapters of the book.

Chapter 2 illustrates instances of socialization, emphasizing the processes through which an individual's male identity and sense of masculinity are constructed. Examples of gender socialization include school, sports, and peers. The chapter also includes selections that reveal some problematic aspects of male socialization for males and females.

The next two chapters examine men's experiences as friends, husbands, and fathers. Cumulatively, the fifteen readings make this a major emphasis of *Men and Masculinity.* In depicting men's intimate relationships, selections include articles on Japanese and Iranian men, Puerto Rican fathers, and a gay male couple. Thus, even within the limited available space, the book portrays some of the diversity of men's experiences of intimacy. Because Chapter 4 looks at men's experiences both as partners and parents, it is the longest chapter in the volume.

Readings in Chapter 5 examine aspects of men's experiences of work, especially some of the prices men pay for failing, falling short, or striving to embody the male work role. It includes readings on the predicament of inner-city black men, the discriminatory atmosphere of the workplace for gay men, and chapters that look at a variety instances of "men without work" (e.g., retirement, unemployment, and at-home fatherhood).

Chapter 6, "Men In Trouble," consists of five readings describing some ways in which men's ongoing efforts to embody culturally appropriate versions of dominant masculinity lead to negative consequences. The readings deal with men's experiences of physical and mental health, and with connections among masculinity, suicide, and crime. Each reading demonstrates ways in which, as a result of the expectations imposed on them because they're men, they suffer shortened lives and diminished well-being.

The final chapter, "Movements of Men and the Politics of Masculinity," consists of four selections that reflect different perspectives on masculinity and men's lives. Readings look at the mythopoetic men's movement and the Christian Promise Keepers, black men, working-class masculinity, and some frustrations of a feminist man. Analytically, readings consider how circumstances such as class, race, and religious affiliation differentiate men's lives and the issues they deem important.

STYLE AND FEATURES

There are four key features that make *Men and Masculinity: A Text Reader* an attractive and useful volume. First, as a reader, most of the volume consists of selected articles and/or chapters from recent research into areas of men's lives. In selecting the readings, I reviewed books, journals, and popular magazines to find recent, relevant, and readable analyses and illustrations of aspects of men's experiences in their many roles and relationships. The research and theories about men and masculinity are ample and growing. I selected the readings because they readably represent some significant slices of men's lives.

In fact, the book is designed to be accessible to students with limited backgrounds in gender courses or in sociology. Important concepts are introduced and illustrated in the chapter introductions, making the readings that follow more relevant and comprehensible for even the inexperienced undergraduate reader. I have compiled readings and original material of a style and at a level that increase the probability that students will read the book.

Second, while mostly a collection of readings, *Men and Masculinity* includes longer chapter introductions than is typical of many edited collections. Although these discussions fall intentionally short of full-length textbook chapters (which might range between thirty and fifty pages), they do more than merely introduce the subsequent readings. Indeed, they place the readings into a broader context, where their substantive and theoretical relevance are established as their main points are highlighted. These introductions are intended to supplement, not replace, textbook coverage, thus, they neither rehash the theories and ideas covered by more general gender texts, nor require students and instructors to cover two overlapping textbooks. Still, by supplying some sense of important points not represented in readings, this anthology achieves wider coverage than is often possible in edited collections.

Third, to give flesh and feeling to the more analytical issues introduced in articles or introductions, the reader includes an assortment of narrative accounts drawn from my reading and research on men's lives. Selected for both their significance and their "readability," these accounts, along with the many fine qualitative research articles, illustrate the experiential dimension of otherwise more analytical issues.

Fourth, along with their purchase of this volume students get access to *InfoTrac College Edition,* Wadsworth's electronic database on the World Wide Web. *InfoTrac College Edition* contains literally hundreds of thousands of articles for students to locate, most of which can be read right on screen. This enables students to pursue topics that interest them further, to locate bibliographies for research projects, and to remain more up to date than any published volume can keep them. *InfoTrac College Edition* has been incorporated into this book at the end of each chapter introduction, where students will find additional readings recommended for follow-up on topics covered in chapter readings.

ACKNOWLEDGMENTS

Many people deserve to be recognized for the roles they played in the completion and production of this book. I thank the following reviewers, whose comments and reactions were encouraging and whose suggestions were helpful:

Marina Abalakina, New Mexico State University

Lynn Davidman, Brown University

Kevin Delaney, Temple University

Judith Howard, University of Washington

Mary Blair Loy, Washington State University

Elizabeth W. Markson, Boston University

Jodi O'Brien, Seattle University

Mark A. Winton, University of Central Florida

A number of people at Wadsworth Publishing Company deserve thanks. Susan Badger, President of Wadsworth, started me thinking about this project when she visited my campus four years ago and first mentioned the idea for such a book. Eve Howard, Publisher, has shepherded this project from its initial prospectus through its multiple revisions. Her helpful advice and her confidence in the project were especially appreciated. I also wish to thank Dee Dee Zobian, Assistant Editor, for her advice, patience and persistence, in seeing that the various pieces all came together as and (close to) when needed. Before Dee Dee, Ari Levenfeld was similarly helpful. I depended much on them throughout the process of writing, revising, editing, and publishing. Thanks, too, to the permissions department at Wadsworth. Rachel Youngman of Hockett Editorial Service did a very effective job copyediting and processing the volume. Jerilyn Emori, in production at Wadsworth, is to be thanked for managing the production process and providing us with an early bound book date.

At Ohio Wesleyan University, a number of people deserve thanks for providing feedback, support, or sympathy through the juggling act of writing and teaching. The Faculty Personnel Committee and Provost William Louthan generously provided me with a special leave during which this project got started. Jan Smith, Mary Howard, Akbar Mahdi, Jim Peoples, John Durst, and Janice Schroeder make my department a warm and supportive place to work. I appreciate their interest and support. Thanks, too, to Akbar, Jim, John, Kim Dolgin, and Gary DeCoker, for their contributions to this volume. Janice Schroeder was especially helpful through the frenzy of activity accompanying the difficult tasks of mailing, duplicating, faxing, rescheduling appointments, and so on, that were necessary with this project. She also helped me find humor in the midst of chaos. I also appreciate the assistance

I received from the staff in the mail room and duplicating, and from Tom Green, Assistant Director, Beeghly Library. I benefited greatly from the efforts of some students: John Morgan and Richard Nightingale enthusiastically agreed to contribute their ideas and allow me to use their experiences and insights as selections in this book; Tim Stanos and Zach Markowitz were helpful in assisting me with my efforts to solicit, organize, and complete copyright permissions.

I also express my appreciation to my family: My parents, Kalman and Eleanor Cohen, and sisters, Laura Cohen and Lisa Merrill, have been supportive and sympathetic. My children, Dan and Allison, are joys to be around and inspirations to me in both my personal and professional life. Someday I hope they get to experience the same pleasure they have given me. Finally, I owe the greatest debt to my wife, Susan J. Cohen. Through a quarter of a century together, she has been my best friend. She shoulders more than she should with less credit and recognition than she deserves. Without her this book would not have been possible. More important, without her it would not have been worth doing.

Men and Masculinity

1

Just Because They're Men

This book explores what men experience in their lives *just because they're men*. The assembled readings illustrate men's experiences in different domains and relationships in their lives. Along the way, we will consider both the broad diversity among men and the similarities and differences between men's and women's lives.

As more than two decades of very fine research and writing have shown, culturally and historically specific ideas about what men are supposed to do and be like affect men. Just as women face culturally produced models of femininity, men are measured against standards of acceptable attitudes and actions for them as men. Such notions of permissible masculinity or femininity are part of the culture into which we all are socialized. None of us completely escapes them.

Have you ever stopped to consider how much your life has been shaped by your being male or female? If you are female, chances are you probably have thought about such issues. The opposite is true for males. Aided by a visible feminist movement and a vocal societal discussion of how women's experiences and opportunities are produced and shaped, we are much more aware of how and why women's lives are what they are. But women are not the only gender. Although men's lives are affected by gender, there is a tendency among men to assume that it's not gender, it's life (i.e., "life is just like that"). Over several years of studying, teaching, and researching gender issues, I have learned otherwise.

In short, men experience what they do, play the roles that they play, and have the kinds of relationships that they form *because they are men*. Think about it. Men are measured against cultural ideas regarding expected and accepted masculinity that lead them to develop particular qualities, behave in distinctive ways, make certain choices, and maintain specific kinds of relationships. Conversely, they are expected to avoid particular situations and suppress certain emotions that might otherwise cause them to fail to measure up to the predominant model of masculinity.

Talking about a "predominant model" of masculinity acknowledges the existence of "multiple masculinities," and the eventual dominance of specific versions of masculinity over others (Connell 1995; Kimmel and Messner 1998; Messerschmidt 1993). Models of masculinity are both dynamic and culturally variable. They change over time (Kimmel 1996), differ across space (Gilmore 1990), and—within a given time and place—are challenged for cultural dominance by those who advocate other versions of masculinity (Kimmel 1994; Connell 1995). Out of a multiplicity of possible masculinities, one comes to be accepted as the model from which men are expected to derive their sense of what "being a man" means (Connell 1995; Messerschmidt 1993). As Robert Connell observed, "At any given time, one form of masculinity rather than others is culturally exalted" (Connell 1995). The triumphant "hegemonic masculinity" then affects both those who accept and those who reject it, because both will be measured against its mandates.

Men's lives are very different from women's lives, in part because through their upbringing they are exposed to the many expectations about how they should act, speak, think, and feel. Moving through life, they confront both opportunities and obstacles associated with their gender. Emphasizing "upbringing" and "opportunities and obstacles" indicates that men's lives are products of the socialization they've experienced, the statuses they occupy, and the subsequent roles they play. Men's lives are shaped by cultural and social structural forces that act upon them because they're men. The readings that follow illustrate the inevitable importance of gender in men's lives.

LIVING GENDERED LIVES

To heighten your sensitivity to the topics and ideas that follow in this book, let's try a little exercise that I use with students in my gender classes. Close your eyes and think carefully about what you believe life must be like for the "opposite sex." Then complete the following statements:

1. "The best thing(s) about being a (male/female) in this society must surely be . . ."
2. "The worst thing(s) about being (male/female) in this society must surely be . . ."

There is one catch: Males reading these questions must answer them about being female, and females must answer about being male.

Every time I use this classroom exercise it generates lively, sometimes heated, occasionally comical discussion. If you are like most of my students, your responses are something like those that follow. Understand that these are not the "correct answers." They are just the ones my students most often mention, summarized and rephrased here.

<u>Women's Views of the Best and Worst Things About Being Male</u>

Best

- Higher pay; access to higher status positions
- Respect
- Freedom of movement; less fear about safety and less concern about the possibility of rape
- No monthly periods, no PMS, no cramps
- Not having to worry about pregnancy, childbirth, and childcare
- Less concern for one's appearance

Worst

- Need to be stoic and emotionally strong, and tendency to be inexpressive
- Pressure to be the breadwinner or devote oneself to work
- Depleted emotional and physical health; earlier death
- Restricted intimacy (including relations with children and loss of custody after divorce)

<u>Men's Views of the Best and Worst Things About Being Female</u>

Best

- Freedom to show feelings, especially feelings of vulnerability or of affection toward friends and loved ones
- Ability to choose—though within limits—whether to seek employment outside the home or raise children
- Depth of connection felt in relationships
- Giving birth
- Longer life span

Worst

- Being sexually harassed, assaulted, or objectified
- Being judged by and related to so much in terms of one's appearance
- Being underpaid, occupationally segregated, discriminated against
- Having so much of the responsibility for children, families, and households
- Having to think so much about one's safety and potential vulnerability
- Being trivialized in conversation and patronized in relationships

No matter how closely your own ideas matched those above, the task itself is an instructive one. Importantly, it is often difficult to do it well. Some are unwilling to take it seriously, preferring lighter, more comical answers, like longer or shorter bathroom lines (which may not be comical under desperate circumstances). Beyond that, though, people often find it hard to look at life through the eyes and experiences of others whose lives seem fundamentally unlike their own. This difficulty extends beyond gender; one could do a similar exercise using race, class, or sexual orientation as the relevant variable. As noted above, we come to take for granted that life is "just this way," without seeing the extent to which others live quite different lives. We think this way especially when we find ourselves in a privileged position—a member of a superior class, majority religion, accepted sexual orientation, or dominant gender. Those who are on the "wrong side" of such hierarchies possess a sharper awareness of the differences and their causes and consequences (Goode 1992; Kimmel and Messner 1998).

This exercise also reminds us that we are all both benefited and harmed because of our being male or female. In fact, comparing or combining the "best" and "worst" lists reveals an interesting cross-gender consensus. Specifically, men's ideas of the "best" things about being female (e.g., the ability to express feelings) are echoed in what women report must be the "worst" things about being male (i.e., not experiencing these things). Thus, both genders are aware of the same differences and perceive them similarly: Having to be emotionally strong, stoic, and inexpressive are among the "worst things" about being male. Similarly, the things women perceive as the "best" about being male (e.g., higher pay) resurface in men's answers about the worst things about being female (e.g., being underpaid or discriminated against). Some differences are exaggerated or oversimplified, but, when pressed, we identify a host of ways in which we "win" or "lose" simply because we are male or female. The readings that follow examine some ways that men in this society are benefited and harmed just because they are men.

To further your sense of the constraining and enabling effects of gender, answer this last question that I also ask my students:

3. Imagine and describe how your life would differ if you had been born and were living as the opposite sex.[1]

Try to get a good picture in your head of yourself and of your life as this "other." Would you be sitting wherever you are now sitting reading this book? Would you be enrolled in the same courses, at the same college? Would you have the same interests, participate in the same activities, be more artistic, more athletic, more confident, or more cautious? Who would your friends be, and what would you do together? Would your life goals be the same as they are now? All these questions illustrate the magnitude of gender's effects in our lives. If you found yourself imagining a different life than what you have lived thus far, what would have created those differences? My guess is that you would stress (a) the ways you were raised, and (b) the chances you got or didn't get (to play certain sports, go to certain schools, take certain classes, and so on). These capture the importance of the different socialization processes and opportunities directed at and, ultimately, differentiating men's and women's lives.

We should avoid exaggerating the differences between men and women, because in many ways the genders are more similar than different (Basow 1992). Still, the readings that follow depict some of what men experience as boys, friends, husbands, fathers, workers, and so forth, simply because they're men and not women. Chapter introductions briefly compare men's and women's experiences; the readings then illustrate some of men's experiences.

Having emphasized differences between genders, we should note how much variation there is among men. Because gender is differently conceived across cultures, men in the United States confront different expectations and opportunities than do men in other societies. The range of cultural diversity is broad, including the number of recognized gender categories, what men and women are expected to do or are presumed to be like, and how young males and females become adult members of their respective genders within their cultural contexts. These issues are nicely addressed in the readings by James Peoples (Reading 1) and Judith Lorber (Reading 2).

Throughout this volume, you will encounter men from a variety of cultures and subcultures, including men from Asia, Pacific Islands, North and South America, and the Middle East. Their lives take shape because of their attempts to embody specific cultural expectations of them as men. We will look explicitly at other important differences that exist among American men. Obviously, being a white, middle-class, middle-aged, heterosexual male is different from being a black male, a working-class or upper-class male, an elderly male, or a gay male. All men don't share the same fate or fortune. Thus, another major emphasis of this book is to illustrate some distinctions in men's lives. In Chapter 1, Alfredo Mirandé illustrates diversity among men by depicting how our conceptions of masculinity vary between groups. Subsequent readings highlight other differences among men across the boundaries of race, social class, ethnicity, age, and sexual orientation. Ideally, this will broaden our eventual understanding of men's lives.

ON THE COMPLEXITY OF GENDER

As indicated in Lorber's "Night to His Day" (Reading 2), gender is comprised of several separate components. As men or women, we simultaneously experience the biological, psychological, social, cultural, political, and economic realities associated with being men or women. Of these, the biological characteristics that designate one's sex category (e.g., chromosomes, hormones, and genitals) represent only one important component of an individual's experience of being gendered.

Of obvious consequence, physical characteristics determine whether one is labeled male or female at birth. This *gender assignment* then initiates a lifelong experience of being assigned roles and confronting expectations based on one's sex category. Additionally, meaningful affective, cognitive, or behavioral consequences may result from some of the physiological differences between the sexes, such as prenatal or postpubertal hormone levels, which affect neuroanatomy, cognitive ability, and longevity (Hoyenga and Hoyenga 1993).

Further interesting attempts have been made to connect behavioral and/or temperamental differences between the sexes to the different biological imperatives and reproductive strategies that evolved among humans. In seeking to maximize their reproductive success, males and females are driven in different directions: males toward spreading their genes by impregnating many females but investing less into each offspring; females toward making sure that each of the offspring receives sufficient care, protection, and provision. We can't say for certain whether and how these different circumstances might explain other gender differences (e.g., in such things as infidelity, violence, competitiveness, father's involvement), but theoretically such connections may exist (Hoyenga and Hoyenga 1993). Men's lives, then, are shaped in part "because they have an abundance of testosterone," just as they are shaped "because they have an XY chromosome," "because they possess a penis," or "because they impregnate (but don't menstruate, gestate, or lactate)."

Of course, whatever role physiological factors play, they are joined by other, more social and cultural processes, which shape our experiences as women or men. Social scientists refer to these as *gender.* Gender consists of a variety of components, beginning with *gender identity,* the conception you have of yourself as a male or a female. Gender identity may share center stage with other elements of your self-identity (e.g., race or class), but it plays a prominent role in the drama of your life. How you see yourself, the relationships that you enter and enjoy, the attitudes and aspirations that you form, the labor that you pursue, and the leisure that you seek, are all guided by your sense of yourself as a female or a male.

Gendered roles reflect the more sociological dimension of gender. Men's lives are shaped by the social statuses they occupy and the expectations they confront as men. Sociologists use the concept of *role* to refer to prescriptions and proscriptions (i.e., do's and don'ts) for behaviors that accompany the various statuses that we occupy (student, parent, teammate, and so on). Although sounding a bit deterministic, "gender roles" cover the many expectations about which behaviors are appropriate, acceptable, or anticipated from a person in a given setting or situation because that person is male or female. These expectations include interests, temperaments, and behaviors.

Even the most ordinary activities—how we walk, sit, tilt our heads, smile, make and maintain or avoid and avert eye contact—are "scripted" differently for males and females. *Gender display* points to the variety of ways in which we reveal, through our verbal and nonverbal demeanor, that we fit in with masculine or feminine ideals. This highlights the interactionist dimension of gender, in which gender is, in part, a performance and a "social construction" (Goffman 1976; West and Zimmerman 1987).

Gendered roles are socially acquired through processes of socialization and interaction. There are sanctions accompanying the various prescriptive and proscriptive norms connected to gender. We are rewarded when we do and punished when we don't act accordingly. These sanctions vary and change, as do the expectations themselves. The organization of this volume emphasizes men's experiences performing a variety of roles (most notably friend, husband, father, worker, athlete, patient, and criminal). These are not exhaustive of all the roles men play in their lives, but they encompass much of male experience.

At the cultural level, we can speak of *gender ideals* to refer to the shared beliefs or models of gender that a majority of society accepts as appropriate masculinity or femininity. There are a variety of useful versions of such cultural constructions of American masculinity (Brannon 1976; Gerzon 1992; Doyle 1994; Kimmel 1996; Connell 1995). Two of the better examples are introduced by Michael Kimmel in "Masculinity as Homophobia" (Reading 3). As Kimmel reports, one of the more popular versions of American masculinity is Robert Brannon's four-pronged description. Brannon (1976) argued that men in the United States are expected to embody characteristics that fall under four broad and powerful themes: display *no sissy stuff;* be a *big wheel* and *sturdy oak,* and, when necessary, *give 'em hell.* These speak, respectively, to the fact that men learn to see masculinity as entailing the avoidance of anything feminine and the display of one's competitiveness, dependability, stoicism, and aggressiveness.

Kimmel's own model of "marketplace masculinity" is a second excellent depiction of how we conceive of masculinity in this society at this point in time. In "marketplace masculinity," or "self-made manhood," men's identities are derived "entirely from . . . activities in the public sphere, measured by accumulated wealth and status, by geographic and social mobility" (p. 17) and by their successful displays of attributes such as aggression and competitiveness. It requires that men constantly prove and display their fitness in such visible tests of their manhood (Kimmel 1996, 23). Both Kimmel and Brannon capture the gender expectations against which most American men are measured.

Finally, the most political dimension of gender is *gender stratification,* the differential access men and women have to desired rewards and resources such as occupations, prestige, wealth, or power. As with stratified social classes (upper, upper-middle, lower-middle, and so on), males and females have different levels of economic well-being and access to power. Compared with women, men earn more, move faster up the occupational ladder, and dominate political office. In interpersonal relationships they enjoy greater decision-making influence and exercise more control over their female partners than those partners exercise over them. More of women's experiences result from what men have done or might do than vice versa. In all of these ways, men are the more privileged and powerful gender.

However, unlike the relationship between social classes, men and women are more diversified. Some women are much better off economically than many men. Although the more desirable occupations and positions of power may be monopolized by men, most men neither hold such occupations nor perceive themselves to be particularly powerful. Although they benefit from the fact that this society has been dominated and shaped by men, men do not all directly or equally enjoy high levels of wealth and power. This drives us back into the intersections of race, class, sexuality, age, and gender. Certain men (e.g., white more than black, upper and upper-middle-class more than working and lower class, heterosexual more than gay) reap more benefits from the nature of sex stratification in our society. Other men pay great costs, despite being men, within this same sex-stratified society. Acknowledging that "men pay too" for the kind of society in which we live neither discounts nor denies women's costs. Simply, men and women pay in measurable if different ways for their positions in a male-dominated society. Similarly, recognizing some of the costs paid by men does not diminish the fact that, as a category, men enjoy relative advantages over women.

Many of the readings that follow reflect the fact that men's lives are also constrained by efforts to embody conventional notions of masculinity. Men die younger; commit more suicide, homicide, and crime; suffer less intimate relationships with their friends and families (especially their children); and identify with or are identified narrowly within their roles as workers. Although men's rewards and costs are further divided depending on matters like race, class, or sexuality, all men pay in ways for their various privileges. We will examine some of the benefits men claim and the costs they pay in their health and well-being, and the positive and negative experiences they have in domains such as friendship, family, and work.

WHAT FOLLOWS

Throughout this book you will find a variety of accounts that depict men's experiences in a variety of domains. These selections come from published and unpublished research across a range of disciplines. Some were originally chapters in books or articles in journals, others are research original to this volume. Alongside the scholarship you'll read from sociology, anthropology, and psychology, you will find many more personal narratives and reflections on areas of men's experiences covered by the more analytical material. These help remind us that though it exists on cultural, economic, and political levels as well, gender is deeply and profoundly personal. Covering the major domains of friendship, family, and work, and looking specifically into areas that are central in men's lives, we will examine how men experience these areas *just because they're men.*

Expanded Contents: Searching the Literature

Beyond the many selections in this volume, students can also access additional writings on men and masculinity through *InfoTrac College Edition,* Wadsworth Publishing Company's electronic database on the World Wide Web. By visiting Wadsworth's Virtual Society web site (http://www.sociology.wadsworth.com) and logging on with the password that came with the purchase of this book, you have twenty-four-hour access to more than 800,000 articles from more than 600 scholarly journals and popular periodicals. Enter the desired search term (i.e., *masculinity*) to gain access to numerous substantive articles (not just abstracts) on men and masculinity. Your search can be further organized by subtopic or subdivision (e.g., *research, public opinion, personal narratives*), either to follow up on topics that you found interesting or to investigate areas of men's lives that are not included in this volume.

InfoTrac College Edition greatly expands your opportunity to learn about men and masculinity. Although I encourage you to search through this literature as you see fit, at the end of the introduction to each chapter of this book you will find titles of recommended articles related to the readings in that chapter. These are by no means the only relevant articles to read or the only connections to make between *InfoTrac College Edition* and readings in this volume. Instead, they are examples that I thought were nicely connected to the selections in this book. In doing your own searches, you will undoubtedly locate many others. If you make effective use of this database, the resulting coverage of men and masculinity will be fresher and broader than would otherwise be possible in any single book.

NOTE

1. After reading the selections by Lorber and Peoples you may have a much deeper awareness of the limitations associated with a notion as taken for granted as "opposite sex." This construct presumes (a) that there are only two possible categories, male and female, and (b) that occupants of these two categories are polar opposites. These ideas are not universal.

REFERENCES

Basow, S. *Gender Stereotypes and Roles.* Pacific Grove, CA: Brooks/Cole, 1992.

Brannon, R. "The Male Sex Role: Our Culture's Blueprint of Manhood and What It's Done for Us Lately." In *The Forty-Nine Percent Majority: The Male Sex Role,* edited by Deborah S. David and Robert Brannon, 1–45. Reading, MA: Addison-Wesley, 1976.

Doyle, J. *The Male Experience.* Dubuque, IA: William C. Brown, 1994.

Connell, R. *Masculinities.* Berkeley: University of California Press, 1995.

Gerzon, M. *A Choice of Heroes.* Boston: Houghton Mifflin, 1992.

Gilmore, D. *Manhood in the Making.* New Haven: Yale University Press, 1990.

Goffman, E. "Gender Display." In *Studies in the Anthropology of Visual Communication, vol. 3, 69–77. 1976.*

Goode, W. "Why Men Resist." In *Rethinking the Family,* edited by B. Thorne with M. Yalom, 287–310. Boston: Northeastern University Press, 1992.

Hoyenga, K., and K. Hoyenga. *Gender Related Differences: Origins and Outcomes.* Needham Heights, MA: Allyn and Bacon, 1993.

Kilmartin, C. *The Masculine Self.* New York: Macmillan, 1994.

Kimmel, M. *Manhood in America: A Cultural History.* New York: The Free Press, 1996.

Kimmel, M., and M. Messner. *Men's Lives.* New York: Macmillan, 1998.

Kimmel, M. "Masculinity as Homophobia" in *Theorizing Masculinities,* edited by H. Brod, Newbury Park, CA: Sage, 1994.

Lorber, J. *Paradoxes of Gender.* New Haven, CT: Yale University Press, 1994.

Messerschmidt, J. *Masculinities and Crime.* Lanham, MD: Rowman and Littlefield, 1993.

West, C., and Zimmerman, D. "Doing Gender." *Gender & Society* 1, no. 2 (June 1987): 125–151.

FOR ADDITIONAL READING

For another example of how definitions of masculinity (and femininity) vary between cultures, read "Masculinity and Femininity in Japanese Culture: A Pilot Study," by Yoko Sugihara and Emiko Katsurada. This study, originally published in *Sex Roles: A Journal of Research,* April 1999 vol. 40 #7–8, examines masculinity and femininity among 265 Japanese college students. Surprisingly, despite Japan's tradition of sharply differentiated gender roles, no significant differences were found between the male and female students in either masculine or feminine qualities (as measured by the Bem Sex Role Inventory).

For further consideration of variation *within* the United States, Pedro Noguera's article, "Reconsidering the 'Crisis' of the Black Male in America," from the Summer 1997 issue of *Social Justice* (vol. 24, #2), looks at the status of black males, and at the intersections of race, class, and gender in shaping those experiences. Among other things, he illustrates why no unilateral characterization of the status and well-being of black men is adequate, as there is considerable variation among African American men, especially across socioeconomic lines.

Finally, for additional reading on aspects of hegemonic masculinity, read "Negotiating the Male Body: Men, Masculinity, and Cultural Ideals," by Chris Wienke (from *The Journal of Men's Studies,* Spring 1998, vol. 6 no. 3). Wienke examines the "muscularity" and fitness associated with the ideal male body, connecting it to dominant constructions of masculinity and the popular culture sources that help disseminate these images. Drawing upon twenty intensive interviews with male college students, he then explores how these images affect men, looking both at men who do and don't "measure up" to the idealized standards.

WADSWORTH VIRTUAL SOCIETY RESOURCE CENTER

http://sociology.wadsworth.com

Visit *Virtual Society* to obtain current updates in the field, surfing tips, career information, and more.

INFOTRAC COLLEGE EDITION: EXERCISE

Go to the *InfoTrac College Edition* web site to find further readings on the topics discussed in this chapter.

1

The Cultural Construction of Gender and Manhood

JAMES G. PEOPLES

Like scholars in other fields, social scientists are affected by the current events, mass media, political ideologies, and intellectual fashions that take place in our own countries. As society changes, anthropologists, psychologists, and sociologists change also, to accommodate new interests, problems, ways of thinking, and political realities. Thus, the types of research projects we tackle, as well as the approaches and assumptions that guide our teaching and interactions with students, are in part a response to our own social and cultural experiences.

Since the 1960s and 1970s, one of these "experiences" has been the feminist movement, one of the most important political and ideological movements of the late twentieth century. True, a few North Americans still long for the days when a man was a man and a woman knew her place, but in the last three decades feminism has changed the lives of millions of women and men in a multitude of ways. More than ever before, we see women/wives in boardrooms and men/husbands in kitchens. These kinds of changes are often called *social changes:* They represent changes in the way both sexes relate to one another and to society at large.

Accompanying these social ("relational") changes are changes in the ways people think about the sexes. Many changes in attitudes about and perceptions of women and men are revealed in popular culture and, especially, in the popular media. It is now quite common for women to play the hero in American action/adventure films, while in many romantic comedies we see Sensitive New Age Guys learning how to get in touch with their feminine side. These kinds of "ideational" changes are *cultural:* They represent changes in how women and men view themselves, how members of each sex view the other, and how everyone views the views of everyone else about the sexes. In other words, our (cultural) *conceptions* of maleness and femaleness, of masculinity and femininity, of manhood and womanhood, and even of boyhood and girlhood have altered, just as have the *actual* (social) *relationships* between females and males.

These and other changes in cultural conceptions about and social relations between the sexes have had many impacts on scholarly research in the social sciences, biological sciences, and humanities. Some argue that the raw, biological differences between women and men hardly matter beyond the two unavoidable facts of reproductive contributions (a woman supplies the egg, a man the sperm) and lactational capabilities (women can, men can't). Biologically, men and women are different, such scholars argue. But out of the raw material of these biological distinctions, people living in different times and places fashion their conceptions of maleness and femaleness in an enormous variety of ways.

Certainly, different cultures reveal varying conceptions of masculinity and femininity. Thus, on the Micronesian island of Chuuk, young men

Previously unpublished. Used by permission.

in their late teens and twenties commonly get drunk and demonstrate their manliness to women, as well as to one another, by fighting (Marshall 1979). In contrast, among the Semai, a tribal people of Malaysia, men see themselves as nonviolent and rarely engage in physical violence. Two men engaged in a quarrel seek a mediator and try to avoid each other, for even to look at each other causes embarrassment (Dentan 1968, 56–7).

Cultural anthropologists, who compare cultures in order to learn about such differences, have investigated hundreds of peoples from all parts of the world. One goal of their research is to address issues such as the extent to which biological differences between the sexes are recognized and made relevant for social life in various cultures. Anthropologists have found that in all known cultures a person's sex is relevant to his/her self-identity, social roles, access to valuable resources, and overall status. Everywhere, your sex matters for who you think you are, what you have, how you interact, and what you can become. But studies that compare cultures to one another have also shown that a person's sex matters in different ways and to different degrees in different cultures. Thus, your sex is much less relevant to your self-identity and status if you are an Aka pygmy of the central African rain forest (Hewlett 1992) than if you are a Yanomamo of the Amazonian rain forest (Chagnon 1997).

So cultures vary in how much importance they attach to whether a person is female or male and in the specific behaviors they expect from females and males. To emphasize such cultural variations, anthropologists and sociologists use the phrase *cultural construction of gender.* This concept relies on a conceptual distinction between *sex* and *gender.* An individual's sex is determined by biological inheritance—your genitals, associated hormonal frequencies, and secondary sexual characteristics (breasts, body size and musculature, and the like) are greatly affected by your genetic makeup. In contrast, gender is culturally defined, not strictly biologically determined. How males and females perceive and define themselves and each other, what it means to be a woman or a man, what roles are seen as appropriate for men and women—these and many other dimensions of femaleness and maleness are learned during socialization rather than fixed at birth. They are culturally variable, not constant across all cultures. Gender, then, in contrast to sex, is "culturally constructed," meaning that different cultures develop different ideas about males and females and use these ideas to define manhood and womanhood.

To illustrate some of the ways in which cultures vary in how they construct gender, consider some of the indigenous peoples of New Guinea. Lying just north of Australia, New Guinea is the second largest island in the world. Because of its history of colonization, the eastern half of the island now comprises the independent nation of Papua New Guinea, while the western half (called Irian Jaya) is a province of Indonesia. The indigenous people of New Guinea are today—as they were in the past—enormously diverse ethnically and linguistically. More than 500 languages are spoken on the island, and there are a roughly equivalent number of ethnic groups, each with their own customs and traditions. The modern nation of Papua New Guinea is, then, among the world's most culturally heterogenous countries.

One reason for such extreme cultural diversity is the rugged terrain of the interior of the island. The high mountains and river valleys of New Guinea's interior were among the last regions on earth to be contacted by Westerners. The first white explorers of the 1930s were astounded to discover millions of people living in the island's remote valleys. Nearly everywhere, New Guinea highlanders lived mainly by growing root crops and raising pigs. Women usually were responsible for most everyday work in the gardens and the care of pigs. Men helped with such tasks also, but more of their time was spent in protecting their community from attack by enemies and in organizing large-scale exchanges of pigs and traditional valuables with their neighbors. Trade networks often connected the people of one mountain valley with those of the valley just over the ridge. It was sometimes the case that people of one valley would simultaneously trade with, intermarry with, and be at war with people in the neighboring valley.

Among many interesting cultural features, highland peoples of the New Guinea interior are known for the unusual ways in which their cultures constructed gender. (Although not considered in this chapter, note that most lowland and coastal New Guinea people are different in their constructions of gender. In fact, anthropologist Maria Lepowsky [1993] describes one lowland people, the Vanatinai, who are remarkably egalitarian in regard to gender and who lack all the cultural beliefs I am about to describe.) Many of the hundreds of tribal peoples living in the highlands believe that, in many situations, women and men must be separated. The reason for such separation—and in some contexts *segregation* is a more appropriate word—is a widespread belief that females can pollute men. Above all else, men fear contact with women's sexual anatomy and, especially, with female menstrual discharges. Such contacts can cause men to sicken, grow old before their time, or even die. Anthropologist Mervyn Meggitt was one of the first outsiders to report on New Guinea beliefs about female contamination. Working among a people known as the Enga, he describes their beliefs about menstrual pollution in these words:

> Men regard menstrual pollution as truly dangerous. They believe that contact with it or a menstruating woman will, in the absence of counter-magic, sicken a man and cause persistent vomiting, turn his blood black, corrupt his vital juices so that his skin darkens and wrinkles as his flesh wastes, permanently dull his wits, and eventually lead to a slow decline and death. (Meggitt 1970, 129)

Beliefs such as these are common—but not universal—in the New Guinea highlands.

Obviously, if a culture inculcates in men a fear of certain kinds of contact with women, there will be many implications for the lives of people of both sexes. For example, many restrictions are placed on women's activities in most highland cultures. A woman must be careful to avoid accidentally making her husband sick by unintentionally polluting his food. In many cultures, women must travel on separate paths from men, lest a man unknowingly step on the secretions left by some woman and thus become polluted. During their menstrual period, women must temporarily leave the community and live in a special structure, often called a menstrual house. A woman who remained in the company of adult men during her period could be punished, for she was endangering male health. For their part, careful men would keep their distance from menstrual houses, lest they become contaminated and, hence, endangered by contact with menstrual discharges.

Social separation between the sexes, however, is not limited to just a few days per month. Commonly, wives and husbands do not even live in the same house! Rather, the married men of the hamlet or village live together in a large structure, appropriately called the *men's house.* Their wives and young children live in a separate house. Wives work hard in the gardens that supply food for themselves, their husbands, and their children. A wife harvests and cooks the food, and daily brings meals to her husband in the men's house. In most cultures, however, a woman may not actually enter the men's house, for this would be dangerous to the men who stay there. Men do visit and spend much time with their wives and children, of course, but in many cultures there are separate spaces in the wife's house for men and women. Men are afraid to sit in their wife's space, and a wife who was found in a place reserved for men could be punished.

The lives of males also are affected by how their culture constructs the genders. For one thing, men in many highland New Guinea cultures claim to be ambivalent about—or even afraid of—having sex. Sex commonly is said to weaken men by depleting their supply of semen, which in many cultures is seen as an important component of a man's "life force." Sex with a menstruating woman is almost universally feared and avoided, but sex with a woman at any time is potentially dangerous since it places the man in intimate contact with the most polluting part of the female anatomy. For another thing, men often have to go through painful bloodletting rituals to cleanse their bodies from female substances, as we shall see later. And, finally, after a certain age young boys are believed to be endangered by continued contact with their mothers. At around ten years of age (depending on the specific culture) boys are taken away from their mothers—by force if necessary—and removed from contact with all females for prolonged periods. They are instructed during special initiation ceremonies that women are dangerous and that interactions with females must be carefully controlled.

Obviously, highland New Guinea peoples are constructing gender differently than do most other cultures. To illustrate these constructions, we shall now consider three specific peoples in more detail. The first example is a people known as the Hua. Hua construct their gender categories on the basis of different substances inside female and male bodies, in addition to biological differences between the sexes. The second, a people called Awa, exemplify a common theme in New Guinea cultures: Young boys must be made into men by undergoing traumatic and painful initiation rituals in the company of their age-mates. The third people, known as the Sambia, believe that boys must acquire semen in order to grow up and become fully masculine, and they have rituals that accomplish this goal.

HUA: CROSS-CUTTING CATEGORIES

One New Guinea people illustrating the cultural construction of gender are the Hua. According to Anna Meigs (1976, 1984, 1990), who conducted anthropological fieldwork among the Hua in the 1970s, Hua culture constructs gender in ways that are unfamiliar to Westerners. Like the Enga and many other New Guinea cultures, the Hua believe in feminine pollution and contamination of males, which makes certain kinds of contact between the sexes dangerous to men. Women must take precautions to ensure that they do not unintentionally harm their husbands or other men by careless behavior. Men also must be careful to avoid situations that will harm their health and decrease their longevity.

The cultural basis for the male fear of females is the Hua belief that human bodies contain a life-giving substance (analogous, perhaps, to a "vital essence") that they call *nu*. Hua people regard *nu* as a physical substance, not a mystical or spiritual force. *Nu* exists in several forms. It is solid (flesh and bone), gaseous (breath), and liquid (blood, sweat, semen, female sexual secretions). *Nu* can be transferred from one person to another in many ways. Transmission of any form of *nu* between individuals may be either harmful/weakening or healthful/invigorating, depending on the nature of the relationship between the donor and the recipient. When a woman serves a meal to her husband, for example, the *nu* from her hands and secretions adheres to the food and is ingested by the man, to the detriment of his health. A woman transmits harmful, polluting *nu* to her husband when they have sexual intercourse. Intercourse is also damaging to a man because he contributes his scarce *nu* (breath, semen) to a woman during sex. She thereby gains strength and vitality at his expense. In these and other ways, when a woman comes into direct or indirect physical contact with a man, the *nu* from her body can pollute and weaken him, Hua believe.

Hua say that *nu* is present to varying degrees in different kinds of foods. Some foods are rich in *nu*, whereas others have less *nu*. By controlling their food intake, people can regulate the amount of *nu* in their bodies. Hua think that various events in individuals' lives require regulation of the amount of *nu* that is present in their bodies. For example, during pregnancy women need extra *nu*, so they eat lots of foods that are rich in *nu*. (Broadly speaking, these are fast-growing foods with high moisture content.) Conversely, women who want to reduce their menstrual flow do so by avoiding *nu*-rich foods.

As for men, they tend not to have enough *nu* for complete growth and maturation. Therefore, during certain stages of boyhood they are encouraged to eat foods that are believed to contain lots of *nu*. But at other times, such as when they are undergoing initiation ceremonies, boys are supposed to rigorously avoid *nu*-rich foods, because such foods will weaken them. In fact, during initiation ceremonies there are strict taboos not only on the eating of foods high in *nu* but also on eating foods grown in gardens tended by women or foods prepared by women. These taboos help to keep the vulnerable boys free of feminine pollution.

In the Hua construction of gender, another difference between men and women is in the amount of *nu* a body naturally contains. Female bodies normally contain an excess of *nu*. In Hua belief, excessive *nu* makes women grow faster and age more slowly than men, but it also makes women unattractively moist. Male bodies naturally contain a smaller amount of *nu*. So men have difficulty in growing (which is why boys sometimes need to eat *nu*-rich foods) and in maintaining youthful vitality later in life, but they are attractively dry and hard. Hua explain many of the differences between the sexes by relative amounts of *nu:* Men are stronger and fiercer because they are drier, for example.

In sum, Hua culture constructs gender categories (male/female) on the basis of beliefs about the quantity of *nu* in a person's body, as well as on the basis of the outwardly observable physical differences between the two sexes. Two Hua beliefs have especially important implications: (1) *nu* helps make men and women different, and (2) *nu* is transferable between the sexes. According to the logic of these beliefs, people who are biologically female can become "like men," and people who are biologically male can become "like women." Thus, a man who has eaten *nu*-rich foods and who has picked up lots of *nu* from women over a lifetime of sexual intercourse can become "like a woman." For their part, women can lose *nu* through various means. Whenever a woman menstruates, she loses *nu,* so after menopause a woman has had so many periods that she is nonpolluting relative to younger women. Each childbirth also significantly drains her body of *nu*. And each time she prepares food, some *nu* from her body adheres to it, resulting in further losses. Thus, over her life, a woman tends to lose *nu,* and if this loss is great enough she can become "like a man."

But *nu* is what makes women polluting to men. And the greater the difference in the amount of *nu* between a woman and a man, the more dangerous that woman is to the man. Since men tend to gain *nu* as they age, and since women can lose large quantities of *nu* as they grow older, as both sexes age women tend to become less and less a threat to men. Thus, older men are less afraid of their female age-mates than are younger men of their female age-mates.

In fact, some older women lose so much *nu* that most restrictions on their behavior are lifted. Specifically, at each childbirth a fertile woman loses such massive amounts of *nu* that a postmenopausal woman who has given birth to three or more children is no longer a danger to men. The bodies of postmenopausal women who have given birth to three or more children have been so drained of feminine substances that they "are understood to have been defeminized," as Meigs (1990, 109) phrases it. Without these feminine substances such women are "like men."

How much "like men"? Enough "like men" that these (biological) females undergo "male" initiation ceremonies and can move into the men's house! Hua have a special verbal category to refer to such women. They are called *kakora,* which Meigs translates as "initiated person." The people who are *kakora* include:

- Males in their teens through about middle age who have been initiated. During their initiation, such men have avoided female foods as well as contact with women. They therefore have minimal quantities of *nu* at this stage of their lives.
- Postmenopausal women with three or more children. Each childbirth has drained them of most feminine substances, so they no longer pollute men.

People who are *kakora*—whether (genitally) male or female—are eligible to live in the men's house and to obtain the secret knowledge learned during "male" initiation ceremonies.

There is a corresponding term for people whose bodies contains lots of feminine substances.

They are *figapa,* translated by Meigs as "uninitiated person." *Figapa* includes:

- Children of both sexes, for they have recently been in intimate contact with their mothers.
- Women in their childbearing years, for their bodies contain maximum amounts of female substances, as proven by their monthly menstruation.
- Postmenopausal women who have had two or fewer children, for their bodies are not sufficiently drained of female substances.
- Elderly men, for feminine *nu* has been transmitted to them by a lifetime of assorted activities (e.g., sexual activity, receiving food from women) that exposed them to contact with women.

These *figapa* people—whether or not their genitals are female genitals—are all "like women" and "uninitiated" because their bodies are rich in substances culturally defined as feminine.

Note that Hua culture constructs the categories of *kakora* and *figapa* in such a way that they cross-cut the biological categories of male and female. At certain culturally defined stages in a person's life, a woman may become like a man, and vice versa. Although the *kakora/figapa* distinction is relevant only in certain ritual contexts (not considered here), it is noteworthy that people who are biologically female or male can be categorized with the opposite gender for certain purposes.

The Hua clearly illustrate the cultural construction of gender. The physical distinctions between women and men (genitals, beards, breasts, and so on) are differences that are recognized in all known cultures, including Hua. But, as the Hua show, various cultures use the raw material (metaphorically speaking) of these distinctions to construct varying beliefs about the specific ways in which the sexes differ. Thus, Hua believe that men and women differ not only in observable physical ways but also in the quantity of *nu* each naturally has. These beliefs affect the attitudes each sex holds about the other and the behavior each sex adopts toward the other. Thus Hua men in some contexts and at some ages are fearful about the possibility of feminine pollution and therefore minimize their contact with women, their intake of food prepared by women, and their sexual activities with women. Finally, the logic of Hua beliefs about the relation between *nu* and sexual identity make it possible for them to have a gender distinction (*kakora/figapa*) that cross-cuts male/female.

But this volume is specifically about males, men, manhood, and masculinity. How about the cultural construction of manhood and masculinity in highland New Guinea cultures? For this, we turn to two other peoples.

AWA: THE MAKING OF MEN

Our second example is the Awa, a New Guinea people numbering about 1,500. As is the case with many other peoples in the region, Awa customs require all boys to undergo a lengthy and traumatic series of rituals, generally known as male initiation rituals. Anthropologists Philip Newman and David Boyd (1982) conducted ethnographic fieldwork among the Awa in the 1960s and 1970s, and describe the lengthy and complex rituals the Awa believe are necessary to make a man. The following account is based upon their description.

Male initiation rituals are customary in many places in the world, not just New Guinea. In most, initiation is a necessary rite of passage, meaning that it symbolically transforms a male from the social status of "boyhood" into the new status of "manhood," with all its responsibilities. Psychologically, initiations both make the boys aware of their new status and help prepare them for it. Socially, initiations announce to the families and community that a group of boys have become men and should be regarded and treated as such.

But in Awa belief, male initiation rituals do more than just prepare a boy for manhood and publicly announce his change of status. Awa believe that the rituals actually produce the physical changes in the boys' bodies that lead to their maturity. Growing up requires drying out: From an amorphous, wet, and soft state, the physical body becomes well formed, dry, and hard. Many of the

Awa initiation rituals are intended to rid male bodies of excessive moisture, helping them to mature physically.

Awa also say that the rituals impart important masculine qualities like courage, strength, aggressiveness, and tolerance of pain. As we shall see, many parts of the rituals are traumatic, but Awa think they are necessary for the boys to learn what they need to know to become men, and to acquire the mental and physical qualities of manhood. These qualities are desirable not just for the boys as individuals, but for the community as a whole, because Awa traditionally were engaged in warfare with their neighbors and, therefore, needed male warriors to survive.

A third purpose of the initiation rituals is to protect boys from feminine contamination. In the course of their initiation, males learn the ritual procedures that will help them cleanse their bodies from female pollution and reduce some of the dangers of sexual relations with women. These dangers include sickness, bodily weakness, lethargy, and even pregnancy. The Awa believe that certain behaviors may lead a married man to become pregnant. "Male pregnancy" may happen to a man if he has sex too often, if he has intercourse with a menstruating woman, or if he penetrates the vagina too deeply during intercourse. The blood and water from a woman's womb will move up his urethra and mix with the semen inside his body, potentially producing a fetus. This will make the man sick, of course, but the Awa have ritual procedures that remove the female substances before harm occurs.

Awa male initiations, then, have three main themes: attaining physical maturity, acquiring masculine qualities, and protecting the initiates from women. In their construction of masculinity, making a boy into a man involves, among other things, drying out his body, teaching him male secrets and endurance, and protecting his health.

The rituals begin when, in their early teens, boys are taken away from their mothers—forcibly if necessary—and undergo a series of complex rituals that last well into their adult years. From infancy, boys have lived with their mother in her house. When several boys in a neighborhood are between about twelve and fourteen, they must leave their mothers and female things behind and join adult men in the men's house. They undergo a series of initiation rituals that take place in five stages.

The first stage of the ritual sequence is the induction of a group ("cohort") of boys into the men's house. This stage is quite intimidating. The boys are beaten with stinging nettles to toughen them. A coarse vine is rubbed vigorously across the inside of their thighs. Their food and water intake is restricted. They are scared by the sounds of special flutes, which they are told are made by spirits. They are taunted and ridiculed by adult men, many of whom are their relatives. But more traumas are to come.

In the second stage, about a year later, the boys experience the first cleansing of their bodies from female pollution. At a secluded site they are forcibly bled and made to vomit. Small bundles of sharp-edged swordgrass are jabbed into their nostrils until blood flows. Two small cuts are made in the glans of their penis. A vine is looped and thrust down their throat, inducing them to vomit. The purpose of these acts is to protect their general physical health and to promote their physical maturity. Bloodletting from the nose and penis helps remove female substances from the boys' bodies and, along with forced vomiting, helps to dry out their bodies, which is necessary for their physical maturity. In addition, like the Hua (see above), Awa believe that it is healthier for men to be drier inside than women.

The third stage occurs when the young men are between about eighteen and twenty. Once again they are purged and dried out by nosebleeding, penis cutting, and vomiting. They are now told certain secrets, known to all adult men but kept secret from all women and young boys. Soon after passing through the third stage, their elders teach the young men that too much contact with women is dangerous. They learn that pollution from female menstruation can overstimulate their growth and age them prematurely. They are warned to avoid sexual intercourse altogether until they are married, because it would be dangerous for them to have sex before they have learned how to protect themselves. After completing this stage, the young men are viewed as fully grown, but not yet ready for marriage and procreation.

About five years later—by now the men are in their twenties and ready for marriage—they go through the fourth stage of their initiation, known as the "sweat ceremony." The initiates sit next to the fire in the men's house and, with minimal water to drink, sweat profusely for a week or more. During this process, elders lecture them about their upcoming responsibilities as husbands and fathers. When the initiates emerge from the men's house after the sweat ceremony, they receive new clothes and body ornaments, including a pair of boar's tusks, which they wear in their noses as a symbolic indicator of their adult status. Prior to the sweat ceremony, the fathers and adult male relatives of the young men have arranged their marriages to women of other groups. The brides are brought to the hamlets of their future husbands after the sweat ceremony.

The fifth and final stage—appropriately called "severe penis cutting"—occurs a few days after the sweat ceremony. Once again, there are prohibitions on drinking and on eating certain kinds of foods. The noses of the young men are bled, even more profusely than before. Small wedges of flesh are cut from both sides of their penises, producing deep gashes and severe bleeding. The experience of pain, as in earlier stages, helps to strengthen the men and also stimulates their "life force." Once a man has been through this final stage, he has completed the passage into manhood and is ready to assume the responsibilities of marriage and fatherhood.

At this final stage of initiation, in fact, all the adult men present (not just the initiates) participate in a painful, but necessary, ritual. They expose their penises to two puncture wounds made by tiny stone-tipped arrows. Awa say this treatment is necessary to remove female substances that have entered the men's body through their penises during sexual intercourse with their wives. Most men have this done several times a year to energize them and remove female substances from their bodies.

For Awa, making a man is a lengthy process that takes place over the course of a decade or more. For a group of boys to achieve manhood requires the active participation of older men, who teach them ritual secrets, who force them to undergo physically painful and psychologically traumatic experiences, who instruct them in their future responsibilities, and who—of course—have been through the same lengthy process themselves. For Awa, older men are needed to build new men out of boys.

Our third example from New Guinea reveals another way in which older men make new men. It illustrates that masculinity in some cultures can be achieved in ways that some individuals in other cultures consider most unmasculine.

SAMBIA: SEXUALITY AND MASCULINITY

One of the most unusual ways in which male initiations make men is found among the Sambia, a people numbering around 2,000. Gilbert Herdt began visiting the Sambia in 1974, and has revisited them several times since then. The following discussion is based mostly upon his richly detailed monograph (1987).

Like the Enga, Hua, Awa, and many other peoples of New Guinea, Sambia men fear pollution from contact with women. Contact with female genitals and menstrual fluids is especially dangerous. Because of their fears, males older than about ten live in a men's house (off-limits to women) until they have reached a certain stage of maturity. Even after they begin to live with their wives, the two sexes tend to maintain separate spaces in the house. To minimize their chances of polluting men, women travel on separate trails and move into menstrual huts during their monthly periods. Like the Awa, Sambia men practice nosebleeding to rid their bodies of female contaminants.

The Sambia live by farming root crops, hunting, and pig herding. Until "tribal" warfare was outlawed by the national government of Papua New Guinea, Sambia men had to be constantly prepared to defend themselves and their settlements from attack. Perhaps because of the need for successful warriors, their culture places a heavy value on male "strength," meaning in the Sambia context prowess as a warrior, bravery, mental toughness, and spiritual potency. According to Sambia beliefs, a male acquires the strength that is so important in making him a man of consequence and a warrior from only one thing: semen. Semen is what hardens a man's muscles, gives him fortitude, makes him fierce in battle, and renders him domineering and aggressive. In short, semen makes men *masculine.*

In Sambia constructions of gender, girls grow up naturally: They develop breasts, reach menarche, and eventually are able to bear children without having to undergo any rituals. Boys, however—who are weak and unmanly because they still lack semen—will not grow up unless they undergo a series of initiation rituals. The intention of the rituals is to "masculinize" the boys (to use fieldworker Herdt's word). Because no boy will become a man unless he passes through the six stages of the ritual, all boys must participate. Those who are too afraid to do so willingly are forced to take part.

Sambia initiations are unusual because of one of their beliefs: Although males are born with a repository for semen (a "semen sack") inside their bodies, the male body cannot manufacture semen. The semen sack will remain empty and the boy will never be masculinized unless it is filled up. Filling up the sack of the young boys is one purpose of the first two stages of the rituals. During the first stage (ages about seven to ten) and the second stage (ages about eleven to fourteen), the boys perform fellatio on young but sexually mature bachelors. They must swallow the semen, for the goal is to nourish their masculinity. Swallowing the semen (no anal intercourse occurs) begins their masculinization. During the actual ritual proceedings, which last several days, the boys live together in a forest culthouse, ingesting semen from bachelors. The bachelors, of course, have been through the first two initiation stages during which they received their supply of semen, which they are now duty-bound to share with the younger boys. In this custom of *ritualized homosexuality,* oral sex is seen as essential for male physiological maturity and for the development of the essential qualities of manhood. The ingested semen is distributed throughout the body, maturing the skin and bones, hardening the muscles, making the body hair grow, and culminating in the changes Westerners call puberty.

The boys also endure nosebleeding. In the first stage of their initiation, elders ram sharp cane grasses into their nostrils to release the pollution the young boys have acquired during their years of association with their mothers and other women. The ritual nosebleeding occurs in later stages of the rituals as well, for masculinization requires ridding the body of feminine substances. The youngsters also are thrashed with sticks and their bodies are rubbed with stinging plants. This is believed to toughen them.

The fourth stage of a man's initiation occurs when he marries the woman chosen for him by his older relatives. By then, the man is in his late teens or early twenties, but his wife usually is far younger, and in fact has not yet reached sexual maturity. Until she has her first menstruation, the couple do not live together. The man remains in the men's house and continues to donate semen to first- and second-stage initiates. Ideally, he has no sex with any female, but remains homosexual. However, in the company of his age-mates, he hunts, goes to war, and makes gardens with the help of his young wife.

The fifth stage of a man's initiation occurs when his wife has her first period and goes into menstrual seclusion. Only then, when the man is in his early twenties, does he engage in heterosexual intercourse. In fact, one main goal of the fifth-stage rituals is to teach the man to protect himself from feminine pollution by learning to ritually bleed his own nose. He is told to thrust sharp grasses into his nostrils after each of his wife's menstrual periods, which removes the pollution from his body and thus makes it possible for him to minimize the dangers of intercourse with a woman. However, he continues to engage in homosexual relations in the men's house and is, in effect, bisexual.

During the fifth stage, then, a man has genital sex with his wife. It is likely to be his first heterosexual experience. As mentioned, a man cannot make more semen for himself, and Sambia believe that ejaculation depletes his supply and reduces his masculinity. (Interestingly, sex with a woman is seen as depleting a man's semen much more than homosexual fellatio.) But, Sambia believe, there is an alternative way for a man to replace his semen, although he does not learn about it until the proper time. During the fifth stage, he is told of the existence of a certain tree that exudes a milky semen-like sap, hitherto kept secret from him. He learns that by drinking the sap of this tree he can replace the semen that he loses from intercourse. Throughout the rest of his life, he will drink this substance so that he can retain his masculinity and health even while engaging in sex.

Homosexuality continues in the men's house, even when the formal rituals are not in process. But it is important to realize that homosexual acts are part of a ritual process—the process of masculinization, which can only be achieved by imparting semen. Once masculinization is complete, once a man has fulfilled his obligation to masculinize boys, and once he has learned how to have sex with women without harming himself or reducing his masculinity, homosexuality should end. Its end is marked by the sixth (and final) stage of initiation, which occurs when the man's wife has their first child. The husband then moves out of the men's house and in with his wife. The homosexual and bisexual period of his life is over. Men who continue homosexual relations after going through the sixth and final stage are looked down upon.

Aside from the fact that they illustrate one of the most unusual ways in which the masculine identity is culturally constructed through undergoing initiation rituals, it is interesting to contrast Sambia notions of homosexual behavior with those of modern North Americans. Many of us see gay men as the antithesis of masculinity, whereas the Sambia view participation in homosexual relationships as necessary to make adult men who are fully masculine. Also, many North Americans (both gays and straights) view the homosexual identity as something a person either "is" or "is not," whereas the Sambia seem to view it as a role that men either "do" or "do not" adopt, depending on the stage of their lives. Finally, many people tend to see homosexuality as unnatural (or sinful), whereas the Sambia view it as an essential part of growing up and its practice in ritual contexts as one of an adult man's duties to youngsters in his group.

CONCLUSION

Setting aside the details about the beliefs and practices of highland New Guinea cultures such as Hua, Awa, and Sambia, what do these peoples teach us? Certainly, they reveal through example that cultures construct gender in different ways. They reveal this so effectively precisely because their cultural constructions of gender—and specifically of manhood and masculinity—are so very different from those of the West (or, for that matter, of most of the East). The more different some culture's "constructions" of reality (gender, the natural world, or whatever) are from your own, the more obvious it is to you that *their* constructions are entirely cultural.

But suppose you were yourself Hua, Awa, or Sambia, reading a chapter like this one about a culture like the one you are living in right now. Then whose construction of gender, manhood, and masculinity would so obviously be entirely cultural?

REFERENCES

Chagnon, Napoleon A. *Yanomamo.* 5th ed. Fort Worth: Harcourt Brace, 1997.

Dentan, Robert. *The Semai: A Nonviolent People of Malaya.* New York: Holt, 1968.

Herdt, Gilbert. *The Sambia: Ritual and Gender in New Guinea.* New York: Holt, 1987.

Hewlett, Barry. *Intimate Fathers.* Ann Arbor, MI: Michigan University, 1992.

Lepowsky, Maria. *Fruit of the Motherland: Gender in an Egalitarian Society.* New York: Columbia University, 1993.

Marshall, Mac. *Weekend Warriors: Alcohol in a Micronesian Culture.* Mountain View, CA: Mayfield, 1979.

Meggitt, Mervyn. "Male-Female Relationships in the Highlands of Australian New Guinea." In *Cultures of the Pacific,* edited by Thomas G. Harding and Ben J. Wallace, 125–43. New York: Free Press, 1970.

Meigs, Anna. "Male Pregnancy and the Reduction of Sexual Opposition in a New Guinea Highlands Society." *Ethnology* 15 (1976): 393–408.

———. *Food, Sex, and Pollution: A New Guinea Religion.* New Brunswick: Rutgers University, 1984.

———. "Multiple Gender Ideologies and Statuses." In *Beyond the Second Sex,* edited by Peggy R. Sanday and Ruth G. Goodenough, 99–112. Philadelphia: University of Pennsylvania, 1990.

Newman, Phillip L., and David J. Boyd. "The Making of Men: Ritual and Meaning in Awa Male Initiation." In *Rituals of Manhood: Male Initiation in Papua New Guinea,* edited by Gilbert Herdt, 239–85. Berkeley: University of California, 1982.

2

"Night to His Day": The Social Construction of Gender

JUDITH LORBER

[Gethenians] do not see each other as men or women. This is almost impossible for our imagination to accept. What is the first question we ask about a newborn baby?

URSULA LE GUIN (1969, 94)

Talking about gender for most people is the equivalent of fish talking about water. Gender is so much the routine ground of everyday activities that questioning its taken-for-granted assumptions and presuppositions is like thinking about whether the sun will come up. Gender is so pervasive that in our society we assume it is bred into our genes. Most people find it hard to believe that gender is constantly created and re-created out of human interaction, out of social life, and is the texture and order of that social life. Yet gender, like culture, is a human production that depends on everyone constantly "doing gender." . . .

And everyone "does gender" without thinking about it. Today, on the subway, I saw a well-dressed man with a year-old child in a stroller. Yesterday, on a bus, I saw a man with a tiny baby in a carrier on his chest. Seeing men taking care of small children in public is increasingly common—at least in New York City. But both men were quite obviously stared at—and smiled at, approvingly. Everyone was doing gender—the men who were changing the role of fathers and other passengers, who were applauding them silently. But there was more gendering going on that probably fewer people noticed. The baby was wearing a white crocheted cap and white clothes. You couldn't tell if it was a boy or a girl. The child in the stroller was wearing a dark blue T-shirt and dark print pants. As they started to leave the train, the father put a Yankee baseball cap on the child's head. Ah, a boy, I thought. Then I noticed the gleam of tiny earrings in the child's ears, and as they got off, I saw the little flowered sneakers and lace-trimmed socks. Not a boy after all. Gender done.

Gender is such a familiar part of daily life that it usually takes a deliberate disruption of our expectations of how women and men are supposed to act to pay attention to how it is produced. Gender signs and signals are so ubiquitous that we usually fail to note them—unless they are missing or ambiguous. Then we are uncomfortable until we have successfully placed the other person in gender status; otherwise, we feel socially dislocated. In our society, in addition to man and woman, the status can be *transvestite* (a person who dresses in opposite-gender clothes) and *transsexual* (a person who has had sex-change surgery). Transvestites and transsexuals carefully construct their gender status by dressing, speaking, walking, gesturing in the ways prescribed for women or men—whichever they want to be taken for—and so does any "normal" person.

For the individual, gender construction starts with assignment to a sex category on the basis of what the genitalia look like at birth. Then babies are dressed or adorned in a way that displays the category because parents don't want to be constantly asked whether their baby is a girl or a boy. A sex category becomes a gender status through naming, dress, and the use of other gender markers. Once a child's gender is evident, others treat those in one gender differently from those in the other, and the children respond to the different treatment by feeling different and behaving differently. As soon as they can talk, they start to refer to themselves as members of their gender. Sex doesn't come into play until puberty, but by that time, sexual feelings and desires and practices have been shaped by gendered norms and expectations. Adolescent boys and girls approach and avoid each other in an elaborately scripted and gendered mating dance. Parenting is gendered, with different expectations for mothers and for fathers, and people of different genders work at different kinds of jobs. The work adults do as mothers and fathers and as low-level workers and high-level bosses shapes women's and men's life experiences, and these experiences produce different feelings, consciousness, relationships, skills—ways of being what we call feminine or masculine. All of these processes constitute the social construction of gender.

Gendered roles change—today fathers are taking care of little children, girls and boys are wearing unisex clothing and getting the same education, women and men are working at the same jobs. Although many traditional social groups are quite strict about maintaining gender differences, in other social groups they seem to be blurring. Then why the one-year-old's earrings? Why is it still so important to mark a child as a girl or a boy, to make sure she is not taken for a boy or he for a girl? What would happen if they were? They would, quite literally, have changed places in their social world.

To explain why gendering is done from birth, constantly and by everyone, we have to look not only at the way individuals experience gender but at gender as a social institution. As a social institution, gender is one of the major ways that human beings organize their lives. Human society depends on a predictable division of labor, a designated allocation of scarce goods, assigned responsibility for children and others who cannot care for themselves, common values and their systematic transmission to new members, legitimate leadership, music, art, stories, games, and other symbolic productions. One way of choosing people for the different tasks of society is on the basis of their talents, motivations, and competence—their demonstrated achievements. The other way is on the basis of gender, race, ethnicity—ascribed membership in a category of people. Although societies vary in the extent to which they use one or the other of these ways of allocating people to work and to carry out other responsibilities, every society uses gender and age grades. Every society classifies people as "girl and boy children," "girls and boys ready to be married," and "fully adult women and men," constructs similarities among them and differences between them, and assigns them to different roles and responsibilities. Personality characteristics, feelings, motivations, and ambitions flow from these different life experiences so that the members of these different groups become different kinds of people. The process of gendering and its outcome are legitimated by religion, law, science, and the society's entire set of values.

In order to understand gender as a social institution, it is important to distinguish human action from animal behavior. Animals feed themselves and their young until their young can feed themselves. Humans have to produce not only food but shelter and clothing. They also, if the group is going to continue as a social group, have to teach the children how their particular group does these tasks. In the process, humans reproduce gender, family, kinship, and a division of labor—social institutions that do not exist among animals. Primate social groups have been referred to as families, and their mating patterns as monogamy, adultery, and harems. Primate behavior has been used to prove the universality of sex differences—as built into our evolutionary inheritance. . . . But animals' sex differences are not at all the same as humans' gender differences; animals' bonding is not kinship; animals' mating is not ordered by marriage; and animals' dominance hierarchies are not the equivalent of human stratification systems. Animals group on sex and age, relational categories that are physiologically, not socially, different. Humans create gender and age-group categories that are socially, and not necessarily physiologically, different. . . .

Mating, feeding, and nurturant behavior in animals is determined by instinct and imitative learning and ordered by physiological sex and age. . . . In humans, these behaviors are taught and symbolically reinforced and ordered by socially constructed gender and age grades. Social gender and age statuses sometimes ignore or override physiological sex and age completely. Male and female animals (unless they physiologically change) are not interchangeable; infant animals cannot take the place of adult animals. Human females can become husbands and fathers, and human males can become wives and mothers, without sex-change surgery. . . . Human infants can reign as kings or queens.

Western society's values legitimate gendering by claiming that it all comes from physiology—female and male procreative differences. But gender and sex are not equivalent, and gender as a social construction does not flow automatically from genitalia and reproductive organs, the main physiological differences of females and males. In the construction of ascribed social statuses, physiological differences such as sex, stage of development, color of skin, and size are crude markers. They are not the source of the social statuses of gender, age grade, and race. Social statuses are carefully constructed through prescribed processes of teaching, learning, emulation, and enforcement. Whatever genes, hormones, and biological evolution contribute to human social institutions is materially as well as qualitatively transformed by social practices. Every social institution has a material base, but culture and social practices transform that base into something with qualitatively different patterns and constraints. The economy is much more than producing food and goods and distributing them to eaters and users; family and kinship are not the equivalent of having sex and procreating; morals and religions cannot be equated with the fears and ecstasies of the brain; language goes far beyond the sounds produced by tongue and larynx. No one eats "money" or "credit"; the concepts of "god" and "angels" are the subjects of theological disquisitions; not only words but objects, such as their flag, "speak" to the citizens of a country.

Similarly, gender cannot be equated with biological and physiological differences between human females and males. The building blocks of gender are *socially constructed statuses.* Western societies have only two genders, "man" and "woman." Some societies have three genders—men, women, and *berdaches* or *hijras* or *xaniths.* Berdaches, hijras, and xaniths are biological males who behave, dress, work, and are treated in most respects as social women; they are therefore not men, nor are they female women; they are, in our language, "male women." There are African and American Indian societies that have a gender status called *manly hearted woman*—biological females who work, marry, and parent as men; their social status is "female men." . . . They do not have to behave or dress as men to have the social responsibilities and prerogatives of husbands and fathers; what makes them men is enough wealth to buy a wife.

Modern Western societies' *transsexuals* and *transvestites* are the nearest equivalent of these crossover genders, but they are not institutionalized as third genders. . . . Transsexuals are biological males and females who have sex-change operations to alter their genitalia. They do so in order to bring their physical anatomy in congruence with the way they want to live and with their own sense of gender identity. They do not become a third gender; they change genders. Transvestites are males who live as women and females who live as men but do not intend to have sex-change surgery. Their dress, appearance, and mannerisms fall within the range of what is expected from members of the opposite gender, so that they "pass." They also change genders, sometimes temporarily, some for most of their lives. Transvestite women have fought in wars as men soldiers as recently as the nineteenth century; some married women, and others went back to being women and married men once the war was over. Some were discovered when their wounds were treated; others not until they died. In order to work as a jazz musician, a man's occupation, Billy Tipton, a woman, lived most of her life as a man. She died recently at seventy-four, leaving a wife and three adopted sons for whom she was husband and father, and musicians with whom she had played and traveled, for whom she was "one of the boys." . . . There have been many other such occurrences of women passing as men to do more prestigious or lucrative men's work. . . .

Genders, therefore, are not attached to a biological substratum. Gender boundaries are breachable, and individual and socially organized shifts from one gender to another call attention to "cultural, social, or aesthetic dissonances." These odd or deviant or third genders show us what we ordinarily take for granted—that people have to learn to be women and men. . . .

FOR INDIVIDUALS, GENDER MEANS SAMENESS

Although the possible combinations of genitalia, body shapes, clothing, mannerisms, sexuality, and roles could produce infinite varieties in human beings, the social institution of gender depends on the production and maintenance of a limited number of gender statuses and of making the members of these statuses similar to each other. Individuals are born sexed but not gendered, and they have to be taught to be masculine or feminine. As Simone de Beauvoir said: "One is not born, but rather becomes, a woman . . . ; it is civilization as a whole that produces this creature . . . which is described as feminine." . . .

Children learn to walk, talk, and gesture the way their social group says girls and boys should. Ray Birdwhistell, in his analysis of body motion as human communication, calls these learned gender displays *tertiary* sex characteristics and argues that they are needed to distinguish genders because humans are a weakly dimorphic species—their only sex markers are genitalia. . . . Clothing, paradoxically, often hides the sex but displays the gender.

In early childhood, humans develop gendered personality structures and sexual orientations through their interactions with parents of the same and opposite gender. As adolescents, they conduct their sexual behavior according to gendered scripts. Schools, parents, peers, and the mass media guide young people into gendered work and family roles. As adults, they take on a gendered social status in their society's stratification system. Gender is thus both ascribed and achieved. . . .

People go along with the imposition of gender norms because the weight of mortality as well as immediate social pressure enforces them. Consider how many instructions for properly gendered behavior are packed into this mother's admonition to her daughter: "This is how to hem a dress when you see the hem coming down and so to prevent yourself from looking like the slut I know you are so bent on becoming." . . .

Gender norms are inscribed in the way people move, gesture, and even eat. In one African society, men were supposed to eat with their "whole mouth, wholeheartedly, and not, like women, just with the lips, that is halfheartedly, with reservation and restraint." . . . Men and women in this society learned to walk in ways that proclaimed their different positions in the society:

> The manly man . . . stands up straight into the face of the person he approaches, or wishes to welcome. Ever on the alert, because ever threatened, he misses nothing of what happens around him. . . . Conversely, a well-brought-up woman . . . is expected to walk with a slight stoop, avoiding every misplaced movement of her body, her head or her arms, looking down, keeping her eyes on the spot where she will next put her foot, especially if she happens to have to walk past the men's assembly. . . .

Many cultures go beyond clothing, gestures, and demeanor in gendering children. They inscribe gender directly into bodies. In traditional Chinese society, mothers bound their daughters' feet into three-inch stumps to enhance their sexual attractiveness. Jewish fathers circumcise their infant sons to show their covenant with God. Women in African societies remove the clitoris of prepubescent girls, scrape their labia, and make the lips grow together to preserve their chastity and ensure their marriageability. In Western societies, women augment their breast size with silicone and reconstruct their faces with cosmetic surgery to conform to cultural ideals of feminine beauty. Hanna Papanek . . . notes that these practices reinforce the sense of superiority or inferiority in adults who carry them out as well as in the children on whom they are done: The genitals of Jewish fathers and sons are physical and psychological evidence of their common dominant religious and familial status; the genitals of African mothers and daughters are physical and psychological evidence of their joint subordination.

Sandra Bem . . . argues that because gender is a powerful "schema" that orders the cognitive world, one must wage a constant, active battle for a child not to fall into typical gendered attitudes and behavior. In 1972, *Ms. Magazine* published Lois Gould's fantasy of how to raise a child free of gender-typing. The experiment calls for hiding the child's anatomy from all eyes except the parents' and treating the child as neither a girl nor a boy. The child, called X, gets to do all the things boys *and* girls do. The experiment is so successful that all the children in X's class at school want to look and behave like X. At the end of the story, the creators of the experiment are asked what will happen when X grows up. The scientists' answer is that by then it will be quite clear what X is, implying that its hormones will kick in and it will be revealed as a female or male. That ambiguous, and somewhat contradictory, ending lets Gould off the hook; neither she nor we have any idea what someone brought up totally androgynously would be like sexually or socially as an adult. The hormonal input will not create gender or sexuality but will only establish secondary sex characteristics; breasts, beards, and menstruation alone do not produce social manhood or womanhood. Indeed, it is at puberty, when sex characteristics become evident, that most societies put pubescent children through their most important rites of passage, the rituals that officially mark them as fully gendered—that is, ready to marry and become adults.

Most parents create a gendered world for their newborn by naming, birth announcements, and dress. Children's relationships with same-gendered and different-gendered caretakers structure their self-identifications and personalities. Through cognitive development, children extract and apply to their own actions the appropriate behavior for those who belong in their own gender, as well as race, religion, ethnic group, and social class, rejecting what is not appropriate. If their social categories are highly valued, they value themselves highly; if their social categories are low status, they lose self-esteem. . . . Many feminist parents who want to raise androgynous children soon lose their children to the pull of gendered norms. . . . My son attended a carefully nonsexist elementary school, which didn't even have girls' and boys' bathrooms. When he was seven or eight years old, I attended a class play about "squares" and "circles" and their need for each other and noticed that all the girl squares and circles wore makeup, but none of the boy squares and circles did. I asked the teacher about it after the play, and she said, "Bobby said he was not going to wear makeup, and he is a powerful child, so none of the boys would either." In a long discussion about conformity, my son confronted me with the question of who the conformists were, the boys who followed their leader or the girls who listened to the woman teacher. In actuality, they both were, because they both followed same-gender leaders and acted in gender-appropriate ways. (Actors may wear makeup, but real boys don't.)

For human beings there is no essential femaleness or maleness, femininity or masculinity, womanhood or manhood, but once gender is ascribed, the social order constructs and holds individuals to strongly gendered norms and expectations. Individuals may vary on many of the components of gender and may shift genders temporarily or permanently, but they must fit into the limited number of gender statuses their society recognizes. In the process, they re-create their society's version of women and men: "If we do gender appropriately, we simultaneously sustain, reproduce, and render legitimate the institutional arrangements. . . . If we fail to do gender appropriately, we as individuals—not the institutional arrangements—may be called to account (for our character, motives, and predispositions)." . . .

The gendered practices of everyday life reproduce a society's view of how women and men should act. . . . Gendered social arrangements are justified by religion and cultural productions and backed by law, but the most powerful means of sustaining the moral hegemony of the dominant gender ideology is that the process is made invisible; any possible alternatives are virtually unthinkable. . . .

FOR SOCIETY, GENDER MEANS DIFFERENCE

The pervasiveness of gender as a way of structuring social life demands that gender statuses be clearly differentiated. Varied talents, sexual preferences, identities, personalities, interests, and ways of interacting fragment the individual's bodily and social experiences. Nonetheless, these are organized in Western cultures into two and only two socially and legally recognized gender statuses, "man" and "woman." In the social construction of gender, it does not matter what men and women actually do; it does not even matter if they do exactly the same thing. The social institution of gender insists only that what they do is *perceived* as different.

If men and women are doing the same tasks, they are usually spatially segregated to maintain gender separation, and often the tasks are given different job titles as well, such as executive secretary and administrative assistant. . . . If the differences between women and men begin to blur, society's "sameness taboo" goes into action. . . . At a rock and roll dance at West Point in 1976, the year women were admitted to the prestigious military academy for the first time, the school's administrators "were reportedly perturbed by the sight of mirror-image couples dancing in short hair and dress gray trousers," and a rule was established that women cadets could dance at these events only if they wore skirts. . . . Women recruits in the U.S. Marine Corps are required to wear makeup—at a minimum, lipstick and eye shadow—and they have to take classes in makeup, hair care, poise, and etiquette. This feminization is part of a deliberate policy of making them clearly distinguishable from men Marines. Christine Williams quotes a twenty-five-year-old woman drill instructor as saying: "A lot of the recruits who come here don't wear makeup; they're tomboyish or athletic. A lot of them have the preconceived idea that going into the military means they can still be a tomboy. They don't realize that you are a *Woman* Marine." . . .

If gender differences were genetic, physiological, or hormonal, gender bending and gender ambiguity would occur only in hermaphrodites, who are born with chromosomes and genitalia that are not clearly female or male. Since gender differences are socially constructed, all men and all women can enact the behavior of the other, because they know the other's social script: "'Man' and 'woman' are at once empty and overflowing categories. Empty because they have no ultimate, transcendental meaning. Overflowing because even when they appear to be fixed, they still contain within them alternative, denied, or suppressed definitions." . . . Nonetheless, though individuals may be able to shift gender statuses, the gender boundaries have to hold, or the whole gendered social order will come crashing down.

Paradoxically, it is the social importance of gender statuses and their external markers—clothing, mannerisms, and spatial segregation—that makes gender bending or gender crossing possible—or even necessary. The social viability of differentiated gender statuses produces the need or desire to shift statuses. Without gender differentiation, transvestism and transsexuality would be meaningless. You couldn't dress in the opposite gender's clothing if all clothing were unisex. There would be no need to reconstruct genitalia to match identity if interests and lifestyles were not gendered. There would be no need for women to pass as men to do certain kinds of work if jobs were not typed as "women's work" and "men's work." Women would not have to dress as men in public life in order to give orders or aggressively bargain with customers. . . .

GENDER RANKING

Most societies rank genders according to prestige and power and construct them to be unequal, so that moving from one to another also means moving up or down the social scale. Among some North American Indian cultures, the hierarchy was male men, male women, female men, female women. Women produced significant durable goods (basketry, textiles, pottery, decorated leather goods), which could be traded. Women also controlled what they produced and any profit or wealth they earned. Since women's occupational realm could lead to prosperity and prestige, it was fair game for young men—but only if they became women in gender status. Similarly, women in other societies who amassed a great deal of wealth were allowed to become men—"manly hearts." According to Harriet Whitehead . . . :

> Both reactions reveal an unwillingness or inability to distinguish the sources of prestige—wealth, skill, personal efficacy (among other things)—from masculinity. Rather there is the innuendo that if a person performing female tasks can attain excellence, prosperity, or social power, it must be because that person is, at some level, a man. . . . A woman who could succeed at doing the things men did was honored as a man would be. . . . What seems to have been more disturbing to the culture—which means, for all intents and purposes, to the men—was the possibility that women, within their own department, might be onto a good thing. It was into this unsettling breach that the berdache institution was hurled. In their social aspect, women were complimented by the berdache's imitation. In their anatomic aspect, they were subtly insulted by his vaunted superiority. . . .

In American society, men-to-women transsexuals tend to earn less after surgery if they change occupations; women-to-men transsexuals tend to increase their income. . . . Men who go into women's fields, like nursing, have less prestige than women who go into men's fields, like physics. Janice Raymond, a radical feminist, feels that transsexual men-to-women have advantages over female women because they were not socialized to be subordinate or oppressed throughout life. She says:

> We know that we are women who are born with female chromosomes and anatomy, and that whether or not we were socialized to be so-called normal women, patriarchy has treated and will treat us like women. Transsexuals have not had this same history. No man can have the history of being born and located in this culture as a woman. He can have the history of *wishing* to be a woman and of *acting* like a woman, but this gender experience is that of a transsexual, not of a woman. Surgery may confer the artifacts of outward and inward female organs but it cannot confer

the history of being born a woman in this society. . . .

GENDER AS PROCESS, STRATIFICATION, AND STRUCTURE

As a social institution, gender is a process of creating distinguishable social statuses for the assignment of rights and responsibilities. As part of a stratification system that ranks these statuses unequally, gender is a major building block in the social structures built on these unequal statuses.

As a *process,* gender creates the social differences that define "woman" and "man." In social interaction throughout their lives, individuals learn what is expected, see what is expected, act and react in expected ways, and thus simultaneously construct and maintain the gender order: "The very injunction to be a given gender takes place through discursive routes: to be a good mother, to be a heterosexually desirable object, to be a fit worker, in sum, to signify a multiplicity of guarantees in response to a variety of different demands all at once." . . . Members of a social group neither make up gender as they go along nor exactly replicate in rote fashion what was done before. In almost every encounter, human beings produce gender, behaving in the ways they learned were appropriate for their gender status, or resisting or rebelling against these norms. Resistance and rebellion have altered gender norms, but so far they have rarely eroded the statuses.

Gendered patterns of interaction acquire additional layers of gendered sexuality, parenting, and work behaviors in childhood, adolescence, and adulthood. Gendered norms and expectations are enforced through informal sanctions of gender-inappropriate behavior by peers and by formal punishment or threat of punishment by those in authority should behavior deviate too far from socially imposed standards for women and men.

Everyday gendered interactions build gender into the family, the work process, and other organizations and institutions, which in turn reinforce gender expectations for individuals. Because gender is a process, there is room not only for modification and variation by individuals and small groups but also for institutionalized change. . . .

As part of a *stratification* system, gender ranks men above women of the same race and class. Women and men could be different but equal. In practice, the process of creating difference depends to a great extent on differential evaluation. As Nancy Jay says: "That which is defined, separated out, isolated from all else is A and pure. Not-A is necessarily impure, a random catchall, to which nothing is external except A and the principle of order that separates it from Not-A." . . . From the individual's point of view, whichever gender is A, the other is Not-A; gender boundaries tell the individual who is like him or her, and all the rest are unlike. From society's point of view, however, one gender is usually the touchstone, the normal, the dominant, and the other is different, deviant, and subordinate. In Western society, "man" is A, "wo-man" is Not-A. (Consider what a society would be like where woman was A and man Not-A.)

The further dichotomization by race and class constructs the gradations of a heterogeneous society's stratification scheme. Thus, in the United States, white is A, African American is Not-A; middle class is A, working class is Not-A, and "African-American women occupy a position whereby the inferior half of a series of these dichotomies converge." . . . The dominant categories are the hegemonic ideals, taken so for granted as the way things should be that white is not ordinarily thought of as a race, middle class as a class, or men as a gender. The characteristics of these categories define the Other as that which lacks the valuable qualities the dominants exhibit.

In a gender-stratified society, what men do is usually valued more highly than what women do because men do it, even when their activities are very similar or the same. In different regions of southern India, for example, harvesting rice is men's work, shared work, or women's work: "Wherever a task is done by women it is considered easy, and where it is done by [men] it is considered difficult." . . . A gathering and hunting society's survival usually depends on the nuts, grubs, and small animals brought in by the women's foraging trips, but when the men's hunt is successful, it is the occasion for a celebration. Conversely, because they are the superior group, white men do not have to do the "dirty work," such as housework; the most inferior group does it, usually poor women of color. . . .

Freudian psychoanalytic theory claims that boys must reject their mothers and deny the feminine in themselves in order to become men: "For boys the major goal is the achievement of personal masculine identification with their father and sense of secure masculine self, achieved through superego formation and disparagement of women." . . . Masculinity may be the outcome of boys' intrapsychic struggles to separate their identity from that of their mothers, but the proofs of masculinity are culturally shaped and usually ritualistic and symbolic. . . .

The Marxist feminist explanation for gender inequality is that by demeaning women's abilities and keeping them from learning valuable technological skills, bosses preserve them as a cheap and exploitable reserve army of labor. Unionized men who could be easily replaced by women collude in this process because it allows them to monopolize the better paid, more interesting, and more autonomous jobs: "Two factors emerge as helping men maintain their separation from women and their control of technological occupations. One is the active gendering of jobs and people. The second is the continual creation of sub-divisions in the work processes, and levels in work hierarchies, into which men can move in order to keep their distance from women."

Societies vary in the extent of the inequality in social status of their women and men members, but where there is inequality, the status "woman" (and its attendant behavior and role allocations) is usually held in lesser esteem than the status "man." Since gender is also intertwined with a society's other constructed statuses of differential evaluation—race, religion, occupation, class, country of origin, and so on—men and women members of the favored groups command more power, more prestige, and more property than the members of the disfavored groups. Within many social groups, however, men are advantaged over women. The more economic resources, such as education and job opportunities, are available to a group, the more they tend to be monopolized by men. In poorer groups that have few resources (such as working-class African Americans in the United States), women and men are more nearly equal, and the women may even outstrip the men in education and occupational status. . . .

As a *structure,* gender divides work in the home and in economic production, legitimates those in authority, and organizes sexuality and emotional life. . . . As primary parents, women significantly influence children's psychological development and emotional attachments, in the process reproducing gender. Emergent sexuality is shaped by heterosexual, homosexual, bisexual, and sadomasochistic patterns that are gendered—different for girls and boys, and for women and men—so that sexual statuses reflect gender statuses.

When gender is a major component of structured inequality, the devalued genders have less power, prestige, and economic rewards than the valued genders. In countries that discourage gender discrimination, many major roles are still gendered; women still do most of the domestic labor and child rearing, even while doing full-time paid work; women and men are segregated on the job and each does work considered "appropriate"; women's work is usually paid less than men's work. Men dominate the positions of authority and leadership in government, the military, and the law; cultural productions, religions, and sports reflect men's interests.

In societies that create the greatest gender difference, such as Saudi Arabia, women are kept out of sight behind walls or veils, have no civil rights, and often create a cultural and emotional world of their own. . . . But even in societies with less rigid gender boundaries, women and men spend much of their time with people of their own gender because of the way work and family are organized. This spatial separation of women and men reinforces gendered differences, identity, and ways of thinking and behaving. . . .

Gender inequality—the devaluation of "women" and the social domination of "men"—has social functions and a social history. It is not the result of sex, procreation, physiology, anatomy, hormones, or genetic predispositions. It is produced and maintained by identifiable social processes and built into the general social structure and individual identities deliberately and purposefully. The social order as we know it in Western societies is organized around racial, ethnic, class, and gender inequality. I contend, therefore, that the continuing purpose of gender as a modern social institution is to construct women as a group to be the subordinates of men as a group. The life of everyone placed in the status "woman" is "night to his day—that has forever been the fantasy. Black to his white. Shut out of his system's space, she is the repressed that ensures the system's functioning." . . .

THE PARADOX OF HUMAN NATURE

To say that sex, sexuality, and gender are all socially constructed is not to minimize their social power. These categorical imperatives govern our lives in the most profound and pervasive ways, through the social experiences and social practices of what Dorothy Smith calls the "everyday/everynight world." . . . The paradox of human nature is that it is *always* a manifestation of cultural meanings, social relationships, and power politics; "not biology, but culture, becomes destiny." . . . Gendered people emerge not from physiology or sexual orientation but from the exigencies of the social order, mostly, from the need for a reliable division of the work of food production and the social (not physical) reproduction of new members. The moral imperatives of religion and cultural representations guard the boundary lines among genders and ensure that what is demanded, what is permitted, and what is tabooed for the people in each gender is well known and followed by most. . . . Political power, control of scarce resources, and, if necessary, violence uphold the gendered social order in the face of resistance and rebellion. Most people, however, voluntarily go along with their society's prescriptions for those of their gender status, because the norms and expectations get built into their sense of worth and identity . . . the way we see and hear and speak, the way we fantasy, and the way we feel . . .

3

Masculinity as Homophobia: Fear, Shame, and Silence in the Construction of Gender Identity

MICHAEL S. KIMMEL

"Funny thing," [Curley's wife] said. "If I catch any one man, and he's alone, I get along fine with him. But just let two of the guys get together an' you won't talk. Jus' nothin' but mad." She dropped her fingers and put her hands on her hips. "You're all scared of each other, that's what. Ever' one of you's scared the rest is goin' to get something on you."

JOHN STEINBECK
Of Mice and Men (1937)

We think of manhood as eternal, a timeless essence that resides deep in the heart of every man. We think of manhood as a thing, a quality that one either has or doesn't have. We think of manhood as innate, residing in the particular biological composition of the human male, the result of androgens or the possession of a penis. We think of manhood as a transcendent tangible property that each man must manifest in the world, the reward presented with great ceremony to a young novice by his elders for having successfully completed an arduous initiation ritual. In the words of poet Robert Bly (1990), "the structure at the bottom of the male psyche is still as firm as it was twenty thousand years ago" (p. 230).

In this chapter, I view masculinity as a constantly changing collection of meanings that we construct through our relationships with ourselves, with each other, and with our world. Manhood is neither static nor timeless; it is historical. Manhood is not the manifestation of an inner essence; it is socially constructed. Manhood does not bubble up to consciousness from our biological makeup; it is created in culture. Manhood means different things at different times to different people. We come to know what it means to be a man in our culture by setting our definitions in opposition to a set of "others"—racial minorities, sexual minorities, and, above all, women.

Our definitions of manhood are constantly changing, being played out on the political and social terrain on which the relationships between women and men are played out. In fact, the search for the transcendent, timeless definition of manhood is itself a sociological phenomenon—we tend to search for the timeless and eternal during moments of crisis, those points of transition when old definitions no longer work and new definitions are yet to be firmly established.

This idea that manhood is socially constructed and historically shifting should not be understood as a loss, that something is being taken away from men. In fact, it gives us something extraordinarily valuable—agency, the capacity to act. It gives us a sense of historical possibilities to replace the despondent resignation that invariably attends timeless, ahistorical essentialisms. Our behaviors are

From Harry Brod (ed.), *Theorizing Masculinities*, Sage Publications, pp. 119–141. ©1994 by Michael Kimmel.
Reprinted by permission of the author.

not simply "just human nature," because "boys will be boys." From the materials we find around us in our culture—other people, ideas, objects—we actively create our worlds, our identities. Men, both individually and collectively, can change.

In this chapter, I explore this social and historical construction of both hegemonic masculinity and alternate masculinities, with an eye toward offering a new theoretical model of American manhood.[1] To accomplish this I first uncover some of the hidden gender meanings in classical statements of social and political philosophy, so that I can anchor the emergence of contemporary manhood in specific historical and social contexts. I then spell out the ways in which this version of masculinity emerged in the United States, by tracing both psychoanalytic developmental sequences and a historical trajectory in the development of marketplace relationships. . . .

MASCULINITY AS HISTORY AND THE HISTORY OF MASCULINITY

The idea of masculinity . . . is the product of historical shifts in the grounds on which men rooted their sense of themselves as men. To argue that cultural definitions of gender identity are historically specific goes only so far; we have to specify exactly what those models were. In my historical inquiry into the development of these models of manhood[2] I chart the fate of two models for manhood at the turn of the 19th century and the emergence of a third in the first few decades of that century.

In the late 18th and 19th centuries, two models of manhood prevailed. The *Genteel Patriarch* derived his identity from landownership. Supervising his estate, he was refined, elegant, and given to casual sensuousness. He was a doting and devoted father, who spent much of his time supervising the estate and with his family. Think of George Washington or Thomas Jefferson as examples. By contrast, the *Heroic Artisan* embodied the physical strength and republican virtue that Jefferson observed in the yeoman farmer, independent urban craftsman, or shopkeeper. Also a devoted father, the Heroic Artisan taught his son his craft, bringing him through ritual apprenticeship to status as master craftsman. Economically autonomous, the Heroic Artisan also cherished his democratic community, delighting in the participatory democracy of the town meeting. Think of Paul Revere at his pewter shop, shirtsleeves rolled up, a leather apron—a man who took pride in his work.

Heroic Artisans and Genteel Patriarchs lived in casual accord, in part because their gender ideals were complementary (both supported participatory democracy and individual autonomy, although patriarchs tended to support more powerful state machineries and also supported slavery) and because they rarely saw one another: Artisans were decidedly urban and the Genteel Patriarchs ruled their rural estates. By the 1830s, though, this casual symbiosis was shattered by the emergence of a new vision of masculinity, *Marketplace Manhood.*

Marketplace Man derived his identity entirely from his success in the capitalist marketplace, as he accumulated wealth, power, status. He was the urban entrepreneur, the businessman. Restless, agitated, and anxious, Marketplace Man was an absentee landlord at home and an absent father with his children, devoting himself to his work in an increasingly homosocial environment—a male-only world in which he pits himself against other men. His efforts at self-making transform the political and economic spheres, casting aside the Genteel Patriarch as an anachronistic feminized dandy—sweet, but ineffective and outmoded, and transforming the Heroic Artisan into a dispossessed proletarian, a wage slave.

As Tocqueville would have seen it, the coexistence of the Genteel Patriarch and the Heroic Artisan embodied the fusion of liberty and equality. Genteel Patriarchy was the manhood of the traditional aristocracy, the class that embodied the virtue of liberty. The Heroic Artisan embodied democratic community, the solidarity of the urban shopkeeper or craftsman. Liberty and democracy, the patriarch and the artisan, could, and did, coexist. But Marketplace Man is capitalist man, and he makes both freedom and equality problematic, eliminating the freedom of the aristocracy and proletarianizing the equality of the artisan. In one sense, American history has been an effort to restore, retrieve, or reconstitute the virtues of Genteel Patriarchy and Heroic Artisanate as they were being transformed in the capitalist marketplace.

Marketplace Manhood was a manhood that required proof, and that required the acquisition of tangible goods as evidence of success. It reconstituted itself by the exclusion of "others"—women, nonwhite men, nonnative-born men, homosexual men—and by terrified flight into a pristine mythic homosocial Eden where men could, at last, be real men among other men. The story of the ways in which Marketplace Man becomes American Everyman is a tragic tale, a tale of striving to live up to impossible ideals of success leading to chronic terrors of emasculation, emotional emptiness, and a gendered rage that leave a wide swath of destruction in [their] wake.

MASCULINITIES AS POWER RELATIONS

Marketplace Masculinity describes the normative definition of American masculinity. It describes his characteristics—aggression, competition, anxiety—and the arena in which those characteristics are deployed—the public sphere, the marketplace. If the marketplace is the arena in which manhood is tested and proved, it is a gendered arena, in which tensions between women and men and tensions among different groups of men are weighted with meaning. These tensions suggest that cultural definitions of gender are played out in a contested terrain and are themselves power relations.

All masculinities are not created equal; or rather, we are all *created* equal, but any hypothetical equality evaporates quickly because our definitions of masculinity are not equally valued in our society. One definition of manhood continues to remain the standard against which other forms of manhood are measured and evaluated. Within the dominant culture, the masculinity that defines white, middle class, early middle-aged, heterosexual men is the masculinity that sets the standards for other men, against which other men are measured and, more often than not, found wanting. Sociologist Erving Goffman (1963) wrote that in America, there is only "one complete, unblushing male":

> a young, married, white, urban, northern heterosexual, Protestant father of college education, fully employed, of good complexion, weight and height, and a recent record in sports. Every American male tends to look out upon the world from this perspective. . . . Any male who fails to qualify in any one of these ways is likely to view himself . . . as unworthy, incomplete, and inferior. (p. 128)

This is the definition that we will call "hegemonic" masculinity, the image of masculinity of those men who hold power, which has become the standard in psychological evaluations, sociological research, and self-help and advice literature for teaching young men to become "real men" (Connell, 1987). The hegemonic definition of manhood is a man *in* power, a man *with* power, and a man *of* power. We equate manhood with being strong, successful, capable, reliable, in control. The very definitions of manhood we have developed in our culture maintain the power that some men have over other men and that men have over women.

Our culture's definition of masculinity is thus several stories at once. It is about the individual man's quest to accumulate those cultural symbols that denote manhood, signs that he has in fact achieved it. It is about those standards being used against women to prevent their inclusion in public life and their consignment to a devalued private sphere. It is about the differential access that different types of men have to those cultural resources that confer manhood and about how each of these groups then develop their own modifications to preserve and claim their manhood. It is about the power of these definitions themselves to serve to maintain the real-life power that men have over women and that some men have over other men.

This definition of manhood has been summarized cleverly by psychologist Robert Brannon (1976) into four succinct phrases:

1. "No Sissy Stuff!" One may never do anything that even remotely suggests femininity. Masculinity is the relentless repudiation of the feminine.
2. "Be a Big Wheel." Masculinity is measured by power, success, wealth, and status. As the current saying goes, "He who has the most toys when he dies wins."
3. "Be a Sturdy Oak." Masculinity depends on remaining calm and reliable in a crisis, holding emotions in check. In fact, proving you're a man depends on never showing your emotions at all. Boys don't cry.
4. "Give 'em Hell." Exude an aura of manly daring and aggression. Go for it. Take risks.

These rules contain the elements of the definition against which virtually all American men are measured. Failure to embody these rules, to affirm the power of the rules and one's achievement of them is a source of men's confusion and pain. Such a model is, of course, unrealizable for any man. But we keep trying, valiantly and vainly, to measure up. American masculinity is a relentless test.[3] The chief test is contained in the first rule. Whatever the variations by race, class, age, ethnicity, or sexual orientation, being a man means "not being like women." This notion of anti-femininity lies at the heart of contemporary and historical conceptions of manhood, so that masculinity is defined more by what one is not rather than who one is.

MASCULINITY AS THE FLIGHT FROM THE FEMININE

Historically and developmentally, masculinity has been defined as the flight from women, the repudiation of femininity. Since Freud, we have come to understand that developmentally the central task that every little boy must confront is to develop a secure identity for himself as a man. As Freud had it, the oedipal project is a process of the boy's renouncing his identification with and deep emotional attachment to his mother and then replacing her with the father as the object of identification. Notice that he reidentifies but never reattaches. This entire process, Freud argued, is set in motion by the boy's sexual desire for his mother. But the father stands in the son's path and will not yield his sexual property to his puny son. The boy's first emotional experience, then, the one that inevitably follows his experience of desire, is fear—fear of the bigger, stronger, more sexually powerful father. It is this fear, experienced symbolically as the fear of castration, Freud argues, that forces the young boy to renounce his identification with mother and seek to identify with the being who is the actual source of his fear, his father. In so doing, the boy is now symbolically capable of sexual union with a motherlike substitute, that is, a woman. The boy becomes gendered (masculine) and heterosexual at the same time.

Masculinity, in this model, is irrevocably tied to sexuality. The boy's sexuality will now come to resemble the sexuality of his father (or at least the way he imagines his father)—menacing, predatory, possessive, and possibly punitive. The boy has come to identify with his oppressor; now he can become the oppressor himself. But a terror remains, the terror that the young man will be unmasked as a fraud, as a man who has not completely and irrevocably separated from mother. It will be other men who will do the unmasking. Failure will de-sex the man, make him appear as not fully a man. He will be seen as a wimp, a Mama's boy, a sissy.

After pulling away from his mother, the boy comes to see her not as a source of nurturance and love, but as an insatiably infantilizing creature, capable of humiliating him in front of his peers. She makes him dress up in uncomfortable and itchy clothing, her kisses smear his cheeks with lipstick, staining his boyish innocence with the mark of feminine dependency. No wonder so many boys cringe from their mothers' embraces with groans of "Aw, Mom! Quit it!" Mothers represent the humiliation of infancy, helplessness, dependency. "Men act as though they were being guided by (or rebelling against) rules and prohibitions enunciated by a moral mother," writes psychohistorian Geoffrey Gorer (1964). As a result, "all the niceties of masculine behavior—modesty, politeness, neatness, cleanliness—come to be regarded as concessions to feminine demands, and not good in themselves as part of the behavior of a proper man" (pp. 56, 57).

The flight from femininity is angry and frightened, because mother can so easily emasculate the young boy by her power to render him dependent, or at least to remind him of dependency. It is relentless; manhood becomes a lifelong quest to demonstrate its achievement, as if to prove the unprovable to others, because we feel so unsure of it ourselves. Women don't often feel compelled to "prove their womanhood"—the phrase itself sounds ridiculous. Women have different kinds of gender identity crises; their anger and frustration, and their own symptoms of depression, come more from being excluded than from questioning whether they are feminine enough.[4]

The drive to repudiate the mother as the indication of the acquisition of masculine gender identity has three consequences for the young boy. First, he pushes away his real mother, and with her the traits of nurturance, compassion, and tenderness she may have embodied. Second, he suppresses those traits in himself, because they will reveal his incomplete separation from mother. His life becomes a lifelong project to demonstrate that he possesses none of his mother's traits. Masculine identity is born in the renunciation of the feminine, not in the direct affirmation of the masculine, which leaves masculine gender identity tenuous and fragile.

Third, as if to demonstrate the accomplishment of these first two tasks, the boy also learns to devalue all women in his society, as the living embodiments of those traits in himself he has learned to despise. Whether or not he was aware of it, Freud also described the origins of sexism—the systematic devaluation of women—in the desperate efforts of the boy to separate from mother. We may *want* "a girl just like the girl that married dear old Dad," as the popular song had it, but we certainly don't want to *be like* her.

This chronic uncertainty about gender identity helps us understand several obsessive behaviors. Take, for example, the continuing problem of the school-yard bully. Parents remind us that the bully is the *least* secure about his manhood, and so he is constantly trying to prove it. But he "proves" it by choosing opponents he is absolutely certain he can defeat; thus the standard taunt to a bully is to "pick on someone your own size." He can't, though, and after defeating a smaller and weaker opponent, which he was sure would prove his manhood, he is left with the empty gnawing feeling that he has not proved it after all, and he must find another opponent, again one smaller and weaker, that he can again defeat to prove it to himself.[5]

One of the more graphic illustrations of this lifelong quest to prove one's manhood occurred at the Academy Awards presentation in 1992. As aging, tough guy actor Jack Palance accepted the award for best supporting actor for his role in the cowboy comedy *City Slickers,* he commented that people, especially film producers, think that because he is 71 years old, he's all washed up, that he's no longer competent. "Can we take a risk on this guy?" he quoted them as saying, before he dropped to the floor to do a set of one-armed push-ups. It was pathetic to see such an accomplished actor still having to prove that he is virile enough to work and, as he also commented at the podium, to have sex.

When does it end? Never. To admit weakness, to admit frailty or fragility, is to be seen as a wimp, a sissy, not a real man. But seen by whom?

MASCULINITY AS A HOMOSOCIAL ENACTMENT

Other men: We are under the constant careful scrutiny of other men. Other men watch us, rank us, grant our acceptance into the realm of manhood. Manhood is demonstrated for other men's approval. It is other men who evaluate the performance. Literary critic David Leverenz (1991) argues that "ideologies of manhood have functioned primarily in relation to the gaze of male peers and male authority" (p. 769). Think of how men boast to one another of their accomplishments—from their latest sexual conquest to the size of the fish they caught—and how we constantly parade the markers of manhood—wealth, power, status, sexy women—in front of other men, desperate for their approval.

That men prove their manhood in the eyes of other men is both a consequence of sexism and one of its chief props. "Women have, in men's minds, such a low place on the social ladder of this country that it's useless to define yourself in terms of a woman," noted playwright David Mamet. "What men need is men's approval." Women become a kind of currency that men use to improve their ranking on the masculine social scale. (Even those moments of heroic conquest of women carry, I believe, a current of homosocial evaluation.) Masculinity is a *homosocial* enactment. We test ourselves, perform heroic feats, take enormous risks, all because we want other men to grant us our manhood.

Masculinity as a homosocial enactment is fraught with danger, with the risk of failure, and with intense relentless competition. "Every man you meet has a rating or an estimate of himself which he never loses or forgets," wrote Kenneth Wayne (1912) in his popular turn-of-the-century advice book. "A man has his own rating, and instantly he lays it alongside of the other man" (p. 18). Almost a century later, another man remarked to psychologist Sam Osherson (1992) that "[b]y the time you're an adult, it's easy to think you're always in competition with men, for the attention of women, in sports; at work" (p. 291).

MASCULINITY AS HOMOPHOBIA

If masculinity is a homosocial enactment, its overriding emotion is fear. In the Freudian model, the fear of the father's power terrifies the young boy to renounce his desire for his mother and identify with his father. This model links gender identity with sexual orientation: The little boy's identification with father (becoming masculine) allows him to now engage in sexual relations with women (he becomes heterosexual). This is the origin of how we can "read" one's sexual orientation through the successful performance of gender identity. Second, the fear that the little boy feels does not send him scurrying into the arms of his mother to protect him from his father. Rather, he believes he will overcome his fear by identifying with its source. We become masculine by identifying with our oppressor.

But there is a piece of the puzzle missing, a piece that Freud, himself, implied but did not follow up.[6] If the pre-oedipal boy identifies with mother, he *sees the world through mother's eyes.* Thus, when he confronts father during his great oedipal crisis, he experiences a split vision: He sees his father as his mother sees his father, with a combination of awe, wonder, terror, *and desire.* He simultaneously sees the father as he, the boy, would like to see him—as the object not of desire but of emulation. Repudiating mother and identifying with father only partially answer his dilemma. What is he to do with that homoerotic desire, the desire he felt because he saw father the way that his mother saw father?

He must suppress it. Homoerotic desire is cast as feminine desire, desire for other men. Homophobia is the effort to suppress that desire, to purify all relationships with other men, with women, with children of its taint, and to ensure that no one could possibly ever mistake one for a homosexual. Homophobic flight from intimacy with other men is the repudiation of the homosexual within—never completely successful and hence constantly reenacted in every homosocial relationship. "The lives of most American men are bounded, and their interests daily curtailed by the constant necessity to prove to their fellows, and to themselves, that they are not sissies, not homosexuals," writes psychoanalytic historian Geoffrey Gorer (1964). "An interest or pursuit which is identified as a feminine interest or pursuit becomes deeply suspect for men" (p. 129).

Even if we do not subscribe to Freudian psychoanalytic ideas, we can still observe how, in less sexualized terms, the father is the first man who evaluates the boy's masculine performance, the first pair of male eyes before whom he tries to prove himself. Those eyes will follow him for the rest of his life. Other men's eyes will join them—the eyes of role models such as teachers, coaches, bosses, or media heroes; the eyes of his peers, his friends, his workmates; and the eyes of millions of other men, living and dead, from whose constant scrutiny of his performance he will never be free. "The tradition of all the dead generations weighs like a nightmare on the brain of the living," was how Karl Marx put it over a century ago (1848/1964, p. 11). "The birthright of every American male is a chronic sense of personal inadequacy," is how two psychologists describe it today (Woolfolk & Richardson, 1978, p. 57).

That nightmare from which we never seem to awaken is that those other men will see that sense of inadequacy, they will see that in our own eyes we are not who we are pretending to be. What we call masculinity is often a hedge against being revealed as a fraud, an exaggerated set of activities that keep others from seeing through us, and a frenzied effort to keep at bay those fears within ourselves. Our real fear "is not fear of women but of being ashamed or humiliated in front of other men, or being dominated by stronger men" (Leverenz, 1986, p. 451).

This, then, is the great secret of American manhood: *We are afraid of other men.* Homophobia is a central organizing principle of our cultural definition of manhood. Homophobia is more than the irrational fear of gay men, more than the fear that we might be perceived as gay. "The word 'faggot' has nothing to do with homosexual experience or even with fears of homosexuals," writes David Leverenz (1986). "It comes out of the depths of manhood: a label of ultimate contempt for anyone who seems sissy, untough, uncool" (p. 455). Homophobia is the fear that other men will unmask us, emasculate us, reveal to us and the world that we do not measure up, that we are not real men. We are afraid to let other men see that fear. Fear makes us ashamed, because the recognition of fear in ourselves is proof to ourselves that we are not as manly as we pretend, that we are, like the young man in a poem by Yeats, "one that ruffles in a manly pose for all his timid heart." Our fear is the fear of humiliation. We are ashamed to be afraid.

Shame leads to silence—the silences that keep other people believing that we actually approve of the things that are done to women, to minorities, to gays and lesbians in our culture. The frightened silence as we scurry past a woman being hassled by men on the street. That furtive silence when men make sexist or racist jokes in a bar. That clammy-handed silence when guys in the office make gay-bashing jokes. Our fears are the sources of our silences, and men's silence is what keeps the system running. This might help to explain why women often complain that their male friends or partners are often so understanding when they are alone and yet laugh at sexist jokes or even make those jokes themselves when they are out with a group.

The fear of being seen as a sissy dominates the cultural definitions of manhood. It starts so early. "Boys among boys are ashamed to be unmanly," wrote one educator in 1871 (cited in Rotundo, 1993, p. 264). I have a standing bet with a friend that I can walk onto any playground in America where 6-year-old boys are happily playing and by asking one question, I can provoke a fight. That question is simple: "Who's a sissy around here?" Once posed, the challenge is made. One of two things is likely to happen. One boy will accuse another of being a sissy, to which that boy will respond that he is not a sissy, that the first boy is. They may have to fight it out to see who's lying. Or a whole group of boys will surround one boy and all shout "He is! He is!" That boy will either burst into tears and run home crying, disgraced, or he will have to take on several boys at once, to prove that he's not a sissy. (And what will his father or older brothers tell him if he chooses to run home crying?) It will be some time before he regains any sense of self-respect.

Violence is often the single most evident marker of manhood. Rather it is the willingness to fight, the desire to fight. The origin of our expression that one has a chip on one's shoulder lies in the practice of an adolescent boy in the country or small town at the turn of the century, who would literally walk around with a chip of wood balanced on his shoulder—a signal of his readiness to fight with anyone who would take the initiative of knocking the chip off (see Gorer, 1964, p. 38; Mead, 1965).

As adolescents, we learn that our peers are a kind of gender police, constantly threatening to unmask us as feminine, as sissies. One of the favorite tricks when I was an adolescent was to ask a boy to look at his fingernails. If he held his palm toward his face and curled his fingers back to see them, he passed the test. He'd looked at his nails "like a man." But if he held the back of his hand away from his face, and looked at his fingernails with arm outstretched, he was immediately ridiculed as a sissy.

As young men we are constantly riding those gender boundaries, checking the fences we have constructed on the perimeter, making sure that nothing even remotely feminine might show through. The possibilities of being unmasked are everywhere. Even the most seemingly insignificant thing can pose a threat or activate that haunting terror. On the day the students in my course "Sociology of Men and Masculinities" were scheduled to discuss homophobia and male-male friendships, one student provided a touching illustration. Noting that it was a beautiful day, the first day of spring after a brutal northeast winter, he decided to wear shorts to class. "I had this really nice pair of new Madras shorts," he commented. "But then I thought to myself, these shorts have lavender and pink in them. Today's class topic is homophobia. Maybe today is not the best day to wear these shorts."

Our efforts to maintain a manly front cover everything we do. What we wear. How we talk. How we walk. What we eat. Every mannerism, every movement contains a coded gender language. Think, for example, of how you would answer the question: How do you "know" if a man is homosexual? When I ask this question in classes or workshops, respondents invariably provide a pretty standard list of stereotypically effeminate behaviors. He walks a certain way, talks a certain way, acts a certain way. He's very emotional; he shows his feelings. One woman commented that she "knows" a man is gay if he really cares about her; another said she knows he's gay if he shows no interest in her, if he leaves her alone.

Now alter the question and imagine what heterosexual men do to make sure no one could possibly get the "wrong idea" about them. Responses typically refer to the original stereotypes, this time as a set of negative rules about behavior. Never dress that way. Never talk or walk that way. Never show your feelings or get emotional. Always be prepared to demonstrate sexual interest in women that you meet, so it is impossible for any woman to get the wrong idea about you. In this sense, homophobia, the fear of being perceived as gay, as not a real man, keeps men exaggerating all the traditional rules of masculinity, including sexual predation with women. Homophobia and sexism go hand in hand.

The stakes of perceived sissydom are enormous—sometimes matters of life and death. We take enormous risks to prove our manhood, exposing ourselves disproportionately to health risks, workplace hazards, and stress-related illnesses. Men commit suicide three times as often as women. Psychiatrist Willard Gaylin (1992) explains that it is "invariably because of perceived social humiliation," most often tied to failure in business:

> Men become depressed because of loss of status and power in the world of men. It is not the loss of money, or the material advantages that money could buy, which produces the despair that leads to self-destruction. It is the "shame," the "humiliation," the sense of personal "failure." . . . A man despairs when he has ceased being a man among men. (p. 32)

In one survey, women and men were asked what they were most afraid of. Women responded that they were most afraid of being raped and murdered. Men responded that they were most afraid of being laughed at (Noble, 1992, pp. 105–106).

HOMOPHOBIA AS A CAUSE OF SEXISM, HETEROSEXISM, AND RACISM

Homophobia is intimately interwoven with both sexism and racism. The fear—sometimes conscious, sometimes not—that others might perceive us as homosexual propels men to enact all manner of exaggerated masculine behaviors and attitudes to make sure that no one could possibly get the wrong idea about us. One of the centerpieces of that exaggerated masculinity is putting women down, both by excluding them from the public sphere and by the quotidian put-downs in speech and behaviors that organize the daily life of the American man. Women and gay men become the "other" against which heterosexual men project their identities, against whom they stack the decks so as to compete in a situation in which they will always win, so that by suppressing them, men can stake a claim for their own manhood. Women threaten emasculation by representing the home, workplace, and familial responsibility, the negation of fun. Gay men have historically played the role of the consummate sissy in the American popular mind because homosexuality is seen as an inversion of normal gender development. There have been other "others." Through American history, various groups have represented the sissy, the non-men against whom American men played out their definitions of manhood, often with vicious results. In fact, these changing groups provide an interesting lesson in American historical development.

At the turn of the 19th century, it was Europeans and children who provided the contrast for American men. The "true American was vigorous, manly, and direct, not effete and corrupt like the supposed Europeans," writes Rupert Wilkinson (1986). "He was plain rather than ornamented, rugged rather than luxury seeking, a liberty loving common man or natural gentleman rather than an aristocratic oppressor or servile minion" (p. 96). The "real man" of the early 19th century was neither noble nor serf. By the middle of the century, black slaves had replaced the effete nobleman. Slaves were seen as dependent, helpless men, incapable of defending their women and children, and therefore less than manly. Native Americans were cast as foolish and naive children, so they could be infantilized as the "Red Children of the Great White Father" and therefore excluded from full manhood.

By the end of the century, new European immigrants were also added to the list of the unreal men, especially the Irish and Italians, who were seen as too passionate and emotionally volatile to remain controlled sturdy oaks, and Jews, who were seen as too bookishly effete and too physically puny to truly measure up. In the mid-20th century, it was also Asians—first the Japanese during the Second World War, and more recently, the Vietnamese during the Vietnam War—who have served as unmanly templates against which American men have hurled their gendered rage. Asian men were seen as small, soft, and effeminate—hardly men at all.

Such a list of "hyphenated" Americans—Italian-, Jewish-, Irish-, African-, Native-, Asian-, gay—composes the majority of American men. So manhood is only possible for a distinct minority, and the definition has been constructed to prevent the others from achieving it. Interestingly, this emasculation of one's enemies has a flip side—and one that is equally gendered. These very groups that have historically been cast as less than manly were also, often simultaneously, cast as hypermasculine, as sexually aggressive, violent rapacious beasts, against whom "civilized" men must take a decisive stand and thereby rescue civilization. Thus black men were depicted as rampaging sexual beasts, women as carnivorously carnal, gay men as sexually insatiable, southern European men as sexually predatory and voracious, and Asian men as vicious and cruel torturers who were immorally disinterested in life itself, willing to sacrifice their entire people for their whims. But whether one saw these groups as effeminate sissies or as brutal uncivilized savages, the terms with which they were perceived were gendered. These groups become the "others," the screens against which traditional conceptions of manhood were developed.

Being seen as unmanly is a fear that propels American men to deny manhood to others, as a way of proving the unprovable—that one is fully manly. Masculinity becomes a defense against the perceived threat of humiliation in the eyes of other men, enacted through a "sequence of postures"—things we might say, or do, or even think, that, if we thought carefully about them, would make us ashamed of ourselves (Savran, 1992, p. 16). After all, how many of us have made homophobic or sexist remarks, or told racist jokes, or made lewd comments to women on the street? How many of us have translated those ideas and those words into actions, by physically attacking gay men, or forcing or cajoling a woman to have sex even though she didn't really want to because it was important to score?

POWER AND POWERLESSNESS IN THE LIVES OF MEN

I have argued that homophobia, men's fear of other men, is the animating condition of the dominant definition of masculinity in America, that the reigning definition of masculinity is a defensive effort to prevent being emasculated. In our efforts to suppress or overcome those fears, the dominant culture exacts a tremendous price from those deemed less than fully manly: women, gay men, nonnative-born men, men of color. This perspective may help clarify a paradox in men's lives, a paradox in which men have virtually all the power and yet do not feel powerful (see Kaufman, 1993).

Manhood is equated with power—over women, over other men. Everywhere we look, we see the institutional expression of that power—in state and national legislatures, on the boards of directors of every major U.S. corporation or law firm, and in every school and hospital administration. Women have long understood this, and feminist women have spent the past three decades challenging both the public and the private expressions of men's power and acknowledging their fear of men. Feminism as a set of theories both explains women's fear of men and empowers women to confront it both publicly and privately. Feminist women have theorized that masculinity is about the drive for domination, the drive for power, for conquest.

This feminist definition of masculinity as the drive for power is theorized from women's point of view. It is how women experience masculinity. But it assumes a symmetry between the public and the private that does not conform to men's experiences. Feminists observe that women, as a group, do not hold power in our society. They also observe that individually, they, as women, do not feel powerful. They feel afraid, vulnerable. Their observation of the social reality and their individual experiences are therefore symmetrical. Feminism also observes that men, as a group, *are* in power. Thus, with the same symmetry, feminism has tended to assume that individually men must feel powerful.

This is why the feminist critique of masculinity often falls on deaf ears with men. When confronted with the analysis that men have all the power, many men react incredulously. "What do you mean, men have all the power?" they ask. "What are you talking about? My wife bosses me around. My kids boss me around. My boss bosses me around. I have no power at all! I'm completely powerless!"

Men's feelings are not the feelings of the powerful, but of those who see themselves as powerless. These are the feelings that come inevitably from the discontinuity between the social and the psychological, between the aggregate analysis that reveals how men are in power as a group and the psychological fact that they do not feel powerful as individuals. They are the feelings of men who were raised to believe themselves entitled to feel that power, but do not feel it. No wonder many men are frustrated and angry.

This may explain the recent popularity of those workshops and retreats designed to help men to claim their "inner" power, their "deep manhood," or their "warrior within." Authors such as Bly (1990), Moore and Gillette (1991, 1992, 1993a, 1993b), Farrell (1986, 1993), and Keen (1991) honor and respect men's feelings of powerlessness and acknowledge those feelings to be both true and real. "They gave white men the semblance of power," notes John Lee, one of the leaders of these retreats (quoted in *Newsweek*, p. 41). "We'll let you run the country, but in the meantime, stop feeling, stop talking, and continue swallowing your pain and your hurt." (We are not told who "they" are.)

Often the purveyors of the mythopoetic men's movement, that broad umbrella that encompasses all the groups helping men to retrieve this mythic deep manhood, use the image of the chauffeur to describe modern man's position. The chauffeur appears to have the power—he's wearing the uniform, he's in the driver's seat, and he knows where he's going. So, to the observer, the chauffeur looks as though he is in command. But to the chauffeur himself, they note, he is merely taking orders. He is not at all in charge.[7]

Despite the reality that everyone knows chauffeurs do not have the power, this image remains appealing to the men who hear it at these weekend workshops. But there is a missing piece to the image, a piece concealed by the framing of the image in terms of the individual man's experience. That missing piece is that the person who is giving the orders is also a man. Now we have a relationship *between* men—between men giving orders and other men taking those orders. The man who identifies with the chauffeur is entitled to be the man giving the orders, but he is not. ("They," it turns out, are other men.)

The dimension of power is now reinserted into men's experience not only as the product of individual experience but also as the product of relations with other men. In this sense, men's experience of powerlessness is *real*—the men actually feel it and certainly act on it—but it is not *true,* that is, it does not accurately describe their condition. In contrast to women's lives, men's lives are structured around relationships of power and men's differential access to power, as well as the differential access to that power of men as a group. Our imperfect analysis of our own situation leads us to believe that we men need *more* power, rather than leading us to support feminists' efforts to rearrange power relationships along more equitable lines.

Philosopher Hannah Arendt (1970) fully understood this contradictory experience of social and individual power:

> Power corresponds to the human ability not just to act but to act in concert. Power is never the property of an individual; it belongs to a group and remains in existence only so long as the group keeps together. When we say of somebody that he is "in power" we actually refer to his being empowered by a certain number of people to act in their name. The moment the group, from which the power originated to begin with . . . disappears, "his power" also vanishes. (p. 44)

Why, then, do American men feel so powerless? Part of the answer is because we've constructed the rules of manhood so that only the tiniest fraction of men come to believe that they are the biggest of wheels, the sturdiest of oaks, the most virulent repudiators of femininity, the most daring and aggressive. We've managed to disempower the overwhelming majority of American men by other means—such as discriminating on the basis of race, class, ethnicity, age, or sexual preference.

Masculinist retreats to retrieve deep, wounded, masculinity are but one of the ways in which American men currently struggle with their fears and their shame. Unfortunately, at the very moment that they work to break down the isolation that governs men's lives, as they enable men to express those fears and that shame, they ignore the social power that men continue to exert over women and the privileges from which they (as the middle-aged, middle-class white men who largely make up these retreats) continue to benefit—regardless of their experiences as wounded victims of oppressive male socialization.

Others still rehearse the politics of exclusion, as if by clearing away the playing field of secure gender identity of any that we deem less than manly—women, gay men, nonnative-born men, men of color—middle-class, straight, white men can reground their sense of themselves without those haunting fears and that deep shame that they are unmanly and will be exposed by other men. This is the manhood of racism, of sexism, of homophobia. It is the manhood that is so chronically insecure that it trembles at the idea of lifting the ban on gays in the military, that is so threatened by women in the workplace that women become the targets of sexual harassment, that is so deeply frightened of equality that it must ensure that the playing field of male competition remains stacked against all newcomers to the game.

Exclusion and escape have been the dominant methods American men have used to keep their fears of humiliation at bay. The fear of emasculation by other men, of being humiliated, of being seen as a sissy, is the leitmotif in my reading of the history of American manhood. Masculinity has become a relentless test by which we prove to other men, to women, and ultimately to ourselves, that we have successfully mastered the part. The restlessness that men feel today is nothing new in American history; we have been anxious and restless for almost two centuries. Neither exclusion nor escape has ever brought us the relief we've sought, and there is no reason to think that either will solve our problems now. Peace of mind, relief from gender struggle, will come only from a politics of inclusion, not exclusion, from standing up for equality and justice, and not by running away.

NOTES

1. Of course, the phrase "American manhood" contains several simultaneous fictions. There is no single manhood that defines all American men; "America" is meant to refer to the United States proper, and there are significant ways in which this "American manhood" is the outcome of forces that transcend both gender and nation, that is, the global economic development of industrial capitalism. I use it, therefore, to describe the specific hegemonic version of masculinity in the United States, that normative constellation of attitudes, traits, and behaviors that became the standard against which all other masculinities are measured and against which individual men measure the success of their gender accomplishments.
2. Much of this work is elaborated in *Manhood: The American Quest* (in press).
3. Although I am here discussing only American masculinity, I am aware that others have located this chronic instability and efforts to prove manhood in the particular cultural and economic arrangements of Western society. Calvin, after all, inveighed against the disgrace "for men to become effeminate," and countless other theorists have described the mechanics of manly proof. (See, for example, Seidler, 1994.)
4. I do not mean to argue that women do not have anxieties about whether they are feminine enough. Ask any woman how she feels about being called aggressive; it sends a chill into her heart because her femininity is suspect. (I believe that the reason for the enormous recent popularity of sexy lingerie among women is that it enables women to remember they are still feminine underneath their corporate business suit—a suit that apes masculine styles.) But I think the stakes are not as great for women and that women have greater latitude in defining their identities around these questions than men do. Such are the ironies of sexism: The powerful have a narrower range of options than the powerless, because the powerless can *also* imitate the powerful and get away with it. It may even enhance status, if done with charm and grace—that is, not threatening. For the powerful, any hint of behaving like the powerless is a fall from grace.
5. Such observations also led journalist Heywood Broun to argue that most of the attacks against feminism came from men who were shorter than 5 ft. 7 in. "The man who, whatever his physical size, feels secure in his own masculinity and in his own relation to life is rarely resentful of the opposite sex" (cited in Symes, 1930, p. 139).
6. Some of Freud's followers, such as Anna Freud and Alfred Adler, did follow up on these suggestions. (See especially Adler, 1980.) I am grateful to Terry Kupers for his help in thinking through Adler's ideas.
7. The image is from Warren Farrell, who spoke at a workshop I attended at the First International Men's Conference, Austin, Texas, October 1991.

REFERENCES

Adler, A. (1980). *Cooperation between the sexes: Writings on women, love and marriage, sexuality and its disorders* (H. Ansbacher & R. Ansbacher, Eds. & Trans.). New York: Jason Aronson.

Arendt, H. (1970). *On revolution.* New York: Viking.

Bly, R. (1990). *Iron John: A book about men.* Reading, MA: Addison-Wesley.

Brannon, R. (1976). The male sex role—and what it's done for us lately. In R. Brannon & D. David (Eds.), *The forty-nine percent majority* (pp. 1–40). Reading, MA: Addison-Wesley.

Connell, R. W. (1987). *Gender and power.* Stanford, CA: Stanford University Press.

Farrell, W. (1986). *Why men are the way they are.* New York: McGraw-Hill.

Farrell, W. (1993). *The myth of male power: Why men are the disposable sex.* New York: Simon & Schuster.

Gaylin, W. (1992). *The male ego.* New York: Viking.

Goffman, E. (1963). *Stigma.* Englewood Cliffs, NJ: Prentice Hall.

Gorer, G. (1964). *The American people: A study in national character.* New York: Norton.

Kaufman, M. (1993). *Cracking the armour: Power and pain in the lives of men.* Toronto: Viking Canada.

Keen, S. (1991). *Fire in the belly.* New York: Bantam.

Kimmel, M. S. (in press). *Manhood: The American quest.* New York: HarperCollins.

Leverenz, D. (1986). Manhood, humiliation and public life: Some stories. *Southwest Review, 71,* Fall.

Leverenz, D. (1991). The last real man in America: From Natty Bumppo to Batman. *American Literary Review, 3.*

Marx, K., & F. Engels. (1848/1964). The communist manifesto. In R. Tucker (Ed.), *The Marx-Engels reader.* New York: Norton.

Mead, M. (1965). *And keep your powder dry.* New York: William Morrow.

Moore, R., & Gillette, D. (1991). *King, warrior, magician, lover.* New York: HarperCollins.

Moore, R., & Gillette, D. (1992). *The king within: Accessing the king in the male psyche.* New York: William Morrow.

Moore, R., & Gillette, D. (1993a). *The warrior within: Accessing the warrior in the male psyche.* New York: William Morrow.

Moore, R., & Gillette, D. (1993b). *The magician within: Accessing the magician in the male psyche.* New York: William Morrow.

Noble, V. (1992). A helping hand from the guys. In K. L. Hagan (Ed.), *Women respond to the men's movement.* San Francisco: HarperCollins.

Osherson, S. (1992). *Wrestling with love: How men struggle with intimacy, with women, children, parents, and each other.* New York: Fawcett.

Rotundo, E. A. (1993). *American manhood: Transformations in masculinity from the revolution to the modern era.* New York: Basic Books.

Savran, D. (1992). *Communists, cowboys and queers: The politics of masculinity in the work of Arthur Miller and Tennessee Williams.* Minneapolis: University of Minnesota Press.

Seidler, V. J. (1994). *Unreasonable men: Masculinity and social theory.* New York: Routledge.

Symes, L. (1930). The new masculinism. *Harper's Monthly, 161,* January.

Wayne, K. (1912). *Building the young man.* Chicago: A. C. McClurg.

What men need is men's approval. (1993, January 3). *The New York Times,* p. C-11.

Wilkinson, R. (1986). *American tough: The tough-guy tradition and American character.* New York: Harper & Row.

Woolfolk, R. L., & Richardson, F. (1978). *Sanity, stress and survival.* New York: Signet.

4

"Macho": Contemporary Conceptions

ALFREDO MIRANDÉ

MI NOCHE TRISTE

My own *noche triste* occurred when my father returned from location on the film *Capitán de Castilla* (Captain from Castille). I remember that he had been gone for a long time, that he came back from Morelia with a lot of presents, and that at first, I was very happy to see him. Then there was a big fight; my parents argued all night, and they separated shortly thereafter. One night when my mother was very sad and depressed, she went to *el árbol de la noche triste.* As she cried by the tree she thought about how she and Hernán Cortés both had been in the same situation: depressed, weeping, and alone.

After my parents separated, my brothers and I went with my father and moved to Tacubaya to live with his mother, Anita, and her mother (my great-grandmother), Carmela (Mamá Mela). Grandmother Anita, or *Abillá,* as we called her, was a petite, energetic little woman, but Mamá Mela was tall, dark, and stately. In Tacubaya we were also surrounded by family, but now it was my father's family, Mirandé-Salazar. His family was smaller because he was an only child and because his father's two siblings, Concha (Consuelo) and Lupe (Guadalupe), never married or had children. My grandfather, Alfredo, died when I was about two years old, but I remember him.

In Tacubaya we first lived in a big, long house with a large green entrance, *El Nueve* (nine), on a street called Vicente Eguía, before moving to an apartment house, *El Trece* (thirteen), down the street. At *El Trece* we lived in the first apartment, and my great-aunts, Concha and Lupe, lived in *El Seis* (six). Concha had been an elementary school teacher and Lupe was an artist. They were retired but very active; both did a lot of embroidering and Lupe was always painting. I was very fond of *las tías.* To me *las tías* always seemed old and very religious, but I was very close to my aunts and loved them deeply. They wore black shawls and went to church early each morning. When I wasn't playing in the courtyard, I was often visiting with my aunts. They taught me catechism, and Tía Lupe was my *madrina,* or godmother, for my first communion.

I would spend hours with *las tías,* fascinated by their conversation. It seemed that every minute was filled with stories about the Mexican Revolution and about my grandfather, Alfredo. I especially liked it when they spoke about him, as I had been named Alfredo and identified with him. They said he was a great man and that they would be very proud and happy if I grew up to be like him someday. No, it was actually that I had no choice—I was destined to be like him. Because I had the good fortune of being named after Alfredo, I had to carry on his name, and, like him, I too would be a great man someday. I should add that my aunts stressed *man* when they talked about him. In other words, I had a distinct impression that my grandfather and I were linked not only because we were both named Mirandé and Alfredo, but also because we were both men. I did not realize it at the time, but my teachers (who were mostly women)—*las tías,* my *Abillá,* Mamá Mela, and my mother—were socializing me into my "sex role." But I don't remember anyone describing Grandfather Alfredo as "macho." Perhaps my *tías* took his being "macho" for granted, since he was obviously male.

I do not know very much about Alfredo's family, except that his father, Juan, or *Jean,* came to México from France and married a *mexicana,* María. I also learned from my mother that Alfredo was of humble origins and was, in a very real sense, a self-made man who studied and pursued a career as a civil engineer. He was committed to bringing about social justice and distributing the land held by the *hacendados* (landowners) among the Mexican *peones.* As a civilian he served under Emiliano Zapata, making cannons and munitions. According to historian John Womack . . . , Alfredo Mirandé was one of Zapata's key assistants and worked as a spy in Puebla for some time under the code name "Delta." While he was in hiding, my *Abillá* would take in other people's clothes to mend and to launder to earn money so that the family could survive. My grandfather grew to be disillusioned, however, as the Revolution did not fulfill its promise of bringing about necessary economic and social reforms.

My *tías* had a photograph of Alfredo standing proudly in front of a new, experimental cannon that he had built. They related that a foolish and headstrong general, anxious to try out the new cannon, pressured Alfredo to fire it before it was ready. My grandfather reluctantly complied and received severe burns all over his body, almost dying as a result of the explosion. It took him months to recover from the accident.

As I think back, most of the stories they told me had a moral and were designed, indirectly at least, to impart certain values. What I learned from my *tías* and, indirectly, from my grandfather was that although one should stand up for principles, one should attempt to avoid war and personal conflicts, if at all possible. One should also strive to be on a higher moral plane than one's adversaries. Alfredo was intelligent, strong, and principled. But what impressed me most is that he was said to be incredibly just and judicious. Everyone who knew him said he treated people of varying educational and economic levels fairly, equally, and with dignity and respect.

I realize that Alfredo lived in a society and a historical period in which women were relegated to an inferior status. Yet I also know that he and my grandmother shared a special intimacy and mutual respect such as I have never personally encountered. By all accounts they loved and respected each other and shared an incredible life together. I have read letters that my grandfather wrote to my grandmother when they were apart, and they indicate that he held her in very high regard and treated her as an equal partner.

"MACHO": AN OVERVIEW

Mexican folklorist Vicente T. Mendoza suggested that the word "macho" was not widely used in Mexican songs, *corridos* (folk ballads), or popular culture until the 1940s. . . . Use of the word was said to have gained in popularity after Avila Camacho became president. The word lent itself to use in *corridos* because "macho" rhymed with "Camacho."

While "macho" has traditionally been associated with Mexican or Latino culture, the word has recently been incorporated into American popular culture, so much so that it is now widely used to describe everything from rock stars and male sex symbols in television and film to burritos. When applied to entertainers, athletes, or other "superstars," the implied meaning is clearly a positive one that connotes strength, virility, masculinity, and sex appeal. But when applied to Mexicans or Latinos, "macho" remains imbued with such negative attributes as male dominance, patriarchy, authoritarianism, and spousal abuse. Although both meanings connote strength and power, the Anglo macho is clearly a much more positive and appealing symbol of manhood and masculinity. In short, under current usage the Mexican macho oppresses and coerces women, whereas his Anglo counterpart appears to attract and seduce them.

This [reading] focuses on variations in perceptions and conceptions of the word "macho" held by Mexican and Latino men. Despite all that has been written and said about the cult of masculinity and the fact that male dominance has been assumed to be a key feature of Mexican and Latino culture, very little research exists to support this assumption. Until recently such generalizations were based on stereotypes, impressionistic evidence, or the observations of ethnographers such as Oscar Lewis . . . , Arthur Rubel . . . , and William Madsen. . . . These Anglo ethnographers were criticized by noted Chicano folklorist Américo Paredes . . . for the persistent ignorance and insensitivity to Chicano language and culture that is reflected in their work. Paredes contended, for example, that although most anthropologists present themselves as politically liberal and fluent in Spanish, many are only minimally fluent and fail to grasp the nuance and complexity of Chicano language. There is, it seems, good reason to be leery of their findings and generalizations regarding not only gender roles but also all aspects of the Mexican/Latino experience.

Utilizing data obtained through qualitative open-ended questions, I look in this chapter at how Latino men themselves perceive the word "macho" and how they describe men who are considered "*muy machos.*" Although all of the respondents were living in the United States at the time of the interviews, many were foreign-born and retained close ties with Mexican/Latino culture. Since they had been subjected to both Latino and American influences, I wondered whether they would continue to adhere to traditional Mexican definitions of "macho" or whether they had been influenced by contemporary American conceptions of the word.

Specifically, an attempt was made in the interviews to examine two polar views. The prevailing view in the social science literature of the Mexican macho is a negative one. This view holds that the origins of the excessive masculine displays and the cult of masculinity in México and other Latino countries can be traced to the Spanish Conquest, as the powerless colonized man attempted to compensate for deep-seated feelings of inadequacy and inferiority by assuming a hypermasculine, aggressive, and domineering stance. There is a second and lesser-known view that is found in Mexican popular culture, particularly in film and music, one that reflects a more positive, perhaps idyllic, conception of Mexican culture and national character. Rather than focusing on violence and male dominance, this second view associates macho qualities with the evolution of a distinct code of ethics.

Un hombre que es macho is not hypermasculine or aggressive, and he does not disrespect or denigrate women. Machos, according to the positive view, adhere to a code of ethics that stresses humility, honor, respect of oneself and others, and courage. What may be most significant in this second view is that being "macho" is not manifested by such outward qualities as physical strength and virility but by such inner qualities as personal integrity, commitment, loyalty, and, most importantly, strength of character. Stated simply, a man who acted like my Tío Roberto would be macho in the first sense of the word but certainly not in the second. It is not clear how this code of ethics developed, but it may be linked to nationalist sentiments and Mexican resistance to colonization and foreign invasion. Historical figures such as Cuauhtémoc, *El Pipíla, Los Niños Héroes,* Villa, and Zapata would be macho according to this view. In music and film positive macho figures such as Pedro Infante, Jorge Negrete, and even Cantinflas are patriots, but mostly they are *muy hombres,* men who stand up against class and racial oppression and the exploitation of the poor by the rich.

Despite the apparent differences between the two views, both see the macho cult as integral to Mexican and Latino cultures. Although I did not formulate explicit hypotheses, I entered the field expecting that respondents would generally identify with the word "macho" and define it as a positive trait or quality in themselves and other persons. An additional informal hypothesis proposed was that men who had greater ties to Latino culture and the Spanish language would be more likely to identify and to have positive associations with the word. I expected, in other words, that respondents would be more likely to adhere to the positive view of macho.

FINDINGS: CONCEPTIONS OF MACHO

Respondents were first asked the following question: "What does the word 'macho' mean to you?" The interviewers were instructed to ask this and all other questions in a neutral tone, as we wanted the respondents to feel that we really were interested in what they thought. We stressed in the interviews that there were no "right" or "wrong" answers to any of the questions. This first question was then followed by a series of follow-up questions that included: "Can you give me an example (or examples) of someone you think is really macho?"; "What kinds of things do people who are really macho do?"; and "Can a woman be macha?"

Each person was assigned an identification number, and the responses to the above questions were typed on a large index card. Three bilingual judges, two men and one woman, were asked to look at the answers on the cards and to classify each respondent according to whether they believed the respondent was generally "positive," "negative," or "neutral" toward the word "macho." Those respondents classified as "positive" saw the term as a desirable cultural or personal trait or value, identified with it, and believed that it is generally good to be, or at least to aspire to be, macho. But those respondents classified as "negative" by the judges saw it as an undesirable or devalued cultural or personal trait, did not identify with being macho, and believed that it is generally bad or undesirable to be macho. In the third category, respondents were classified as "neutral" if they were deemed to be indifferent or ambivalent or to recognize both positive and negative components of the word "macho." For these respondents, macho was "just a word," or it denoted a particular male feature without imputing anything positive or negative about the feature itself.

Overall there was substantial agreement among the judges. In 86 percent of 105 cases the judges were in complete agreement in their classifications, and in another 12 percent two out of three agreed. In other words, in only two instances was there complete disagreement among the judges in which one judge ranked the respondents positive, another negative, and still another neutral.

One of the most striking findings is the extent to which the respondents were polarized in their views of macho. Most had very strong feelings; very few were neutral or indifferent toward the word. In fact, only 11 percent of the 105 respondents were classified as neutral by our judges. No less surprising is the fact that, contrary to my expectations, very few respondents viewed the word in a positive light. Only 31 percent of the men were positive in their views of macho, compared to 57 percent who were classified as negative. This means, in effect, that more than two-thirds of the respondents believed that the word "macho" had either negative or neutral connotations.

My expectation that those individuals with greater ties to Latino culture would be more likely to identify and to have positive associations with "macho" was also not supported by the data. Of the thirty-nine respondents who opted to be interviewed in Spanish, only 15 percent were seen as having a positive association with macho, whereas 74 percent were negative and 10 percent were neutral toward the term. In contrast, of the sixty-six interviewed in English, 41 percent were classified as positive, 47 percent as negative, and 12 percent as neutral toward the term.

Although negative views of the word "macho" were more prevalent than I had expected, the responses closely parallel the polar views of the word "macho" discussed earlier. Responses classified as "negative" by our judges are consistent with the "compensatory" or "deficit" model, which sees the emphasis on excessive masculinity among Mexicans and Latinos as an attempt to conceal pervasive feelings of inferiority among native men that resulted from the Conquest and the ensuing cultural, moral, and spiritual rape of the indigenous population. Those classified as "positive," similarly, are roughly consistent with an "ethical" model, which sees macho behavior as a positive, nationalist response to colonization, foreign intervention, and class exploitation.

Negative Conceptions of "Macho"

A number of consistent themes are found among the men who were classified as viewing the word "macho" in a negative light. Though I divide them into separate themes to facilitate the presentation of the findings, there is obviously considerable overlap between them.

Negative Theme 1: Synthetic/Exaggerated Masculinity. A theme that was very prevalent in the responses is that machos are men who are insecure in themselves and need to prove their manhood. It was termed a "synthetic self-image," "exaggerated masculinity," "one who acts tough and is insecure in himself," and an "exaggerated form of manliness or super manliness." One respondent described a macho as

> one who acts "bad." One who acts tough and who is insecure of himself. I would say *batos* [dudes] who come out of the *pinta* [prison] seem to have a tendency to be insecure with themselves, and tend to put up a front. [They] talk loud, intimidate others, and disrespect the meaning of a man.

Another person described it as

> being a synthetic self-image that's devoid of content. . . . It's a sort of facade that people use to hide the lack of strong, positive personality traits. To me, it often implies a negative set of behaviors. . . . I have a number of cousins who fit that. I have an uncle who fits it. He refuses to have himself fixed even though he was constantly producing children out of wedlock.

Negative Theme 2: Male Dominance/Authoritarianism. A second, related theme is that of male dominance, chauvinism, and the double standard for men and women. Within the family, the macho figure is viewed as authoritarian, especially relative to the wife. According to one respondent, "They insist on being the dominant one in the household. What they say is the rule. They treat women as inferior. They have a dual set of rules for women and men." Another respondent added:

> It's someone that completely dominates. There are no two ways about it; it's either his way or no way. My dad used to be a macho. He used to come into the house drunk, getting my mother out of bed, making her make food, making her cry.

A Spanish-speaker characterized the macho as follows:

> *Una persona negativa completamente. Es una persona que es irresponsable en una palabra. Que anda en las cantinas. Es no es hombre. Si, conozco muchos de mi tierra; una docena. Toman, pelean. Llegan a la case gritando y golpeando a la señora, gritando, cantando. Eso lo vi yo cuando era chavalillo y se me grabó. Yo nunca vi a mi papá que golpeara a mi mamá* (A completely negative person. In a word, it's a person who is irresponsible. Who is out in the taverns. That's not a man. Yes, I know many from my homeland; a dozen. They drink, fight. They come home yelling and hitting the wife, yelling, singing. I saw this as a child and it made a lasting impression on me. I never saw my father hit my mother).

Negative Theme 3: Violence/Aggressiveness. A third, related theme is macho behavior manifested in expressions of violence, aggressiveness, and irresponsibility, both inside and outside the family. It is "someone that does not back down, especially if they fear they would lose face over the most trivial matters." Another person saw macho as the exaggeration of perceived masculine traits and gave the example of a fictional figure like Rambo and a real figure like former president Ronald Reagan. This person added that it was "anyone who has ever been in a war," and "it's usually associated with dogmatism, with violence, with not showing feelings." A Spanish-speaking man summarized it succinctly as "*el hombre que sale de su trabajo los viern̄s, va a la cantina, gasta el cheque, y llega a su casa gritando, pegándole a su esposa diciendo que él es el macho*" (the man who gets out of work on Friday, goes to a bar, spends his check, and comes home yelling and hitting his wife and telling her that he is the macho [i.e., man]). Still another felt that men who were macho did such things as "drinking to excess," and that associated with the word "macho" was "the notion of physical prowess or intimidation of others. A willingness to put themselves and others at risk, particularly physically. For those that are married, the notion of having women on the side."

One of our Spanish-speaking respondents mentioned an acquaintance who lost his family because he would not stop drinking. "*Él decía, 'La mujer se hizo para andar en la casa y yo pa' andar en las cantinas'*" (He used to say, "Woman was made to stay at home and I was made to stay in taverns"). This respondent also noted that men who are real machos tend not to support their families or tend to beat them, to get "dandied up," and to go out drinking. Another said that they "drink tequila" and "have women on their side kissing them."

Negative Theme 4: Self-Centeredness/*Egoísmo*. Closely related is the final theme, which views someone who is macho as being self-centered, selfish, and stubborn, a theme that is especially prevalent among respondents with close ties to México. Several men saw machismo as *un tipo de egoísmo* (a type of selfishness) and felt that it referred to a person who always wanted things done his way—*a la mía*. It is someone who wants to impose his will on others or wants to be right, whether he is right or not. It is viewed, for example, as

> *un tipo de egoísmo que nomás "lo mío" es bueno y nomás mis ideas son buenas. Como se dice, "Nomás mis chicharrones truenan." . . . Se apegan a lo que ellos creen. Todo lo que ellos dicen está correcto. Tratan que toda la gente entre a su manera de pensar y actuar, incluyendo hijos y familia* (A type of selfishness where only "mine" is good and only my ideas are worthwhile. As the saying goes, "Whatever I say goes." . . . They cling to their own beliefs. Everything they say is right. They try to get everyone, including children and family, to think and act the way they do).

Some respondents who elaborated on the "self-centeredness" or *egoísta* theme noted that some men will hit their wives "just to prove that they are machos," while others try to show that they "wear the pants" by not letting their wives go out. One person noted that some men believe that wives and daughters should not be permitted to cut their hair because long hair is considered "a sign of femininity," and another made reference to a young man who actually cut off a finger in order to prove his love to his sweetheart.

Because the word "macho" literally means a "he-mule" or a "he-goat," respondents often likened macho men to a dumb animal such as a mule, goat, or bull: "Somebody who's like a bull, or bullish"; "The man who is strong as though he were an animal"; "It's an ignorant person, like an animal, a donkey or mule"; and "It's a word that is outside of that which is human." One person described a macho as

> the husband of the mule that pulls the plow. A macho is a person who is dumb and uneducated. *Hay tienes a* [There you have] Macho Camacho [the boxer]. He's a wealthy man, but that doesn't make a smart man. I think he's dumb! . . . They're aggressive, and they're harmful, and insensitive.

Another respondent said, "Ignorant, is what it means to me, a fool. They're fools, man. They act bully type." Another similarity linked it to being "ignorant, dumb, stupid," noting that they "try to take advantage of their physical superiority over women and try to use that as a way of showing that they are right."

Given that these respondents viewed "macho" in a negative light, it is not surprising to find that most did not consider themselves macho. Only eight of the sixty men in this category reluctantly acknowledged that they were "somewhat" macho. One said, "Yes, sometimes when I drink, I get loud and stupid," and another, "Yes, to an extent because I have to be headstrong and bullish as a teacher."

Positive Conceptions of Macho: Courage, Honor, and Integrity

As previously noted, only about 30 percent of the respondents were classified as seeing macho as a desirable cultural or personal trait or value, and those who did so were much more apt to conduct the interview in English. Some 82 percent of the men who had positive conceptions were interviewed in English.

As was true of men who were classified as negative toward the word "macho," several themes were discernible among those classified as positive. And as with the negative themes, they are separate but overlapping. A few respondents indicated that it meant "masculine" or "manly" (*varonil*), a type of masculinity (*una forma de masculinidad*), or male. The overriding theme, however, linked machismo to internal qualities like courage, valor, honor, sincerity, respect, pride, humility, and responsibility. Some went so far as to identify a distinct code of ethics or a set of principles that they saw as being characteristic of machismo.

Positive Theme 1: Assertiveness/Standing Up for Rights. A more specific subtheme is the association of machismo with being assertive, courageous, standing up for one's rights, or going "against the grain" relative to other persons. The following response is representative of this view:

> To me it means someone that's assertive, someone who stands up for his or her rights when challenged. . . . Ted Kennedy because of all the hell he's had to go through. I think I like [Senator] Feinstein. She takes the issues by the horns. . . . They paved their own destiny. They protect themselves and those that are close to them and attempt to control their environment versus the contrast.

It is interesting to note that this view of being macho can be androgynous. Several respondents mentioned women who exemplified "macho qualities" or indicated that these qualities may be found among either gender. Another man gave John Kennedy and Eleanor Roosevelt as examples and noted that people who are macho

> know how to make decisions because they are confident of themselves. They know their place in the world. They accept themselves for what they are and they are confident in that. They don't worry about what others think. . . . They know what to do, the things that are essential to them and others around them.

A Spanish-speaking respondent added:

> *En respecto a nuestra cultura es un hombre que defiende sus valores, en total lo físico, lo emocional, lo psicológico. En cada mexicano hay cierto punto de macho. No es arrogante, no es egoísta excepto cuando tiene que defender sus valores. No es presumido* (Relative to our culture, it's a man that stands up for what he believes, physically, emotionally, and psychologically. Within every Mexican there is a certain sense of being macho. He is not arrogant, not egotistic, except when he has to defend his values. He is not conceited).

Positive Theme 2: Responsibility/Selflessness. A second positive macho theme is responsibility, selflessness, and meeting obligations. In direct opposition to the negative macho who is irresponsible and selfish, the positive macho is seen as having a strong sense of responsibility and as being very concerned with the welfare and well-being of other persons. This second positive macho theme was described in a number of ways: "to meet your obligations"; "someone who shoulders responsibility"; "being responsible for your family"; "a person who fulfills the responsibility of his role . . . irrespective of the consequences"; "they make firm decisions . . . that take into consideration the well-being of others." According to one respondent,

A macho personality for me would be a person that is understanding, that is caring, that is trustworthy. He is all of those things and practices them as well as teaches them, not only with family but overall. It encompasses his whole life.

It would be a leader with compassion. The image we have of Pancho Villa. For the Americans it would be someone like Kennedy, as a strong person, but not because he was a womanizer.

Positive Theme 3: General Code of Ethics. The third theme we identified embodies many of the same traits mentioned in the first and second themes, but it differs in that respondents appear to link machismo not just to such individual qualities as selflessness but to a general code of ethics or a set of principles. One respondent who was married to an Israeli woman offered a former defense minister of Israel as exemplifying macho qualities. He noted that

It's a man responsible for actions, a man of his word. . . . I think a macho does not have to be a statesman, just a man that's known to stand by his friends and follow through. A man of action relative to goals that benefit others, not himself.

Another said that it means living up to one's principles to the point of almost being willing to die for them. One of the most extensive explications of this code of ethics was offered by the following respondent:

To me it really refers to a code of ethics that I use to relate values in my life and to evaluate myself in terms of my family, my job, my community. My belief is that if I live up to my code of ethics, I will gain respect from my family, my job, and my community. Macho has nothing to do with how much salsa you can eat, how much beer you can drink, or how many women you fuck!

They have self-pride, they hold themselves as meaningful people. You can be macho as a farmworker or judge. It's a real mixture of pride and humility. Individualism is a part of it—self-awareness, self-consciousness, responsibility.

Positive Theme 4: Sincerity/Respect. The final positive theme overlaps somewhat with the others and is often subsumed under the code of ethics or principles. A number of respondents associated the word "macho" with such qualities as respect for oneself and others, acting with sincerity and respect, and being a man of your word. One of our interviewees said,

Macho significa una persona que cumple con su palabra y que es un hombre total. . . . Actúan con sinceridad y con respeto (Macho means a person who backs up what he says and who is a complete man. . . . They act with sincerity and respect).

Another mentioned self-control and having a sense of oneself and the situation.

Usually they are reserved. They have kind of an inner confidence, kind of like you know you're the fastest gun in town so you don't have to prove yourself. There's nothing to prove. A sense of self.

Still another emphasized that physical prowess by itself would not be sufficient to identify one as macho. Instead, "It would be activities that meet the challenge, require honor, and meet obligations." Finally, a respondent observed:

Macho to me means that you understand your place in the world. That's not to say that you are the "he-man" as the popular conception says. It means you have respect for yourself, that you respect others.

Not surprisingly, all of the respondents who viewed machismo in a positive light either already considered themselves to have macho qualities or saw it as an ideal they hoped to attain.

Neutral Conceptions of Macho

Twelve respondents could not be clearly classified as positive or negative in their views of "macho." This so-called neutral category is somewhat of a residual one, however, because it includes not only men who were, in fact, neutral but also those who gave mixed signals and about whom the judges could not agree. One said that "macho" was just a word that didn't mean anything; another said that it applied to someone strong like a boxer or a wrestler, but he did not know anyone who was macho, and it was not clear whether he considered it to be a positive or negative trait. Others were either ambivalent or pointed to both positive and negative components of being macho. A street-wise young man in his mid-twenties, for example, indicated that

> The word macho to me means someone who won't take nothing from no one. Respects others, and expects a lot of respect from others. The person is willing to take any risk. . . . They always think they can do anything and everything. They don't take no shit from no one. They have a one-track mind. Never want to accept the fact that women can perform as well as men.

Significantly, the judges were divided in classifying this respondent; one classified him as negative, another as positive, and the third as neutral. The fact is that rather than being neutral, this young man identifies both positive ("respects others and self") and negative ("never want to accept the fact that women can perform as well as men") qualities with being macho.

Another person observed that there were at least two meanings of the word—one, a brave person who is willing to defend his ideals and himself, and the other, a man who exaggerates his masculinity—but noted that "macho" was not a term that he used. Another respondent provided a complex answer that distinguished the denotative (i.e., macho) and connotative (i.e., machismo) meanings of the term. He used the word in both ways, differentiating between being macho or male, which is denotative, and machismo, which connotes male chauvinism. He considered himself to be macho but certainly not *machista*.

> *Ser macho es ser valiente o no tener miedo. La connotación que tiene mal sentido es poner los intereses del hombre adelante de los de la mujer o del resto de la familia. Representa egoísmo. . . . Macho significa varón, hombre, pero el machismo es una manera de pensar, y es negativo* (To be macho is to be brave or to not be afraid. The connotation that is negative is to put the interests of the man ahead of those of the woman or the rest of the family. It represents selfishness. . . . Macho means male, man, but machismo is a way of thinking, and it is negative).

Another person similarly distinguished between being macho and being *machista*.

> *Pues, en el sentido personal, significa el sexo masculino y lo difiere del sexo femenino. La palabra machismo existe solamente de bajo nivel cultural y significa un hombre valiente, borracho y pendenciero* (Well, in a personal sense, it means the masculine gender and it distinguishes it from the feminine. The word machismo exists only at a low cultural level and it means a brave man, a drunkard, and a hell-raiser).

Six of the twelve respondents who were classified as neutral considered themselves to be at least somewhat macho.

REGIONAL AND SOCIOECONOMIC DIFFERENCES IN CONCEPTIONS OF MACHO

Conceptions of the word "macho" do not vary significantly by region, but there are significant differences according to socioeconomic status. Men with more education, with a higher income, and in professional occupations were more likely to have a positive conception of the word. This is not to suggest that they are necessarily more *machista*, or chauvinistic, but that they simply see the word in a more positive light. Almost half (42 percent) of the respondents who were professionals associated the word "macho" with being principled or standing up for one's rights, whereas only 23 percent of nonprofessionals had a positive conception of the word.

Place of birth and language were also significantly associated with attitudes toward machismo, but, ironically, those respondents who were born in the United States and those who were interviewed in English were generally more positive toward the word "macho." Forty-two percent of those born in the United States have positive responses, compared with only 10 percent of those who were foreign-born.

An English-speaking respondent said that "macho equals to me chivalry associated with the Knights of the Round Table, where a man gives his word, defends his beliefs, etc." Another noted that machos were people who "stand up for what they believe, try things other people are afraid to do, and defend the rights of others." But one Mexican man saw it as the opposite—"*Mexicanos que aceptan que la mujer 'lleve los pantalones,' irresponsables, les dan mas atención a sus aspectos sociales que a sus responsabilidades*" (Mexicans who accept that the woman 'wear the pants,' they are irresponsible, these men pay more attention to their social lives than to their responsibilities).

REGIONAL AND SOCIOECONOMIC DIFFERENCES IN "HOW MACHOS ACT"

After defining the word "macho," respondents were asked to give an example of how people who are macho act or behave. The answers ranged from drinking to excess, acting "bad" or "tough," being insecure in themselves, to having a "synthetic self-image," a code of ethics, and being sincere and responsible. Because responses typically were either negative or positive rather than neutral or indifferent, they were grouped into two broad categories.

Regional differences were not statistically significant, although southern Californians were more likely than Texans or northern Californians to see macho behavior as aggressive or negative and to associate it with acting tough, drinking, or being selfish.

The general pattern that was observed with regard to occupation, education, and income was that professionals, those with more education, and those with higher incomes were less likely to associate the word "macho" with negative behaviors such as drinking and trying to prove one's masculinity.

Place of birth and the language in which the interview was conducted were also related to the type of behavior that was associated with the word "macho." Men born in the United States and those who opted to conduct the interview in English were significantly more likely to associate such positive behaviors as being responsible, honorable, or respectful of others with people they considered to be macho.

CONCLUSION

These data provide empirical support for two very different and conflicting models of masculinity. The compensatory model sees the cult of virility and the Mexican male's obsession with power and domination as futile attempts to mask feelings of inferiority, powerlessness, and failure, whereas the second perspective associates being macho with a code of ethics that organizes and gives meaning to behavior. The first model stresses external attributes such as strength, sexual prowess, and power; the second stresses internal qualities like honor, responsibility, respect, and courage.

Although the findings are not conclusive, they have important implications. First, and most importantly, the so-called Mexican/Latino masculine cult appears to be a more complex and diverse phenomenon than is commonly assumed. But the assumption that being macho is an important Mexican cultural value is seriously called into question by the findings. Most respondents did not define macho as a positive cultural or personal trait or see themselves as being macho. Only about one-third of the men in the sample viewed the word "macho" positively. If there is a cultural value placed on being macho, one would expect that those respondents with closer ties to Latino culture and the Spanish language would be more apt to identify and to have positive associations with macho, but the opposite tendency was found to be true. Respondents who preferred to be interviewed in English were much more likely to see macho positively and to identify with it, whereas the vast majority of those who elected to be interviewed in Spanish viewed it negatively.

A major flaw of previous conceptualizations has been their tendency to treat machismo as a unitary phenomenon. The findings presented here suggest that although Latino men tend to hold polar conceptions of macho, these conceptions may not be unrelated. In describing the term, one respondent observed that there was almost a continuum between a person who is responsible and one who is chauvinistic. If one looks more closely at the two models, moreover, it is clear that virtually every trait associated with a negative macho trait has its counterpart in a positive one. Some of the principal characteristics of the negative macho and the positive counterparts are highlighted in Table 4.1.

Table 4.1 Negative and Positive Macho Traits

Negative	Positive
Bravado	Brave
Cowardly	Courageous
Violent	Self-defensive
Irresponsible	Responsible
Disrespectful	Respectful
Selfish	Altruistic
Pretentious	Humble
Loud	Soft-spoken
Boastful	Self-effacing
Abusive	Protective
Headstrong/bullish	Intransigent
Conformist	Individualistic
Chauvinistic	Androgynous
Dishonorable	Honorable
External qualities	Internal qualities

The close parallel between negative and positive macho traits is reminiscent of Vicente T. Mendoza's distinction between genuine and false macho. According to Mendoza, the behavior of a genuine machismo is characterized by true bravery or valor, courage, generosity, stoicism, heroism, and ferocity; the negative macho simply uses the appearance of semblance of these traits to mask cowardliness and fear. . . .

From this perspective much of what social scientists have termed "macho" behavior is not macho at all, but its antithesis. Rather than attempting to isolate a modal Mexican personality type or determining whether macho is a positive or a negative cultural trait, social scientists would be well served to see Mexican and Latino culture as revolving around certain focal concerns or key issues such as honor, pride, dignity, courage, responsibility, integrity, and strength of character. Individuals, in turn, are evaluated positively or negatively according to how well they are perceived to respond to these focal concerns. But because being macho is ultimately an internal quality, those who seek to demonstrate outwardly that they are macho are caught in a double bind. A person who goes around holding his genitals, boasting about his manliness, or trying to prove how macho he is would not be considered macho by this definition. In the final analysis it is up to others to determine the extent to which a person lives up to these expectations and ideals.

It is also important to note that to a great extent, the positive internal qualities associated with the positive macho are not the exclusive domain of men but extend to either gender. One can use the same criteria in evaluating the behavior of women and employ parallel terminology such as *la hembra* (the female) and *hembrismo* (femaleness). *Una mujer que es una hembra* (a woman who is a real "female") is neither passive and submissive nor physically strong and assertive, for these are external qualities. Rather, *una hembra* is a person of strong character who has principles and is willing to defend them in the face of adversity. Thus, whereas the popular conception of the word "macho" refers to external male characteristics such as exaggerated masculinity or the cult of virility, the positive conception isolated here sees being macho as an internal, androgynous quality.

2

Making Men Out of Them: Male Socialization in Childhood and Adolescence

As a father of a son and a daughter, I have had much opportunity to witness the emergence of gender differences. When my children were very young, I liked to watch especially closely when they played. Sometimes, I'd find myself standing just outside their doors or pausing between rooms to listen to their conversations with their friends or with each other. Other times, I'd observe, with a wince or a smile, their reactions to certain jokes, television shows, books, or movies. I am still occasionally surprised by how deeply Dan, 15, and Allison, 12, display an awareness of whether a particular game, movie, shirt, or song is "for girls" or "for boys," but I was stunned by how young they were when such gender consciousness first surfaced. I'd wonder, "Where did this come from?" How did they develop such sensitivity to gender? Could these ideas have come from me or their mother? Hadn't we always tried to do it "right"? We were neither early nor enthusiastic buyers of toy guns. Both Dan and Allison took gymnastics and played soccer, and both continue to play baseball or softball. And we must have read them Stan and Jan Berenstain's *He Bear, She Bear* a hundred times each, enough so that I can still recite much of it verbatim ("You can be anything, you see, whether you are He or She . . . So many things to do and be, He Bear, She Bear, you and me . . ."). So what happened?

In graduate school, I had a friend who would liken life to "one big gender school." We'd use that shorthand expression to depict the many ways and places in which we learn to be male or female, feminine or masculine. Now that I live with two of its sharper students, who are not about to let themselves flunk out, I realize how broad the gender curriculum is; there is so much one learns, from so many teachers. You'll never graduate from "gender school," though your performance on its various tests will be evaluated and "graded" throughout your life. This chapter illustrates some of the ways in which boys and young men are taught and tested as they are socialized toward manhood.

SOME SUBTLETIES OF SOCIALIZATION

The study of men and masculinity has been dominated by an emphasis on gender socialization. *Socialization* is a complex, gradual, and cumulative process through which we learn the attitudes, values, beliefs, and behaviors appropriate to our social positions. As one of the most life-defining social positions of the many that we occupy, one's gender determines how others treat us, what they make of us, and what they attempt to make us into.

Within this perspective, we are born male or female, but we are *made* into men or women. Furthermore, without dismissing the importance of biological (i.e., hormonal, chromosomal, and anatomical) characteristics, the greater sources of masculine identities, male roles, and ideals of manhood are found in what happens after we're born, not in what we "bring with us" at birth. After all, the wide-ranging cultural differences in men's experiences of masculinity and manhood are not especially well explained by strictly biological influences. The readings in this section were selected to illustrate some instances and consequences of male socialization.

Socialization is a more variable process than might first be imagined. It is a mixture of messengers and messages, some intended, some not, some explicit, some tacit, through which we make males masculine. Referring to some socialization as unintended doesn't make it any less significant in shaping our lives. This kind of socialization results from much of what we are exposed to through the mass media and some of what we learn at home or school. The goals of artists, producers, and sponsors are more explicitly oriented around entertaining us and selling us products than teaching us about gender, but the latter may still result from their efforts (Renzetti and Curran 1995).

At school, gender messages are communicated through examples used in class discussions or textbooks, and the relative representation of female and male faculty or administrators. At home, as children observe the division of family responsibilities (e.g., mommy reads bedtime stories at night, daddy is gone for longer stretches of the day), whether parents intend it or not, they are socializing their children toward certain gender ideas. In fact, children may derive gender messages that are completely counter to the ones parents or teachers intend to construct and convey by the things they say.

Of course, much socialization is more intentional. Beginning when children are very young, parents—especially fathers—take a keen interest in seeing that their children—especially their sons—are appropriately "gendered." Their actions are conscious attempts to help "genderize" them. Peers, in witnessing and weighing what we do and how we do it, are among the most deliberate agents of gender socialization. From the threatening taunts of the male peer group ("What are you, anyway, a sissy, wimp, fag, girl?") to the competition for popularity that pushes boys and girls toward conformity, peer groups are powerful factors influencing how we turn out. Although parents and peers have different objectives, each works with a "gender agenda."

One can draw a second, stylistic distinction to differentiate how socialization occurs. Some socialization matches the clichéd parental admonition, "Come on, son, stiff upper lip. Big boys don't cry." Some children's literature, occasional classroom lectures, much peer and parent proscription (i.e., "don't do that" or "don't be that way") and prescription (i.e., "do this" or "be that") are so explicit. Other socialization is more subtle or tacit. It may even be entirely nonverbal, as seen in the absence or underrepresentation of characters in stories, textbooks, or television programming, or in the overlap of age and gender hierarchies (or occupational and gender statuses) in characters. And again, just by observing the division of daily family tasks, real, though subtle gender reinforcement occurs.

Things unsaid may also communicate powerful messages. For example, meaning is conveyed when a father commends his son's baseball performance and talks at length about the relative strengths and weaknesses of the boy's Little League team, but fails to compliment his daughter's equally successful softball performance or treat her team with the same seriousness. He may not say to his daughter, "I don't care about your softball," but he implies that he doesn't think it is as important as the game her brother plays (or that it is not as important that she plays as it is that her brother plays).

Social scientists are divided over the exact mechanism that best accounts for gender acquisition (parental modeling, cognitive development, differential reinforcement, and so on), and whether such socialization is ultimately necessary and beneficial (i.e., "functional") or exploitative. Most agree, though, that we are exposed to a range of "socialization messages," which act with consequence in making boys into men and girls into women.

More recent social constructionist perspectives downplay socialization, objecting especially to its more deterministic implications. They point out that we don't passively react to gender messages, that the messages themselves vary across time and space, and that gendered behavior is the product of active construction, sometimes even in direct and deliberate opposition to the dominant ideas within our society. What they less often criticize is the notion that we learn that certain behaviors or characteristics are associated more with males than females, and vice versa. We are taught these connections by a number of influential others, including our families and friends, teachers and textbooks, and the various mass media that we consume daily.

Boys versus Girls

From birth on, boys and girls receive different reactions and are given different messages about what they can and cannot do or should or should not be like. This begins as soon as parents start to relate to their newborns. How we dress, hold, soothe, play with, and praise or punish our children are actions that all influence the types of characteristics they acquire and display. The furnishings and accessories of their daily lives reveal our gendered ideas about what they should like and be like.

Broadly speaking, boys learn that being male means being strong, competitive, aggressive, inexpressive, and courageous. They further learn that *masculinity* is tightly connected to their (hetero)-sexuality. As they grow up, American males learn a variety of sexual myths (e.g., sex, not intimacy, is all that counts; you can never get enough sex) that distort what they know about sexuality and may ultimately contribute to a variety of sexual dysfunctions (Doyle 1995). Additional connections between gender and sexual behavior are often learned within the male peer group. For example, Nathan McCall's "Trains" (Reading 8) illustrates the linkages often made between masculinity, dominance, sex, and aggressiveness. McCall shows how some males learn to use sex (i.e., coercive sex) to display gender (dominance).

Tommi Avicolli's "He Defies You Still" (Reading 9) demonstrates another powerful message boys learn about the connections between masculinity and sexuality: To be fit as a man, one must prefer women. In other words, acceptable masculinity presumes heterosexuality. Feminine males are seen as suspect males and, potentially, as "failed heterosexuals." As a consequence, males develop a contempt for aspects of themselves and each other that might cause anyone to wonder about their masculinity and, hence, their sexuality. Aware of the boundaries around "acceptable" male behavior, they then patrol those borders seeking to identify, isolate, and intimidate transgressors. Lest anyone wonder about them, they also monitor their own behavior, showing only acceptable levels of closeness, intimacy, emotional expressiveness, weakness, vulnerability, and sensitivity. Men, overall, pay a price for their acceptance of such a narrow range of acceptable behaviors. An even greater price may be paid by those males whose sexuality is demonized by taunting teammates, classmates, and others. As Avicolli shows, gay males are also reminded of the contempt felt by other groups of males toward males "like them."

Boys learn that traits such as strength, competitiveness, inexpressiveness, and aggressiveness are expected of them as males. The competitiveness of sports, for example, initiates and reinforces a competitive demeanor overall. The restraint suggested by mandates like "boys don't cry" may carry over into men's style of relating to others in even their most intimate relationships. The tolerated aggressiveness of boyhood may carry over into a kind of "push/shove" mentality: When "push comes to shove," real men have the capacity to push and shove, or hit, punch, kick, and, as displayed by boxer Mike Tyson, bite.

These masculine qualities are seen in the overall theme of male dominance. Male dominance is exercised both relative to other males but, especially, relative to women. Cross-sex displays of dominance surface in a number of readings to come. For example, the excerpt from Bernard Lefkowitz's *Our Guys: The Glen Ridge Rape and the Myth of the Perfect Suburb* (Reading 15), and Boswell and Spade's article on fraternities and rape (Reading 35), both illustrate how this sense of dominance is aggressively and destructively displayed by groups of young men. Even in more benign ways, over issues of watching television (Walker, Reading 16) or caring for children (Walzer, Reading 18), some of what results within intimate relationships reflects men's dominance in such relationships.

If the above illustrates what boys learn, where do they learn it? The answer is "everywhere." In buying their toys, picking their clothes, decorating their rooms, reading them books, enrolling them in art or music lessons, registering them for sports programs, parents shape the types of people their children become. In doing so, parents reveal their concern that their children be appropriately gendered.

Let's look at toys. Children's toys teach skills, reflect adult-appropriate roles, entertain, nurture and reinforce attributes. They also communicate gender messages. One sees the careful construction of gender appropriateness in the marketing of toys in ads and catalogues and in the toys themselves (Renzetti and Curran 1995). Boys' toys encourage the development of attributes like exploration and aggression, foster the development of spatial ability, and promote science and math achievement (Renzetti and Curran 1995; Hoyenga and Hoyenga 1993).

Toy companies and toy stores may be very gender conscious in what they manufacture and market, but toys don't come home by themselves. Toys are advertised to children and parents, and purchased by parents (or children through their parents), who make their choices from a wide variety of possibilities. The patterns of parents' choices suggests a high degree of deliberation and concern for the suitability of toys for their children, especially their sons (Basow 1992). Parents are joined by other outsiders who, intentionally or not, explicitly or subtly, help shape us to the gender expectations they have for us. Some of these "outside influences" are depicted in the readings that follow.

School and Sport

In this chapter, we focus mostly upon actors and domains outside the home, including school and sports. School is a major part of socialization, especially of young children, and it retains importance well beyond the many years during which we are students. Through formal and informal content, schools fit us to the cultural and institutional needs of our society at the particular place and time in which we live. Through curricular materials that differently represent and encourage women and men, and through interactions with peers, teachers, coaches, administrators, and counselors, schools also construct and reinforce gender differences.

Teachers are especially significant in how they differently reinforce and respond to male and female students. Teachers praise boys' abilities more, call on boys more, help boys find and correct their errors more, and propose more challenging academic situations to them (Sadker and Sadker 1994). But, as shown by William Pollack (Reading 6), boys also suffer from their school experiences in meaningful ways that girls don't. For example, the attention boys receive is not always positive—they are subject to more discipline and receive more of the teacher's anger than do girls, even when one controls for the disruptiveness of their behavior. Furthermore, their academic performance often suffers, as indicated by their rates of failing, acting up, and/or dropping out (Sadker and Sadker 1994; Renzetti and Curran 1995; Pollack 1998). Thus, both girls and boys "pay" for what school does to them, and both would benefit from greater gender equality.

School also brings us into contact with same-age peers of both sexes. These groups become the audiences who witness and the critics who review our gender performances. Selectively employing labels like "sissy" and "tomboy," children monitor each other and reinforce appropriate behaviors (Thorne 1993). Transgressions, like the labels, have quite different impacts on females and males. Whereas "tomboy" (or "jock") may be taken positively as a sign of a girl's athleticism or assertiveness, "sissy" (or "fag," "wuss," or any other contemporary equivalent) is never meant or received as at all complimentary.

Much of what boys do, they do to avoid being "sissified." In the article by Jonah Blank, "The Kid No One Noticed" (Reading 10), the pressure to measure up to acceptable masculinity in his Paducah, Kentucky, high school is seen as a major part of what drove Michael Carneal to murder three classmates and wound five others. Although most efforts to measure up to masculinity are much less extreme, they, too, illustrate how much of what boys do they do to fit the mold.

Sport Both in and outside of school, sports constitute a particularly large part of male upbringing in the United States. Most boys face the expectation that they will play some kind of sports, and the most popular boys tend to be those who play best (Coleman 1976; Gerzon 1992; Messner 1992). The adulation paid to the successful 1999 United States Women's World Cup soccer team showed the extent to which sports have opened to women. Female athletes, like these deserving champions, benefited from two decades' worth of legislative policy, social pressure, and parental support that have increased women's opportunities to play sports if they choose. However, there has been no equivalent movement to free men from the obligation to play sports if they prefer not to (Doyle 1995; Fine 1987; Stein and Hoffman 1978). "Real boys" play ball.

As sociologist Michael Messner states, "sport remains the single most important element of the peer-status system of U.S. adolescent males. . . . boys are, to a greater or lesser extent, judged according to their ability, or lack of ability, in competitive sport" (1992, 24). Those who lack ability or interest in sports find that these become signs of masculine inadequacy (Whitson 1990). They face a kind of ascribed feminization or, in the homophobic world within which boys reach manhood, homosexualization. Boys who don't play well are said to "play like girls," or like "sissies" or "fags." Thus, it is a high-stakes game boys enter, whether they step on or avoid the various ball fields and courts of their youth.

Sport would be important even if it were only a masculine province in which boys were sifted and sorted as better or worse athletes, or even as better or worse boys, on the basis of their athletic abilities. It is more, however, than something most boys, for at least a time, do with each other. Participation in competitive sports is a large component of socialization into masculinity. In playing, one is exposed to the values and expectations that help make one appropriately masculine. Team sports act as "an adolescent male's '*rite de passage*' through which boys become men" (Doyle 1995). Sports such as baseball, soccer, football, basketball, rugby, and lacrosse celebrate and glorify qualities that constitute hegemonic masculinity, thus adding legitimacy to that version of masculinity. Toughness, the ability to withstand pain, physical strength, aggressiveness, and an avoidance of feminine activities and values are all transmitted via men's involvement in sports. Reading 7, by Sabo and Panepinto, illustrates how this is accomplished in the sport of football.

Research on gender socialization through children's books, mass media influences, and school experiences shows that gender messages are cumulative and consistent. No single encounter or exposure will result in gendered behaviors. However, there is a consistency in the messages conveyed across these many early socialization influences that is almost inescapable (Basow 1992).

Boys versus Boys

Variations in male socialization by culture, class, and race abound. American boys lack the formalized rites of passage into manhood that characterize male socialization in many other cultures (Gilmore 1990). David Gilmore (Reading 5) describes the Samburu and Masai ritualized transitions from boyhood through *moranhood* to manhood, publicly initiated through the painful act of circumcision. Expected to display the stoic, tough, fearless qualities associated with their upcoming roles, Samburu initiates are to remain silent and motionless through what may be four minutes of agonizing, unanesthetized cutting. Following this test are numerous other rituals that the young men must perform if they hope to ascend through the status of *moran* and eventually achieve the status of *lee* (worthy man).

Within American society other, less extreme variations in male socialization exist. Evidence suggests that black parents socialize their children in less gender-divided and sex-segregated ways, stressing hard work, independence and achievement for both their sons and their daughters. Young males, like young females, are raised to value domestic equality (sharing tasks), and face the expectation that they will be nurturant, emotionally expressive, and assertive (Renzetti and Curran 1995).

Class variations surface as well. Some research indicates that there is less gender stereotyping as one moves up the economic ladder. Whereas working-class parents may "differentiate between the sexes more sharply than middle-class families," middle-class families seem to differentiate somewhat less (Basow 1992). They will, for example, give stereotypical male toys to sons *and* daughters (Basow 1992).

Race and class variations in socialization do not stop at home. African American boys reportedly receive greater peer reinforcement for qualities such as street smarts, athletic prowess, and sexual competence than their white (middle-class) counterparts (Basow 1992). In school, black males face especially difficult circumstances and receive the most unfavorable teacher treatment when compared with white males, white females, or black females (Sadker and Sadker 1994; Basow 1992). They receive the most recommendations for special education and are subjected to low expectations by teachers. Teachers describe black males as having the worst work habits, and they predict lower levels of academic success for them, regardless of their actual behavior (Basow 1992).

Socialization differs, too, for heterosexual and gay males. Although no particular set of socialization experiences is necessarily responsible for sexual orientation (McCaghy 1997), gay males have certain socialization experiences because they are gay. Gay males are targets for a variety of reactions including rejection and contempt, as can be seen in Avicolli's painful personal narrative (Reading 9). They learn early that males like them may become the targets for harassment and hostility. Those who are confused by or in denial of their sexual orientation witness what happens to others who are suspected or known to be gay. Even in the "enlightened" campus environments surrounding colleges and universities, gay men (and lesbians) are the recipients of "the most severe hostilities," including violence or the threat of violence (Levin and McDevitt 1993). There are life-shaping consequences to such experiences.

SOCIALIZATION AS BEGINNING, NOT END

Before closing, we ought to carefully put socialization in its place. Those who prefer it as more acceptable than a biological explanation of gendered behavior have occasionally fallen into the same sort of trap as those whose emphasis on biology they criticize and reject. Dismissing the more controversial assertion that we are limited by our physical characteristics, socialization theorists argue that we are products of our early experiences and environments. Being differently exposed to environmental influences makes males and females develop different characteristics and qualities. Once we develop them, we are, in ways similar to biologically based arguments, limited (Risman and Schwartz 1989).

As a number of scholars warn, there is a danger in overemphasizing socialization in explaining social behavior (Risman and Schwartz 1989; Gerson 1993; West and Zimmerman 1987). Alternative perspectives, like microstructural theory, emphasize the importance of such things as situations and circumstances in creating gendered behavior through the opportunities they create, the obstacles they impose, and the occasions they provide for individuals to "do gender" (Risman and Schwartz 1989). Childhood socialization may shape men's (and women's) initial life plans or outlooks, but the range of later life experiences, including the relationships one forms and the constraints imposed or opportunities afforded by one's work, can redirect people's lives in unexpected directions (Risman 1989; Gerson 1993).

Furthermore, as more interactionist theories have demonstrated, gendered behavior is not guaranteed, despite whatever early exposure to external influences we have had. Individuals, in the various social situations in which they find themselves, must accomplish or "do gender," and their performances will be evaluated by those with whom they are interacting.

REFERENCES

Basow, S. *Gender Stereotypes and Roles.* Pacific Grove, CA: Brooks/Cole, 1992.

Coleman, J. "Athletics in High School." In *The 49% Majority,* edited by D. David and R. Brannon, 264–269. Reading, MA: Addison-Wesley, 1976.

Doyle, J. *The Male Experience.* 3rd ed. Dubuque, IA: Wm. C. Brown Communications, 1995.

Fine, G. *With the Boys: Little League Baseball and Preadolescent Culture.* Chicago: University of Chicago Press, 1987.

Gerson, K. *No Man's Land: Men's Changing Commitment to Family and Work.* New York: Basic Books, 1993.

Gerzon, M. *A Choice of Heroes.* Boston: Houghton Mifflin, 1992.

Hoyenga, K., and K. Hoyenga. *Gender Related Differences: Origins and Outcomes.* Needham Heights, MA: Allyn and Bacon, 1993.

Lefkowitz, B. *Our Guys: The Glen Ridge Rape and the Myth of the Perfect Suburb.* Berkeley: University of California Press, 1997.

Levin, J., and J. McDevitt. *Hate Crimes: The Rising Tide of Bigotry and Bloodshed.* New York: Plenum, 1993.

McCaghy, C., and T. Capron. *Deviant Behavior: Crime, Conflict, and Interest Groups.* Needham Heights, MA: Allyn and Bacon, 1997.

Messner, M. *Power at Play: Sports and the Problem of Masculinity.* Boston: Beacon Press, 1992.

Pollack, W. *Real Boys: Rescuing Our Sons from the Myths of Boyhood.* New York: Henry Holt, 1998.

Renzetti, C., and D. Curran. *Women, Men, and Society.* Boston: Allyn and Bacon, 1995.

Risman, B. "Can Men 'Mother'? Life as a Single Father." In *Gender in Intimate Relationships,* edited by B. Risman and P. Schwartz, 155–164. Belmont, CA: Wadsworth, 1989.

Risman, B., and P. Schwartz. *Gender in Intimate Relationships: A Microstructural Approach.* Belmont, CA: Wadsworth, 1989.

Sadker, M., and D. Sadker. *Failing at Fairness: How America's Schools Cheat Girls.* New York: Charles Scribner's Sons, 1994.

Stein, P., and S. Hoffman. "Sports and Male Role Strain." *Journal of Social Issues* 34 (1978): 136–150.

Thorne, B. *Gender Play: Girls and Boys in School.* New Brunswick, NJ: Rutgers University Press, 1993.

West, C., and D. Zimmerman. "Doing Gender." *Gender and Society* 1987: 1(2): 125–151.

Whitson, D. "Sport in the Social Construction of Masculinity." In *Sport, Men, and the Gender Order: Critical Feminist Perspectives,* edited by M. Messner and D. Sabo, 19–29. Champaign, IL: Human Kinetics Press, 1990.

FOR ADDITIONAL READING

The preceding discussion and the readings that follow illustrate some key aspects of male socialization. As broad as such socialization is, however, much has gone uncovered. Thus, the following articles are recommended to students who might like to read further. For a short overview of parents' roles in the gender socialization process, read Susan D. Witt's article, "Parental Influence on Children's Socialization to Gender Roles," from the journal *Adolescence.* For considerations of some media effects on male socialization, read Tracy L. Dietz's article, "An Examination of Violence and Gender Role Portrayals in Video Games: Implications for Gender Socialization and

Aggressive Behavior," from *Sex Roles: A Journal of Research* (March 1998, vol. 38, no. 5–6) or "Marketing Masculinity: Gender Identity and Popular Magazines" from the same journal (July 1998, vol. 39, no. 1–2). Finally, for an analysis of how sports can produce *nontraditional* masculinity, read "Alternative Masculinity and Its Effect on Gender Relations in the Subculture of Skateboarding," by Becky Beal, from *Journal of Sport Behavior.*

WADSWORTH VIRTUAL SOCIETY RESOURCE CENTER

http://sociology.wadsworth.com

Visit *Virtual Society* to obtain current updates in the field, surfing tips, career information, and more.

INFOTRAC COLLEGE EDITION: EXERCISE

Go to the *InfoTrac College Edition* web site to find further readings on the topics discussed in this chapter.

5

Markers to Manhood: Samburu

DAVID GILMORE

It has always been the prime function of mythology and rite to supply the symbols that carry the human spirit forward, in counteraction to those other constant human fantasies that tend to tie it back.

JOSEPH CAMPBELL
The Hero with a Thousand Faces

Living in different continents, pursuing different goals, the Trukese brawlers, the Mehinaku wrestlers, the American he-men, and the Spanish macho-men have little in common other than their passionate concern for demonstrating manhood. But they are alike in one other way: they all navigate a pathway to manliness that is without clear signposts. To attain their goal they inch forward by trial and error, following sometimes vague injunctions laid down by their cultural script. Along the way, they must avoid pitfalls and temptations that tie them back, in Campbell's terms above, to childhood.

Their voyage to manhood, then, resembles what Erik Erikson in his *Childhood and Society* (1950) referred to as epigenetic stages of psychosocial development. By this he meant step-by-step sequences of growth which, when transversed, confer upon the individual simultaneously an ego-identity, or sense of self, and a cultural identity appropriate to his or her time and place. As Erikson showed by comparing a variety of cultures, each society provides its own unique materials and rewards by which incremental steps are encouraged. The process itself remains similar with the same end, but the details of the passage differ widely.

Many traditional societies are highly structured in the way they acknowledge adulthood for both sexes, providing rigid chronological watersheds as, for example, ceremonies or investments, complete with magical incantations and sacred paraphernalia. Others, more changeful, such as Western industrial societies, are loosely structured with little public recognition of progress and certainly no magic. Some, as Erikson showed for modern America, offer a bewildering range of options for men at every stage of life, creating problems of diffuseness and ambiguity—a "dilemma" . . . or a "crisis of masculinity" . . . — that men must resolve in their own way to reach their culture's goal. The result, as in contemporary America, can often be a "makeshift masculinity." . . .

A number of primitive societies provide collective rites of passage that usher youths through sequential stages to an unequivocal manhood. Such rites "dramatize" . . . the masculine transition through a clear-cut process of ritual investiture complete with emblems and culminating in the public conferral of an adult status that equals manhood. The basic framework for interpreting these male rituals of initiation was provided long ago by Arnold van Gennep in his classic *The Rites of Passage* (1960), first published in 1908. The underlying theme of such passage rites, according to van Gennep, is a change in status and identity: the boy "dies" and is "reborn" a man, each stage accompanied by appropriate symbolization.

For van Gennep, rites of passage represent the death of childhood. Van Gennep claimed that this death-rebirth theme is played out in three stages: separation, transition, and incorporation. During the first stage, the boy severs relations with childhood, often literally, by renouncing the mother or being forcibly taken away from her. In the transition stage, he is sent away to a new place in the bush or is otherwise isolated where he remains in limbo, a "liminal" (or transitional) status when he is neither boy nor man but something in-between. Finally, he emerges a "man," through the vigorous exit ceremonies such as those we will examine here.

Such prolonged and collective rites of passage are found mainly in primitive (or preliterate) societies. Peasants and urban peoples rarely celebrate adulthood for either sex through elaborate or sacred ritualization, usually opting for tacit recognitions of individual growth. . . . [T]he Mediterranean and Middle Eastern peoples, in particular, are generally left to their own devices, as are modern Americans and most other Westerners. An exception is the Jews (if the Jews indeed can be so classified), who present an interesting case in point of ritual dramatization. Let me consider them for a moment as we turn toward the issue of ritualized manhood.

As is well known, the bar mitzvah ceremony among the Jews confers an official manhood status, usually at age thirteen. A centerpiece of religious and ethnic identity, it is both a chronological marker and a ritual of initiation, dramatized publicly. The ceremony is individual, rather than collective, but often the entire congregation participates, adding a community flavor. Because Jewish culture in the United States has a familiar, close-to-home quality, the bar mitzvah has not been the subject of much sociological interest here, especially since, in modern secular Jewish-American culture, the rite resembles nothing more profound than an exercise in extravagant partying. Still, observation of the rite can tell us much about the nature of the masculine transition under advanced, monotheistic, literate conditions, because underlying everything is the traditional involvement with the acquisition of male powers.

One feature of the bar mitzvah often forgotten in the modern vulgarization of the ceremony is that it entails, among other things, a specific test of the initiate's mettle. The youth must memorize and publicly recite important passages from the Torah, in Hebrew, in the presence of both the congregation and a ritual specialist, or rabbi, who has instructed him and who certifies his performance. The initiate can either pass or fail in this; the whole procedure depends upon his public performance. This examination element is present even in the watered-down Reform ceremony in which circuslike entertainments overshadow the serious religious purpose. But how comes this arcane recital to be a test of cultural competence? Can this memorization and recitation of sacred texts be compared to the physical testing of the other peoples we have discussed?

On the surface there is a tenuous similarity to the strong thought of the Trukese weekend pugilists, which also implies a competence in male "thinking," and certainly there is an element of risk in the public exposure. Yet the deeper homologies are obscured by the surface variability of expression, its doctrinal rather than physical outcome. The similarity lies in a form of testing that measures cultural fitness and prepares the individual for making contributions to specific, and historically valid, group goals.

The test for the Jewish boy is one of memory and comprehension, the successful mastery of sacred texts; but the abilities being measured are clearly intellectual skills of a particular sort, especially scholarly potential. For example, in my own experience—typical of secular suburban culture in the northeastern United States—the process amounted to a purely academic performance, rather like taking an examination in school, or so it seemed to me. This performance constituted, in this milieu, a critical arena of personal achievement with all its attendant anxieties and possibilities of both public humiliation and triumph. During the memorization and recital, the boy is, as van Gennep argued, in a limbo or transitional status; afterward, if successful, he reemerges a "man" and childhood is dead, a victim again of manly competence.

Stereotypically, the Jewish people are not associated with the phallic posturing of the more warlike peoples we have discussed above. Their avenue of success has always been intellectual, to live by one's wits. Viewed in this light, we can see that the Jewish youth achieves a transition to a culturally constituted manhood in a way that serves to measure publicly those skills by which the group itself excels and by which it survives and prospers. This "ethnic niche" approach . . . to male ritual lets us see manhood imagery in the light of a broader theory of economic adaptation of the group within a wider context of social relations. Not all challenges are physical and not all success can be measured against an external opponent.

Yet, despite their lack of a warlike tradition, the Jews do have a manhood concept that shares certain features with that of more bellicose peoples. Even secular, assimilated Jewish-American culture, one of the few in which women virtually dominate men, has a notion of manhood. A wife who is satisfied with her husband will say that he is a "mensch," which in both Yiddish and its parent tongue, German, means a real man; and her mother, the ultimate judge of manly virtue, might even agree. Being a mensch means being competent, dependable, economically secure, and most of all helpful and considerate to dependents. In modern middle-class Jewish culture, the mensch is a take-charge kind of personality, a firm, dependable pillar of support. He has fathered presentable children, provided economic support for his family, and given his wife what she needs (or, rather, wants).

The opposite of the mensch is the bungler, the failure who lets others take advantage of him. He is alternately called, in all the richness of Yiddish, a schlemiel (jerk), a schnook (dope), a nudnik (nincompoop), a schmendrick (nitwit), a schnorrer (toady), or most commonly a schmuck ("prick," but literally a useless bauble, used to mean bumbling or incompetent). All these synonyms (and there are many others: putz, nebbish, schmo, schmeggegie) convey the sense of masculine inferiority or disgrace measured in degrees of inadequacy and comic ineptitude in culturally assigned jobs and tasks. The schlemiel or the schmuck is a failure in the Jewish tradition of what a man should be (one rarely hears these terms used to describe women). He is the subject of endless ridicule. He has let his family down by being incompetent. He represents the Yiddish equivalent of the Mehinaku trash yard man or the New Guinea rubbish man: the antihero, the non-man. He has failed the test of manhood.

But to return to the question of more exotic rituals of passage . . . we will look closely at [a very different society that awards] manhood through ritual testing of male competence: the pastoralist Samburu of East Africa . . . [who could not] be confused with suburban Jews. . . .

A STRUCTURED ASCENSION TO MANHOOD

The British anthropologist Paul Spencer . . . has studied the East African Samburu tribe for many years, beginning in the 1950s. The following account is drawn mainly from his classic monograph, *The Samburu* (1965), and from a worthy companion study, *Nomads in Alliance* (1973). The Samburu are a black Nilotic pastoral people who live in an upland plateau between Lake Rudolf and the Uaso Ngiro river in northern Kenya. They were colonized like many of their neighboring tribes by the British at the turn of the century but not so severely treated or forcibly acculturated as some other tribes. In language and culture, they are very similar to the more famous and colorful Masai (or Maasai), to whom they are historically related (we will take a peek at the Masai later), and to a host of neighboring East African cattle pastoralists such as the Rendille, Jie, Turkana, and Nuer, among many others.

Like these other pastoral tribes, the Samburu still live mainly off their herds of cattle, sheep, goats, plus a few donkeys, as their ancestors did, using everything but the squeal. It is upon the value of cattle in particular that the emphasis is squarely placed in traditional Samburu culture, as it is among the Dodoth of Uganda, . . . at times almost to the exclusion of small stock. Possession of cattle here is the mark of a man of substance, a big man. A man who has cattle is important, they say. "He can have many wives and many sons to look after his herds." . . . Cattle represent both the principal source of nourishment and the main trade item; they are the central cultural value—aside from wives and children. It should therefore come as no surprise that their manhood imagery revolves around competence in dealing with herds, as it did among the Dodoth.

In the Samburu culture, cattle represent liquid wealth—an all-purpose currency for trading and for thus acquiring other good things. The Samburu are as obsessed with this cattle wealth as the Jews are with learning, the Trukese with fighting and accumulating consumer goods, and the Mehinaku with fishing and wrestling. The most deeply held ideal among them is the notion that each man should have his own herd and ultimately be able to manage it independently and see to its increase. The key term in this equation is *independent,* meaning both economic and social freedom: no debts, no lords, no masters. But aside from possession, a man must also be in full, uncompromised control of his stock; he must be viewed as an autonomous entrepreneur of herds, exercising administrative control, without even a whiff of dependency about him. The Samburu are insistent upon this point, regarding independent husbandry as the very basis of adult male status.

Accumulated wealth in the form of meat, managed and augmented through careful and unencumbered stock-breeding, allows a man to marry often in this polygynous society and to achieve the coveted status of "worthy man" by adding wives, wealth, and children to his lineage. A man rich in cattle is a worthy elder and "a man of respect," but only if he is also "generous to the point of self denial." . . . That is, Samburu men accumulate herds so that, bursting with pride, they can give them away, engaging in a "battle" of generosity in the tribal competition to distribute food to the people. . . . To manage this greedy generosity, a man must be free to trade and breed cattle and to slaughter on his own initiative.

In the past, before the British came in the early years of this century, the Samburu, like the fierce Masai, were courageous warriors as well as stockmen, but warfare has virtually ceased since the arrival of the Pax Britannica. One of the main goals of East African warfare and raiding formerly was to capture cattle, and the Samburu warriors ranged far and wide, battling and raiding the powerful Masai and the tough Boran (to whom they are also distantly related) in order to build up their herds. In return, they were raided and preyed upon by their equally opportunistic neighbors. Accordingly, the principal way for a Samburu youth to garner approval was to travel far and wide in order to kill people of other tribes and to steal their stock for prestige, "without any political justification." . . . In turn, they were attacked by others, so every Samburu man was expected to be a brave warrior. The survival of the tribe depended upon it.

It appears, then, from Spencer's description that warfare had an overt economic as well as prestige motivation. When all this abruptly ceased after the imposition of peace by the British, the Samburu were not to be denied, and they continued to engage in small-scale skirmishing with the purpose not so much of killing but of rustling cattle, as did their neighbors. Because cattle rustling is the easiest and most expeditious means of acquiring stock and of increasing herd size in a short time, it remains a critical coefficient of male social value. Manhood thus retains a military quality, and all "worthy" men are expected to engage in successful marauding, for otherwise they cannot be rich and cannot sponsor the festivals and feasts that nourish and enrich the tribe. So previously, while the youths engaged in warfare "with a definite role to play in the survival of the tribe," . . . today, with the tribe's survival no longer at stake, the main context for the display of courage and manly virtue is in the accumulation of stock through raiding, an economic end of lavish gifting in feasts and rituals, in which personal courage provides the means of expression. Especially for young men (called *moran,* an important term we will return to shortly), who have little other opportunity to amass animals for breeder stock, raiding and rustling represent the principal means of achieving manhood and all its social rewards: respect, honor, wives, children. The only way to win a wife is by showing one's potential as a herdsman to potential parents-in-law.

In some respects the Samburu, like the Ethiopian Amhara, have a set of fiercely held gender ideals. . . . Aside from personal autonomy and force, they, too, have strong notions of masculine "honor" (*nkanyit*) and a corollary ideal of "prestige" (Samburu term not given). These notions are related, but not identical. Honor is tied up with meeting expectations as a member of a corporate group defined by genealogy or affinity—whether lineage, clan, club, or age-set—and thereby protecting the group's collective reputation. Prestige is tied up with matters of personal status within any particular social group, . . . therefore containing an element of intragroup rivalry. As Spencer describes it, questions of degrees of masculine honor imply satisfying tribal ideals, not necessarily leading to intratribal competition. . . . However, both notions relate strongly to displays of physical courage, and both imply a touchy male vanity that will tolerate no slights. Both notions also imply a fierce defense of the social unit of which one is a member. Honor is clearly male reputation measured by degrees of control over and protection of units of collective identity. . . .

These traditional male values, despite modernization and government suppression, remain vibrant. . . . Spencer says the following: "Today, honour and prestige are still treasured values of the *moran* [youth], and account for much of their behavior. A moran who has been slighted in some way—perhaps he has been taunted or his mistress has been seduced—is expected to retaliate against the offender, and hence it is his *honor* which is involved. But when he exceeds these expectations in some way, asserting himself more than is necessary, it becomes also a matter of personal prestige." . . .

Strong stress on personal prestige and group honor is a social value we would expect from a society with a proud military tradition and an acephalous, segmentary political organization. But the Samburu, like the Andalusians and many other less militaristic peoples, round out this corporate "honor" with other associated male values that relate directly to notions of cultural competence and sexual prowess. The composite image of the worthy man includes a heavy emphasis on economic self-sufficiency and productivity. Material self-denial and an associated "giver" status are elevated to moral paramountcy. The ideal among the Samburu is that a man "should aim at increasing his herd and the size of the family he founds until he can form a homestead in its own right." . . . In other words, he is expected to reproduce the homestead, the basic kinship unit of the tribe, by the sweat of his brow and to make it grow and prosper so that it becomes more than self-sufficient, to be fruitful and multiply, as in Jehovah's exhortation to the Israelites. A man must produce both cattle and children in equal measure, protecting them both from predators and interlopers, and he must expend his social energies in economic reinvestment rather than personal consumption that might deplete his family's precious patrimony. The Samburu man is always under public pressure to measure up in this regard. But his worthiness comes not only from accumulating but also from lavish gifting during feasts. This donor function looms larger in Samburu male imagery than sexual valor.

An institutionalized competition of tribal generosity spurs men to try to best each other in a kind of potlatching of meat. At every feast, each man tries to give the most and take the least. "At most feasts, for instance, younger men would insist on giving their seniors the best pieces of meat, and affines would be treated in a similar way. Even among age mates, whose equality was beyond question, there was a continual battle of politeness in which the man who seemed to eat the least and encourage his neighbors to eat the most was the moral victor, the truly worthy man." A selfish man, an "eater" of his own herds, is described as *laroi*—a term used among the Samburu for small, paltry things, undersized or fragile objects, and for tools that do not work effectively, like a leaky water bucket or a Land Rover that keeps breaking down. The *laroi* man, then, is childlike (undersized), mean, deficient, wasteful, inefficient. . . . He takes more than he gives. Like the trash yard man, the rubbish man, the schlemiel, he is despised on this account, and this dislike can amount to an "unvoiced curse bringing with it disaster" upon his useless person. . . . Conversely, the Samburu sense of masculinity is a kind of moral invisible hand that guides the activities of self-respecting individuals toward the collective end of capital accumulation. The goal is to become a tribal patron, creating dependencies in others, involving oneself in battles of generosity with other men in which everyone indirectly benefits. . . . In this equation, subordination to another man is considered by the Samburu "an ineradicable stain upon personal honor." . . .

AGE-SETS

What makes the Samburu famous in the ethnographic literature is their system of age-sets and colorful attendant rituals by which the male life cycle is advanced and celebrated. In this, they are similar to their related East African tribes, but because of Spencer's superb documentation they serve as an exemplar of the ritualized masculine transition common to all. Samburu males must pass through a complicated series of age-sets and age-grades by which their growing maturity and responsibility as men in the light of these tribal values are publicly acknowledged. Basically, there are three main age-grades, which are subdivided into a number of chronological age-sets. These main groupings, the age-grades, are those of, first, boy; then *moran,* or adolescent/young adult (Spencer gives moran as both singular and plural); and finally elder.

A celebrated feature of East African cattle societies, the special status of moran is a long, drawn-out transitional period of testing; it is the one with which I am most concerned here. Previously adapted toward preparing boys for a stoic warrior's life, and today for success in rustling, moranhood begins at about age fourteen or fifteen and lasts for about twelve years, divided into numerous age-sets. It is initiated by a circumcision procedure described below, which begins the separation stage.

After passing the test of circumcision, the initiates are spatially isolated as a group, being removed from the confines of the village to a preselected place in the bush where they will live for the next decade or so, perfecting their skills and functions as elders. In addition, during moranhood there are a number of subsidiary rituals and tests that each youth must perform. The most important of these are the arrow ceremony, the naming ceremony, and, finally, the bull ceremony, all involving tests of skill, endurance, or competence. The bull ceremony involves the boy's first sacrifice of his own ox and the distribution of its meat. This represents the exit ceremony of moranhood, eventually ushering the young man into full status as *lee,* or worthy man, . . . corresponding to van Gennep's incorporation stage. But no moran may marry or father children until he kills his first ox, so everything, again, revolves around the moran's ability to acquire and provide meat. Reflecting this, the ritualization of manhood culminates in a very specific act of tribal generosity indicating a future of successful industry. Rather than describing all the minor details of the age-grade series, which the reader may find clearly elaborated in Spencer's classic studies, . . . I will only point out the main features of this transitional period as they relate to the question of the meaning of manhood.

MORANHOOD

The first test for the boys entering moranhood is that of the traumatic circumcision procedure. A trial of bravery and stoicism, the operation is intensely painful, and no anesthetics are used; nor is anything done to lessen the anticipatory terror of the initiate, suggesting that the purpose is specifically one of testing. Each youth, placed on view before his male relatives and prospective in-laws, must remain motionless and silent during the cutting, which may last four minutes or more. Even an involuntary twitch would be interpreted as a sign of fear. The word used for such a minor movement is *a-kwet,* literally "to run," which Spencer glosses as "to flinch." . . . If the boy makes the slightest movement or sound, there is a collective gasp of shock and dismay; he is forever shamed as a coward and will be excluded from joining his age-set in the march toward adult status. No other initiate would want to form an age-mate relationship with a boy who "runs," for this boy will bear a stigma of inferiority for the rest of his life. Besides being cruelly ostracized himself, . . . he brings ruin upon his entire lineage forever. In the Samburu proverb, all of his people must publicly "eat their respect" . . . because one of their boys has run.

This use of an oral metaphor to express collective shame is of comparative interest here, reminding one not only of the New Guinea rubbish man, whose main fault is to "eat" the land without producing, but also of the role of oral or "hunger" imagery and of similar physical tropes in the construction of masculine identity that others have noted elsewhere. . . . The Samburu metaphor conjures a frequent image that relates prestige to the function of giving others food to eat, in this case, of provisioning with cattle. Instead of providing the stuff of respect—the meat of worthiness for others to eat, that is, to admire—the shamed lineage members are forced to swallow (and to choke on) their pride. In producing a boy who has "run," they become consumers rather than donors of a value symbolically likened to meat: the substance of honor. The entire lineage suffers from this inversion of valences.

The food references continue, made concrete in associated moran proscriptions on meat and milk, both primary foods. During the early stages of the moranhood, the young novice must ritually pass by his own mother's hut for the last time. At this dramatic and stylized juncture, he swears publicly before her and before all those assembled that he will no longer eat any meat "seen by any married woman," . . . that is, a woman of his mother's social status. Such meat is henceforth *menong'*, despised food. Furthermore, the boy swears to refrain from drinking milk from any source inside the village settlement, and he also forswears drinking milk altogether if it has been produced by certain categories of people related to the women he may later marry. This self-denial is important because milk, either alone or mixed with cow's blood, is a major source of food for the Samburu. A much loved delicacy in all its forms, it is naturally hard to resist. Giving it up is a real sacrifice. Doing so demonstrates a maturity of resolve that amounts to a symbolic disavowal of dependency.

These frustrating food taboos are of major cultural significance to the Simburu; they see such abstemiousness as "a determining criterion of moranhood." . . . Although this is missed by Spencer, whose interests lie elsewhere, essentially what the young initiates are communicating through these oral renunciations is the symbolic forswearing of the nourishing of mothers and potential affines. It is an act of self-denial by which the boy enacts a personal transformation from receiver to giver of sustenance. The milk prohibitions communicate two linked declarations: first, the boy renounces all food dependency on his own mother; second, he renounces the juvenescent wish to seek substitute oral gratification among affines and older women, people whom he must later support through his own economic activities. This double abnegation of maternal nurturing reenacts both ritually and psychodynamically the trauma of weaning, for it conveys a public confirmation that he has renounced the breast voluntarily in favor of delayed gratifications of work-culture. All women will henceforth be treated as receivers rather than givers of food; the boy will no longer need mothering.

Aside from food and ritual taboos, the main ideal of moranhood has to do with acquiring cattle, which, as we have seen, is the main source of tribal wealth. Much of moranhood is spent in learning how to manage a herd and how to procure cattle, as this skill will enable a moran to create and sustain a new homestead. The boys learn husbandry early: they are led out to their isolated bush settlements by elders, who exhort them, "go to your cattle camps . . . to your herds . . . so that you may become rich elders." . . . For the impecunious youths, this entails mainly raiding and stealing, which in turn demand raw courage and steely fortitude. As Spencer notes, stock theft, preferably from some other tribe, remains an ideal of moranhood today, "and many are tempted to try it." . . . Rustlers may be captured, beaten, jailed, or even killed, but, if successful, rustling bestows manhood on the young moran. This naturally makes him attractive to the girls, who find such exploits manly. The girls are "enthralled with the notion that [the] moran should be manly in every way." . . .

A PARALLEL: THE MASAI

As I mentioned before, the Samburu experience is repeated among a host of other East African cattle-herding tribes. A colorful parallel is the Masai, sometimes held up as the apotheosis of the East African cattle-warrior complex. Let us take a brief comparative look at their ideas of masculinity. Here I use data provided by a female anthropologist, Melissa Llewelyn-Davies (1981), to avoid male bias; in addition I use data from the autobiography of a Masai man, Tepilit Ole Saitoti (1986), to avoid ethnocentrism—charges that might be made, though not substantiated I think, about Spencer.

The Masai, living in the hills of the Kenya-Tanzania border, have the same age-set institution and moran transition as the Samburu. Among these famous warriors, also, the qualities of masculinity are created by passage through this critical threshold, which they call "the door to manhood." . . . For the Masai, manhood is a status that does not come naturally, but rather is an elaborated idea symbolically constructed as a series of tests and confirmations during the moran period, which lasts for approximately the same length of time as among the Samburu. . . . Paramount in this formative period is the demonstration of physical courage, which is the sine qua non of the Masai warrior. Physical courage, however, is needed not only against human enemies but also against the destructive forces of nature, especially wild animals and predators such as lions, rhinos, and elephants, which are particularly active in Masailand. Whenever there is danger, the moran mobilize to meet it head on:

> Moran are, first of all, warriors. In the event of an attack upon a village, all able-bodied men present will attempt to defend it. But the responsibilities of moran in this regard are greater than those of others. . . . They are always the first to be called upon for the execution of dangerous tasks such as dealing with the wild animals that sometimes harass people and livestock; indeed lion hunts are an important and highly ritualized feature of moranhood. . . . Thus moran are associated with death and killing of enemies, both human and animal. . . .

As guardians of their people, the Masai moran therefore must be very brave. They live unaided in the forests; they hunt lions, scare away rhinos, and they face death on raids. The message comes across early: mothers constantly exhort their small sons to be brave "like moran." . . . Induction into moranhood begins only when the Masai boy has shown indications of the necessary qualities. In his autobiography, Tepilit Ole Saitoti . . . writes of how he nagged his father to let him be initiated into manhood, only to be told to wait. His cautious father was looking for signs of readiness in the boy. Finally, one dark night Tepilit bravely confronted and killed a huge lioness that had attacked his family's cattle and threatened the children. After this, his father relented: "Two months after I had killed the lioness, my father summoned all of us together. In the presence of all his children he said, 'We are going to initiate Tepilit into manhood. He has proven before all of us that he can now save children and cattle.' " . . .

During their apprenticeship, the boys undergo the same painful circumcision as the Samburu. The initiate must not so much as blink an eye. Shortly before his initiation, the apprehensive Tepilit is told, "You must not budge; don't move a muscle or even blink. You can face only one direction until the operation is completed. The slightest movement on your part will mean you are a coward, incompetent and unworthy to be a Masai man." . . . Although Masai girls are also circumcised (excision of the clitoris and the labia minora), bravery during the operation is not an important matter for them. No lasting stigma attaches to the many who do cry out or even try to escape. Girls are not expected to be brave; their self-control is "of little public significance," as the feminist Llewelyn-Davies writes. . . . Tepilit again: "It is common to see a woman crying and kicking during circumcision. Warriors are usually summoned to help hold her down." . . .

The Masai conception of masculinity has much to do with physical courage, but it encompasses much more. Llewelyn-Davies in fact argues that Masai manhood is firmly based not only on the warrior ethic but also on the idea of economic independence, and that Masai manhood is grounded in property accumulation. . . . A moran must not only learn the art of war but must also learn to be an autonomous herdsman and to create disposable wealth. His masculine potential is based as much on the accrual of wealth and its control and distribution as it is upon military prowess. She argues that moranhood must be seen as a process by which "propertyless boys" become "men" by gaining rights to property (*in jus*) and by giving this property away, creating rights over dependents (*in rem*). In contrast, women are permanently and inherently "dependent." . . . It is this contrast that defines both gender statuses. This notion of maleness as dominance over property and persons is tied to a generalized stress on reproductive success and fertility. The aspiration of all men is conceptualized as being "the indefinite increase, not merely of his herds and flocks, but of the number of his human dependents." . . .

Given the stress on prolific procreation, there is a macho sexual element here. Sexual competence is important to the Masai male image. A man must be aggressive in courtship and potent in sex. Tepilit tells the story of his friend, a young moran warrior, who boldly approaches a much older woman:

> He put a bold move on her. At first the woman could not believe his intention, or rather was amazed by his courage. The name of the warrior was Ngengeiya, or Drizzle.
>
> "Drizzle, what do you want?"
>
> "To make love to you."
>
> "I am your mother's age."
>
> "It's either her or you."
>
> This remark took the woman by surprise. She had underestimated the saying, "There is no such thing as a young warrior." When you are a warrior, you are expected to perform bravely in any situation. Your age and size are immaterial.
>
> "You mean you could really love me like a grown-up man?"
>
> "Try me, woman."
>
> He moved in on her. Soon the woman started moaning with excitement, calling out his name. "Honey Drizzle, Honey Drizzle, you *are* a man." In a breathy, stammering voice, she said, "A real man."

A real man is brave in all risky situations. Like the audacious Honey Drizzle, he performs under pressure.

As among the Samburu, a generalized emphasis on procreativity is ritually formalized during the stage of life when a moran is expected to start procuring food. A high point of the boy's moranhood is the sacrifice of his first ox. The major portion of the meat is then given to the boy's mother, an act that is described as a thank-you to her for having reared and fed him as a boy. This ritual feeding of his mother symbolizes the boy's status reversal from consumer to provider of meat and thus indicates a responsible adult manhood. The tests of moranhood are thus "crucially important to masculine identity." . . . As the central concept upon which the tribe's prosperity rests, however, moranhood embodies even more to the Masai: "Moranhood thus embodies more than an abstract ideal of masculinity. It also displays Masai culture at its most brilliant; it is crucial to the image of the culture as Masai represent it to themselves and to the outside world." . . . For the Masai, as for the Samburu, the idea of manhood contains also the idea of the tribe, an idea grounded in a moral courage based on commitment to collective goals. Their construction of manhood encompasses not only physical strength or bravery but also a moral beauty construed as selfless devotion to national identity. It embodies the central understanding that the man is only the sum of what he has achieved and that what he has achieved is nothing more or less than what he leaves behind.

6

Inside the World of Boys: Behind the Mask of Masculinity

WILLIAM POLLACK

"I get a little down," Adam confessed, "but I'm very good at hiding it. It's like I wear a mask. Even when the kids call me names or taunt me, I never show them how much it crushes me inside. I keep it all in."

THE BOY CODE: "EVERYTHING'S JUST FINE"

Adam is a fourteen-year-old boy whose mother sought me out after a workshop I was leading on the subject of boys and families. Adam, she told me, had been performing very well in school, but now she felt something was wrong.

Adam had shown such promise that he had been selected to join a special program for talented students, and the program was available only at a different—and more academically prestigious—school than the one Adam had attended. The new school was located in a well-to-do section of town, more affluent than Adam's own neighborhood. Adam's mother had been pleased when her son had qualified for the program and even more delighted that he would be given a scholarship to pay for it. And so Adam had set off on this new life.

At the time we talked, Mrs. Harrison's delight had turned to worry. Adam was not doing well at the new school. His grades were mediocre, and at midterm he had been given a warning that he might fail algebra. Yet Adam continued to insist, "I'm fine. Everything's just fine." He said this both at home and at school. Adam's mother was perplexed, as was the guidance counselor at his new school. "Adam seems cheerful and has no complaints," the counselor told her. "But something must be wrong." His mother tried to talk to Adam, hoping to find out what was troubling him and causing him to do so poorly in school. "But the more I questioned him about what was going on," she said, "the more he continued to deny any problems."

Adam was a quiet and rather shy boy, small for his age. In his bright blue eyes I detected an inner pain, a malaise whose cause I could not easily fathom. I had seen a similar look on the faces of a number of boys of different ages, including many boys in the "Listening to Boys' Voices" study. Adam looked wary, hurt, closed-in, self-protective. Most of all, he looked alone.

One day, his mother continued, Adam came home with a black eye. She asked him what had happened. "Just an accident," Adam had mumbled. He'd kept his eyes cast down, she remembered, as if he felt guilty or ashamed. His mother probed more deeply. She told him that she knew something was wrong, something upsetting was going on, and that—whatever it was—they could deal with it, they could face it together. Suddenly, Adam erupted in tears, and the story he had been holding inside came pouring out.

From William Pollack, *Real Boys.*

Adam was being picked on at school, heckled on the bus, goaded into fights in the schoolyard. "Hey, white trash!" the other boys shouted at him. "You don't belong here with *us*!" taunted a twelfth-grade bully. "Why don't you go back to your own side of town!" The taunts often led to physical attacks, and Adam found himself having to fight back in order to defend himself. "But I never throw the first punch," Adam explained to his mother. "I don't show them they can hurt me. I don't want to embarrass myself in front of everybody."

I turned to Adam. "How do you feel about all this?" I asked. "How do you handle your feelings of anger and frustration?" His answer was, I'm sad to say, a refrain I hear often when I am able to connect to the inner lives of boys.

"I get a little down," Adam confessed, "but I'm very good at hiding it. It's like I wear a mask. Even when the kids call me names or taunt me, I never show them how much it crushes me inside. I keep it all in."

"What do you do with the sadness?" I asked.

"I tend to let it boil inside until I can't hold it any longer, and then it explodes. It's like I have a breakdown, screaming and yelling. But I only do it inside my own room at home, where nobody can hear. Where nobody will know about it." He paused a moment. "I think I got this from my dad, unfortunately."

Adam was doing what I find so many boys do: he was hiding behind a mask, and using it to hide his deepest thoughts and feelings—his real self—from everyone, even the people closest to him. This mask of masculinity enabled Adam to make a bold (if inaccurate) statement to the world: "I can handle it. Everything's fine. I am invincible."

Adam, like other boys, wore this mask as an invisible shield, a persona to show the outside world a feigned self-confidence and bravado, and to hide the shame he felt at his feelings of vulnerability, powerlessness, and isolation. He couldn't handle the school situation alone—very few boys or girls of fourteen could—and he didn't know how to ask for help, even from people he knew loved him. As a result, Adam was unhappy and was falling behind in his academic performance.

Many of the boys I see today are like Adam, living behind a mask of masculine bravado that hides the genuine self to conform to our society's expectations; they feel it is necessary to cut themselves off from any feelings that society teaches them are unacceptable for men and boys—fear, uncertainty, feelings of loneliness and need.

Many boys, like Adam, also think it's necessary that they handle their problems alone. A boy is not expected to reach out—to his family, his friends, his counselors, or coaches—for help, comfort, understanding, and support. And so he is simply not as close as he could be to the people who love him and yearn to give him the human connections of love, caring, and affection every person needs.

The problem for those of us who want to help is that, on the outside, the boy who is having problems may seem cheerful and resilient while keeping inside the feelings that don't fit the male model—being troubled, lonely, afraid, desperate. Boys learn to wear the mask so skillfully—in fact, they don't even know they're doing it—that it can be difficult to detect what is really going on when they are suffering at school, when their friendships are not working out, when they are being bullied, becoming depressed, even dangerously so, to the point of feeling suicidal. The problems below the surface become obvious only when boys go "over the edge" and get into trouble at school, start to fight with friends, take drugs or abuse alcohol, are diagnosed with clinical depression or attention deficit disorder, erupt into physical violence, or come home with a black eye, as Adam did. Adam's mother, for example, did not know from her son that anything was wrong until Adam came home with an eye swollen shut; all she knew was that he had those perplexingly poor grades.

THE GENDER STRAITJACKET

Many years ago, when I began my research into boys, I had assumed that since America was revising its ideas about girls and women, it must have also been reevaluating its traditional ideas about boys, men, and masculinity. But over the years my research findings have shown that as far as boys today are concerned, the old Boy Code—the outdated and constricting assumptions, models, and rules about boys that our society has used since the nineteenth century—is still operating in force. I have been surprised to find that even in the most progressive schools and the most politically correct communities in every part of the country and in families of all types, the Boy Code continues to affect the behavior of all of us—the boys themselves, their parents, their teachers, and society as a whole. None of us is immune—it is so ingrained. I have caught myself behaving in accordance with the code, despite my awareness of its falseness—denying sometimes that I'm emotionally in pain when in fact I am; insisting that everything is all right, when it is not.

The Boy Code puts boys and men into a gender straitjacket that constrains not only them but everyone else, reducing us all as human beings, and eventually making us strangers to ourselves and to one another—or, at least, not as strongly connected to one another as we long to be.

OPHELIA'S BROTHERS

In Shakespeare's *Hamlet,* Ophelia is lover to the young prince of Denmark. Despondent over the death of his father, Hamlet turns away from Ophelia. She, in turn, is devastated and she eventually commits suicide. In recent years, Mary Pipher's book on adolescent girls, *Reviving Ophelia,* has made Ophelia a symbolic figure for troubled, voiceless adolescent girls. But what of Hamlet? What of Ophelia's brothers?

For Hamlet fared little better than Ophelia. Alienated from himself, as well as from his mother and father, he was plagued by doubt and erupted in uncontrolled outbursts. He grew increasingly isolated, desolate, and alone, and those who loved him were never able to get through to him. In the end, he died a tragic and unnecessary death.

The boys we care for, much like the girls we cherish, often seem to feel they must live semi-inauthentic lives, lives that conceal much of their true selves and feelings, and studies show they do so in order to fit in and be loved. The boys I see—in the "Listening to Boys' Voices" study, in schools, and in private practice—often are hiding not only a wide range of their feelings but also some of their creativity and originality, showing in effect only a handful of primary colors rather than a broad spectrum of colors and hues of the self.

The Boy Code is so strong, yet so subtle, in its influence that boys may not even know they are living their lives in accordance with it. In fact, they may not realize there is such a thing until they violate the code in some way or try to ignore it. When they do, however, society tends to let them know—swiftly and forcefully—in the form of a taunt by a sibling, a rebuke by a parent or a teacher, or ostracism by classmates.

But, it doesn't have to be this way. I know that Adam could have been saved a great deal of pain if his parents and the well-meaning school authorities had known how to help him, how to make him feel safe to express his real feelings, beginning with the entirely natural anxiety about starting at a new school. This could have eased the transition from one school to a new one, rather than leaving Adam to tough it out by himself—even though Adam would have said, "Everything's all right." . . .

BEHIND THE MASK OF MASCULINITY: SHAME AND THE TRAUMA OF SEPARATION

Just as Adam and his parents unwittingly adhered to the Boy Code, most parents and schools do the same. It has been ingrained in our society for so long, we're unaware of it. One educational expert recently suggested that the way to achieve equality in schooling would be by "teaching girls to raise their voices and boys to develop their ears." Of course boys should learn to listen. They should also speak clearly, in their own personal voices. I believe, however, that it's not boys who cannot hear us—it is we who are unable to hear them.

Researchers have found that at birth, and for several months afterward, *male infants are actually more emotionally expressive than female babies.* But by the time boys reach elementary school much of their emotional expressiveness has been lost or has gone underground. Boys at five or six become less likely than girls to express hurt or distress, either to their teachers or to their own parents. Many parents have asked me what triggers this remarkable transformation, this squelching of a boy's natural emotional expressiveness. What makes a boy who was open and exuberant unwilling to show the whole range of his emotions?

Recent research points to two primary causes for this change, and both of them grow out of assumptions about and attitudes toward boys that are deeply ingrained in the codes of our society.

The first reason is the use of shame in the toughening-up process by which it's assumed boys need to be raised. Little boys are made to feel ashamed of their feelings, guilty especially about feelings of weakness, vulnerability, fear, and despair.

The second reason is the separation process as it applies to boys, the emphasis society places on a boy's separating emotionally from his mother at an unnecessarily early age, usually by the time the boys are six years old and then again in adolescence.

The use of shame to "control" boys is pervasive. . . . Boys are made to feel shame over and over, in the midst of growing up, through what I call society's shame-hardening process. The idea is that a boy needs to be disciplined, toughened up, made to act like a "real man," be independent, keep the emotions in check. A boy is told that "big boys don't cry," that he shouldn't be "a mama's boy." If these things aren't said directly, these messages dominate in subtle ways in how boys are treated—and therefore how boys come to think of themselves. Shame is at the heart of how others behave toward boys on our playing fields, in schoolrooms, summer camps, and in our homes. A number of other societal factors contribute to this old-fashioned process of shame-hardening boys. . . .

The second reason we lose sight of the real boy behind a mask of masculinity, and ultimately lose the boy himself, is the premature separation of a boy from his mother and all things maternal at the beginning of school. Mothers are encouraged to separate from their sons, and the act of forced separation is so common that it is generally considered to be "normal." But I have come to understand that this forcing of early separation is so acutely hurtful to boys that it can only be called a trauma—an emotional blow of damaging proportions. I also believe that it is an unnecessary trauma. Boys, like girls, will separate very naturally from their mothers, if allowed to do so at their own pace.

As if the trauma of separation at age six were not wrenching enough, boys often suffer a second separation trauma when they reach sexual maturity. As a boy enters adolescence, our society becomes concerned and confused about the mother-son relationship. We feel unsure about how intimate a mother should be with her sexually mature son. We worry that an intense and loving relationship between the two will somehow get in the way of the boy's ability to form friendships with girls his own age. As a result, parents—encouraged by the society around them—may once again push the boy away from the family and, in particular, the nurturing female realm. Our society tells us this is "good" for the boy, that he needs to be pushed out of the nest or he will never fly. But I believe that the opposite is true—that a boy will make the leap when he is ready, and he will do it better if he feels that there is someone there to catch him if he falls.

This double trauma of boyhood contributes to the creation in boys of a deep wellspring of grief and sadness that may last throughout their lives.

MIXED MESSAGES: SOCIETY'S NEW EXPECTATIONS FOR BOYS

But there is another problem too: society's new expectations for boys today are in direct conflict with the teachings of the Boy Code—and we have done little to resolve the contradiction. We now say that we want boys to share their vulnerable feelings, but at the same time we expect them to cover their need for dependency and *hide* their natural feelings of love and caring behind the mask of masculine autonomy and strength. It's an impossible assignment for any boy, or, for that matter, any human being.

THE SILENCE OF LOST BOYS

Often, the result of all this conflation of signals is that the boys decide to be silent. They learn to suffer quietly, in retreat behind the mask of masculinity. They cannot speak, and we cannot hear. It's this silence that is often confusing to those of us concerned about the well-being of boys because it fools us into thinking that all is well, when much may be awry—that a boy doesn't need us, when in fact he needs us very much.

The good news is that we now know of many ways that we can help boys, and they are based on various patterns we now understand about typical boy behavior. Understanding these patterns, these ways of a real boy's life, will, I believe, help us raise boys of all ages in more successful and authentic ways. For the truth is that once we help boys shed the straitjacket of gender—once we hear and understand what a real boy says, feels, and sees—the silence is broken and replaced by a lively roar of communication. The disconnection quickly becomes reconnection. And once we reconnect with one boy, it can lead to stronger bonds with all the males in our lives—our brothers and fathers and husbands and sons. It can also help boys to connect again with their deepest feelings, their true selves.

LIVING WITH HALF A SELF—THE "HEROIC" HALF

Until now, many boys have been able to live out and express only *half* of their emotional lives—they feel free to show their "heroic," tough, action-oriented side, their physical prowess, as well as their anger and rage. What the Boy Code dictates is that they should suppress all other emotions and cover up the more gentle, caring, vulnerable sides of themselves. In the "Listening to Boys' Voices" study, many boys told me that they feel frightened and yearn to make a connection but can't. "At school, and even most times with my parents," one boy explained, "you can't act like you're a weakling. If you start acting scared or freaking out like a crybaby, my parents get mad, other kids punch you out or just tell you to shut up and cut it out." One mother told me what she expected of her nine-year-old son. "I don't mind it when Tony complains a little bit," she said, "but if he starts getting really teary-eyed and whiny I tell him just to put a lid on it. It's for his own good because if the other boys in the area hear him crying, they'll make it tough for him. Plus, his father really hates that kind of thing!"

Boys suppress feelings of rejection and loss also. One sixteen-year-old boy was told by his first girlfriend, after months of going together, that she didn't love him anymore. "You feel sick," confessed Cam. "But you just keep it inside. You don't tell anybody about it. And, then, maybe after a while, it just sort of goes away."

"It must feel like such a terrible burden, though, being so alone with it," I remarked.

"Yep," Cam sighed, fighting off tears. "But that's what a guy has to do, isn't it?"

Jason, age fifteen, recently wrote the following in an essay about expressing feelings:

> If something happens to you, you have to say: "Yeah, no big deal," even when you're really hurting. . . . When it's a tragedy—like my friend's father died—you can go up to a guy and give him a hug. But if it's anything less . . . you have to punch things and brush it off. I've punched so many lockers in my life, it's not even funny. When I get home, I'll cry about it.

I believe, and my studies indicate, that many boys are eager to be heard and that we, as parents and professionals, must use all our resources to reach out and help them. As adults, we have both the power and perspective to see through the boys' false front of machismo, especially when we know enough to expect it and to understand it for what it is—a way to look in-charge and cool.

A four-year-old boy shrugs and tries to smile after he is hit in the eye with a baseball, while blinking back tears of pain. A ten-year-old boy whose parents have just divorced behaves so boisterously and entertainingly in class he's branded the "class clown," but underneath that bravado is a lot of suffering; he longs for the days when his parents were together and he didn't need that kind of attention. A fourteen-year-old flips listlessly through a sports magazine while his school counselor discusses the boy's poor conduct. When the counselor warns the boy that his behavior may well lead to failure and suspension from school—trying to discipline through shame, through a threat of rejection—the boy retorts, "So what?"

Unfortunately, at times we all believe the mask because it fits so well and is worn so often it becomes more than just a barrier to genuine communication or intimacy. The tragedy is that the mask can actually become impossible to remove, leaving boys emotionally hollowed out and vulnerable to failure at school, depression, substance abuse, violence, even suicide.

BOYS TODAY ARE FALLING BEHIND

While it may seem as if we live in a "man's world," at least in relation to power and wealth in adult society we do not live in a "boy's world." Boys on the whole are not faring well in our schools, especially in our public schools. It is in the classroom that we see some of the most destructive effects of society's misunderstanding of boys. Thrust into competition with their peers, some boys invest so much energy into keeping up their emotional guard and disguising their deepest and most vulnerable feelings, they often have little or no energy left to apply themselves to their schoolwork. No doubt boys still show up as small minorities at the top of a few academic lists, playing starring roles as some teachers' best students. But, most often, boys form the majority of the bottom of the class. Over the last decade we've been forced to confront some staggering statistics. From elementary grades through high school, boys receive lower grades than girls. Eighth-grade boys are held back 50 percent more often than girls. By high school, boys account for two thirds of the students in special education classes. Fewer boys than girls now attend and graduate from college. Fifty-nine percent of all master's degree candidates are now women, and the percentage of men in graduate-level professional education is shrinking each year.

So, there is a gender gap in academic performance, and boys are falling to the bottom of the heap. The problem stems as much from boys' lack of confidence in their ability to perform at school as from their actual inability to perform.

When eighth-grade students are asked about their futures, girls are now twice as likely as boys to say they want to pursue a career in management, the professions, or business. Boys experience more difficulty adjusting to school, are up to ten times more likely to suffer from "hyperactivity" than girls, and account for 71 percent of all school suspensions. In recent years, girls have been making great strides in math and science. In the same period, boys have been severely lagging behind in reading and writing.

BOYS' SELF-ESTEEM—AND BRAGGING

The fact is that *boys' self-esteem as learners is far more fragile than that of most girls.* A recent North Carolina study of students in grades six to eight concluded that "Boys have a much lower image of themselves as students than girls do." Conducted by Dr. William Purkey, this study contradicts the myth that adolescent boys are more likely than girls to see themselves as smart enough to succeed in society. Boys tend to brag, according to Purkey, as a "shield to hide deep-seated lack of confidence." It is the mask at work once again, a façade of confidence and bravado that boys erect to hide what they perceive as a shameful sense of vulnerability. Girls, on the other hand, brag less and do better in school. It is probably no surprise that a recent U.S. Department of Education study found that among high school seniors fewer boys than girls expect to pursue graduate studies, work toward a law degree, or go to medical school.

What we really need for boys is the same upswing in self-esteem as learners that we have begun to achieve for girls—to recognize the specialized academic needs of boys and girls in order to turn us into a more gender-savvy society.

Overwhelmingly, recent research indicates that girls not only outperform boys academically but also feel far more confident and capable. Indeed the boys in my study reported, over and over again, how it was not "cool" to be too smart in class, for it could lead to being labeled a nerd, dork, wimp, or fag. As one boy put it, "I'm not stupid enough to sit in the front row and act like some sort of teacher's pet. If I did, I'd end up with a head full of spitballs and then get my butt kicked in." Just as girls in coeducational environments have been forced to suppress their voices of certainty and truth, boys feel pressured to hide their yearnings for genuine relationships and their thirst for knowledge. To garner acceptance among their peers and protect themselves from being shamed, boys often focus on maintaining their masks and on doing whatever they can to avoid seeming interested in things creative or intellectual. To distance themselves from the things that the stereotype identifies as "feminine," many boys sit through classes without contributing and tease other boys who speak up and participate. Others pull pranks during class, start fights, skip classes, or even drop out of school entirely.

SCHOOLS AND THE NEED FOR GENDER UNDERSTANDING

Regrettably, instead of working with boys to convince them it is desirable and even "cool" to perform well at school, teachers, too, are often fooled by the mask and believe the stereotype; and this helps to make the lack of achievement self-fulfilling. If a teacher believes that boys who are not doing well are simply uninterested, incapable, or delinquent, and signals this, it helps to make it so. Indeed when boys feel pain at school, they sometimes put on the mask and then "act out." Teachers, rather than exploring the emotional reasons behind a boy's misconduct, may instead apply behavioral control techniques that are intended somehow to better "civilize" boys.

Sal, a third-grader, arrived home with a note from his teacher. "Sal had to be disciplined today for his disruptive behavior," the teacher had written. "Usually he is a very cooperative student, and I hope this behavior does not repeat itself."

Sal's mother, Audrey, asked her son what he had done.

"I was talking out of turn in class," he said.

"That's it?" she asked. "And how did your teacher discipline you?"

"She made me stay in during recess. She made me write an essay about why talking in class is disruptive and inconsiderate." Sal hung his head.

"I was appalled," recalls Audrey. "If the teacher had spent one minute with my child, trying to figure out why he was behaving badly, this whole thing could have been avoided." The teacher had known Sal to be "a very cooperative student." It seems that, the night before, Sal had learned that a favorite uncle had been killed in a car crash. "I told my son that I understood that he was having a really hard day because of his uncle, but that, even so, it's wrong to disrupt class. He was very relieved that I wasn't mad," Audrey said. "The episode made me think about how boys get treated in school. I think the teacher assumed that Sal was just 'being a boy.' And so, although what he really needed was a little understanding and extra attention, instead she humiliated him. It reminded me to think about how Sal must be feeling when something like this happens, because he often won't talk about what's bothering him unless we prompt him to."

As a frequent guest in schools across the country, I have observed a practice I consider to be inappropriate, even dangerous—and based on a misunderstanding of boys. Elementary school teachers will offer the boys in their class a special "reward"—such as a better grade, an early recess, or an extra star on their good-behavior tally sheet—if the boys will *not* raise their hand more than once per class period. They find that some boys are so eager to talk and so boisterous in clamoring to be called on that their behavior disrupts the order of the classroom.

High school teachers sometimes adopt the same practice with their adolescent boy students, particularly those who act up or talk out of turn in class. The teachers will let the boys leave early or take a short break from class if they demonstrate that they can keep quiet and "behave." In other words, instead of trying to look behind the behavior to the real boy, to what is going on inside him, teachers assume a negative, and ask these boys to make themselves even *more* invisible and to suppress their genuine selves further. Ironically, they're asking boys to act more like the old stereotype of the passive, "feminine" girl. The teachers may get what they want—a quiet classroom—but at what cost? Such approaches silence boys' voices of resistance and struggle and individuality, and serve to perpetuate boys' attention-seeking acts of irreverence.

We need to develop a new code for real boys, gender-informed schools, and a more gender-savvy society where both boys and girls are drawn out to be themselves.

If we want boys to become more empathic, we must be more empathic toward them.

THE POTENCY OF CONNECTION—A NEW CODE FOR BOYS AND GIRLS

Growing up as a boy brings its own special difficulties, but the good news is that boys can and do overcome them when and if they feel connected to their families, friends, and communities. My research demonstrates that despite society's traumatizing pressure on boys to disconnect from their vulnerable inner selves, many, if not most, boys maintain an inner wellspring of emotional connectedness, a resilience, that helps to sustain them. Sometimes these affective ties are formed with special male friends—boys' "chumships." Boys may also forge empathic and meaningful friendships with girls and young women, relationships that are often platonic.

The fact is that boys experience deep subliminal yearnings for connection—a *hidden yearning for relationship*—that makes them long to be close to parents, teachers, coaches, friends, and family. Boys are full of love and empathy for others and long to stay "attached" to their parents and closest mentors. These yearnings, in turn, can empower parents and professionals to become more deeply connected to the boys in their lives, much as Professor Carol Gilligan at Harvard and researchers at the Stone Center Group at Wellesley College have so eloquently advocated we do for girls. This intense power to connect of parents and others is part of the "potency of connection" that needs to be at the heart of a revised real-boy code. Through the potency of connection a boy can be helped to become himself, to grow into manhood in his own individual way—to be fully the "real boy" we know he is.

7

Football Ritual and the Social Reproduction of Masculinity

DONALD F. SABO
JOE PANEPINTO

This [reading] focuses on the relationship between the football coach and his players in order to understand how football ritual contributes to the social reproduction of masculinity. The masculinity-validating dimensions of football ritual have always been one of the game's prominent cultural features. . . . Football's historical prominence in sport media and folk culture has sustained a hegemonic model of masculinity that prioritizes competitiveness, asceticism, success (winning), aggression, violence, superiority to women, and respect for and compliance with male authority. The ability to persuasively initiate and legitimate this model of masculinity has, in part, allowed men who inhabit positions of power and wealth to "reproduce the social relationships that generate their dominance." . . . Therefore, in addition to reinforcing sex inequality, football's reproduction of hegemonic masculinity has also helped maintain class inequality by masking and legitimating the processes that advantage some groups of males and marginalize or disadvantage others.

Football ritual is seen here as a locus of interaction between structural, cultural, and psychological processes that engenders players and prepares them for life within the "sex-gender system." The sex-gender system refers to a "set of social relations which has a material base and in which there are hierarchical relations between men and solidarity among them, which enables them in turn to dominate women." . . . Hence, the sex-gender system has two major, interdependent structural dimensions. The first is sex inequality, which allows for male domination of women. The second is the intermale dominance hierarchy, which fosters solidarity among males, conformity to hegemonic models of masculinity, and acceptance of status inequality among male groups.

The coach-player relationship is the epicenter of football ritual. Much previous research on coaches has explored the value orientations of coaches, their personality traits, and the coach role (Gould & Martens, 1979; Sage, 1974). More recent work focuses on the gender of coaches; Fine's (1987) and Chambliss's (1988) respective studies of Little League baseball coaches and elite swimming coaches offer valuable, detailed analyses of the men's athletic experiences. What is missing from these studies, however, is a theoretical framework that facilitates a critical feminist analysis of the coach-player relationship, that is, one that contextualizes this relationship within the larger sex-gender system.

In this study, we used a "bottom-up" perspective to analyze the coach-player relationship. We modeled our approach after Effron (1971), whose

From Michael Messner and Donald Sabo (eds.), *Sport, Men, and the Gender Order: Critical Feminist Perspectives*, pp. 115–125, ©1990 by Human Kinetics Books. Reprinted by permission of the author and publisher.

literary analysis of Cervantes's epic novel was done through the eyes of Sancho Panza rather than Don Quixote. Likewise, we sought to explicate coach-player relations by interviewing players, not coaches. Our discussions with former football players were also guided from the outset by a feminist-informed anthropological framework that assumed that football ritual resembles primitive male initiation rites in fundamental ways.

THEORY: FOOTBALL AS MALE INITIATION RITE

Ritual is organized action. La Fountaine (1985) asserts that ritual derives from social structure, and through the allocation of roles and the shaping of individual identity, ritual occasions "mobilize this structure in action" (p. 11). Lukes (1975) emphasizes that ritual is "rule governed activity of a symbolic character which draws the attention of its participants to objects of thought and feeling which they hold to be of special significance" (p. 291). Hence ritual can be seen as a dynamic process that at once reproduces the structure and cultural ethos of a community and, at the same time, enables the community to enmesh itself in its own identity. "We engage in rituals," Leach (1976) says, "in order to transmit messages to ourselves" (p. 45).

A ritual can be described as patriarchal when it contains elements of gender socialization that promote and express institutionalized patterns of both sex segregation and male dominance. The most thorough anthropological treatment of patriarchal ritual is found in studies of male initiation rites (Benedict, 1959; Godelier, 1986; Herdt, 1982b; Moore, 1986). Through these rites, F. Young (1965) states, boys "learn the definition of the male situation maintained by the adult males" (p. 30). Masculinity rites in male-dominated societies contain several common elements:

- **Man-Boy Relationships.** Masculinity rites entail ongoing interaction between two key groups: older men, or "officiants," and younger men, or "initiates." Officiants are socially visible occupants of midlevel statuses who define and orchestrate rituals that recruit and motivate selected initiates to engage in appropriate behaviors, beliefs, and values. Officiants are not the architects of ritual; rather, they attend to its social construction in light of a larger, long-standing cultural blueprint.

- **Conformity and Control.** Officiants use a variety of methods to induce conformity to the rules and requirements of ritual. In order to control their tribal charges, officiants may instruct, pressure, console, chide, implore, threaten, punish, or trick to achieve compliance and conformity from initiates.

- **Social Isolation.** Male initiates are socially isolated from family and other tribal groups, especially girls and women. Whereas the virtues of men and masculinity are extolled, women and femininity are ignored or denigrated. Among the Sambia of New Guinea, for example, boys are told by adults that female spirits wish to kill and eat them. Such threats frighten boys who, in turn, turn to officiants for guidance and protection (Herdt, 1982a). Hence, tribal teachings about gender distill from and are amplified by structured sex segregation.

- **Deference to Male Authority.** The initiation process at once introduces boys to the wider male status hierarchy and acclimates them to male authority. The relationship between officiant and initiate conveys information about hierarchical rank and authority. Knowledge or special skills are usually taught according to some hierarchical scheme, and initiates are encouraged to pass from one stage or level to another, moving ever closer to the achievements of elder, higher status males.

- **Pain.** Initiation rites are filled with the infliction of pain on initiates. For example, Zuni boys from New Mexico are flogged with yucca whips, Nandi boys from East Africa face ingenious tortures linked to circumcision, and in some North American tribes torture is self-inflicted (Benedict, 1959). Courage and ability to endure pain not only set initiates apart from uninitiated boys and from women, these qualities place initiates above these other groups as well. Sex inequality and the intermale dominance hierarchy are thus reproduced.

Football ritual contains the above elements of primitive male initiation rites (Sabo, 1987). First, football is a social theater with an all-male, intergenerational cast. The older-coach/younger-player relationship develops over many years and, at least in part, is defined as a testing ground for adult manhood. Second, though the individual styles of coaches may vary from authoritarian to facilitative, they exert a great deal of control over their players and insist on conformity (Coakley, 1986). Third, football ritual unfolds in sex-segregated contexts such as the locker room and playing field. Coaches and players most often train, travel, eat, and recreate in all-male settings. If women are present, they are usually in subservient positions vis-à-vis men (i.e., cheerleaders, stewardesses, fans, and mothers who clean uniforms and serve meals at banquets). Fourth, football is also hierarchically structured. The most obvious status difference is between coach and players, but additional rankings exist between head coach and assistant coaches, first-team and second-team players, and stars and average players. Authority is concentrated almost totally in the coach, and players are expected to obey the rules. And finally, football ritual is filled with pain. As one former professional stated, "From the moment training camp opens until the season is long over, [players] do not have a day that is free of some degree of injury and pain." . . .

In summary, we drew from the anthropological literature in order to identify the ideal typical elements of primitive male initiation rites in male-dominated societies. This was done for two reasons. Theoretically, it enabled us to understand the coach-player (officiant-initiate) relationship within the contexts of football ritual and the larger sex-gender system. Methodologically, the resulting ideal type helped us decide what directions to explore in the interviews.

METHOD

We conducted 25 in-depth, semistructured interviews with a convenient sample of former football players. Seventeen had played through high school and college, though six did not complete their degrees. One had professional experience, and seven were strictly high school players. All but one of the men were white and came from either working-class or middle-class backgrounds. There were 20 face-to-face interviews and 5 telephone interviews. Following the ideal/typical contours of primitive male initiation rites detailed above, interviews generally focused on

- the coach-player relationship,
- conformity,
- social isolation,
- male authority, and
- pain.

In contrast to traditional approaches to interviewing, we assumed that interaction between social scientist and interviewee was intersubjective in character rather than occurring between an objective, unbiased, or detached observer-interviewer and the object of scientific scrutiny. . . . Both interviewers are former intercollegiate football players themselves. This fact was immediately communicated with our subjects, and furthermore, disclosures about our own experiences on the gridiron were part of the interview.

RESULTS

The interviews revealed many similarities between the coach-player relationship and the officiant-initiate relationship within primitive masculinity rites.

Coach-Player Relationships

The football coach fulfilled many of the requirements of the officiant role. Coaches and players spent many hours together learning football lore, rehearsing complex maneuvers, developing physical skills, and performing in games before the community. Most interviewees stated that football was the most important aspect of their lives. Relationships with coaches, therefore, were primary, and players perceived them this way. Some remembered coaches with deep affection and admiration. A former high school player said, "He was like a god to me, like he was 8 feet tall." Another explained, "My father died when I was young and Coach Johnson, I think, became kind of a father to me. I loved and admired him." Most players had mixed feelings toward coaches. They used descriptive terms ranging from "good man," "incredible guy," and "a real leader" to "a crazy man," "a real prick," and "mean son of a bitch."

> He wasn't exactly a god, but we sometimes thought of him that way. . . . As a freshman and sophomore, I'd have done anything for him. Anything! By my junior year, though, I thought he was much too inflexible.
>
> Over the years you learn that coaches are out for themselves, especially at the college level. You're their meal ticket, their glory ride. They're the boss, you're the employee.

Regardless of sentiment or circumstances, all players agreed that coaches exerted an important influence on their "growing up."

Conformity and Control

Coaches exerted a great deal of influence over their athletes' lives. In addition to teaching the basic rules of the game itself, coaches imposed training rules such as curfews, exercise regimens, dietary and dating restrictions, study programs, and sometimes clothing regulations. They not only taught boys how to train, tackle, and block, but also to dress and act like football players and gentlemen on and off the field.

Conformity and control were secured in several ways. Often coaches, particularly coaches with winning reputations, exercised astonishing amounts of personal control over the athletes simply through what Crosset . . . calls the "promise of grandeur." As one player put it, "The formula was 'follow my orders and win games.'" Although during the interviews some players expressed anger at coaches' pressures or demands to conform, others saw this as a necessary part of the game and life. Indeed, as one stated, "We wanted to conform. He taught us that you have to learn to work with a team if you're going to succeed at anything." Others felt that coaches helped them stay out of trouble, cut back on drug use, and take school more seriously.

Some coaches manipulated in-group/out-group tensions to insure conformity. One player offered, "We had been losing for so long that we would do anything he asked us to do because he made us believe we would win. . . . He acted like our best friend—turned us against other groups and the administration so that we would never look at him as the source of our problem." Another offered, "One of his favorite sayings was '90% of the people you meet are going to be assholes. You, gentlemen, are the 10%.'"

Ridicule was another tactic used to induce conformity. Athletes would be "chewed out" during practices, on the sidelines at games, or during team reviews of game films. Ridicule was often tinged by homophobia and misogyny. One coach hung a bra in a players' locker to signify that player wasn't tough enough. In order to inflame aggression or compliance, coaches called players "pussies" or "limp wrists" and told them "go home and play with your sisters" or "start wearing silk panties." These messages affirm asymmetric and opposed categories of gender (i.e., masculine vs. feminine).

Most coaches capitalized on the developmental urges of their preadolescent and adolescent charges to forge autonomous, individuated identities distinct from adult prescriptions. Coaches defined the challenges and struggles of the game in highly individualistic or self-fulfilling terms. Boys were encouraged to "excel," "stick it out," or "go the distance" for themselves, to perceive football as a test of personal resolve, toughness, and allegiance to peers. As one former college player observed,

> When I started out in grade school, the game was just fun. Later in high school it became a big quest to prove myself. I never was quite sure what I was proving (I'm still not), but I worked like crazy and did almost everything the coach said.

Conformity was thus cloaked by rugged individualism but, in effect, independence striving became accommodation. As Mailer (1968) observed, "There are negative rites of passage as well. Men learn in a negative rite to give up the best things they were born with, and forever."

Social Isolation

One form of social isolation is sex segregation. As Fine's (1987) study of Little League baseball players shows, a "distinctive maleness" is engendered by homosocial leisure settings. Football coaches arranged for and enforced homosociality in many ways. Practices were closed to cheerleaders and inquiring moms. Boys were often told to avoid too much contact with or attachment to girls; deeper involvement was said to promote distraction, siphon energy, and erode team loyalty. Two married college players said their coach exhorted them to avoid sexual relations with their wives for 3 days before games; it was argued that sex would dull their competitive edge.

Players who developed more committed relationships with women reported being chided by coaches and teammates. The message was that being "pussy-whipped" is no asset to individual athletic excellence and team success. Indeed, the demands placed on many college players made pursuing or maintaining serious relationships unfeasible.

> We were figuring it out one night before an away-game at Boston College. During the season, we went to classes 5 days a week, we had 3-hour practices 5 days a week followed by team meals. Friday night was psych-up time and Saturday was game day. Sunday we reviewed game films for 3 hours and had a team meeting, not to mention that we were sore as hell and couldn't move worth a damn. The only time we could chase girls was Saturday nights after home games, and, even then, the coaches said they'd prowl the bars to catch somebody drinking or breaking curfew. The only thing we had time for was going to class, playing ball, and jerking off.

As is the case with primitive masculinity rites, the structured exclusion of women tended to exaggerate masculine traits and abilities and devalue feminine ones. One former college and professional player suggested,

> Football is a macho game, the ultimate expression of macho in America. You look at yourself as the ultimate physical male and so look at other genders differently . . . if they can't compete with me, they can't be on the same level as me.

As Bryson (1983) argues, there are two ways sport ritual leads to male dominance and the "inferiorization" of females: first, by linking maleness to highly valued and visible skills, and second, by linking maleness with the positively sanctioned use of aggression/force/violence. Football does both.

Deference to Male Authority

Primitive ritual fixes and affirms boundaries and involves ideas about hierarchical order. So, too, the coach-player interaction helped develop a "hierarchical awareness" in players or what Chodorow (1978) would label "positional identity." How did this awareness take root?

Coaches occupy the highest status position within the social organization of the team. The coach-officiant also links players to the larger community and sex-gender hierarchy. High school coaches, for example, are fixed in a complex web of school boards, principals, parents, and community leaders (Sage, 1987b). Boys learn about the structure and workings of both the sex and male-dominance hierarchies through observations of their coaches' interactions with the wider community. Players recognized that coaches were answerable to powers who reside outside the ritualized contours of the game (e.g., high school or college coaches defer to administrative and booster club leaders; professional coaches cater to ownership or local government officials). As one player described it,

> I was a guard so, when we won, my line coach was happy. When we lost, he'd catch hell from Coach Parker [the head coach] who would have to answer to the principal. . . . Our school also had a long winning streak and a reputation in the city and state to uphold.

Coaches also taught boys useful lessons in how to comport themselves with school officials, the media, government, and important fans. One college player recalled how "we were taught how to speak to journalists, what to say about our opponents and teammates." Another stated,

> Football helped me overcome shyness, helped me handle myself with important people. In high school I once met the bishop who presented our team with an award. I had to make a speech.

Like their primitive counterparts, coach-officiants used social isolation to enhance their control and authority. Consider this example detailed by an interviewee. Upon their arrival to campus, players were instructed to drag their mattresses over to the un-iced hockey rink and place them along the boards. Sixty players slept on the floor of the rink without electricity or alarm clocks. Each morning, the players were collectively awakened by the snapping on of the house lights and the voice of the head coach coming across the public address system. One player recalled,

> The physical things he made us do did not bother me as much as the mental part. I can't walk into the rink to this day without thinking about it. . . . He knew what he was doing, he wanted to establish who was the boss.

Players often described their coaches as military officers. The coaches were said to be tough, like drill sergeants who whipped people into shape. Some coaches seemed to demand complete obedience. Most of the interviewees expressed some disbelief at their own tendencies to obey orders blindly and follow lockstep along any path laid out by their coaches:

> I still like the guy but I can't believe what I did. It was like a form of blind obedience. . . . I just thought that these were the steps we have to take and the system is probably right. . . . He was shaping my mind and I knew it. At first when I heard him call himself El Supremo, I thought he was only kidding. Now, looking back on it, I think that maybe he really believed it.

In summary, the coach-player relationship was a vehicle for boys to learn about and adjust to life within the intermale dominance hierarchy of football and hierarchical organization in general. Some players wholeheartedly accepted the coach's authority as legitimate whereas others secretly defied or resented it. Both groups, however, adapted to and at least outwardly conformed to the coach's authority.

Infliction of Pain

Coaches orchestrated and rationalized a variety of pain-inducing experiences. Physical pain was inflicted by injury, playing while hurt, excessive conditioning, contact drills, and fighting. Five interviewees reported being punched or slapped by coaches, though this was not normal practice. Emotional pain was inflicted by verbal criticism, public ridicule for errors or inadequacies, and humiliation by benching or demoting players.

Interviewees enthusiastically shared stories of coaching techniques that "drove us to the limits." Coaches assured players that pain is an inevitable part of the game, and interviewees seemed to have bought the axiom "jock, stock, and knee brace." Athletes reported participating in many pain-inducing activities that were organized (i.e., ritualized) by coaches as part of the training regimen or as a form of punishment.

> He'd make us run wind sprints until half the team dropped.
>
> At preseason camp, to sort out who was meanest, he'd stick two guys in the pit and whoever got out first was better. It wasn't

really football, because guys would just punch and scratch the hell out of each other until somebody managed to crawl out.

If we had a poor practice, we'd run the tires until somebody puked or dry heaved, or he'd have us beat the shit out of each other in tackling drills or something like that.

Coaches also taught players to inflict pain on their teammates and opponents. In order to prepare boys to play the game, coaches necessarily teach players to physically "punish the other guy." Despite the inherent "logic" of such training tactics within the game, the net result is that players (much like ancient gladiators) are ritualistically accomplices in one another's physical brutalization.

Players talked about learning to play with pain as one of football's greatest lessons. A former Big East player capsulized his coach's rationale as "If you can get through this, you can get through anything." Some expressed disbelief that they were able to get through their ordeals at all.

Physically, few coaches could get away with what he did to us. I'm surprised that the trainer didn't do anything about it. Our bodies were completely abused.

We all worshipped the ground he walked on, but he was also a huge prick. He beat the shit out of us and we loved him for it. It sounds weird now, but that's the way it was.

Coaches de-emphasized the degree of physical suffering and probability of serious injury. Players were taught to deny pain. Coaches encouraged boys to "toughen up," to "learn to take your knocks," and "to sacrifice your body." Boys were expected to "run until you puke" and "push until it aches." Permanent injury and debilitation sometimes resulted. Coaches' responses to players' pains and injuries varied between mild empathy to studied indifference. A high school player recalled,

It was during a scrimmage and I heard Joe's nose pop from 10 feet away. The blood was shooting right out of his face. Coach Benson came over, took one look at him, yelled for the trainer, and said so everybody could hear, "OK, he's done. Get a warm body in here."

Many interviewees suspected that there may have been other agendas beneath the painful training practices. For example, one player felt that the physical abuse served to psychologically condition players to accept the coach's authority. The player mused that "the more exhausted people are, the more accepting." Another observed that there was a point where the infliction of pain became an end in itself. On his college team, painful episodes became so ritualized that their importance overshadowed the stated purpose for practice—improved performance. As one former college player reflected, "It was counterproductive to the real goal of playing; the punishment for losing interfered with our ability to remedy our losing ways."

Pain infliction facilitates the domination of the mind through the body (Scarry, 1985). In primitive masculinity rites, pain infliction is the central element that bonds initiates together, induces a psychological change in them, and distinguishes them from women and male underlings. In football ritual, we suggest, pain appears to cement hierarchical distinctions between males, fuse the players' allegiance to one another, set men apart from and above women, and solidify the coach's authority within the intermale dominance hierarchy (Sabo, 1986).

FOOTBALL AS MASCULINITY RITUAL

Officiants of primitive masculinity rites are, in part, the social-psychological managers of boys' gender identity development. Officiants arrange for significant actors within the community to engage one another in ways that convey gender expectations and prepare boys for adult life. So, too, did coaches interpret the meaning of the game for the player-initiates: a "manly" enterprise, "preparation for life," "more than just a game," "a sport that requires great sacrifice," or "a game you'll grow to love."

Many of the meanings that coaches reportedly attached to football revolved around hegemonically masculine themes: distinctions between boys and men, physical size and strength, avoidance of feminine activities and values, toughness, aggressiveness, violence, and emotional self-control. Sometimes the coach's masculine counsels were overt: "Football is the closest thing to war you boys will ever experience. It's your chance to find out what manhood is really all about." More often than not, however, the processes of gender learning were covert or expressed mainly through the structured interplay of power and role relations between coach and players, players and women, or among teammates themselves. Indeed, interviewees were often at a loss for words when asked how their coaches had helped shape their identities as men. This oddity or contradiction may be partially explained by La Fountaine's (1985) observation that, despite the significance of certain meanings or values for the community, they are "rarely recognized by the participants for whom ritual has its own purpose" (p. 12).

In contrast to this reticence about gender, all interviewees espoused the view that football helps make for success in later life. They saw football as "training for life" and held that both good and bad experiences with coaches were "necessary lessons" for success. Indeed, the coach's emphasis on hard work and competition was repeatedly cited as the most valuable lesson of the football experience. As one 37-year-old sales representative put it, "Football taught me how to define goals and work my ass off to achieve them. There are no free rides in this life." A law student explained,

> In spite of what he did, he taught us benefits we'll carry for the rest of our lives. [We] have to learn to be competitive, because it is a competitive world. You have to be a real tough bastard to get to your goal.

In summary, we discovered that the coach-player relationship simultaneously served as a training ground for hegemonic masculinity and a fountainhead of achievement ideology. Indeed, the masculine strivings of players were inextricably tied to their mobility aspirations. Despite their relative successes or failures in life, players came away from football with an abiding belief in the American Dream ideology (i.e., hard work and personal sacrifice lead to economic success). Football's messages about manliness, male dominance, and women's place in the world were ultimately filtered through the men's perceived relations to the class system. In the final analysis, it was impossible to discern where "homo patriarchus" ended and "homo economicus" began.

SUMMARY

Much Marxist-inspired analysis of sport attempts to show how sport ideology and athletic socialization transmit status or class position. Such efforts are consistent with general reproduction theory, which explains "how societal institutions perpetuate (or reproduce) the social relationships and attitudes needed to sustain the existing relations of production in a capitalist society" (J. MacLeod, 1987, p. 9). For social reproduction theorists like Bourdieu (1977, 1978) or Althusser (1971), however, the theoretical attention paid to social class relations has dwarfed the analysis of gender and sex inequality. In contrast, previous feminist analyses of men have categorically treated them as an undifferentiated monolith, the collective dominators of women. Men's myriad relations to one another within the intermale dominance system, and how and why their relationships to women vary throughout the class system, remain underresearched and unexplicated.

This study shows that the football coach-player relationship can be understood as a nexus of patriarchal ritual that reproduces hegemonic forms of masculinity as well as competitive behavior and achievement ideologies that are more closely tied to class inequality. Football is a type of male interaction that perpetuates male privilege through dominance bonding (Farr, 1988). In-depth interviews helped us to discern "the extent to which, and in what ways, sport is involved in the mediation of ideas and beliefs, some of which become linked to the interests of classes and dominant groups, and others of which are concealed" (J. A. Hargreaves, 1982, p. 15). Our research shows that it is the patriarchal aspects of football ritual that have been concealed.

The interviews also demonstrate that gender dynamics represent only one aspect of the coach-player relationship. At this stage of its development in the American political economy, football ritual appears to be just as much a source of achievement ideology as it is gender initiation. The emphasis on meritocratic ideologies is particularly evident in corporate advertising and mass media portrayals of the game. Indeed, the cultural efficacy of the football player and coach as living symbols of manliness may be waning while other models for masculinity are ascendant. As a former Baltimore Colt from the 1960s era put it, "When we played men were really men not like these bums today with their briefcases and goddamn stock portfolios" (Donovan, 1987). In truth, pro-football players have joined labor, which is now a third-class power in America; big biceps make as little sense in the picket line as in the corporate boardroom. NCAA Division I coaches, like their professional counterparts, are becoming fixtures in upper level management, and coaches are increasingly identified more with winning and fiscal outcomes than with shaping the manly destinies of young subordinates.

The lessons born of boys' manly struggles with coaches, teammates, and opponents, therefore, are simultaneously linked to the reproduction of larger systems of both gender and class relations. The football coach is not simply a man doing a job. Although it is true that, increasingly, aspects of the coach's role are defined by the occupational structure of the postindustrial capitalist order, he is also a man among men, a cultural figurehead, an agent of social reproduction within a homosocial world who introduces boys to the ever-changing sex-gender system and its hegemonic model of masculinity.

8

Trains

NATHAN McCALL

It was the first day of summer vacation. I was fourteen years old and had just completed the eighth grade, marking the end of my junior high school days. I was sitting at home, watching TV, when the telephone rang. "Hello," I said.

"Yo, Nate, this is Lep!"

"Yo, Lep, what's up?"

"We got one. She phat as a motherfucka! Got nice titties, too! We at Turkey Buzzard's crib. You better come on over and get in on it!"

"See you in a heartbeat."

When I got to Turkey Buzzard's place a few blocks away, Bimbo, Frog Dickie, Shane, Lep, Cooder, almost the whole crew—about twelve guys in all—were already there, grinning and joking like they had stolen something. Actually, they *had* stolen something: They were holding a girl captive in one of the back bedrooms. Turkey Buzzard's parents were away at work. I learned that the girl was Vanessa, a black beauty whose family had recently moved into our neighborhood, less than two blocks from where I lived. She seemed like a nice girl. When I first noticed her walking to and from school, I had wanted to check her out. Now it was too late. She was about to have a train run on her. No way she could be somebody's straight-up girl after going through a train.

Vanessa was thirteen years old and very naive. She thought she had gone to Turkey Buzzard's crib just to talk with somebody she had a crush on. A bunch of the fellas hid in closets and under beds. When she stepped inside and sat down, they sprang from their hiding places and blocked the door so that she couldn't leave. When I got there, two or three dudes were in the back room, trying to persuade her to give it up. The others were pacing about in the living room, joking and arguing about the lineup, about who would go first.

Half of them were frontin', pretending they were more experienced than they really were. Some had never even had sex before, yet they were trying to act like they knew what to do. I fronted, too. I acted like I was eager to get on Vanessa, because that's how everybody else was acting.

I went back to the room and joined the dudes trying to persuade Vanessa to let us jam her. She wouldn't cooperate. She said she was a virgin. That forced us to get somebody to play the crazy-man role, act like he was gonna go off on her if she didn't give it up. That way, she'd get scared and give in. That's how the older boys did it. We figured it would work for us.

Lep played the heavy. He started talking loud enough for her to hear. His eyes got wide, like Muhammad Ali's used to do when he was talking trash. As guys pretended to hold Lep back, he struggled wildly, like he was fighting to get into the room. "If that bitch don't give me some, she ain't never leavin' this house!"

Frog played the good guy, acting like he was fighting to hold Lep off. "Come on, man, let the girl go. She said she don't wanna do nothing, so you can't make her."

In the ruckus, Turkey Buzzard stood over a deathly silent Vanessa as she sat on the bed, horrified, looking at the doorway where the staged struggle was going on. Buzzard talked soothingly to Vanessa, trying to convince her that he, too, was a nice guy who was on her side. "Look, baby, if you let one of us do it, then the rest of them will be satisfied and they'll let you go. But if you don't let at least one of us do it, then them other dudes gon' get mad and they ain't gonna wanna let you leave. . . . Lep is crazy. We can't keep holdin' him off like that. If we let him come in this room, it's gonna be all over for you."

Vanessa seemed in a daze, like she couldn't believe what was happening to her. She looked up at Buzzard, glanced at the doorway, then looked back at Buzzard. I stood off to the side, studying her. I could see the wheels turning in her head. She knew she was cornered. She had never been in a situation like this before and she didn't have a clue how to handle it. I could tell by the way she kept looking at Buzzard, searching his face for something she could trust, that she was on the verge of cracking. She *had* to trust Buzzard. There was really no choice. It was either that or risk having that crazy Lep burst into the room and pounce on her like he was threatening to do. I could see her thinking about it, adding things up in her mind, trying to figure out if there was something she could do or say to get herself out of that jam. Then a look of resignation washed over her face. It was a sad, fearful look.

She looked so sad that I started to feel sorry for her. Something in me wanted to reach out and do what I knew was right—do what we all instinctively knew was right: Lean down, grab Vanessa's hand, and lead her from that room and out of that house; walk her home and apologize for our temporary lapse of sanity; tell her, "Try, as best as possible, to forget any of this ever happened."

But I couldn't do that. It was too late. This was our first train together as a group. All the fellas were there and everybody was anxious to show everybody else how cool and worldly he was. If I jumped in on Vanessa' behalf, they would accuse me of falling in love. They would send word out on the block that when it came to girls, I was a wimp. Everybody would be talking at the basketball court about how I'd caved in and got soft for a bitch. There was no way I was gonna put that pressure on myself. I thought, *Vanessa got her stupid self into this. She gonna have to get herself out.*

Turkey Buzzard put his hand on her shoulder and said, "What you gonna do, girl? You gonna let one of us do it?"

Vanessa's eyes filled with water. Her lips parted momentarily, as if she intended to speak, but no words came out. Instead, she swallowed hard and nodded her head up and down, indicating that she'd give in. Slowly, she lay back on the bed, like Buzzard told her, and closed her eyes tight. Moving quickly, Buzzard slipped her pants off. When he grabbed the waistband of her panties, she rose up suddenly and grabbed his hands, as if she'd changed her mind. She gripped his hands tight for a second, looking pleadingly into his eyes. Then she turned and looked at Frog and me as if she wanted one of us to come and rescue her. I looked away. Buzzard said, almost in a whisper, "C'mon, girl, it's gonna be all right. It'll be over in a minute."

Sitting upright, Vanessa searched his face again, then released his hand, lay back on the bed, and cupped her hands over her eyes.

By then, the fellas in the living room had grown quiet. They knew she was about to give in, and the sense of anticipation—and fear—rose. Some guys, curious, crept to the doorway and kneeled low so that she couldn't see them peeping into the room. Others stayed in the living room, smiling and whispering among themselves. "Buzzard got her pants off. He got the pants."

I don't remember who went first. I think it was Buzzard. When Vanessa tried to get up after the first guy finished, another was there to climb on top. Guys crowded into the room and hovered, wide-eyed, around the bed, like gawkers at a zoo. Then another went, and another, until the line of guys climbing on and off Vanessa became blurred.

After about the fifth guy had gone, I still hadn't taken my turn. I could have gone before then, but I was having a hard time mustering the heart to make a move. I was in no great hurry to have sex in front of a bunch of other dudes.

The whole scene brought to mind a day a few years earlier when I was sitting in the gym bleachers at my junior high school, watching an intramural basketball game. Scobie-D and some older hoods came and sat near me. At one point, Scobe got on another dude's case for failing to make a move on a girl who'd given him some play. I'll never forget it: Scobe stood up and announced loudly to the rest of the guys, "That motherfucka is scared a' pussy!" The other dude hung his head in shame.

While hovering near Vanessa, I remembered how Scobe had disgraced that guy. I wasn't about to let that happen to me. I wasn't about to let it be said that I was scared of pussy. I took a deep breath and tried to relax and free my mind. I knew I couldn't get an erection if I wasn't relaxed. That would be even more embarrassing. I didn't want to pull my meat out unless I had an erection. It would look small. Somebody might see it, shriveled up and tiny, and start laughing. Imitating some of the others in the room, I took one hand and cuffed it between my legs, caressing my meat, trying to coax it to harden so I could pull it out proudly and take my turn. Then Lep said, "Nate, have you gone yet?"

I said, "Naw, man. I'm gettin' ready to go now."

I moved forward. My heart pounded like crazy, so crazy that I feared that if somebody looked closely enough they could see it beating against my frail chest. I thought, *It's my turn. I gotta go now.*

As several other guys hung around nearby, I went and stood over Vanessa. She was stretched out forlornly on the bed with a pool of semen running between her legs. She stayed silent and kept her hands cupped over her eyes, like she was hiding from a bad scene in a horror movie. With my pants still up, I pulled down my zipper, slid on top of her and felt the sticky stuff flowing from between her legs. Half-erect and fumbling nervously, I placed myself into her wetness and moved my body, pretending to grind hard. After a few miserable minutes, I got up and signaled for the next man to take his turn.

While straightening my pants, I walked over to a corner, where two or three dudes stood, grinning proudly. Somebody whispered, "That shit is *good,* ain't it?"

I said, "Yeah, man. That shit is good." Actually, I felt sick and unclean.

After the last man had taken his turn, Vanessa got up, put on her pants, and went into the bathroom, holding her panties balled up in her hands. She came out and stood in the living room, waiting for somebody to open the front door and let her out. By then, the fellas and I were lounging around, stretched out on Buzzard's mother's couch, sitting on the floor or wherever else we could find a spot. Vanessa looked solemnly around the room at each of us. Nobody said a word. Then Lep led her out and walked with her down the street.

I felt sorry for Vanessa, knowing she would never be able to live that down. I think some of the other guys felt sorry, too. But the guilt was short-lived. It was eclipsed in no time by the victory celebration we held after she left. We burst into cheers and slapped five with each other like we'd played on the winning baseball team. We joked about who was scared, whose dick was small, and who didn't know how to put it in. Everybody had a story to tell: "Beamish couldn't even get his dick hard! He had to go in a corner and beat his meat!"

"Did you see the bed sink when Bimbo climbed his big, elephant ass up there? I thought he was gonna crush the poor girl!"

"She acted like she didn't even wanna grind when I got on it! I had to teach the bitch how to grind. . . . "

That train on Vanessa was definitely a turning point for most of us. We weren't aware of what it symbolized at the time, but that train marked our real coming together as a gang. It certified us as a group of hanging partners who would do anything and everything together. It sealed our bond in the same way some other guys consummated their alliances by rumbling together in gang wars against downtown boys. In so doing, we served notice—mostly to ourselves—that we were a group of up-and-coming young cats with a distinct identity in a specific portion of Cavalier Manor that we intended to stake out as our own.

After that first train, we perfected the art of luring babes into those kinds of traps. We ran a train at my house when my parents were away. We ran many at Bimbo's crib because both his parents worked. And we set up one at Lep's place and even let his little brother get in on it. He couldn't have been more than eight or nine. He probably didn't even have a sex drive yet. He was just imitating what he saw us do, in the same way we copied older hoods we admired.

One night, when I was sitting at home by myself, thinking about it, it occurred to me that the whole notion of running trains was weird. Even though it involved sex, it didn't seem to be about sex at all. Like almost everything else we did, it was a macho thing. Using a member of one of the most vulnerable groups of human beings on the face of the earth—black females—it was another way for a guy to show the other fellas how cold and hard he was.

It wasn't until I became an adult that I figured out how utterly confused we were. I realized that we thought we loved sisters but that we actually hated them. We hated them because they were black and we were black and, on some level much deeper than we realized, we hated the hell out of ourselves.

I didn't understand all that at the time. I don't think any of us understood. But by then, we had started doing a whole lot of crazy things we didn't understand. . . .

9

He Defies You Still: The Memoirs of a Sissy

TOMMI AVICOLLI

You're just a faggot
No history faces you this morning
A faggot's dreams are scarlet
Bad blood bled from words that scarred[1]

SCENE ONE

A homeroom in a Catholic high school in South Philadelphia. The boy sits quietly in the first aisle, third desk, reading a book. He does not look up, not even for a moment. He is hoping no one will remember he is sitting there. He wishes he were invisible. The teacher is not yet in the classroom so the other boys are talking and laughing loudly.

Suddenly, a voice from beside him:

"Hey, you're a faggot, ain't you?"

The boy does not answer. He goes on reading his book, or rather pretending he is reading his book. It is impossible to actually read the book now.

"Hey, I'm talking to you!"

The boy still does not look up. He is so scared his heart is thumping madly; it feels like it is leaping out of his chest and into his throat. But he can't look up.

"Faggot, I'm talking to you!"

To look up is to meet the eyes of the tormentor.

Suddenly, a sharpened pencil point is thrust into the boy's arm. He jolts, shaking off the pencil, aware that there is blood seeping from the wound.

"What did you do that for?" he asks timidly.

"Cause I hate faggots," the other boy says, laughing. Some other boys begin to laugh, too. A symphony of laughter. The boy feels as if he's going to cry. But he must not cry. Must not cry. So he holds back the tears and tries to read the book again. He must read the book. Read the book.

When the teacher arrives a few minutes later, the class quiets down. The boy does not tell the teacher what has happened. He spits on the wound to clean it, dabbing it with a tissue until the bleeding stops. For weeks he fears some dreadful infection from the lead in the pencil point.

SCENE TWO

The boy is walking home from school. A group of boys (two, maybe three, he is not certain) grab him from behind, drag him into an alley and beat him up. When he gets home, he races up to his room, refusing dinner ("I don't feel well," he tells his mother through the locked door) and spends the night alone in the dark wishing he would die. . . .

These are not fictitious accounts—I *was* that boy. Having been branded a sissy by neighborhood children because I preferred jump rope to baseball and dolls to playing soldiers, I was often taunted with "hey sissy" or "hey faggot" or "yoo hoo honey" (in a mocking voice) when I left the house.

To avoid harassment, I spent many summers alone in my room. I went out on rainy days when the street was empty.

From *Radical Teacher,* 24, 1985, pp. 4–5. Reprinted with permission.

I came to like being alone. I didn't need anyone, I told myself over and over again. I was an island. Contact with others meant pain. Alone, I was protected. I began writing poems, then short stories. There was no reason to go outside anymore. I had a world of my own.

In the schoolyard today
they'll single you out
Their laughter will leave your ears ringing
like the church bells
which once awed you. . . .[2]

School was one of the more painful experiences of my youth. The neighborhood bullies could be avoided. The taunts of the children living in those endless repetitive row houses could be evaded by staying in my room. But school was something I had to face day after day for some two hundred mornings a year.

I had few friends in school. I was a pariah. Some kids would talk to me, but few wanted to be known as my close friend. Afraid of labels. If I was a sissy, then he had to be a sissy, too. I was condemned to loneliness.

Fortunately, a new boy moved into our neighborhood and befriended me; he wasn't afraid of labels. He protected me when the other guys threatened to beat me up. He walked me home from school; he broke through the terrible loneliness. We were in third or fourth grade at the time.

We spent a summer or two together. Then his parents sent him to camp and I was once again confined to my room.

SCENE THREE

High school lunchroom. The boy sits at a table near the back of the room. Without warning, his lunch bag is grabbed and tossed to another table. Someone opens it and confiscates a package of Tastykakes; another boy takes the sandwich. The empty bag is tossed back to the boy who stares at it, dumbfounded. He should be used to this; it has happened before.

Someone screams, "faggot," laughing. There is always laughter. It does not annoy him anymore.

There is no teacher nearby. There is never a teacher around. And what would he say if there were? Could he report the crime? He would be jumped after school if he did. Besides, it would be his word against theirs. Teachers never noticed anything. They never heard the taunts. Never heard the word, "faggot." They were the great deaf mutes, pillars of indifference; a sissy's pain was not relevant to history and geography and god made me to love honor and obey him, amen.

SCENE FOUR

High school religion class. Someone has a copy of *Playboy*. Father N. is not in the room yet; he's late, as usual. Someone taps the boy roughly on the shoulder. He turns. A finger points to the centerfold model, pink fleshy body, thin and sleek. Almost painted. Not real. The other asks, mocking voice, "Hey, does she turn you on? Look at those tits!"

The boy smiles, nodding meekly; turns away.

The other jabs him harder on the shoulder, "Hey, whatsamatter, don't you like girls?"

Laughter. Thousands of mouths; unbearable din of laughter. In the arena: thumbs down. Don't spare the queer.

"Wanna suck my dick? Huh? That turn you on, faggot!"

The laughter seems to go on forever . . .

Behind you, the sound of their laughter
echoes a million times
in a soundless place
They watch you walk/sit/stand/breathe. . . .[3]

What did being a sissy really mean? It was a way of walking (from the hips rather than the shoulders); it was a way of talking (often with a lisp or in a high-pitched voice); it was a way of relating to others (gently, not wanting to fight, or hurt anyone's feelings). It was being intelligent ("an egghead" they called it sometimes); getting good grades. It means not being interested in sports, not playing football in the street after school; not discussing teams and scores and playoffs. And it involved not showing fervent interest in girls, not talking about scoring with tits or *Playboy* centerfolds. Not concealing naked women in your history book; or porno books in your locker.

On the other hand, anyone could be a "faggot." It was a catch-all. If you did something that didn't conform to what was the acceptable behavior of the group, then you risked being called a faggot. If you didn't get along with the "in" crowd, you were a faggot. It was the most commonly used put-down. It kept guys in line. They became angry when somebody called them a faggot. More fights started over someone calling someone else a faggot than anything else. The word had power. It toppled the male ego, shattered his delicate facade, violated the image he projected. He was tough. Without feeling. Faggot cut through all this. It made him vulnerable. Feminine. And feminine was the worst thing he could possibly be. Girls were fine for fucking, but no boy in his right mind wanted to be like them. A boy was the opposite of a girl. He was not feminine. He was not feeling. He was not weak.

Just look at the gym teacher who growled like a dog; or the priest with the black belt who threw kids against the wall in rage when they didn't know their Latin. They were men, they got respect.

But not the physics teacher who preached pacifism during lectures on the nature of atoms. Everybody knew what he was—and why he believed in the anti-war movement.

My parents only knew that the neighborhood kids called me names. They begged me to act more like the other boys. My brothers were ashamed of me. They never said it, but I knew. Just as I knew that my parents were embarrassed by my behavior.

At times, they tried to get me to act differently. Once my father lectured me on how to walk right. I'm still not clear on what that means. Not from the hips, I guess, don't "swish" like faggots do.

A nun in elementary school told my mother at Open House that there was "something wrong with me." I had draped my sweater over my shoulders like a girl, she said. I was a smart kid, but I should know better than to wear my sweater like a girl!

My mother stood there, mute. I wanted her to say something, to chastise the nun; to defend me. But how could she? This was a nun talking—representative of Jesus, protector of all that was good and decent.

An uncle once told me I should start "acting like a boy" instead of like a girl. Everybody seemed ashamed of me. And I guess I was ashamed of myself, too. It was hard not to be.

SCENE FIVE

Priest: Do you like girls, Mark?

Mark: Uh-huh.

Priest: I mean *really* like them?

Mark: Yeah—they're okay.

Priest: There's a role they play in your salvation. Do you understand it, Mark?

Mark: Yeah.

Priest: You've got to like girls. Even if you should decide to enter the seminary, it's important to keep in mind God's plan for a man and a woman. . . . [4]

Catholicism of course condemned homosexuality. Effeminacy was tolerated as long as the effeminate person did not admit to being gay. Thus, priests could be effeminate because they weren't gay.

As a sissy, I could count on no support from the church. A male's sole purpose in life was to father children—souls for the church to save. The only hope a homosexual had of attaining salvation was by remaining totally celibate. Don't even think of touching another boy. To think of a sin was a sin. And to sin was to put a mark upon the soul. Sin—if it was a serious offense against god—led to hell. There was no way around it. If you sinned, you were doomed.

Realizing I was gay was not an easy task. Although I knew I was attracted to boys by the time I was about eleven, I didn't connect this attraction to homosexuality. I was not queer. Not I. I was merely appreciating a boy's good looks, his fine features, his proportions. It didn't seem to matter that I didn't appreciate a girl's looks in the same way. There was no twitching in my thighs when I gazed upon a beautiful girl. But I wasn't queer.

I resisted that label—queer—for the longest time. Even when everything pointed to it, I refused to see it. I was certainly not queer. Not I.

We sat through endless English classes, and history courses about the wars between men who were not allowed to love each other. No gay history was ever taught. No history faces you this morning. You're just a faggot. Homosexuals had never contributed to the human race. God destroyed the queers in Sodom and Gomorrah.

We learned about Michelangelo, Oscar Wilde, Gertrude Stein—but never that they were queer. They were not queer. Walt Whitman, the "father of American poetry," was not queer. No one was queer. I was alone, totally unique. One of a kind. Were there others like me somewhere? Another planet, perhaps?

In school, they never talked of the queers. They did not exist. The only hint we got of this other species was in religion class. And even then it was clouded in mystery—never spelled out. It was sin. Like masturbation. Like looking at *Playboy* and getting a hard-on. A sin.

Once a progressive priest in senior year religion class actually mentioned homosexuals—he said the word—but was into Erich Fromm, into homosexuals as pathetic and sick. Fixated at some early stage; penis, anal, whatever. Only heterosexuals passed on to the nirvana of sexual development.

No other images from the halls of the Catholic high school except those the other boys knew: swishy faggot sucking cock in an alley somewhere, grabbing asses in the bathroom. Never mentioning how much straight boys craved blow jobs, it was part of the secret.

It was all a secret. You were not supposed to talk about the queers. Whisper maybe. Laugh about them, yes. But don't be open, honest; don't try to understand. Don't cite their accomplishments. No history faces you this morning. You're just a faggot faggot no history just a faggot

EPILOGUE

The boy marching down the Parkway. Hundreds of queers. Signs proclaiming gay pride. Speakers. Tables with literature from gay groups. A miracle, he is thinking. Tears are coming loose now. Someone hugs him.

> You could not control
> the sissy in me
> nor could you exorcise him
> nor electrocute him
> You declared him illegal illegitimate
> insane and immature
> But he defies you still.[5]

NOTES

1. From the poem "Faggot" by Tommi Avicolli, published in *GPU News,* Sept. 1979.
2. Ibid.
3. Ibid.
4. From the play *Judgment of the Roaches* by Tommi Avicolli, produced in Philadelphia at the Gay Community Center, the Painted Bride Arts Center and the University of Pennsylvania; aired over WXPN-FM, in four parts; and presented at the Lesbian/Gay Conference in Norfolk, VA, July, 1980.
5. From the poem "Sissy Poem," published in *Magic Doesn't Live Here Anymore* (Philadelphia: Spruce Street Press, 1976).

10

The Kid No One Noticed

JONAH BLANK

Paducah, Ky.—When Michael Carneal warned friends last Thanksgiving to stay away from their high school lobby, it was not, he now says, because he knew that a tragedy would occur there once the long weekend came to an end. "Just about every day I told people that something was going to happen on Monday." He had developed a habit of making frequent but empty threats, he says, after logging onto a Web site called "101 Ways to Annoy People."

This week, Michael Carneal pleads guilty but mentally ill to three counts of murder and five of attempted murder—the result of a threat that proved anything but empty. When he gunned down eight classmates at a prayer circle in the lobby of Heath High School last December 1, he was a frail 14-year-old, a little over 5 feet tall and weighing 110 pounds. Now, at 15, he is a few inches taller and 20 pounds heavier. He has spent the past 10 months in juvenile detention, which he prefers to high school. He likes the food, sleeps well, and, he says, "people respect me now."

Of all the school shootings that made headlines in America over the past year, the Paducah killings may be the most baffling. Nearly every theory trotted out at the time of the tragedy now seems hollow: the obsession of a gun nut or the revenge of a bully's victim, atheistic nihilism or the influence of violent movies, the traumas of a dysfunctional childhood or the ravages of criminal insanity—all important social problems but, in this case, each a dead end. What's striking about Michael Carneal is how ordinary he is. But he had an extraordinary craving for "respect."

U.S. News has obtained a copy of the psychiatric report prepared as evidence for his trial—an evaluation by doctors who spent several days interviewing him, his family, and five of his friends. "Michael Carneal was not mentally ill nor mentally retarded at the time of the shootings," the doctors found. His lawyers agree that he was not legally insane at the time but say he is mentally ill and needs treatment. The Carneal quotes in this [reading] come from the psychiatric report.

In school shootings from Mississippi to Arkansas to Oregon, an inner darkness seems to have preceded the mayhem: membership in a satanic cult, a history of torturing animals, or a fanatical fascination with firearms and explosives. But whatever demons may have lurked in Michael's heart remain well hidden. Examine the psychiatric reports and the police records, talk to anyone in Paducah who is still willing to talk, and a picture gradually comes into focus: Michael Carneal is, and was, insecure, self-centered, and hungry for attention, a boy wrestling with the frustrations of puberty and desiring the approval of his peers—hardly different from millions of kids across America. In some ways, though, he seemed younger than most teenagers. While his contemporaries were listening to gangsta rap, Michael still liked Smurfs. On Heath High's social ladder, he was barely clinging to the lower rungs.

Heroes A few Friday nights ago, while Michael read a Stephen King book in his cell, his ex-schoolmates donned camouflage fatigues. The Heath Pirates were playing a football archrival, the Ballard Bombers, and the kids were decked out in military garb to show their school spirit. Even after a 35-0 defeat, Heath High's heroes were clearly the boys in football gear. They were the ones who would be talked about until the next Friday, the ones for whom the cheerleaders cheered, for whom the band played. Until the shooting, Michael was a band member, a skinny freshman with a baritone horn.

Once he wrote a secret story, a tale in which a shy kid named Michael was picked on by "preps"—the popular kids—but was saved by a brother with a gun. "Michael" gave the corpses of the slain preps to his mother as a gift. The story might have set off alarms, but it remained hidden until after the shootings.

The actual Michael Carneal had no heroic brother, no fictional alter ego to save him from a threat that seemed quite real. He felt alienated, pushed around, picked on. "I didn't like to go to school," he said. "I didn't feel as if anyone really liked me." But he cited little evidence of bullying. Once in middle school, someone pulled his pants down; friends say such things were happening to kids all the time. In his own view, however, he was a castaway, at the mercy of cruel Pirates.

He was never very close to his father, a lawyer. His older sister, Kelly, got much of the family's attention. A popular girl, she became the school valedictorian half a year after the shootings. "He tries to be as good as me," she told psychiatrists, "and he can never size up." Michael compensated by becoming a class clown, what one friend called an "energetic prankster who would get attention any way he could." Michael later told doctors he stole CDs, sold parsley to a classmate as marijuana, and downloaded Internet pornography, which he passed around the school.

He was not a gun enthusiast; he may have handled firearms only a few times in his life. His friends noticed no pronounced interest in violent movies, music, TV, or video games. He dismissed the media-spread notion that his rampage was inspired by a scene in the movie *The Basketball Diaries,* which he described as boring. "I don't know why it happened, but I know it wasn't a movie."

One possible element was hardly mentioned at the time: unrequited love, or unrequited lust. Gwen Hadley, mother of one of the victims, confirms to *U.S. News* what many students had suspected: "Michael was in love with my daughter Nicole, and Nicole had no interest in him." Michael told his doctors he liked Nicole but said they never dated. Gwen Hadley says he had phoned Nicole almost nightly in the weeks before the shootings, ostensibly to discuss chemistry.

Their photos tell the whole story: Nicole was tall, pretty, and, in a school where many a girl seems to be a bottle blond, she had enough self-assurance to remain a defiant brunet. But Michael, with his thick glasses, had the owlish aspect of a pubescent Steve Forbes. Nicole played in the band with Michael and did not treat him with disdain. On a few occasions, they did homework together in his home.

Michael told psychiatrists he had begun to date but had never so much as kissed a girl. He was upset when students called him a "faggot," and he almost cried when a school gossip sheet labeled him gay. Classmates now say kids toss such taunts around freely, but, for Michael, the barbs stung. Alone in his room, with his Internet porn for company, the lack of a girlfriend was a source of physical frustration and social embarrassment. When he fired into the prayer circle, his first bullet struck Nicole Hadley.

The why may never be fully explained, but the what of those final days is now tragically clear. On Thanksgiving, Michael stole two shotguns, two semiautomatic rifles, a pistol, and 700 rounds of ammunition from a neighbor's garage. He hauled the cache home on his bike and sneaked it through a bedroom window. "I was feeling proud, strong, good, and more respected," he would tell the psychiatrists. "I had accomplished something. I'm not the kind of kid who accomplishes anything. This is the only adventure I've ever had."

The next day, he stole two more shotguns, these from his parents' bedroom. A day later, on Saturday, he took "the best" guns to a friend's house; the boys admired the longarms and shot targets with a pistol. On Sunday, after church and homework, he wrapped his arsenal in a blanket.

He did not plan to shoot anyone, he said, but just wanted to show the guns off. "Everyone would be calling me and they would come over to my house or I would go to their house. I would be popular. I didn't think I would get into trouble."

On Monday, he put the bundle into the trunk of his sister's Mazda. In his backpack was a Ruger .22 pistol. He rode with Kelly to school and found his friends, as usual, hanging out in the front lobby. A few feet away, the morning prayer circle was ending. His moment had arrived: He announced what was in the blanket and waited for the adulation he was certain would follow.

The boys talked about the guns for a minute or two but were not especially impressed. In western Kentucky, firearms are a part of everyday life. As the discussion turned to new CDs, Michael reached into his backpack. He put plugs in his ears and rammed the ammo clip into his Ruger. "I pulled it out," he recalled, "and nobody noticed." Witnesses thought they were merely watching the class clown. Three shots killed three girls who had played beside Michael in the band: Nicole Hadley, Kayce Steger, and Jessica James. The remaining five shots wounded five other students.

"I had guns." When he opened fire, Michael said, he was trying to get people to notice him: "I don't know why I wasn't bluffing this time. I guess it was because they ignored me. I had guns, I brought them to school, I showed them to them, and they were still ignoring me." He didn't think the light-calibered Ruger could kill, he said, nor did he aim at anyone in particular. But after another student—a football player—talked him into dropping the gun, the meaning of his actions sank in. "I said, 'Kill me, please. Please kill me.' I wanted to die. I knew what I had done."

Now, the families of his victims are searching for a sliver of meaning. At a two-day quilting bee, where they stitched together panels of sympathy sent from as far away as France and Japan, mothers and fathers grasped at the hope of preventing the next schoolyard shooting.

"We've got to teach parents to love and respect their kids," said Jessica James's father, Joe, who took up quilting especially for the occasion. "We don't have homes anymore," added his wife, Judy. "We only have places where kids come to sleep." Yet the Carneal family, by all external appearances and the judgment of several psychiatrists, was in no way dysfunctional.

The families see Michael as a cold, deliberate killer and believe the sure threat of harsh punishment might have prevented the tragedy. But Michael, who faces a life sentence with no chance of parole for 25 years, seems not to have pondered the consequences of his actions. His best explanation for pulling the trigger is pathetically childish: "I just wanted the guys to think I was cool."

Gwen Hadley, for all her efforts at positive renewal, knows that her daughter won't be the last schoolgirl cut down for no reason. "It's going to happen again," she said. "This can happen anywhere, at any time." Then she put another stitch in the quilt of hope.

3

Men as Friends

This chapter is the first of two to examine men's experiences within intimate relationships. In the literature on intimacy, from friendship through love and family, there is a recurring theme highlighting men's supposed shortcomings. Unlike women, who are raised to relate more easily and deeply with others and who develop a greater capacity for disclosing and sharing their inner selves, men are said to maintain greater emotional distance, even as they experience their closest relationships.

Francesca Cancian argued that there is a gender bias in cultural constructions of love that may distort our understanding of how both men and women love. Defining or "seeing" love in largely expressive terms ignores other aspects of women's and men's intimacy. For example, much of what women do as expressions of love (especially for spouses and children) consists of instrumental tasks associated with nurturing and caregiving. Likewise, if men believe they "show" or express love by what they do more than by what they say, conceptualizing or recognizing love in terms of things said renders men's sincere attempts to show intimacy invisible, and leaves them looking especially inadequate as intimate partners (Cancian 1985a).

Cancian's critique of "the feminization of love" pertains equally well to friendship. We tend to conceptualize "real" or "true" friendship by such qualities as emotional support and self-disclosure. Friends share their inner lives with each other; they tell each other how they feel, including how they feel about each other. Close friends, especially best friends, can, should, and do tell each other just about anything. This notion of friendship draws heavily upon a female measuring rod, and may understate the real intimacy that men share with male friends if such closeness is expressed in other ways. Scott Swain's study of "covert intimacy" among men (Reading 13) strongly suggests this possibility.

MEN VERSUS WOMEN

Whether one elects to criticize or defend the behavioral patterns found among male friends, readings in this chapter illustrate how, in number and nature, men and women live in different friendship worlds. Compared with women, men have numerous friends but lack "close friends." Further, unlike women's time with friends, men spend their time with friends doing things but revealing little (Dolgin, Reading 11). Additionally, men display less affection, in either words or touch, than women do toward their friends.

Based on her interviews with women and men about their friends, Karen Walker (Reading 12) warns that such reported gender differences often are exaggerated because of what people say about their friendships more than what they actually experience as friends. In talking about friendship, Walker's informants articulated the more common characteristics of gender differences. In talking about their friends, they revealed more complex, often gender "inappropriate" patterns of relating. Similarly, the narrative by Richard Nightingale (Reading 14) illustrates how individual men often find that characterizations about men's relationships overlook their experiences. His relationship with his "best friend" is both active and affective. With such qualifications in mind, we need to examine the patterns so frequently revealed in the literature on men and friendship (e.g., Bell 1981; Rubin 1985; Kilmartin 1994).

In identifying the factors that shape men's friendships, most researchers point to aspects of men's socialization (McGill 1985; Basow 1992). Some emphasize elements of the dominant cultural construction of masculinity, wherein men are expressive, competitive, rational, and uncomfortable with revealing their innermost feelings, especially feelings of vulnerability or of affection toward other males. Men's friendship styles are efforts to meet these masculine mandates (Bell 1981; Rubin 1985; McGill 1985; Stein 1986).

Some researchers suggest that men and women develop different relationship styles because of differences in how they resolve the developmental task of early childhood identity formation (Rubin 1985). Borrowing from Nancy Chodorow (Chodorow 1978), Rubin emphasizes the consequences for both genders of being "mothered" and having one's closest early relationship be with a female. Women develop "permeable ego boundaries" that are open to relationships with others and retain a strong connection to their mothers. Men are forced to separate from their mothers, identify with absent or less present fathers, and build boundaries around themselves in relation to their most nurturant caregivers (i.e., mothers). This haunts them throughout their later relationships, because it makes them less able to "connect" intimately with others (Rubin 1985). Women experience themselves in the context of relationships, while men—depicted as "selves in separation"—remain oriented more toward independence and task completion (Kilmartin 1994).

Other researchers believe that being "mothered" but not "fathered" has important role model consequences. Without a loving, attentive, nurturing presence from fathers or other male role models, boys come to inhibit their own emotional expressiveness, identifying such behavior as typical of mothers (and women, in general) and thus to be avoided. Because of the relative involvement of mothers versus fathers in caring for young children and the much greater prevalence of single mother versus single father households, boys have fewer available male role models for intimacy. Furthermore, what role models they have are also products of gender socialization and carry a style of relating that results from that socialization.

Whatever the source (e.g., normative masculinity, male developmental needs, or a lack of role models), men later avoid the types of emotional displays and self-disclosures that are more characteristic of women's friendships. Thus, men are portrayed as either lacking the capacity for or fearing the deeper intimacy found among female friends because they behave differently within their relationships. This has potentially disturbing implications for men's relationships with each other but, as Nightingale and Walker show, it is neither inevitable nor total.

Aside from how men's friendships differ from women's, what can we say about them? In boyhood, male friendships are characterized by a "distinctly male" style of relating to each other, including an orientation toward dominance, competition, and rough-and-tumble play, displayed in groups that are larger and less intimate than those of female friends. Boys interrupt each other more and engage in more bragging, commanding, storytelling, and ridiculing than is typically seen among girls (Kilmartin 1994).

As they age, male friendships remain activity-based; male friends do things together. In Reading 13, Scott Swain suggests that a "covert intimacy" is expressed in such shared activities, especially athletic activities. Via the sharing of sports, joking, doing one another favors, and teasing, men report that they exchange feelings of liking and being liked by their male friends. They come to value the kinds of nonverbal displays of male-male friendship, seeing how closeness can be expressed and felt without having to say anything. If this is true, we shortchange men's relationships when we expect them to be like women's relationships. Swain also illustrates the connections between male friendships and sports. Intricately tied to male socialization, sports also have importance in shaping the experiences many men have of friendship. Sports can occasion male bonding with friends as well as between fathers and sons, adult friends, and co-workers. As children, many males experience their closest male relationships in and around sports. Mutual participation intensifies both the connection and competitiveness between them. Even when not playing, watching and talking about sports teams provide boys with a common bond that reinforces their friendship (and often differentiates them from girls). Later, shared spectatorship often serves as an occasion for father-son sharing. I learned this well as a son; I have used it as a father. Even in the absence of other meaningful or noncontroversial areas of conversation, many fathers and sons retain what connection they have via conversations about favorite sports teams (in my family, across three generations, it's the Yankees). Sports may provide a rich and relatively safe source of bonding (Kilmartin 1994). In adulthood, a similar sort of vicarious involvement can sustain relationships between neighbors, co-workers, and friends, and may even help turn neighbors and co-workers into friends.

Occasionally, as Bernard Lefkowitz illustrates, male friends can bring out the worst in each other (Reading 15). Lefkowitz depicts the power and pull of the male peer group over a particular group of high school males. Sitting at the apex among the cliques in their suburban New Jersey high school, "the Jocks" engaged in several acts of vandalism and violence. Eventually, some were convicted of brutally sexually assaulting a developmentally handicapped young woman in a court case that is the focus of Lefkowitz's powerful book (Lefkowitz 1997). In Reading 15, Lefkowitz describes some of the Jocks' earlier activities as they bonded more tightly through gross displays of violence and contempt for a female acquaintance.

Some research indicates that men are more open and intimate in cross-sex relationships than in their friendships with other men. Wives or romantic partners are often the closest confidants in men's lives. In those relationships, men find themselves reaping the benefits that come from greater disclosure, even if the levels at which they disclose don't always match what their partners desire. Certainly, the tendency to "funnel their intimacy" into one relationship, especially marriage, is consistent with the cultural expectations of marriage as best friendship. But even outside of marriage, the depth of men's disclosure to women stands in contrast to the male-male style, suggesting not so much an inability as an unwillingness or a discomfort with male-male intimacy. In some analyses, the role of homophobia is made paramount in accounting for what men don't do or don't share with male friends (McGill 1985; Reid and Fine 1992). For example, consider this comment Michael McGill received in his research on men's intimacy:

> You have to be wary of the guys who want to get a little too friendly. It's sad, but there is a lot of guilt by association where gays are concerned. Maybe that's why we men aren't any closer to each other than we are; we're all afraid that everybody else will think we're gay. (McGill 1985)

MEN VERSUS MEN

There are culturally variable patterns in men's relationships with other men. Among Asian men, for example, open expression of friendship is not uncommon (Doyle 1995). During the war in Vietnam, American men saw repeated instances of Vietnamese men holding hands while walking together. To the American servicemen, this suggested homosexuality; to the Vietnamese men, it was a way of symbolizing their friendship (Doyle 1995). More open displays of affection between men (kissing in public, gentle touch, and so on) have been commonplace in many other parts of the world, even as they have been absent in the experiences of most American men.

The readings that follow suggest some ways in which male friendships vary among American men. Dolgin and Walker both note some differences that occur across social class lines. By shaping the nature of occupational mobility and the dominance of the job in one's life, social class creates impediments on the time one has for, or the use one makes of, friendship. This appears to be true across racial lines (Franklin 1992). While black males reportedly experience more intimate relationships with male friends than do most white males (Basow 1992), Clyde Franklin's research on working-class and "upwardly mobile" black men's friendships identified social class differences among blacks. Working-class black men use their more intense, emotional, and "extremely close" friendships to buffer themselves against the ways in which they are marginalized by the wider society. Friendships among middle-class black men resemble more the previously depicted pattern found among middle-class white men, and are often sacrificed in the quest to fit the mold required for upward mobility (Franklin 1992).

Karen Walker (Reading 12) suggests that the male pattern of friendship may be a product of structural constraints, especially the occupational constraints faced more often by men than women. She notes, for example, that middle-class women in traditionally male-dominated occupations displayed the same "lack of intimacy" in friendship that is more typically reported of men. Like men, they found themselves without the time or "space" (Allan 1989) for active friendship on top of demanding jobs and family responsibilities. On the other hand, working-class men's (and women's) friendships, though highly sex-segregated, included more of the intimate disclosures typical of women's friendship patterns.

In fact, to understand what happens to men's friendships in the life stages of child rearing and career building, we need to consider both the structure of opportunities and the culture of priorities. Friendships must compete with other roles and relationships for our time and loyalty (Hess 1972; Allan 1989; Cohen 1992). They are often subordinated to both work and family. Our time for friendship is affected by the reality of what we must do in order to survive economically, and by beliefs about what we ought to do and where we ought to be. Neither friendship nor family assert the inflexible temporal claims that work commands (Lewis and Weigert 1981). Although we may wish it were otherwise, jobs make claims on our time that are unavoidable and uncontrollable. If that leaves us in a "time bind" regarding family time (Hochschild 1997), it has more dire consequences for friendship, which lacks the cultural priority that family time possesses. In deciding where we ought to spend the time we more directly control ourselves, the family occupies a prominent place among our priorities. Friendships lose out (Cohen 1992). If men develop a pattern of having more numerous but specialized relationships, like with "lunch buddies" at work, that style may be a response to the structure of their lives as well as their prior socialization into intimacy.

As suggested, patterns of friendship are also affected by one's life cycle stage. In Dolgin's review of literature on men's friendships (Reading 11), she more broadly explores the ways such relationships develop and change throughout the life cycle. One key difference between older and younger men is that in the later years of life men's and women's friendships become more alike (though not identical). Richard Nightingale's (Reading 14) description of his "best friend" also may reflect age effects, as many of the more "masculine" characteristics of friendships surface after men finish school and enter the more time-draining and competitive environment of the workplace, while also adding marital or parental commitments. Like him, however, you may find the usual depiction of men's friendships to overlook contexts in which deeper expressions of intimacy occur.

There are other important dimensions on which men's friendships may vary. Research by Peter Nardi (1992, 1998) indicates how gay men's friendships both fit with and depart from conventional models of male intimacy. Nardi notes that gay men differentiate their friendships from sexual relationships. As an extension of the masculine pattern in which sexual intimacy often precedes emotional intimacy, gay men may initially engage in sexual relationships without a strong intimate attachment. However, once friendship emerges, a sort of "incest taboo" among the "family" of friends is applied (Nardi 1992). Importantly, gay men's friendships also depart from masculine scripts in that they represent deeper levels of emotional intimacy among men (Nardi 1992, 1998). Analytically, such relationships reveal that men possess the potential for more disclosing and emotionally supportive relationships with other men (Nardi 1998). Experientially, such friendships provide solidarity among those who are sexually stigmatized by the wider society and offer gay men opportunities to share and support each other, and reveal their "true selves." Such support is especially important as gay men continue to confront the realities of discrimination and face the dangers associated with hate crimes and AIDS.

REFERENCES

Allan, G. *Friendship: Developing a Sociological Perspective.* Boulder, CO: Westview, 1989.

Basow, S. *Gender Stereotypes and Roles.* Pacific Grove, CA: Brooks/Cole, 1992.

Bell, R. *Worlds of Friendship.* Beverly Hills, CA: Sage, 1981.

Cancian, F. "Gender Politics: Love and Power in the Public and Private Spheres." In *Gender and the Life Course,* edited by A. Rossi, 253–262. Hawthorne, NY: Aldine, 1985.

Chodorow, N. *The Reproduction of Mothering.* Berkeley, CA: University of California Press, 1978.

Cohen, T. "Men's Families, Men's Friends: A Structural Analysis of Constraints on Men's Social Ties." In *Men's Friendships,* edited by P. Nardi, 115–131. Newbury Park, CA: Sage, 1992.

Doyle, J. *The Male Experience.* Dubuque, IA: William C. Brown, 1995.

Franklin, C. "'Hey Home—Yo, Bro:' Friendship Among Black Men." In *Men's Friendships,* edited by R. Nardi, 201–213. Newbury Park, CA: Sage, 1992.

Hess, B. "Friendship." In *Aging and Society.* Vol. 3, *A Sociology of Age Stratification,* edited by M. Riley, M. Johnson, and A. Foner, 357–396. New York: Russell Sage, 1972.

Hochschild, A. *The Time Bind.* New York: Henry Holt, 1997.

Kilmartin, C. *The Masculine Self.* New York: Macmillan, 1994.

Lefkowitz, B. *Our Guys: The Glen Ridge Rape and the Secret Life of the Perfect Suburb.* Berkeley, CA: University of California Press, 1997.

Lewis, J., and A. Weigert. "The Structures and Meanings of Social Time." *Social Forces 60* (1981): 432–462.

McGill, M. *The McGill Report on Male Intimacy.* New York: Henry Holt, 1985.

Nardi, P. "The Politics of Gay Men's Friendships." In *Men's Lives,* 4th ed., edited by M. Kimmel and M. Messner, 250–253. Boston: Allyn and Bacon, 1998.

______ "Sex Friendship and Gender Roles Among Gay Men." In *Men's Friendships,* edited by P. Nardi, 173–185. Newbury Park, CA: Sage, 1992.

Reid, H., and G. Fine. "Self-Disclosure in Men's Friendships: Variations Associated with Intimate Relations." In *Men's Friendships,* edited by P. Nardi. Newbury Park, CA: Sage, 1992.

Rubin, L. *Just Friends: The Role of Friendship in Our Lives.* New York: Harper & Row, 1985.

Stein, P. "Men and Their Friendships." In *Men in Families,* edited by R. Lewis and R. Salt, 261–270. Beverly Hills, CA: Sage, 1986.

FOR ADDITIONAL READING

For a qualitative illustration of friendship patterns among young men, read Niobe Way's article, "Using Feminist Research Methods to Understand the Friendships of Adolescent Boys," from the *Journal of Social Issues* (Winter 1997, vol. 53, no. 4). Way's article is based on longitudinal research with an ethnically diverse sample of nineteen low-income, adolescent males. Their accounts suggest high levels of distrust and a fear of betrayal for which Way considers a variety of explanations. To read more about characteristics of men's friendships, especially the boundaries imposed on disclosure and affection, read "I Love You Man: Overt Expression of Affection in Male-Male Interaction," by Mark Morman and Kory Floyd (*Sex Roles: A Journal of Research,* May 1998, vol. 38, no. 9–10).

WADSWORTH VIRTUAL SOCIETY RESOURCE CENTER

http://sociology.wadsworth.com

Visit *Virtual Society* to obtain current updates in the field, surfing tips, career information, and more.

INFOTRAC COLLEGE EDITION: EXERCISE

Go to the *InfoTrac College Edition* web site to find further readings on the topics discussed in this chapter.

11

Men's Friendships: Mismeasured, Demeaned, and Misunderstood?

KIM G. DOLGIN

Friendships are voluntary relationships that exist primarily for personal satisfaction and enjoyment rather than for the fulfillment of particular goals or tasks (Wiseman 1986). People enjoy their friendships, and as an added bonus derive many benefits from having friends. In addition to providing fun and relaxation, friends give you a sense that someone cares about you and that you are not alone in the world. They allow you to express and receive intimacy, and through that sharing to gain insight into yourself and into others (Jourard 1971). Friends serve as confidants with whom you can sound out your problems, and they often offer both emotional and material support in solving those problems. They are individuals with whom you share interests, and are therefore outlets with whom you can engage in or discuss those activities. Friendships are extremely important to most people, and, ultimately, what makes your friends your friends are the feelings of caring and closeness that you share (Brehm 1985).

To what degree are each of the above characteristics true of *men's* friendships? Many researchers would say that several are not. In particular, virtually all researchers agree that men do not frequently confide in each other, and the majority of authors have questioned men's abilities to form truly intimate, truly close friendships with each other. For example, Kilmartin (1994), following Goldberg (1976), claimed that men have numerous buddies but few true friends; Tognoli (1980) wrote that men alienate themselves from other men; Lewis (1978) went so far as to propose that many men have *never* had a close male friendship that was not plagued with guilt and fear; while Balswick and Peek (1976) deemed men's lack of friends "tragic," and Pleck (1976) said that friendships between adult men are "weak and often lacking." Other scholars vigorously disagree, claiming that men's friendships have been mismeasured and misunderstood (e.g., Cancian 1986; Swain 1989; Wood and Inman 1993) and that they are in fact deeply intimate. One author, writing in the late 1960s, even claimed that, because of the survival pressures associated with our species' early evolutionary history of hunting and tribal warfare, men and only men are imbued with a deep-rooted, genetic predilection to form intense friendships (Tiger 1969)! In any case, since the yardstick against which men's friendships have been measured and found either superior or wanting is *women's* friendships, addressing the quality issue entails examining gender differences in friendship patterns.

Used by permission of the author. Previously unpublished.

GENDER DIFFERENCES IN FRIENDSHIP PATTERNS

A literature review reveals five characteristics of men's friendships that appear to set them apart from women's friendships. First, men's friendships are frequently described as "agentic," "instrumental," and "activity-based." In other words, men *do* things with their friends: they watch sporting events, play games, fix up their cars, or go fishing (Aries and Johnson 1983; Booth and Hess 1974; Candy et al. 1981; Duck 1988; Mazur and Olver 1987; Roberto and Kimboko 1989; Weiss and Lowenthal 1975; Wellman 1992; Winstead 1986). Engaging in these "male bonding" activities fosters a sense of camaraderie (Kilmartin 1994; Sherrod 1989; Swain 1989), and the sense of structure that these activities give helps men relax and enjoy themselves (Farr 1988; Maccoby 1990) and gives them an excuse to spend time together. Wright (1982) captured the essence of men's friendships by describing them as "side by side." Women's friendships, in contrast, are centered more around talking and conversation. (In Wright's words, they are "face to face.")

Second, when men do converse with their friends, the talk is less personal and emotionally revealing than when women talk. Men speak about the activities that they enjoy and the things they have accomplished, whereas women talk about their feelings and about their friendship itself (Bell 1981; Caldwell and Peplau 1982; Rubin 1985). This pattern holds true even among best friend pairs (Davidson and Duberman 1982; Dolgin et al. 1991; Dosser et al. 1986; Reis et al. 1985; Snell et al. 1988; Winstead et al. 1984).

Third, men tend to feel a certain degree of competition with their friends that is lacking in women's friendships, and this has the potential of leading to tension or discomfort (Basow 1992; Garfinkel 1985; O'Neil et al. 1986). Winning and losing are more often a part of men's friendly activities (e.g., golf, cards, and so on) than women's, and even their conversations have more of a one-upmanship quality (Wood 1994). Male friends often choose activities that allow each participant to reciprocally demonstrate his competencies and expertise (Swain 1989).

Fourth, male friends are less affectionate with each other than female friends. Although it is common for women friends to say that they care for each other, this is something that men rarely do. In addition, female friends are physically affectionate: They give each other hugs and touch each other when they speak. Male friends, in contrast, touch each other only in highly ritualized ways (e.g., the high five), and do so most frequently when they are in circumstances in which they are otherwise affirming their masculinity. Male athletes, for example, are allowed to hug each other at the end of a victorious game.

Finally, as an outgrowth of the activity-based nature of their friendships, men tend to have "specialized friendships"; that is, a man tends to have one friend with whom he goes bowling, another with whom he has lunch at work, and another with whom he goes drinking on Friday nights (Barth and Kinder 1988). Women, in contrast, tend to have "generalized friendships": They have one or two best friends with whom they spend most of their time. By extension, at most times in their lives men have more friends than women, though they know each friend less well and spend less time with each of them (Aukett et al. 1988; Blyth and Foster-Clark 1987; Lowenthal et al. 1975; Reisman 1990; see Caldwell and Peplau 1982 for conflicting data on college students).

HOW INTIMATE ARE MEN'S FRIENDSHIPS?

This circles us back to where we began. The sixth and most controversial purported gender difference in friendship is that women's friendships are closer and more intimate than men's. This has been asserted time and time again (e.g., Booth 1972; Booth and Hess 1974; Candy et al. 1981; Huyck and Hoyer 1982; Komarovsky 1976; Reis et al. 1985; Rubin 1985; Strommen 1977; Weiss and Lowenthal 1975; Wheeler et al. 1983; Winstead 1986). This claim is largely, although not exclusively, based upon the fact that women disclose more intimate personal information to their friends than men and are more emotionally revealing to their friends than men (e.g., Aries and Johnson 1983; Caldwell and Peplau 1982; Cozby 1973; Dindia and Allen 1992; Dolgin et al. 1991; Hays 1984). When men do discuss their feelings, they express mostly negative emotions such as anger (Saurer and Eisler 1990) and are unlikely to talk about their fears (Davidson and Duberman 1982) or sadnesses (Rubin 1985). When men do speak about themselves, they are more apt than women to focus upon their strengths, victories, and achievements (Cancian 1986; Rawlins 1992). For example, Dolgin and Minowa (in press) found that the most common disclosures by male college students to their friends involved revealing self-flattering facts to another male; in contrast, the most common statements made by females were unflattering facts about themselves to other females.

Nonverbally, too, men express less affection and intimacy for each other than women. For example, males maintain a greater physical distance from each other than females (Aiello and Aiello 1974; Heshka and Nelson 1972; Patterson and Schaeffer 1977). They are less likely to freely touch each other when they greet (Greenbaum and Rosenfeld 1980), and they make less eye contact when they speak (Dabbs et al. 1980; Exline et al. 1965; Libby 1970).

Those who take issue with the assertion that women's friendships are more intimate than men's friendships dispute the relationship among signs of affection, knowledge of the other, and intimacy, not the fact that men are less self-revealing. Cancian (1986) was perhaps the first to argue, in her words, against the "feminization of love." Concurring, Wood and Imman (1993) reasoned that feminine ways of expressing closeness (such as displays of affection and personal disclosure) have been misinterpreted as defining closeness rather than as a means of demonstrating it. Swain (1989) posited that men display "covert intimacy"; that is, men indirectly communicate affection with each other by using ritualized gestures, by joking, by teasing, and by exchanging favors. Since men all share this code, they are aware that these signs denote intimate feelings. These researchers and others (e.g., Sherrod 1989) hold that men's dislike of personal conversation should not be interpreted as a lack of intimacy. For example, men do not find disclosure as stress reducing as they find diversionary activities (Reisman 1990; Swain 1989), and so they may avoid intimate conversation in favor of those activities for this reason. Also, Reis et al. (1985) did demonstrate that, although they prefer not to, men *can* converse intimately when they choose. Their findings bolster the view that men are *capable* of intimacy (*contra,* for example, Hacker 1981), but express it differently than women.

WHY ARE MEN'S FRIENDSHIPS THE WAY THEY ARE?

Whether or not it is appropriate to use disclosure as a yardstick for intimacy, few would disagree that men are less self-disclosing, especially about their emotions and weaknesses. The question, then, is why? Multiple, interrelated pressures contribute to this self-restraint.

One major contributor is the need to conform to the societally sanctioned male role. A principal facet of this role is that men should be strong and silent, and so its adoption causes one to become emotionally inexpressive (Basow 1992; Dosser et al. 1986; Pleck 1976). Several studies have indicated that androgynous men disclose more than traditionally masculine men, with levels similar to women (Lavine and Lombardo 1984; Balswick 1988); conversely, highly sex-typed men disclose much less, especially to other males (Winstead et al. 1984). Working-class men, who typically embrace the masculine stereotype more strongly than middle-class men, are especially unlikely to shun emotional expression (Hacker 1981). Also, because self-disclosure violates the norm, disclosure by men is not necessarily as advantageous for them as it is for women. Walker and Wright (1976) found that intimate self-disclosure facilitated liking between women more than between men, and Derlega and Chaiken (1976) found that men who reveal weakness are seen as less mentally healthy than women who do.

A related reason that many men are uncommunicative is that they obey the "antifemininity norm" (O'Neil 1981; O'Neil et al. 1986). Whether this tendency comes from modeling, lack of male role models, or a psychoanalytic need to reject mother identification (Chodorow 1978), "real" men avoid doing anything womanly. Since revealing weaknesses and displaying tender feelings are things women do, are part of the female role, these activities must be avoided.

A third reason that men are less comfortable disclosing personal information to each other is because men feel competitive with other men, even their friends (Basow 1992; DePaulo 1982; Farr 1988; Garfinkel 1985; Lewis 1978; O'Neil et al. 1986; Tognoli 1980). Mazur and Olver (1987) reported that men see other men as "dangerous" and threatening to them, and Sattel (1976) viewed men's inexpressiveness as a way of maintaining power over others. As Helgeson et al. (1987) reasoned, hiding their failures allows men to appear strong and powerful. Unfortunately, this competitive jockeying for position within one's own group can spill over and foster hostile feelings about those not in one's own group: My status is elevated not only by raising my place in my hierarchy but also by elevating my hierarchy over others' hierarchies. This "dominance bonding" (Farr 1988) can sometimes—tragically—erupt into hostile behavior toward members of those other groups, for example, women, gays, members of other ethnic groups (see Sanday 1990).

Homophobia also appears to push male friends apart (Clark 1974; Garfinkel 1985; Lehne 1976; Lewis 1978; Morin and Garfinkle 1978; Naifeh and Smith 1984; Tognoli 1980). For fear of their gestures being misinterpreted, men cannot express love for each other and can only touch each other in highly circumscribed ways (Basow 1992), generally in situations when they are otherwise engaging in overtly masculine behavior. (The pat on the rear that football players give each other is the most vivid example of this.) Support for the impact of homophobia comes from studies by Devlin and Cowan (1985) and Kite and Deaux (1986), who found that the degree of a man's homophobia was related to the intimacy he had with his male friends.

Yet another contributor to male emotional inexpressiveness is a lack of role models from whom to copy (Kilmartin 1994; Lewis 1978). Since boys learn the definition of appropriate masculine behavior by watching the men around them, most boys—those with inexpressive fathers—will be undemonstrative themselves. By this logic, boys with affectionate fathers, however, should be more prone to display affection. This was exactly the pattern uncovered by Balswick (1988).

Finally, the reciprocity norm (Norrell 1984), which dictates that one returns as much disclosure as one has received, pressures men to be close mouthed. Men receive less disclosure from others than women (Bell 1981; Buhrke and Fuqua 1987; Reis et al. 1985; Snell et al. 1988). They, in turn, should not be comfortable being highly self-revealing.

One factor that does *not* contribute to male inexpressiveness is a hypersensitivity to intimacy. Since information perceived as very intimate is less discussed than less intimate information, men might be reluctant to converse about themselves because they consider such disclosure to be more intimate than do women. For example, men might believe that discussing one's health is highly intimate, whereas women see it as only moderately intimate. This does not appear to be true (Dolgin, unpublished manuscript; Dolgin and Kim 1994; Reis et al. 1985).

THE DEVELOPMENT OF THE MALE FRIENDSHIP STYLE

The masculine friendship style—specialized, activity-based friendships filled with impersonal talk and competition and lacking in expressed affection (and perhaps intimacy)—does not emerge full-blown when men enter adulthood. This pattern is, rather, a continuation and outgrowth of differences in male and female children's play that begins late in infancy. Since early in childhood children define "friends" as "those they play with," these gender differences in play style lead to gender differences in friendship style that extend into adulthood.

The stage for gendered friendships thus begins to be set when children reach the age of 15–18 months. At this time, boys and girls begin to preferentially play with different toys. This is significant because the toys boys choose (or are given) are less likely to elicit talking than girls' toys and are more likely to lead to physical activity and mock aggression (Caldera et al. 1989; O'brien & Huston 1985). Gender segregation in choice of play partner begins very early, too, at about 27 months, when girls begin to preferentially play with other girls. Girls begin to reject boys (and hence force boys into all-male play groups) because they do not like boys' rough-and-tumble play style (Maccoby 1990). The tendency for males to roughhouse may well have a biological component, since it is not only seen cross-culturally but is also common in male nonhuman primate young (Maccoby 1986). At about 36 months, boys begin to actively prefer male play partners with whom they can wrestle and play with their favorite toys (LaFreniere et al. 1984).

These tendencies become even more pronounced in preschool and kindergarten. Beal (1994) suggests that twin forces pull the genders apart: continuing play style differences and the preschool-age children's needs to formulate gender-role identities. As in infancy, girls don't like to play with boys because boys play more assertively and aggressively (DiPietro 1981). Boys push and shove to get their way (Charlesworth and Dzur 1987) and ignore girls' attempts at influence (Maccoby and Jacklin 1987). Competitive play fighting and chasing—rough-and-tumble play revisited—remain among the favorite games of preschool boys (Humphreys and Smith 1987). Perhaps because they are surrounded by fewer same-sex adult role models than girls (Beal 1994), male preschoolers are less interested in adult attention than female preschoolers, and so they spend relatively less time with adults and more with each other, in all-male groups (Feiring and Lewis 1987). They therefore turn to each other when searching for male role models. There is some evidence that this is especially true for African-American boys (Rashid 1989). Supporting evidence can be found in several studies, which indicate that children who are relatively advanced in their gender-role knowledge are the most likely to play in same-sex groups (Fagot et al. 1986; Martin 1991, cited in Beal 1994). Finally, not only gender differences in play behavior but expectations about friendship's responsibilities begin to emerge at this time. For example, preschool girls are more willing to share with a friend than with an acquaintance, whereas boys do not make this distinction (Birch and Billman 1986).

By middle childhood, children actively avoid interacting with members of the opposite sex (Adler et al. 1992; Gottman 1986), and sex segregation is pronounced even when children are engaged in gender-neutral activities. When excursions into the "enemy camp" do occur in middle childhood, these forays often have romantic overtones. Thorne (1986) terms this "border work," and believes that it serves to further emphasize the boundaries between the sexes. This marked sex segregation is a near-universal phenomenon, occurring in cultures as disparate as those found in Sweden (Tietjen 1982) and western Kenya (Harkness and Super 1985). (See also Whiting and Edwards 1988.)

This segregation is particularly meaningful, because as Maccoby (1986) points out, while boys and girls individually show much overlap in their behaviors, in groups they behave quite differently. Boys tend to play in groups of three or more, whereas girls prefer dyads (Brooks-Gunn and Matthews 1979; Lever 1976; Waldrop and Halverson 1975); or, as Archer (1992) puts it, boys form packs and girls form pairs. Packs allow for elaborate games and team sports, and boys learn to associate fun with being in the company of large groups of peers (Lips 1993). Boys are less exclusive about friendship than girls, and are more willing to include nonfriends in their group activities (Eder and Hallinan 1978). At the same time, packs lack privacy and are not conducive to intimacy.

Boys' packs differ from girls' pairs in another way, too, that foreshadows adult male friendship patterns: The members have different statuses, and the groups are arranged in dominance hierarchies (Maccoby and Jacklin 1987). Boys know and are concerned with their ranking in the hierarchy. This causes *restrictive play* among male friends: Boys play in short bursts punctuated by interruptions and shifts in activity (Benenson and Apostoleris 1993; Boulton and Smith 1990). For example, a boy might begin a scuffle if he is losing a game. In contrast, girls' play sessions tend to be more extended and tranquil. Boys like to win and have their way when they play; girls like to compromise and smooth over (Goodwin 1990).

Because of this, boys and girls develop different styles of speaking to their friends. Boys' style of speaking serves to assert their dominance, attract and maintain an audience, and assert themselves when others have the floor. Girls' speech is directed at creating relationships of closeness and equality, criticizing others in gentle, acceptable ways, and ensuring that they have accurately interpreted other girls' speech (Maltz and Borker 1983). Boys order and command, boast and taunt, while girls use indirect forms, make requests, and pause so that others have a chance to speak (Cook et al. 1985; Goodwin 1990; Maltz and Borker 1983). This difference, too, clearly paves the way for gender differences in adult friendship styles.

Although some studies have shown no gender differences in children's understanding of the meaning of friendship (Bigalow and La Giapa 1975; Furman and Bierman 1983), and no gender differences in children's intimate knowledge of their friends (Diaz and Berndt 1982), other research indicates that boys' and girls' understanding of friendship continues to diverge. Sharing intimacy and feelings becomes a part of girls' friendships earlier than boys' (Bigalow 1977; Reis et al. 1993), and girls are more expressive to each other than boys (Borisoff 1993; Rawlins 1992; Rotenberg and Chase 1992). Foot et al. (1977) found that boys were uncomfortable and tense when asked to interact intimately with one of their friends, whereas girls were not. Girls both expect and receive more kindness, loyalty, commitment, and empathy from their best friends than boys (Clark and Bittle 1992).

By adolescence each of the elements of the typical adult patterns of gendered friendships is firmly in place. Although the rigid single-sexed structure of friendship breaks down (Dunphy 1963), adolescents have predominantly same-sex friends (Douvan and Adelson 1966; Hartup 1983; Savin-Williams 1980), and it is at this age that the issue of intimacy truly begins to distinguish males' and females' friendships (Berndt 1986; Berndt and Perry 1990). Adolescent girls claim that their friendships are more intimate than those of boys (Berndt 1986; Blyth and Foster-Clark 1987; Buhrmester and Furman 1987; Hunter and Youniss 1982; Kon and Losenkov 1978), and they disclose more than boys do (Dimond and Munz 1967; Dolgin and Kim 1994). Boys are often concerned that they would be teased if they disclosed to their friends (Berndt 1990; Youniss and Smollar 1985), and so when boys do talk, it is largely about their achievements (Stapley and Haviland 1989). Adolescent girls are more likely to turn to peers for support than boys (Burke and Weir 1978; Eaton et al. 1991; Lederman 1993), and the support they seek takes a different form than that of boys: Boys seek concrete, material help from their friends, not emotional support, and they expect their friends to stand by them when they are in trouble with authority figures (Douvan and Adelson 1966). Girls begin to feel the stresses associated with female friendships and are more concerned about faithfulness and rejection than boys (Berndt 1982; Kon and Losenkov 1978; Schneider and Coutts 1985).

HOW GREAT ARE THE DIFFERENCES BETWEEN MEN'S AND WOMEN'S FRIENDSHIPS?

The literature summarized above gives the impression that men's and women's friendships are very different from each other. There is little doubt that there are some men whose friendships neatly match the classic pattern, but there are others for whom it holds less true. Many aspects of a man's life affect the form of his friendships. Race, for example, appears to play a role: Several studies indicate that gender differences in friendship patterns may be more pronounced in whites than in blacks (Coates 1987; Stewart and Vauz 1986). Sex-typing, too, contributes to friendship style: Highly masculine, as opposed to androgynous, men are most likely to have typical male friendships (Balswick 1988; Lavine and Lombardo 1984). Socioeconomic status is also a factor. In particular, most men's friends are in the same or in adjacent socioeconomic classes, and men who hold white-collar jobs tend to have more friends than blue-collar workers (Farrell 1985). In addition, marital status affects the time that men have to devote to friendships. Cohen (1992) and Shulman (1975) both report that married men are less involved with friends than single men, and that fathers spend less time with their friends than married men without children; married men also tend to be less close to their friends than single men (Fischer and Phillips 1982) and to disclose less to them (Tschann 1988). In keeping with this pattern, elderly widowed men spend more time with friends than married elderly men (Powers and Bultena 1976). Elderly men's friendships in general differ from younger men's in at least one way: Gender differences in friendship become more attenuated in old age, and men's and women's friendships become more similar toward the end of the life span (Antonucci and Akiyama 1987; Roberto and Scott 1986). (Elderly men do continue to see friends in groups more than women do [Field and Minkler 1988], and are less likely to have intimate relationships outside the family [Connidis and Davies 1992].) Therefore, while some men are unlikely to open up to their buddies, others are more likely to confide in their close friends. In particular, the single status and middle-class backgrounds of many college students—coupled with their close living quarters, shared experiences, and large amounts of leisure time—means that many college men have friendships that are closer than those they will have later in life.

Even leaving demographic variables aside, men's and women's friendships may be more similar than the literature makes them appear. For one thing, much (though not all) of the gender-difference evidence is based upon self-report data. This is problematic for two reasons. First, between-gender differences are usually exaggerated in such data because subjects respond in socially acceptable, gender-stereotypical ways (Basow 1992; Matlin 1996). Second, even if subjects are responding accurately and in good faith, they may mean different things even when using the same terms. When examining the relationships of men and women with their "close friends," for example, we may be comparing apples and oranges (Caldwell and Peplau 1982); this too would magnify between-gender differences. A third limitation of the studies is that effect sizes are rarely reported (see Dolgin et al. 1991). When data are collected from large samples, small, consistent differences in the data sets are often "statistically significant" even when they are of small magnitude. In other words, if 200 women each made one more intimate comment per day to their best friend than the 200 men in the sample—let's say 14 such comments as opposed to 13—the results would be statistically significant *but not large or meaningful.* It is usually true that the magnitude of between-gender differences is small compared to within-gender differences (Matlin 1996). Finally, most of the data have been drawn from white middle- and working-class samples, and these samples may not be representative of Americans at large. Therefore, most of the reported gender differences are probably trends rather than absolutes, and matters of degree rather than dichotomies. (See Duck and Wright 1993; Wright 1988; Wright and Scanlon 1991.)

CROSS-SEX FRIENDSHIPS

The discussion above has focused upon men's friendships with other men. Some men, of course, have women friends as well. Most researchers agree that cross-sex friendships are considerably less common than same-sex friendships (e.g., Bell 1981; Booth and Hess 1974; Swain 1992). Wright and Scanlon (1991) calculated that fewer than half—approximately 40 percent—of noncollege men have at least one cross-sex friend, while college students are more likely to have one. Androgynous men and women are more likely to have cross-sex friends than gender-stereotyped individuals (Ickes and Barnes 1978; Lavine and Lombardo 1984; Lombardo and Lavine 1981).

There are numerous reasons for the relative scarcity of cross-sex friendships. First, adults do not enter adulthood with a developmental history of numerous cross-sex friendships. As was detailed above, girls actively segregate themselves from boys because boys dominate their cross-sex interactions (Maccoby 1990), and boys avoid playing with girls because they fear that doing so will lower their place in the hierarchy (Kilmartin 1994). Second, many adults lead quite sex-segregated lives and rarely interact in any meaningful way with individuals of the other gender (Bell 1981; O'Meara 1989).

A third impediment comes from the sexual overtones and tensions that can underlie these cross-sex relationships (Johnson et al. 1991; O'Meara 1989). Men are more likely to undertake cross-sex friendships because of initial sexual attraction than women; in fact, it is often their primary motivation (Rose 1985). Although men are willing to consider their female friends as potential romantic partners, women are more likely to feel that sexual feelings undermine their cross-sex friendships (Sapadin 1988). Compounding this barrier is the fact that men are likely to perceive flirtation in a behavior that women perceive as merely friendly (Abbey 1982; Abbey and Melby 1986; Saal et al. 1989). In addition, the lack of role models, social pressure, gender inequality and the prevalence of sex-typed behaviors diminish the opportunities for cross-sex friendships (see O'Meara 1989; Swain 1992).

Given all of these obstacles, why do cross-sex friendships exist? Simply because they are pleasurable for both parties. Both men and women report that they welcome getting an "insider perspective" from their cross-sex friends and getting feedback on how someone of the opposite gender thinks (Sapadin 1988). Men get the opportunity to talk about themselves and seek emotional support (Buhrke and Fuqua 1987; Hacker 1981; Wellman 1992; Wheeler et al. 1983). Women report enjoying having friendships that are less emotionally intense than their female-female relationships (Basow 1992), and they often find men more interesting and stimulating than their female friends (Wright and Scanlon 1991).

Still, cross-sex friendships do not appear to be equally satisfying for men and women: On the whole, men are more satisfied with them than women. Men are more likely to name a woman as a close friend than vice versa (Aukett et al. 1988; Reisman 1990), and in addition, while men say that they get more support from their women friends than their men friends, women say they get less support from their men friends (Aukett et al. 1988; Reisman 1990). Both men and women say that their friendships with women are closer and more satisfying than their friendships with men (Aries and Johnson 1983; Buhrke and Fuqua 1987), and being with women decreases men's feelings of loneliness more than being with other men (Wheeler et al. 1983). Female-female friends tend to differ from, and be more intimate and rewarding than, other friendship combinations (Dolgin et al. 1991; Wright and Scanlon 1991).

CONCLUSIONS

And so, where does this leave us? With men's and women's friendships that share many characteristics (Helgeson et al. 1987; Wright 1982). Men and women both find friendship deeply gratifying and important (Mazur 1989; Rawlins 1992; Swain 1989), and we all agree that close friendships involve intimacy, acceptance, trust, and help (Sherrod 1989). Still, there is no denying that true differences exist, especially when married, traditional, masculine sex-typed men are compared with women. It is likely true that gender role identification is more important than maleness *per se* at determining men's friendship quality (Burda et al. 1984; Jones et al. 1990; Wheeler et al. 1983).

The majority of men, however, subscribe to the gender role to some degree, and so the tendencies to do rather than to talk, and to shy away from displays of affection are there to a greater or lesser degree for most men.

Is the male pattern lacking? Are women's friendships closer than men's? Are women's friendships better than men's? Friends play a larger role in women's lives than men's lives (Duck and Wright 1993); this is especially true of married men. Furthermore, men don't expect as much from friends as do women, even in realms having nothing to do with disclosure and intimacy (Elkins and Peterson 1993), and women are more likely to be instrumental with their friends than men are disclosive to theirs (Wright and Scanlon 1991). Because women discuss their friendships, they can monitor how they are going and more easily work on improving them (Winstead 1986). Also, men more often report being lonely than women (Stokes and Levin 1986; Borys and Perlman 1985), and men's feelings of loneliness are more decreased by contact with female friends than with male friends (Wheeler et al. 1983). It therefore seems that the friendships that many men have, while sustaining and gratifying and emotionally satisfying in many ways, are not sufficient to meet the broad spectrum of their emotional needs.

A burden that many men carry is that, whereas women have many confidants, men usually have only one: their wives or romantic partners (Huyck and Hoyer 1982; Veroff et al. 1981). Since they do not link intimacy and romance, women can turn to their friends in times of crisis more than men can (Brown and Fox 1979; Chiriboga et al. 1979; Veroff et al. 1981). Many men, therefore, are in the risky position of having placed all their emotional eggs in one basket. As long as they have a wife or girlfriend to whom they can vent and with whom to be emotionally free, their emotional support needs can be met and they can contentedly banter with their male friends. However, if men are between wives or lovers or girlfriends, or if their relationships with their wives are not good, they likely do not have anyone to whom they can turn for emotional support. In addition, it is unfortunate that men can't reveal weaknesses and seek personal advice from their male friends, because (1) they have more of them and (2) men are likely to have had experiences that women have not (e.g., being turned down when asking someone out on a date, or having been drafted). Women may therefore be less able to relate to or empathize with those experiences. Although the female style of friendship is not costless—for example, women make themselves vulnerable by shared secrets—having no one to confide in is not a risk that women often face.

In sum, although it is impossible to categorically state that men's friendships are less intimate than or less close than women's friendships—since the answer depends upon one's definitions of intimacy and closeness—it does appear that men's same-sex friendships do not meet as great a range of men's needs as women's same-sex friendships do. Although this is a far cry from calling men's friendships "tragic," the inability to be emotionally open to one's friends is certainly not advantageous. As is often the case, the ability to be androgynous and to take the best from both the traditionally masculine and the traditionally feminine styles of behavior seems most adaptive. In the case of male friendships, this means being able to reduce stress by losing yourself in a ball game and by coming to your friends when you are needy.

REFERENCES

Abbey, A. "Sex Differences in Attributions for Friendly Behavior: Do Males Misperceive Females' Friendliness?" *Journal of Personality and Social Psychology* 42 (1982): 830–838.

Abbey, A., and C. Melby. "The Effects of Nonverbal Cues on Gender Differences in Perceptions of Sexual Interest." *Sex Roles* 15 (1986): 283–298.

Adler, P. A., S. J. Kleiss, and P. Adler. "Socialization to Gender Roles: Popularity Among Elementary School Boys and Girls." *Sociology of Education* 65 (1992): 169–187.

Aiello, J. R., and T. D. Aiello. "The Development of Personal Space: Proxemic Behavior of Children 6 Through 16." *Human Ecology* 2 (1974): 177–189.

Antonucci, T.C., and H. Akiyama. "An Examination of Sex Differences in Social Support in Mid- and Late Life." *Sex Roles* 17 (1987): 737–749.

Archer, J. "Childhood Gender Roles: Social Context and Organization." In *Childhood Social Development: Contemporary Perspectives,* edited by H. McGurk. Hove, UK: Erlbaum, 1992.

Aries, E. J., and F. L. Johnson. "Close Friendships in Adulthood: Conversational Context Between Same-Sex Friends." *Sex Roles* 9 (1983): 1183–1196.

Aukett, R., J. Ritchie, and K. Mill. "Gender Differences in Friendship Patterns." *Sex Roles* 19 (1988): 57–66.

Balswick, J. *The Inexpressive Male.* Lexington, MA: D. C. Heath, 1988.

Balswick, J. O., and C. W. Peek. "The Inexpressive Male: A Tragedy of American Society." In *The Forty-Nine Percent Majority: The Male Sex Role,* edited by D. Brannon and R. Brannon, 55–57. Reading, MA: Addison-Wesley, 1976.

Barth, R. J., and B. N. Kinder. "A Theoretical Analysis of Sex Differences in Same-Sex Friendships." *Sex Roles* 19 (1988): 349–363.

Basow, S. *Gender: Stereotypes and Roles.* 3rd ed. Pacific Grove, CA: Brooks-Cole, 1992.

Beal, C. R. *Boys and Girls: The Development of Gender Roles.* New York: McGraw-Hill, 1994.

Bell, R. R. "Friendships of Men and Women." *Psychology of Women Quarterly* 5 (1981): 402–417.

Benenson, J. F., and N. H. Apostoleris. "Gender Differences in Group Interactions in Early Childhood." Paper presented at the biennial meeting of the Society for Research in Child Development, New Orleans, March 1993.

Berndt, T. J. "The Features and Effects of Friendship in Early Adolescence." *Child Development* 53 (1982): 1447–1460.

Berndt, T. J. "Children's Comments About Their Friendships." In *Cognitive Perspectives on Children's Social and Behavioral Development: The Minnesota Symposium on Child Psychology,* vol. 18, edited by M. Perlmutter, 189–212. Hillsdale, NJ: Erlbaum, 1986.

Berndt, T. J. "Intimacy and Competition in the Friendships of Adolescent Boys and Girls." In *Gender Roles Across the Lifespan,* edited by M. Stevenson. Madison, WI: University of Wisconsin Press, 1990.

Berndt, T. J., and T. B. Perry. "Distinctive Features and Effects of Early Adolescent Friendships." In *From Childhood to Adolescence: A Transitional Period,* edited by R. Montemayor, R. Adams, and T. P. Gullotta. Newbury Park, CA: Sage, 1990.

Bigalow, B. J. "Children's Friendship Expectations: A Cognitive Developmental Study." *Child Development* 48 (1977): 246–253.

Bigalow, B. J., and J. J. La Giapa. "Children's Written Descriptions of Friendships: A Multidimensional Analysis." *Developmental Psychology* 11 (1975): 857–858.

Birch, L. L., and J. Billman. "Preschool Children's Food Sharing with Friends and Acquaintances." *Child Development* 57 (1986): 387–395.

Blyth, D. A., and F. S. Foster-Clark. "Gender Differences in Perceived Intimacy with Different Members of Adolescents' Social Networks." *Sex Roles* 17 (1987): 689–718.

Booth, A. "Sex and Social Participation." *American Sociological Review* 37 (1972): 183–193.

Booth, A., and E. Hess. "Cross-Sex Friendship." *Journal of Marriage and the Family* 36 (1974): 38–47.

Borys, S., and D. Perlman. "Gender Differences in Loneliness." *Personality and Social Psychology Bulletin* 11 (1985): 63–74.

Boulton, M., and P. K. Smith. "Rough-and-Tumble Play, Aggression and Dominance: Perceptions and Behavior in Children's Encounters." *Human Development* 33 (1990): 271–282.

Brehm, S. *Intimate Relationships.* New York: Random House, 1985.

Brooks-Gunn, J., and W. S. Matthews. *He and She: How Children Develop Their Sex-Role Identity.* Englewood Cliffs, NJ: Prentice Hall, 1979.

Brown, P., and H. Fox. "Sex Differences in Divorce." In *Gender and Disordered Behavior: Sex Differences in Psychopathology,* edited by E. Gomberg and V. Frank. New York: Brunner/Mazel, 1979.

Buhrke, R. A., and D. R. Fuqua. "Sex Differences in Same- and Cross-Sex Supportive Relationships." *Sex Roles* 17 (1987): 339–352.

Buhrmester, D., and W. Furman. "The Development of Companionship and Intimacy." *Child Development* 58 (1987): 1101–1113.

Burda, P. C., Jr., A. Vaux, and T. Schill. "Social Support Resources: Variation Across Sex and Sex-Role." *Personality and Social Psychology Bulletin* 10 (1984): 119–126.

Burke, R. J., and T. Weir. "Sex Differences in Adolescent Life Stress, Social Support, and Well-Being." *Journal of Psychology* 98 (1978): 277–288.

Caldera, Y. M., A. C. Huston, and M. O'brien. "Social Interactions and Play Patterns of Parents and Toddlers with Feminine, Masculine, and Neutral Toys." *Child Development* 60 (1989): 70–76.

Caldwell, M. A., and L. A. Peplau. "Sex Differences in Same-Sex Friendship." *Sex Roles* 8 (1982): 721–732.

Cancian, F. M. "The Feminization of Love." *Signs: Journal of Women in Culture and Society* 11 (1986): 692–709.

Candy, S. G., L. W. Troll, and S. G. Levy. "A Developmental Exploration of Friendship Functions in Women." *Psychology of Women Quarterly* 5 (1981): 456–472.

Charlesworth, W. R., and C. Dzur. "Gender Comparisons of Preschoolers' Behavior and Resource Utilization in Group Problem Solving." *Child Development* 58 (1987): 191–200.

Chiriboga, D. A., A. Coho, J. A. Stein, and J. Roberts. "Divorce, Stress, and Social Supports: A Study in Help-Seeking Behavior." *Journal of Divorce* 3 (1979): 121–135.

Chodorow, N. *The Reproduction of Mothering: Psychoanalysis and the Sociology of Gender.* Berkeley, CA: University of California Press, 1978.

Clark, D. "Homosexual Encounter in All-Male Groups." In *Men and Masculinity,* edited by J. H. Pleck and J. Sawyer, 88–93. Englewood Cliffs, NJ: Prentice Hall, 1974.

Clark, M. L., and M. L. Bittle. "Friendship Expectations and the Evaluation of Present Friendship in Middle Childhood and Early Adolescence." *Child Study Journal* 22 (1992): 115–135.

Coates, D. L. "Gender Differences in the Structure and Support Characteristics of Black Adolescents' Social Networks." *Sex Roles* 17 (1987): 719–736.

Cohen, T. "Men's Families, Men's Friends: A Structural Analysis of Constraints on Men's Social Ties." In *Men's Friendships,* edited by P. M. Nardi, 115–131. Newbury Park, CA: Sage, 1992.

Connidis, I. A., and L. Davies. "Confidants and Companions: Choices in Later Life." *Journal of Gerontology: Social Sciences* 47 (1992): 115–122.

Cook, A. S., J. J. Fritz, B. L. McCornack, and C. Visperas. "Early Gender Differences in the Functional Use of Language." *Sex Roles* 21 (1985): 909–915.

Cozby, P. "Self-Disclosure: A Literature Review." *Psychological Bulletin* 70 (1973): 73–91.

Dabbs, J. M., Jr., M. S. Evans, C. H. Hopper, and J. A. Purvis. "Self-Monitors in Conversation: What Do They Monitor?" *Journal of Personality and Social Psychology* 39 (1980): 278–284.

Davidson, L. R., and L. Duberman. "Friendship: Communication and Interactional Patterns in Same-Sex Dyads." *Sex Roles* 8 (1982): 809–822.

DePaulo, B. "Social Psychological Processes in Informal Help Seeking." In *Basic Processes in Helping Relationships,* edited by T. A. Wills. New York: Academic Press, 1982.

Derlega, V. J., and A. L. Chaiken. "Norms Affecting Self-Disclosure in Men and Women." *Journal of Consulting and Clinical Psychology* 44 (1976): 376–380.

Devlin, P. K., and G. A. Cowan. "Homophobia, Perceived Fathering, and Male Intimate Relationships." *Journal of Personality Assessment* 49 (1985): 467–473.

Diaz, R. M., and T. J. Berndt. "Children's Knowledge of a Best Friend: Fact or Fancy?" *Developmental Psychology* (1982): 787–794.

Dimond, R. E., and D. C. Munz. "Original Position of Birth and Self-Disclosure in High School Students." *Psychological Reports* 21 (1967): 829–833.

DiPietro, J. A. "Rough and Tumble Play: A Function of Gender." *Developmental Psychology* 17 (1981): 50–58.

Dindia, K., and M. Allen. "Sex Differences in Self-Disclosure: A Meta-Analysis." *Psychological Bulletin* 112 (1992): 106–124.

Dolgin, K. G. "An Apparent Lack of Gender Differences in Perceptions of Disclosure Intimacy." Manuscript in preparation.

Dolgin, K. G., and S. Kim. "Adolescents' Disclosure to Best and Good Friends: The Effects of Gender and Topic Intimacy." *Social Development* 3 (1994): 146–157.

Dolgin, K. G., L. Meyer, and J. Schwartz. "Effects of Gender, Target's Gender, Topic and Self-Esteem on Disclosure to Best and Middling Friends." *Sex Roles* 25 (1991): 311–329.

Dolgin, K. G., and N. Minowa (in press). "Gender Differences in Self-Presentation: A Comparison of the Roles of Flatteringness and Intimacy in Self-Disclosure to Friends." *Sex Roles.*

Dosser, D., J. Balswick, and C. Haverson. "Male Inexpressiveness and Relationships." *Journal of Social and Personal Relationships* 3 (1986): 241–258.

Douvan, E., and J. Adelson. *The Adolescent Experience.* New York: Wiley, 1966.

Duck, S. *Relating to Others.* Chicago: Dorsey Press, 1988.

Duck, S., and P. Wright. "Reexamining Gender Differences in Same-Gender Friendships: A Close Look at Two Kinds of Data." *Sex Roles* 28 (1993): 709–727.

Dunphy, D. C. "The Social Structure of Urban Adolescent Peer Groups." *Sociometry* 26 (1976): 230–246.

Eaton, Y. M., M. L. Mitchell, and J. A. Jolley. "Gender Differences in the Development of Relationships During Late Adolescence." *Adolescence* 26 (1991): 565–568.

Eder, D., and M. T. Hallinan. "Sex Differences in Children's Friendships." *American Sociological Review* 43 (1978): 237–250.

Elkins, L. E., and C. Peterson. "Gender Differences in Best Friendships." *Sex Roles* 29 (1993): 497–508.

Exline, R., D. Gray, and D. Schuette. "Visual Behavior in a Dyad as Affected by Interview Content and Sex of Respondent." *Journal of Personality and Social Psychology* 1 (1965): 201–209.

Fagot, B. I., M. D. Leinback, and R. Hagan. "Gender Labelling and the Adoption of Sex-Typed Behaviors." *Developmental Psychology* 22 (1986): 440–443.

Farr, K. "Dominance Bonding Through the Good Old Boys Sociability Group." *Sex Roles* 18 (1988): 259–277.

Farrell, M. P. "Friendships Between Men." *Marriage and Family Review* 9 (1985): 163–197.

Feiring, C., and M. Lewis. "The Child's Social Network: Sex Differences from Three to Six Years." *Sex Roles* 17 (1987): 621–636.

Field, D., and M. Minkler. "Continuity and Change in Social Support Between Young-Old and Old-Old or Very-Old Age." *Journal of Gerontology: Psychological Sciences* 43 (1988): 100–106.

Fischer, C., and S. L. Phillips. "Who Is Alone? Social Characteristics of People with Small Networks." In *Loneliness: A Source-Book of Current Theory, Research, and Therapy,* edited by L. Peplau and D. Perlman. New York: Wiley-Interscience, 1982.

Foot, H. C., A. J. Chapman, and J. R. Smith. "Individual Differences in Children's Social Responsiveness in Human Situations." In *It's Not a Funny Thing, Humour,* edited by A. J. Chapman and H. C. Foot. Oxford, England: Pergamon Press, 1977.

Furman, W., and K. Bierman. "Developmental Changes in Young Children's Conceptions of Friendship." *Child Development* 54 (1983): 549–556.

Garfinkel, P. *In A Man's World: Father, Son, Brother, Friend, and Other Roles Men Play.* New York: New American Library, 1985.

Goldberg, H. *The Hazards of Being Male: Surviving the Myth of Masculine Privilege.* New York: Nash, 1976.

Goodwin, M. H. "Tactical Uses of Stories: Participation Frameworks Within Girls' and Boys' Disputes." *Discourse Processes* 13 (1990): 33–71.

Gottman, J. M. "The World of Coordinated Play: Same- and Cross-Sex Friendships in Young Children." In *Conversations of Friends: Speculation on Affectional Bonds. Studies in Emotion and Social Interaction,* edited by J. M. Gottman and J. G. Parkes, 139–191. Cambridge: Cambridge University Press, 1986.

Greenbaum, P. E., and H. M. Rosenfeld. "Varieties of Touching in Greetings: Sequential Structure and Sex-Related Differences." *Journal of Nonverbal Behavior* 5 (1980): 13–25.

Hacker, H. "Blabbermouths and Clams: Sex Differences in Self-Disclosure in Same-Sex and Cross-Sex Friendship Dyads." *Psychology of Women Quarterly* 5 (1981): 385–401.

Harkness, S., and C. Super. "The Cultural Context of Gender Segregation in Children's Peer Groups." *Child Development* 56 (1985): 219–224.

Hartup, W. "Peer Relations." In *Handbook of Child Psychology: Vol. 4. Socialization, Personality, and Social Development,* edited by E. M. Hetherington. New York: Wiley, 1983.

Hays, R. B. "The Development and Maintenance of Friendship." *Journal of Social and Personal Relations* 1 (1984): 75–98.

Helgeson, V. S., P. Saver, and M. Dyer. "Prototypes of Intimacy and Distance in Same Sex and Opposite Sex Relationships." *Journal of Social and Personal Relationships* 4 (1987): 195–233.

Heshka, S., and Y. Nelson. "Interpersonal Speaking Distances as a Function of Age, Sex, and Relationship." *Sociometry* 35 (1972): 491–498.

Humphreys, A. P., and P. K. Smith. "Rough-and-Tumble, Friendship, and Dominance in School Children: Evidence for Continuity and Change with Age." *Child Development* 58 (1987): 201–212.

Hunter, F. T., and J. Youniss. "Changes in Functions of Three Relations During Adolescence." *Developmental Psychology* 18 (1982): 806–811.

Huyck, M. H., and W. J. Hoyer. *Adult Development and Aging.* Belmont, CA: Wadsworth, 1982.

Ickes, W., and R. D. Barnes. "Boys and Girls Together—and Alienated: On Enacting Stereotyped Sex Roles in Mixed-Sex Dyads." *Journal of Personality and Social Psychology* 36 (1978): 669–683.

Johnson, C. B., M. S. Stockdale, and F. E. Saal. "Persistence of Men's Misperceptions of Friendly Cues Across a Variety of Interpersonal Encounters." *Psychology of Women Quarterly* 15 (1991): 463–465.

Jones, D. C., N. Bloys, and M. Wood. "Sex Roles and Friendship Patterns." *Sex Roles* 23 (1990): 133–145.

Jourard, S. M. *The Transparent Self.* New York: Van Nostrand, 1971.

Kite, M. E., and K. Deaux. "Attitudes Towards Homosexuality: Assessment and Behavioral Consequences." *Basic and Applied Social Psychology* 7 (1986): 137.

Kilmartin, C. T. *The Masculine Self.* New York: Macmillan, 1994.

Komarovsky, M. *Dilemmas of Masculinity.* New York: Norton, 1976.

Kon, I. S., and V. A. Losenkov. "Friendship in Adolescence: Values and Behavior." *Journal of Marriage and the Family* 40 (1978): 143–155.

LaFreniere, P. J., F. F. Strayer, and R. Gauthier. "The Emergence of Same-Sex Affiliative Preferences Among Preschool Peers: A Developmental/Ethological Perspective." *Child Development* 55 (1984): 1958–1965.

Lavine, L. O., and J. P. Lombardo. "Self-Disclosure: Intimate and Nonintimate Disclosure to Parents and Best Friends as a Function of Bem Sex-Role Category." *Sex Roles* 11 (1984): 760–768.

Lederman, L. C. "Gender and the Self." In *Women and Men Communicating,* edited by L. P. Arliss and D. J. Borisoff. Fort Worth, TX: Harcourt Brace Jovanovich, 1993.

Lehne, G. K. "Homophobia Among Men." In *The 49-Percent Majority,* edited by D. David and R. Brannon, 66–88. Reading, MA: Addison-Wesley, 1976.

Lever, J. "Sex Differences in the Games Children Play." *Social Problems* 23 (1976): 478–487.

Lewis, R. A. "Emotional Intimacy Among Men." *Journal of Social Issues* 34 (1978): 108–121.

Libby, W. L. "Eye Contact and Direction of Looking as Stable Individual Differences." *Journal of Experimental Research in Personality* 4 (1970): 303–312.

Lips, H. M. *Sex and Gender: An Introduction.* 2nd ed. Mountain View, CA: Mayfield, 1993.

Lombardo, J. P., and L. O. Lavine. "Sex-Role Stereotyping and Patterns of Self-Disclosure." *Sex Roles* 7 (1981): 403–411.

Lowenthal, M., M. Thurnher, and D. Chiriboga. *Four Stages of Life: A Comparative Study of Women and Men Facing Transitions.* San Francisco: Jossey-Bass, 1975.

Maccoby, E. E. "Social Groupings in Childhood: Their Relationship to Prosocial and Antisocial Behavior in Boys and Girls." In *Development of Antisocial and Prosocial Behavior,* edited by D. Olweus, J. Block, and M. Radke-Yarrow, 263–284. San Diego: Academic-Press, 1986.

Maccoby, E. E. "Gender and Relationships." *American Psychologist* 45 (1990): 513–520.

Maccoby, E. E., and C. N. Jacklin. "Gender Segregation in Childhood." *Advances in Child Development and Behavior* 20 (1987): 239–287.

Matlin, M. *The Psychology of Women.* 3rd ed. Fort Worth, TX: Harcourt Brace, 1996.

Maltz, D. N., and R. A. Borker. "A Cultural Approach to Male-Female Miscommunication." In *Language and Social Identity,* edited by J. A. Gumperz, 195–216. New York: Cambridge University Press, 1983.

Mazur, E. "Predicting Gender Differences in Same-Sex Friendships from Affiliation Motive and Value." *Psychology of Women Quarterly* 13 (1989): 277–292.

Mazur, E., and R. Olver. "Intimacy and Structure: Sex Differences in Imagery of Same-Sex Relationships." *Sex Roles* 16 (1987): 539–558.

Morin, S., and E. Garfinkle. "Male Homophobia." *Journal of Social Issues* 34 (1978): 29–47.

Naifeh, S., and G. Smith. *Why Can't Men Open Up? Overcoming Men's Fear of Intimacy.* New York: Clarkson N. Potter, 1984.

Norrell, J. E. "Self-Disclosure: Implications for the Study of Parent-Adolescent Interaction." *Journal of Youth and Adolescence* 13 (1984): 163–179.

O'Brien, M., and A. C. Huston. "Development of Sex-Typed Play Behavior in Toddlers." *Developmental Psychology* 52 (1985): 114–121.

O'Meara, J. D. "Cross-Sex Friendship: Four Basic Challenges of an Ignored Relationship." *Sex Roles* 21 (1989): 525–543.

O'Neil, J. M. "Patterns of Gender Role Conflict and Strain: Sexism and Fear of Femininity in Men's Lives." *Personnel and Guidance Journal* 60 (1981): 203–210.

O'Neil, J. M., B. J. Helms, R. K. Gable, L. David, and L. S. Wrightsman. "Gender-Role Conflict Scale: College Men's Fear of Femininity." *Sex Roles* 14 (1986): 335–350.

Patterson, M. L., and R. E. Schaeffer. "Effects of Size and Sex Composition on Interaction Distance, Participation, and Satisfaction in Small Groups. *Small Group Behavior* 8 (1977): 432–433.

Pleck, J. H. "The Male Sex Role: Definitions, Problems, and Sources of Change." *Journal of Social Issues* 32 (1976): 155–162.

Powers, E., and G. Bultena. "Sex Differences in Intimate Friendships of Old Age." *Journal of Marriage and the Family* 38 (1976): 739–747.

Rashid, H. M. "Divergent Paths in the Development of African-American Males: A Qualitative Perspective." *Urban Research Review* 12 (1989): 12–13.

Rawlins, W. K. *Friendship Matters: Communications, Dialectics, and the Life Course.* New York: Aldine DeGruyter, 1992.

Reis, H. T., M. Senchak, and B. Solomon. "Sex Differences in the Intimacy of Social Interaction: Further Examination of Potential Explanations." *Journal of Personality and Social Psychology* 48 (1985): 1204–1217.

Reis, H. T., Y. Lin, M. E. Bennett, and J. B. Nezlek. "Change and Consistency in Social Participation During Early Adulthood." *Developmental Psychology* 29 (1993): 633–645.

Reisman, J. M. "Intimacy in Same-Sex Friendships." *Sex Roles* 23 (1990): 65–82.

Roberto, K. A., and P. J. Kimboko. "Friendships in Later Life: Definitions and Maintenance Patterns." *International Journal of Aging and Human Development* 28 (1989): 9–19.

Roberto, K. A., and J. P. Scott. "Friendships of Older Men and Women: Exchange Patterns and Satisfaction." *Psychology and Aging* 1 (1986): 103–109.

Rose, S. M. "Same-Sex and Cross-Sex Friendships and the Psychology of Homosexuality." *Sex Roles* 12 (1985): 63–74.

Rotenberg, K. J., and N. Chase. "Development of the Reciprocity of Self-Disclosure." *Journal of Genetic Psychology* 153 (1992): 75–86.

Rubin, L. *Just Friends: The Role of Friendship in Our Lives.* New York: Harper and Row, 1985.

Saal, F. E., C. B. Johnson, and N. Weber. "Friendly or Sexy? It May Depend on Whom You Ask." *Psychology of Women Quarterly* 13 (1989): 263–276.

Sanday, P. R. *Fraternity Gang Rape: Sex, Brotherhood, and Privilege on Campus.* New York: New York University Press, 1990.

Sapadin, L. A. "Friendship and Gender: Perspectives of Professional Men and Women." *Journal of Social and Personal Relations* 5 (1988): 387–403.

Sattel, J. "The Inexpressive Male: Tragedy or Sexual Politics?" *Social Problems* 23 (1976): 469–477.

Saurer, M. K., and R. M. Eisler. "The Role of Masculine Gender Role Stress in Expressivity and Social Support Network Factors." *Sex Roles* 23 (1990): 261–271.

Savin-Williams, R. C. "Dominance Hierarchies in Groups of Middle to Late Adolescent Males." *Journal of Youth and Adolescence* 9 (1980): 75–85.

Schneider, F. W., and L. M. Coutts. "Person Orientation of Male and Female High School Students: To the Educational Disadvantage of Males?" *Sex Roles* 13 (1985): 47–63.

Sherrod, D. "The Influence of Gender on Same-Sex Friendships." In *Close Relationships,* edited by C. Hendrick, 164–186. Newbury Park, CA: Sage, 1989.

Shulman, N. "Life-Cycle Variations in Patterns of Close Relationships." *Journal of Marriage and the Family* 37 (1975): 813–821.

Snell, W., Jr., R. Miller, and S. Belk. "Development of the Emotional Self-Disclosure Scale." *Sex Roles* 18 (1988): 59–73.

Stapley, J. C., and J. M. Haviland. "Beyond Depression: Gender Differences in Normal Adolescents' Emotional Experiences." *Sex Roles* 20 (1989): 295–308.

Stewart, D., and A. Vaux. "Social Support Resources, Behaviors, and Perceptions Among Black and White College Students." *Journal of Multicultural Counseling and Development* 14 (1986): 65–72.

Stokes, J., and I. Levin. "Gender Differences in Predicting Loneliness from Social Network Characteristics." *Journal of Personality and Social Psychology* 51 (1986): 1069–1074.

Strommen, E. A. "Friendship." In *Women: A Psychological Perspective,* edited by E. Doelson and J. Gullahorn. New York: Wiley, 1977.

Swain, S. O. "Men's Friendships with Women: Intimacy, Sexual Boundaries, and the Informant Role." In *Men's Friendships,* edited by P. M. Nardi, 153–171. Newbury Park, CA: Sage, 1992.

Swain, S. O. "Covert Intimacy: Closeness in Men's Friendships." In *Gender in Intimate Relationships: A Microstruc-*

tural Approach, edited by B. Risman and P. Schwartz, 71–86. Belmont, CA: Wadsworth, 1989.

Thorne, B. "Girls and Boys Together . . . but Mostly Apart: Gender Arrangements in Elementary Schools." In *Relationships and Development,* edited by W. Hartup and K. Rubin, 167–184. Hillsdale, NJ: Erlbaum, 1986.

Tietjen, A. "The Social Networks of Preadolescent Children in Sweden." *International Journal of Behavioral Development* 5 (1982): 111–130.

Tiger, L. *Men in Groups.* New York: Random House, 1969.

Tognoli, J. "Male Friendship and Intimacy Across the Life Span." *Family Relations* 29 (1980): 273–279.

Tschann, J. "Self-Disclosure in Adult Friendships: Gender and Marital Status Differences." *Journal of Social and Personal Relationships* 5 (1988): 65–81.

Veroff, J., R. Kulka, and E. Douvan. *Mental Health in America: Patterns of Help-Seeking from 1957–1976.* New York: Basic Books, 1981.

Waldrop, M., and C. Halverson. "Intensive and Extensive Peer Behavior: Longitudinal and Cross-Sectional Analysis." *Child Development* 46 (1975): 19–26.

Walker, L. S., and P. H. Wright. "Self-Disclosure in Friendship." *Perceptual and Motor Skills* 42 (1976): 735–742.

Weiss, L., and M. Lowenthal. "Life Course Perspectives on Friendship." In *Four Stages of Life,* edited by M. Lowenthal, M. Thurnher, and D. Chiriboga. San Francisco: Jossey-Bass, 1975.

Wellman, B. "Men in Networks: Private Communities, Domestic Friendships." In *Men's Friendships,* edited by P. M. Nardi, 74–114. Newbury Park, CA: Sage, 1992.

Wheeler, L., H. Reis, and J. Nezlek. "Loneliness, Social Interaction, and Sex Roles." *Journal of Personality and Social Psychology* 45 (1983): 943–953.

Whiting, B. B., and C. P. Edwards. *Children of Different Worlds: The Formation of Social Behavior.* Cambridge, MA: Harvard University Press, 1988.

Winstead, B. A. "Sex Differences in Same-Sex Friendships." In *Friendship and Social Interaction,* edited by V. J. Derlega and B. A. Winstead. New York: Springer-Verlag, 1986.

Winstead, B., V. Derlega, and P. Wong. "Effects of Sex-Role Orientation on Behavioral Self-Disclosure." *Journal of Research in Personality* 18 (1984): 541–553.

Wiseman, J. P. "Friendship: Bonds and Binds in a Voluntary Relationship." *Journal of Social and Personal Relationships* 3 (1986): 191–211.

Wood, J. T. *Gendered Lives: Communication, Gender, and Culture.* Belmont, CA: Wadsworth, 1994.

Wood, J. T., and C. C. Inman. "In a Different Mode: Recognizing Masculine Styles of Communicating Closeness." *Journal of Applied Communication Research* 21 (1993): 279–295.

Wright, P. H. "Men's Friendships, Women's Friendships and the Alleged Inferiority of the Latter." *Sex Roles* 8 (1982): 1–20.

Wright, P. H. "Interpreting Research on Gender Differences in Friendship: A Case for Moderation and a Plea for Caution." *Journal of Social and Personal Relations* 5 (1988): 367–373.

Wright, P. H., and M. B. Scanlon. "Gender Role Orientations and Friendships: Some Attenuation, but Gender Differences Abound." *Sex Roles* 24 (1991): 551–556.

Youniss, J., and J. Smollar. *Adolescent Relationships with Mothers, Fathers, and Friends.* Chicago: University of Chicago Press, 1985.

12

Men, Women, and Friendship: What They Say, What They Do

KAREN WALKER

Using data from 52 in-depth interviews with working-class and professional men and women, I examine gender differences in friendships. Men and women respond to global questions about friendship in culturally specific ways. Men focus on shared activities, and women focus on shared feelings. Responses to questions about specific friends, however, reveal more variation in same-sex friendships than the literature indicates. Men share feelings more, whereas women share feelings less; furthermore, the extent to which they do so varies by class. I argue that conceptualizing gender as an ongoing social construction explains the data better than do psychoanalytic or socialization accounts.

Stereotypes about friendship represent women's friendships as intimate relationships in which sharing feelings and talk are the most prevalent activities. Men's friendships are represented as ones in which sharing activities such as sports dominate interaction. In this [reading], however, I argue that the notions that women share intimate feelings whereas men share activities in their friendships are more accurately viewed as cultural ideologies than as observable gender differences in behavior. Using data from in-depth interviews with working- and middle-class men and women, I show that men and women use these ideologies to depict their friendships and to orient their behavior. Responses to global questions about friendship indicate that men and women think about their friendships in culturally specific ways that agree with stereotypes about men's and women's friendships.

Responses to questions about specific friends, however, reveal more variation in same-sex friendships than the stereotypes or the social scientific literature lead one to expect. When men and women discuss friendship they emphasize the behavior that corresponds to their cultural notions of what men and women are like. Men focus on shared activities, and women focus on shared feelings. When specific friendships are examined, however, it becomes clear that men share feelings more than the literature indicates, whereas women share feelings less than the literature indicates; furthermore, the extent to which they do so varies by class. Employed middle-class women indicate that they are sometimes averse to sharing feelings with friends. Working-class men, on the other hand, report regularly sharing feelings and discussing personal problems.

This approach differs significantly from much of the social scientific literature on friendship by closely examining the link between behavior and ideology and seeing gender as an ongoing construction of social life. Many friendship studies over the past decade have emphasized the extent to which men and women have different kinds of friendships; the conclusions concur with the most prevalent stereotypes. Lillian Rubin (1985) has argued that men bond through shared activities, whereas women share intimate feelings through talk. She ascribes these differences to two

From *Gender & Society,* Vol. 8, No. 2, June 1994, pp. 246–265.

phenomena. First, socialization of children encourages attention to relationships for girls and competition among boys and men. Second, the psychic development of girls leads girls and women to develop permeable ego boundaries and relational, nurturing capacities that encourage them to seek intimacy within friendship with other females. Boys and men, on the other hand, are threatened by having close, intimate friendships. Intimacy threatens their sense of masculinity because it touches that feminine part of their psyche that they were forced to repress in early childhood. According to Rubin, shared activities and competition are compensatory structures for men that prevent them from becoming too intimate.

Other authors differ somewhat on the causes and evaluation of these differences. Scott Swain (1989), for example, argues that perceiving men as deficient in intimate capacities as measured by verbal communication ignores nonverbal intimacy. In a study of intimate friendships among antebellum men, Karen Hansen (1992) has argued that arguments resting on psychic development are essentialist and neglect historical changes in intimate behavior. Other authors attribute observed gender differences in friendships solely to socialization rather than psychic development (Allan 1989; Swain 1989). There is, however, little debate that contemporary men characteristically engage in shared activities, whereas women engage in verbal sharing of feelings with their best friends (Bell 1981; Caldwell and Peplau 1982; Eichenbaum and Orbach 1989; Oliker 1989; Sherrod 1987; Swain 1989).

In contrast to the most commonly developed explanations in friendship studies, some sociologists have argued in recent years that gender is constructed on an ongoing basis in social life (Connell 1987; Leidner 1991; West and Zimmerman 1987). These approaches account not only for the variation within same-sex friendships that I observed but also for the strength of the ideologies about friendship. The process of representing themselves as adhering to cultural norms when they discuss friendship is one way men and women create coherent understandings about themselves as gendered humans; furthermore, because social processes of gender construction emphasize the naturalness of gender differences, individuals rarely question the extent to which those differences exist. Leidner observes that even "the considerable flexibility of notions of proper gender enactment does not undermine the appearance of inevitability and naturalness that continues to support the division of labor by gender" (1991, 158). To this I would add that even when individuals understand gender differences as socially rather than biologically caused, as several of my respondents did, they see those social causes as having shaped their personalities in such a way that change is difficult, if not impossible, to achieve.

In addition to exploring how women and men construct gender-specific understandings of their friendships, I also explore how social class influences men's and women's capacities for conforming to gendered behaviors. Women's or men's material circumstances affect their abilities to conform to gendered norms about friendship. The friendships of women who did not work in the paid labor force or women whose family took priority over their labor market participation generally conformed to a model of intimate friendship. Middle-class men and women who were geographically or occupationally mobile tended to report a lack of intimacy in specific friendships. Working-class women and men who participated in dense social networks and whose resources limited the extent of their social activities often spent time talking to their friends and sharing feelings about events in their lives, thereby creating intimate friendships; thus social class shapes the experiences men and women have with friendship, and it may do so in ways that contradict stereotypes about men's and women's behaviors.

METHOD OF STUDY

My data were gathered from a series of 52 in-depth interviews that I conducted in 1991 and 1992 with working- and middle-class men and women[1] between the ages of 24 and 48. Because I wanted to observe groups of three or four individuals within each group of men and women from a particular class, I asked some of the men and women I interviewed for referrals to their friends, preferably friends who knew each other. All but three respondents, therefore, had at least one other friend in the study. Several had two friends, and two working-class women had three friends in the study. I wished to see what issues were most salient to groups of friends. I also wished to see to what extent friends agree on what their interactions are like.[2]

Many respondents were married; most worked outside the home. Of the 33 women interviewed, 60 percent worked in the labor force either part- or full-time and 75 percent were married or living with men. Of the 19 men, 84 percent were married and 84 percent worked full-time in the labor force. In addition, two working-class men were formally unemployed but working in the informal labor force. Respondents were mostly white but ethnically diverse. Fifteen percent identified themselves as Italian American, 17 percent as Irish Americans, 45 percent as of Northern European descent, and 17 percent as Jewish (all but one of the Jewish respondents were middle class). One respondent was African American, one was Puerto Rican, and one was Arab American.

Most studies have missed the variation this study found between reports about friendship in general and what respondents say actually occurs in specific friendships. With the exception of network studies (which neglect issues of meaning within friendship), studies exploring gender differences in friendship have tended to ask either global questions about friends (Bell 1981; Caldwell and Peplau 1982; Rubin 1985; Swain 1989) or to focus on best friendships (Oliker 1989). Rebecca Adams (1989), in contrast, reports that the most accurate data about friendship are obtained by asking detailed questions about each member of a network and then calculating summary scores. Asking about a person's friendship in general is not as effective because "global questions require respondents to describe friendships as if they were all similar" (1989, 26).

Responses to global questions, however, are not inaccurate attempts by individuals to answer poorly designed questions. There are striking patterns to these responses that vary by gender. Individuals respond to such general questions as "How do you define friendship?" and "Do you think men's friendships differ from women's friendships?" in culturally specific ways. One gains knowledge of the meaning of friendship to individuals by asking global questions. This knowledge does not emerge by asking detailed questions about specific friendships precisely because there is a gap between individuals' notions of their social world and their actions in it. Both global and detailed questions are thus necessary to understand the interplay between ideology and actions.

ACCOUNTS OF FRIENDSHIP

Definitions of Friendship

Many men's and women's general accounts of friendship agree with the notion that men bond through shared activities, whereas women share intimate feelings. One man, a working-class father in his late twenties, said that men share a lot of things together, such as

> fishing, baseball, you know, maybe playing softball in the field, huntin', you know, just getting together and having a rap session, you know, sit back and relax and just talk and you know, work on cars, talk about, you know, there's different things, more you can share with a male because most of them, you know, know what's going on.

Other men also focused on shared activities as a basis for friendship. Many men reported that they watched sports with friends and did athletics such as running or playing basketball. Others focused less on physical activities and more on discussing politics or shared political activism as a basis for friendship.

When asked how they defined friendship, most men commented on its affective dimensions and the importance of trust, but they frequently did so in terms of sharing experiences:

> People I consider friends are people who I'm likely to have warm feelings toward, feel like I have some history with, some events together. (middle-class man)

> How do I define friendship? Someone I can trust I would call a friend. Someone I like to spend some time with, you know, I enjoy certain, I enjoy doing certain things with. (working-class man)

These men shared events, they did things together, and they played games together. They also trusted each other and had warm feelings toward each other, but they did not mention the importance of sharing feelings together.

In contrast, most women responded that friendship was characterized by a shared history, giving and receiving support—primarily emotional support—and being able to talk to each other about things. When asked, "How do you define friendship?" two women responded saying,

> Sharing, caring, being there for each other. (working-class woman)

> It's a special interest in another person, and it shows caring and openness. (middle-class woman)

These responses emphasize support, talk, and sharing feelings. Whereas men also stressed supportiveness in friends, they often indicated that friends "do anything" for them, for example, give financial help or physical support in confrontations with others. Only one man reported that friendship meant being able to "tell anything and everything" to your friends, whereas several women did.

Perceived Differences in Men's and Women's Friendships

In response to questions about whether men's and women's friendships were similar or different, respondents' answers reflected cultural gender ideology. In a conversation immediately preceding our interview, one professional woman, Anna,[3] asked me what my dissertation was about. Told that it was a study of men's and women's friendships, she said that was very interesting: Her friendships were much different than her husband's friendships. Her husband had a friend named Jim whom Tom (Anna's husband) referred to as his "buddy." But even though Jim and Tom had a lot of fun together, Tom could not establish a more intimate friendship with Jim. Jim did not express his feelings. Anna thought this was because Jim was afraid his feelings would overwhelm him.

Although Anna was unique in bringing up the subject of men's and women's differences herself, about 40 percent of the respondents reported that men were not as open as women.[4] When asked "Do you think that men's friendships are different than women's friendships?", one man responded this way:

> Well, we know that they are often. But whether that's a result of social construction or genetics I have no idea. [How are they different?] Well, there's a lot more sense of competition between a lot of men. A lot less openness about personal matters. It's just not as personal as being friends, uh, as your women friends. (middle-class man)

Other responses sounded remarkably similar in focusing on openness, the degree of closeness, and intimacy:

> Men keep more to themselves. They don't open up the way women do. Some women will spill their guts at the drop of the hat. (working-class man)

> I don't think men are as close as women are to each other. I think they're a little more distant with each other. I don't think men tell each other everything. (working-class woman)

Although a focus on intimacy and closeness was the most common response to questions about gender differences in friendships, another 25 percent of the respondents suggested that men and women do different things with their friends. Men play sports and share activities; women talk. Women discuss children and families; men discuss cars, work, and politics. Thus, about 65 percent of the respondents expressed and engaged in the construction of dominant cultural ideology about friendship that women are more intimate and talk in their friendships, whereas men engage in activities and do not share feelings.

Responses of the remaining one-third who gave other answers about men's and women's friendships were fairly evenly divided between those who said that women's friendships showed greater conflict, those who did not know if there were differences, and those who said men's and women's friendships were essentially similar in that loyalty and trust were requisites of all friendships.

Not only did men and women give gendered accounts of friendship, they also assumed that men and women were different. In doing so, they did not necessarily know how men and women differed, but they believed that they did. Individuals sometimes understood the response of someone from the opposite gender as characteristic of that gender if it differed from their own responses. Men and women frequently told me how their spouses were different from them. A lawyer told me how this process worked for him:

> So if Alberta and Deborah, Alberta my friend at work and Deborah my wife, take the same perspective I will say, "Oh, here's another example of women having the same perspective and men having a different perspective."

He assumed that women are different from him and then categorized a shared response by two women as characteristically feminine. Not only did men and women have different cultural notions about how men and women were different, but they created these notions on an ongoing basis using generalizations. Their accounts of their friendships were deeply influenced by these ideas.

EXPERIENCES IN SPECIFIC FRIENDSHIPS

Many respondents gave general accounts of their friendships that agreed with their understandings of what appropriate gendered behavior is. When I asked respondents detailed questions about activities they did with friends and things they talk about, however, I discovered that there was frequently a disparity between general representations that indicated cultural beliefs and specific information.

For instance, the man quoted above who thought that men friends shared activities such as hunting and fishing never went hunting or fishing with friends. His economic resources were severely limited, he worked two jobs, and had little time to socialize except in the evenings. His wife worked part-time in the evenings and on weekends, and while she worked he took care of their two children. Sometimes friends came over, and they watched TV and drank together. Most frequently, however, he socialized with friends at work in a high school where he was a janitor. He and his work friends discussed retirement, their wives, their children, as well as shared interests in sports.

When asked what he had talked about with his closest friend at work that day, he said they had discussed their wives' preferences for marital courtship. One wife liked to have the scene set before sex. She and her husband went out to eat together, they went shopping together, and there was a period in which they both knew what would happen later in the evening. My respondent told his friend that he and his wife liked spontaneity. Preparing for sex through a series of rituals seemed superficial to them. This is the kind of detailed talk about a significant relationship in which women are reputed to engage. It is a far cry from the stereotype of men friends who discuss sports and who do not have intimate discussions. Nor does it fit with my respondent's own perception of what friendships are like.

There were many similar examples. Seventy-five percent of the men reported engaging in nongendered behavior with friends—all of whom reported that they spoke intimately about spouses, other family members, and their feelings. Furthermore, one-third of those men reported that they engaged in other nongendered behavior as well as intimate talk. A lawyer who said that he did not discuss marital issues or personal matters with his friends later reported that he and his friends had talked with some regularity about fertility problems they were having with their wives. They discussed the fact that their wives were much more concerned about their difficulties in conceiving than they were.

Some men reported the impact of divorce on friendship interaction. One professional man told me of at least two men whom he had supported through their divorces, listening to them talk about their feelings. He told me of friends who had helped him through his divorce, men who allowed him to "ventilate." This man repeatedly told me that he was very emotive and expressive and that he complained a lot to friends. That these conversations are disregarded in how men shape their understandings about being men is indicated by what happened later in the interview. He told me that he thought women were more open than men were. He said he had gone to hear Robert Bly talk, and Robert Bly had said that women had an ability to get to the heart of the matter, to really articulate things, whereas men were unable to do so. My respondent said he really thought Bly was right, and he regretted his own inability to be more open. He also said that he never discussed marital problems or issues with his friends. He reported that he and his friends usually discussed sports, even though my respondent did not much care for sports.

Three young lawyers in an open network talked about doing sports with friends—wrestling in college, running as adults, playing basketball. They also talked about going to football games together and discussing sports in telephone calls. These young men seemed to have stereotypically masculine friendships, but when asked for detailed information about how often they saw their friends and under what circumstances, they reported that the most frequent interaction was during the holidays—at parties, going out to dinner, and getting together with their wives. They did get together for one or two sports events a year. In addition, two of the men had run in local foot races together on two occasions. Although these events were symbolically important in terms of these men's images of their friendships, they did not accurately represent the men's common pattern of friendship interaction. Their friendships were largely carried on through bimonthly or sometimes weekly telephone conversations from their offices, during which they said they discussed sports, work, and family. One of these men was involved in a transatlantic relationship. What to do about his relationship had been a serious issue for the month or two prior to our interview. So, when asked what he and specific friends had talked about in their last conversations, he almost always said he talked about what he should do: should he go to England, should he break up with the woman, was she serious in wanting a commitment?

In addition to talking about things that men are reputed not to talk about—feelings and relationships—sometimes men did things that did not fit with their ideas of what men typically did. For instance, both men and women thought of shopping as something women, but not men, do together. Most men, in fact, said they did not shop with friends. The response of Joe, a working-class man, illuminates the meaning of shopping as a gendered activity. He reported that sometimes he went food shopping with a married friend: "Anita gives him the list and we go to the supermarket like two old ladies and we pick out things, 'Well, this one's cheaper than that one so let's get this.'" Joe seemed a little embarrassed by this activity. He laughed softly as he reported these shopping trips. His married friend, whom I also interviewed, denied that he shopped with friends.

Gene, a middle-class man, said that he was concerned about moisturizer, but he did not discuss it with friends. I asked why not, and he said it was "faggotty" to do so, blushed, and laughed with embarrassment. Then he talked about going to visit a gay friend in another city. The morning after he got there he and his friend went out early to buy moisturizer and vegetable pâté at Neiman-Marcus—activities he thought of as stereotypically gay and not things most heterosexual men do (his friend was in the entertainment industry in Hollywood, and his behavior could just as easily have been ascribed to the demands of a lifestyle valuing appearance and style). Gene's use of the term "faggotty" indicated that there are certain things that he did not do as he constructed his masculinity in everyday interaction. This accords with Carrigan, Connell, and Lee's (1987) discussion about the importance of hegemonic heterosexual masculinity in constructing power relations among men. Men who construct alternative forms of masculinity, such as being gay, are perceived as not masculine. Like the working-class man who did not feel like a man when he shopped for food with his friend, Gene did not feel like a man when he discussed moisturizer, and so he did not do it with straight friends.

Given that gay men are perceived as not masculine, it is interesting to note that Gene's friendship with Al, his gay friend, did not appear to threaten Gene's masculinity. Gene's account of how Al came out to him emphasized friendship over sexual orientation. Longtime college friends, Al told Gene he was gay several years after college, and he added that he knew his coming out might end their friendship. Gene reported being "very insulted" that Al thought he might stop being his friend. Whether or not being friends with Al threatened his masculinity was unimportant compared to Gene's public reaffirmation of the norm of loyalty to friends; furthermore, Gene emphasized joking behavior as important to friendship and connected joking to masculinity. Al, it turned out, had an acerbic and imaginative sense of humor that Gene appreciated.

Even in a situation where behavior contradicts gendered norms, some aspect of the interaction often constructs gender in ways consonant with ideology. Whereas the action of one person may contradict ideology, the response of the other may construct gender in ways conforming to dominant ideology. This pattern of emphasizing friendship activities that fit with their ideologies of masculinity is analogous to what Leidner (1991) observed in her study of men and women interactive service workers who emphasized those aspects of their jobs that they thought of as consonant with their gender identities. For instance, Gene and a friend were sitting together one night:

> And he says, "So, how are things going?" And I, I had been through a real bad period. And I said, "Fuck it, I'm going to tell him." So I told him how everything was wrong in my life, and toward the end of it he said, "Oh, that's the last time I ask you how you're feeling." . . . I mean, he was actually offensively unsupportive.

In this instance, Gene violated gender ideology dictating that he not discuss his problems with his friend. His friend responded by letting Gene know that he did not want to hear Gene's problems. By belittling Gene's disclosures, his friend also constructed masculinity in ways conforming to dominant norms. Gene, who tended to reject the legitimacy of the dominant norms and who self-consciously violated those norms, interpreted his friend's behavior as typically male. Gene used this event to describe masculine behavior to me, even though it also included Gene's own transgressions of gender ideology.

The construction of gender, therefore, is a highly complex process that occurs through behavior, through ideology, and through accounts of both. One person may act in ways that contradict gender ideology, and the response may be disapproval, which reinforces the ideology. In addition, there are many events and forms of interaction that occur simultaneously in one situation, and flexibility in the construction of one aspect of gender may not threaten the overall construction of gender if other interactions, such as joking, reinforce gender identity.

Nongendered Behavior Among Women

Similar processes of understanding their own behavior in terms of cultural ideologies and norms occurred among women. Women reported that they could tell their friends anything. They readily volunteered that they talked about their husbands and lovers. Few women volunteered that shared activities were essential to friendship; the middle-class women were more likely to do so than the working-class women—they had more resources with which to engage in shared activities. Few women volunteered that they go to spectator sports with friends. Few women volunteered that they engaged in athletic activities with friends.

Just as there were disparities between men's statements about what their friendships were generally like and what their specific experiences were, there were also disparities between what women said their friendships were like generally and what specific friendships were like. About 65 percent of all women reported engaging in behavior that did not conform to gender ideology. Some women, like some men, occasionally went to spectator sporting events with friends, but, with one exception, I only heard about these activities when I asked directly if they went to sporting events or did athletics with friends. Unlike men, women did not volunteer that they did so. About 25 percent of all women worked out together, went to aerobics classes together, or belonged to local sports teams, but they primarily defined these occasions as other times to see each other and talk; the activities themselves were defined as relatively unimportant.

Another 15 percent of the women respondents said that many friendship interactions occurred while they were doing things with friends. Several middle-class women belonged to musical groups where they met friends. A few working-class women went out to clubs together, leaving their spouses at home. Sharing activities, therefore, sometimes provided the basis for women's friendships just as it provided the basis for some men's friendships.

Finally, 25 percent of the women respondents reported that they considered certain information private and would not discuss it with friends. Ilana, the professional woman quoted above who said that friendship entailed showing openness and caring, reported that there were many things about which she would not talk to friends. For instance, she did not discuss her relationships with men with other women.

Ilana indicated that her reticence was unusual, but it did not force her to reconsider her femininity. Instead, she both acknowledged and denied the masculine stereotype of sharing activities without talk. On the one hand, she said, "There are a lot of male friendships based on doing certain male things together without saying much." On the other hand, she said, "There are also men that surprise me in that it seems to me that they like to gossip and talk about women even more than women may want to talk about men." She went on to report:

> I was going out with someone and we were at a video store and met a friend of his. And it was the first time I ever met this friend, it was only the third time I had ever seen the person I was going out with and immediately they were talking about what happened . . . the other guy was basically stood up by a woman and they were talking about it and, you know, just the whole thing. And then I went to dinner with the person I was going out with and he told me the whole story of this friend, of the whole relationship. . . . And it just didn't seem appropriate somehow.

Ilana, who hesitated to speak with people about personal matters, reconstructed masculinity as gossipy, and she reacted in distaste to her companion's openness; furthermore, she contrasted men's to women's gossip, saying that some men like to gossip about women more than women gossip about men. When Ilana herself did not conform to the gender stereotype of feminine openness she still constructed her behavior in opposition to masculine behavior.

Two academic women reported doing very little intimate sharing with friends. One woman reported that she had no friends whatsoever, although she was on friendly relations with many colleagues. She did not discuss her relationship with her husband or her family, or many intimacies with anyone other than her husband. She also did not talk to people about worries or problems because she said she did not like to burden people. The woman who referred me to her told me that she, too, rarely discussed intimate matters with friends. Like the first woman, she said she did not do so because she did not want to impose her problems on others; she did not think they would be interested.

One group of four working-class women defined openness as an essential component to friendship, and three of the women regarded the fourth woman, Susan, as not open enough, a failing they urged her to overcome. Susan concurred with her friends' opinions that she was not open enough, but she chose silence and reserve as a form of rebellion because she was angry with the woman in the group to whom she was closest. Although Susan's behavior did not conform to the ideal of feminine openness and intimacy, the talk of her friends constructed femininity even as they censured Susan for her secrecy.

Several professional women regretted that they did not have close intimate friendships, or they accepted what they took to be the fact that they did not make good friends. They felt like failures for not having friends with whom they could "say anything." On the one hand, they gave me reasons why they did not have those friendships. On the other hand, they felt as if they were unusual and "bad" friends. This belief in their own failure did not emerge in interviews with men. Whereas some men regretted not being able to open up, no men said they were bad for not doing so. For men, it was characteristically masculine to lack the ability to share intimate feelings with friends.

CLASS AND FRIENDSHIP

Sociologists have noted that social class as well as gender influences friendship patterns (Allan 1989; Willmott 1987). In this study, one way social class influenced behavior was that behavior conflicted with popular ideology, particularly for middle-class women. Middle-class women in traditionally male-dominated occupations such as university professors, lawyers, and doctors most frequently reported a lack of intimacy in friendships. Many had been very mobile. They moved from city to city while they pursued their educations. Some reported having intimate friendships when they were in college or graduate school, but indicated that recent friendships were less likely to be so. Whereas some women regularly corresponded with old friends with whom they were intimate, some did not and therefore reported a lack of intimate friendships. One doctor reported:

> Well, pretty much all my life, although I find this sort of dwindling down now . . . but I think for a long time I always had a couple really close friends, like people I'd see almost every day, talk to a lot, and um, would tell virtually anything and everything to.

The same doctor was asked, "How is it different now?" and responded,

> I don't know if I'm so busy or everybody else, everybody has sort of tunnel vision, you tend not to talk to people or feel you don't have time to talk to people about anything but the essentials. . . . You don't have as many peers, basically . . . you're not in a medical school class of 180 of whom a third are women. That's part of it. . . . Maybe I always did look to my peers who were in the training group or whatever for my friends, and the group has gotten smaller. The other thing is the group has gotten more competitive. I mean, to be blunt, I don't particularly like a lot of people that I'm training with. They are very competitive and they're not very friendly, really. There was only one other woman in my year and I think she viewed me as a rival, not as a potential friend.

In addition to being occupationally mobile, this woman, like many of the middle-class respondents, had been geographically mobile. She had infrequent contact with friends who lived in other cities, and she experienced attrition in groups of older friendships from college and medical school; furthermore, she did not replace friends as quickly as she once did. In addition, drawing on the expectation that friends could be drawn from peers with whom she was relatively equal, she found there were fewer equals in her pool of potential friends. Finally, competition at work dictated against the formation of intimate friendships.

Helen Gouldner and Mary Symons Strong (1987) noted similar experiences among executive women who lacked the time to form friendships off the job and hesitated to form friendships with people at work. Time was a major factor in limiting intimate friendships for the employed middle-class women I interviewed. For many, the development of intimacy required some minimal amount of time. Professional women often lacked the time to give to friends, particularly if they were married and had children. Most of their socializing with friends was done with work colleagues with whom they limited the amount of personal information they exchanged or with their husbands and other couples. Socializing with couples was not very intimate, but were occasions when men and women alike reported that they discussed politics, children, and work.

Professional men had similar friendships in these respects. They, too, had been mobile. Their groups of college friends had shrunk over the years. Friends had not been replaced as quickly as they had been lost. Time was limited for many men, particularly married men with children. In addition, much of their socializing was also done in couples, when intimate sharing was at a minimum. These constraints on professional men's and women's abilities to form long-lasting friendships with extensive interaction produced friendships resembling the masculine model of friendship more than the feminine model, especially when compared to unemployed women or working-class men and women.

When middle-class respondents had intimate friendships, they tended to have them with long-time friends and levels of intimacy varied over time. Periods of crisis, such as divorce, were most likely to be times when men and women had intimate discussions, but those times tended to be limited, and after the troubles subsided, they returned to their old patterns of interaction.

The middle-class women who had friendships conforming to gender stereotypes of intimacy tended to be those women who stayed at home with small children and had neighborhood friends. They had more opportunities to see friends alone, which increased their opportunities for intimate discussions. In addition, collegial norms against sharing intimate details did not exist for these women.

Working-class respondents' lives were structured differently from those of professional men and women, and their friendships were correspondingly different.[5] Only one working-class man was from another state (whereas three-fourths of the middle-class respondents grew up in other states). Working-class men and women tended to know their friends for much longer periods of time—they met in school or in their neighborhoods. They had fewer friends, but they saw them more often. In addition, they frequently saw their friends informally at home or in the neighborhood; as a result, they knew about each other's troubles in a way that professional men and women who socialized occasionally with friends did not. Their social lives were more highly gender segregated, which provided greater opportunities for intimate discussion. Men, as well as women, reported that they talked to their friends about intimate things on a regular basis. Thus nongendered behavior among working-class respondents tended to occur among men who reported frequent intimate discussions ("We're worse than a bunch of girls when it comes to that").

The working-class respondents also had more problems—financial, substance abuse, family, health—and having these problems meant they tended to be topics of conversations with their friends. When professional men and women faced troubles, they also talked to their friends, but their lives were more stable in many respects. As a group the middle-class respondents were affluent, and they reported having few financial difficulties. Middle-class respondents reported that heavy periods of drug and alcohol use tended to have been in their late adolescence and early twenties, times when they were not married. The working-class respondents married young, and drug and alcohol use tended to be a problem because it interfered with marriage and family. Financial problems, drug problems, and marital problems often seemed to occur together among the working-class respondents with each problem exacerbating the others, whereas middle-class respondents tended to deal with a single problem at a time and to emerge from periods of crisis more quickly. To the extent that discussing problems is a mark of intimacy for many individuals, working-class men appeared to have more intimate friendships than professional men and women.

CONCLUSION

I have conceptualized gender as an ongoing social creation rather than as a role individuals learn or a personality type they develop that causes differences in behavior. This approach to gender accounts for many of the differences between how men and women represent their friendships in general and the specific patterns their friendships have. In interviews, men and women gave me general accounts of their friendships based on specific cultural norms of masculinity and femininity. These accounts often came early in the interview when I asked respondents to give me their definition of friendship and to tell me why friendship was important to them, if it was. They also came up toward the end of the interview when I asked them if they thought men's friendships differed from women's friendships. And they often contradicted specific information gathered during the friendship history.

It is not, however, sufficient to understand these processes as the ongoing production of gender. The process of gender construction is universal, but the forms and content of the construction vary. One must, therefore, ask why these particular ideologies about gender differences in friendship dominate rather than others. Earlier ideologies of friendship represented women as incapable of loyalty and true friendship and men as noble friends. Why has a reversal occurred? Asking this question, Barry Wellman (1992) argues that the domestication of the community through suburbanization has decreased the importance of men's public and semipublic social ties, at the same time increasing the importance of private domestic tips, at which women have traditionally excelled; thus men's friendships are beginning to resemble women's friendships in that they are carried out in the home and as "relations of emotional support, companionship, and domestic services" (Wellman 1992, 101). Wellman is effectively addressing a perceived change over time in friendship behavior among men, not the question of why women's friendships are privileged in contemporary culture.

The answer to that lies in the emergence of the women's movement and the self-conscious attempts to valorize women—and the response to that valorization by men. Much of the academic literature about women's friendship patterns is written by feminist social scientists who respond to older cultural ideologies that men were capable of true friendship and women were not (Rubin 1985; Smith-Rosenberg 1975). Rubin's argument, which claims that men are psychically incapable of the kind of intimacy characterizing women's friendships, draws explicitly on Carol Gilligan's and Nancy Chodorow's feminist work. Although it is often difficult to determine the source of cultural ideology, the efflorescence of articles and books about women's friendships and relationships occurs after the beginning of second wave feminism, and it is only then that men begin to respond systematically to those works.

In everyday interaction, these cultural ideologies are reproduced through discussions with friends, through the kind of generalizing people come to regarding gender that I mentioned earlier, through socialization, and through exposure to ideas about gender differences by the media. If men and women accept cultural definitions of reality, and many of them do, they interpret their behavior in light of such depictions and feel either deficient or validated depending on whether they fit or not. Many also orient their behavior toward these prescriptions for behavior. The fact that so much of their behavior does not match the cultural prescriptions, however, indicates the limits of those prescriptions for determining behavior. It also calls into question the adequacy of explanations that conclude that women and men are essentially different. Men and women respond to the demands of their lives in a variety of ways; some of those responses may disaffirm those same men's and women's gender ideologies.

In addition to asking why ideologies about women's openness and men's activities dominate other ideologies, one may also ask why men and women accept the stereotypes when their own behavior differs significantly from them. Several reasons account for this. First, friends sometimes frown on men's and women's gender-inappropriate behaviors. Such disapproval reinforces gender norms, and is interpreted by the one who receives disapproval as an indication that the behavior of the one who disapproves conforms to gender ideologies; therefore, the person whose behavior does not conform comes to see his or her behavior as anomalous. Second, women and men sometimes do not see the disparity between behavior and ideology because they do not reflect on their behavior. Given the enormous number of interactions in which people participate in daily life, it is not surprising that they neglect to reflect on behavior that they do not think meaningful in the construction of their identities.

Finally, friendship interactions are very complex, and men and women do gender in a variety of ways in single interactions. Often their behaviors do not conform to gender stereotypes, but other elements of the interaction construct gender along normative lines. The usual outcome of doing gender, buttressed by life's experiences, is that the actors have no doubts about their gender identities as women or men. In the case of friendship and gender stereotypes, men's and women's flexibility in behavior does not threaten their identities as men or women because so many other aspects of interaction reinforce their identities, which they reify. They are not, therefore, normally faced with a problem of gender identity that they must reflect on and solve either through modifying their behaviors or through modifying their ideologies to better reflect behavior. As I noted earlier, the change in ideology from one valorizing men's friendships to one valorizing women's friendships was initiated by feminists who had identified problems and self-consciously acted to solve them. Their success may rest on the fact that, when asked to reflect on their own experiences, others also recognized the problems in the way men's and women's friendships were perceived and the way they were experienced; but it was first necessary to bring the disparities to people's attention. Even this change did not fundamentally alter the notion that women's friendships are emotionally expressive, whereas men's friendships are not; rather, it altered the *value* of men's and women's friendships.

One must be wary of arguments that posit the early development of gendered personalities that cause differences in friendship; however, provocative, psychoanalytic theories of masculine or feminine personalities cannot account for the variation one sees in men's or women's behaviors. Nor does it account for the disparity between people's general accounts of friendship and their specific experiences, a disparity that makes sense if one understands those general representations as attempts to provide coherence, and understanding, and to find norms by which to live as gendered humans.

NOTES

1. The terms *working class* and *middle class* are used in this article to refer to people who differ in the work they do in the labor force and in their status. Middle-class men and women held professional or managerial jobs. They all had bachelor's degrees, and many had graduate or professional degrees. Most lived in suburbs or in downtown apartments. In contrast, working-class respondents worked in traditional craft, clerical, or low-level service occupations. Most had high school diplomas or less, although one working-class man had a bachelor of arts degree and seven working-class women had one or two years of college. Most lived in twin or row houses in densely populated working-class urban neighborhoods. Individuals not working in the paid labor force at the time of the interview were defined as working or middle class both on the basis of their status and by the kind of work they did when they worked in the labor force; thus the unemployed working-class women had done clerical work or low-level service work (cashiering), whereas the three formally unemployed working-class men reported that they were a tile setter, a plumber, and a cook, respectively. Unemployed middle-class women had worked in mid-level management or professional jobs.

2. Snowball sampling often introduces bias. In this study, I often found that friends had similar life experiences and educational levels, and the fact that friends shared activities with one another limited the effective size of the sample. My ability to generalize, therefore, is limited. Discrepancies between gender ideology and behavior occurred among most individuals, however, and the fact that they did suggests that earlier explanations of gender differences in friendship, as well as the extent of those differences, should be closely examined.
3. All the names of the respondents have been changed.
4. Results about perceived differences in men's and women's friendships are based on responses of 46 individuals. At the beginning of the study, four women and two men were not asked to compare men's and women's friendships.
5. Clyde Franklin makes similar observations about black men: "While working class black men's same-sex friendships are warm, intimate, and holistic, upwardly mobile black men's same-sex friendships are cool, non-intimate, and segmented" (1992, 212).

REFERENCES

Adams, R. G. 1989. Conceptual and methodological issues in studying friendship of older adults. In *Older adult friendship: Structure and process,* edited by R. G. Adams and R. Blieszner. Newbury Park, CA: Sage.

Allan, G. 1989. *Friendship: Developing a sociological perspective.* Boulder, CO: Westview.

Bell, R. 1981. *Worlds of friendship.* Beverly Hills, CA: Sage.

Caldwell, M. A., and L. A. Peplau. 1982. Sex differences in same-sex friendships. *Sex Roles* 8:721–32.

Carrigan, T., B. Connell, and J. Lee. 1987. Hard and heavy: Toward a new sociology of masculinity. In *Beyond patriarchy: Essays by men on pleasure, power, and change,* edited by M. Kaufman. Toronto: Oxford University Press.

Connell, R. W. 1987. *Gender and power.* Stanford, CA: Stanford University Press.

Eichenbaum, L., and S. Orbach. 1989. *Between women: Love, envy, and competition in women's friendships.* New York: Penguin.

Franklin, C. W., II. 1992. Friendship among black men. In *Men's friendships,* edited by P. M. Nardi. Newbury Park, CA: Sage.

Gouldner, H., and M. S. Strong. 1987. *Speaking of friendship: Middle class women and their friends.* New York: Greenwood.

Hansen, K. V. 1992. Our eyes behold each other: Masculinity and intimate friendship in antebellum New England. In *Men's friendships,* edited by P. M. Nardi. Newbury Park, CA: Sage.

Leidner, R. 1991. Serving hamburgers and selling insurance: Gender, work, and identity in interactive service jobs. *Gender & Society* 5:154–77.

Oliker, S. 1989. *Best friends and marriage.* Berkeley: University of California Press.

Rubin, L. 1985. *Just friends: The role of friendship in our lives.* New York: Harper & Row.

Smith-Rosenberg, C. 1975. The female world of love and ritual. *Signs* 1:1–29.

Sherrod, S. 1987. The bonds of men: Problems and possibilities in close male relationships. In *The making of masculinities: The new men's studies,* edited by H. Brod. Boston: Allen and Unwin.

Swain, S. 1989. Covert intimacy: Closeness in men's friendships. In *Gender in intimate relationships: A microstructural approach,* edited by B. Risman and P. Schwartz. Belmont, CA: Wadsworth.

Wellman, B. 1992. Men in networks: Private communities, domestic friendships. In *Men's friendships,* edited by P. M. Nardi. Newbury Park, CA: Sage.

West, C., and D. Zimmerman. 1987. Doing gender. *Gender & Society* 1:125–51.

Willmott, P. 1987. *Friendship networks and social support.* London: Policy Studies Institute.

13

Covert Intimacy: Closeness in Men's Friendships

SCOTT SWAIN

This study is an analysis of college men's intimate behavior in same-sex friendships and their standards for assessing intimacy. It documents the development, causes, and manifestation of a covert style of intimate behavior in men's friendships. Covert intimacy is a private, often nonverbal, context-specific form of communication. The concept of covert intimacy is rooted in the behaviors that men reported as indicative of closeness and intimacy in their friendships with other men.

First, I trace differences in the development of men's and women's adolescent friendships that shape and promote differing styles of intimacy. Next, such contexts are linked to the emergence of the separate worlds of men and women and how such separate worlds and microstructural contexts continue into adulthood. I analyze these separate worlds for the specific behaviors and values that shape intimacy among same-sex friends, and then clarify the distinctive cues and nuances of men's intimate behavior by comparing them to behaviors in male-female platonic friendships and friendships among women. I conclude the study with an assessment of the strengths and limitations of men's covert style of intimacy with men friends and its relationship to the inexpressive male.

THE INEXPRESSIVE MALE, OR SEX-SPECIFIC STYLES OF INTIMACY

Sex-role theorists have characterized men as instrumental (Parsons and Bales, 1955), agentive (Bakan, 1966), and task-oriented (Komarovsky, (1964). Women have been characterized as expressive (Parsons and Bales, 1955), communal (Bakan, 1966), and empathic (Hoffman, 1977; Bem, [Martyna, and Watson,] 1976). Consistent with these theoretical formulations, researchers on the male role have interpreted men's interpersonal behavior as nonintimate and have stressed the restraints and limitations that cultural conceptions of masculinity impose on intimate expression. Examples of this *deficit approach* to men's intimate capabilities are Jack Balswick's "The Inexpressive Male" (1976) and Mirra Komarovsky's concept of men's "trained incapacity to share" (1964).

In recent years many of these generalizations, which were based on slight yet significant sex differences, have been reexamined. The majority of studies that measure interpersonal skills and relationship characteristics report nonsignificant sex differences (Maccoby and Jacklin, 1974; Pleck, 1981; Tavris and Wade, 1984). When studies report significant sex differences, the results have been mixed and sometimes conflicting. In support of the male deficit model, men are reported to be less likely than women to disclose sadness and fears (Rubin, 1983; Allen and Haccoun, 1976; Davidson and Duberman, 1982), less affective and spontaneous with friends than women (Booth, 1972), and less adept than women at nonverbal decoding skills (Henley, 1977). Men are also reported to be more homophobic than women, which may inhibit the use of certain interpersonal skills in men's friendships (Lehne, 1976; Morin and Garfinkle, 1978).

However, the majority of self-disclosure studies reveal nonsignificant sex differences (Cozby, 1973); and related analyses report that men score higher than women on nonverbal decoding skills (Hall and Halberstadt, 1981), rate their friendships as more trusting and spontaneous than do women (Davidson and Duberman, 1982), and value intimacy in friendship as much as do women (Caldwell and Peplau, 1982). In view of such findings and the conflicting results of other related studies, sex differences in interpersonal behavior appear to be minor or not adequately measured. However, notions of the "inexpressive male" continue to persist and guide research on men's interpersonal behavior.

Perhaps the most consistently reported difference in men's and women's friendships is men's preference for joint activities and women's preference for talking (Caldwell and Peplau, 1982; Cancian, 1985). Men's emphasis on instrumental action has been interpreted by past researchers as a less personal and less intimate form of interaction than verbal self-disclosure (Komarovsky; 1964; Rubin, 1983; Davidson and Duberman, 1982). This interpretation may be influenced by researchers' reliance on measuring feminine-typed styles of behavior to assess topics involving love and interpersonal behavior. This bias has been critiqued by Cancian as the "feminization of love" (1986). Researchers concerned with intimacy have assumed that verbal self-disclosure is the definitive referent for intimacy, and have thus interpreted alternative styles that involve instrumental action as a less intimate, or nonintimate, behavior. [Previous definitions of intimacy relied primarily on verbal self-disclosure as an indicator of intimacy. But, the relationship between intimacy and self-disclosure is usually only implied and not specifically defined.]

Caldwell and Peplau (1982) suggest that men and women may place the same value on intimacy in friendships, yet have different ways of assessing intimacy. Men are reported to express a wider range of intimate behaviors, including self-disclosure, while participating in gender-validating activities (Swain, 1984). Men may develop sex-specific contexts, cues, and meanings, which connote feelings and appraisals of intimacy similar to those connoted by self-disclosure for women.

Intimacy is defined in the present study as *behavior in the context of a friendship that connotes a positive and mutual sense of meaning and importance to the participants.* This definition allows respondents to determine what behaviors are meaningful and intimate, and assumes that there may be several avenues that may result in the experience of intimacy.

The results presented here are based on in-depth interviews with fifteen men and five women. The college sample was young and white with a mean age of 22.5 years. This small sample was used to further explore sex-specific friendship behaviors, which have been significantly documented using larger samples of similar populations (Swain, 1984; Caldwell and Peplau, 1982).

The interview protocol was based on two empirical studies; the first was a pilot study ($N = 232$) that measured the relative value of activities in men's and women's same-sex friendships, and the second ($N = 140$) measured the relative importance and meaning that men and women attributed to those activities (Swain, 1984).

Interviews lasted an average of an hour and a half, with a female interviewer working with female subjects and a male interviewer (myself) working with the male subjects. For analysis, we then transcribed and organized the interviews by question and content. A disadvantage of such a focused sample is that the results may not generalize across age groups, or even represent this particular subgroup. However, we selected a private and personal interview setting to collect more detailed data about sensitive information concerning intimacy in friendships than would otherwise be possible when using larger samples and less personal data-collection techniques. Because of the college setting, we expected the sample to have more friends and contact with same-sex friends than men and women from the general population. The advantages of this sample of young adults are their temporal closeness to the development of adolescent friendships and their frequent interaction with friends because of the college environment. This should promote clarity in their recollections of the development of adolescent friendship behaviors and give them a sharpened and more sensitive vantage point from which to describe their current friendships with men and women.

THE DEVELOPMENT OF SEX-SPECIFIC STYLES OF INTIMACY: THE SEPARATE WORLDS OF BOYS AND GIRLS

Men and women grow up in overlapping, yet distinctively different worlds. Sex segregation begins at an early age when boys and girls are differentially rewarded for various play activities. Boys are encouraged to actively participate in the outside environment by parental acceptance of the risks of physical injury and parents' flexible attitude toward personal hygiene and appearance. Torn clothes, skinned knees, and dirty hands are signs of normal growth of healthy boys. If girls choose these same activities, they may be tolerated; however, they may be sanctioned differently. For example, the term "tomboy" is used to distinguish a girl with "boyish" behaviors, and to designate a stage of development that deviates from normative expectations of the female child. Several men attributed the distinctive friendship behaviors of men to this early segregation while growing up. Jim said:

> Well, you do different things. Little boys, they'll play in the dirt and things, whereas a guy and a girl they might play in the house or something. The guys like . . . they don't mind getting dirty. I don't want to stereotype or anything, but it's just the way I see it. The guys are more rugged and things.

Pete responded to the question, "How do you act just around the guys?"

> You'd talk about anything, do anything. You aren't as polite. You don't care as much how you look, how you dress, what you wear, things of that nature. Even if we're just platonic friends, for some reason when you're around girls you're different. In the United States men and women don't share bathrooms together in the public restrooms. That's a good example, right there—obviously men and women are segregated then. That segregation exists in friendships, too.

Separate restrooms are a concrete manifestation of the different realms experienced by boys and girls as they grow up and of the restrictions on crossing over into the other sex's domain. The curiosity that boys and girls experience about what the bathrooms are actually like for the other sex is evidence of this separateness. A boy who is teased and pushed into a girls' bathroom is called a "girl" as he hastily exits. Thus, children internalize sex-segregated boundaries and enforce these restrictions. Evidence of the long-term influence of this segregation is the humiliation and embarrassment an adult feels when accidentally entering the "wrong" bathroom. The association between gender and specific contexts is also suggested by men referring to a woman who is included in a men's poker game as being "just like one of the guys." The separate contexts of men and women continue throughout the life cycle to shape the ways they express intimacy.

The male world is the outside environment of physical activity. Boys share and learn activities with male friends that involve an engagement with this outside world. Social encouragement is evident in such organizations as the Boy Scouts (Hantover, 1978), sports and recreation programs (Coleman, 1976; Stein and Hoffman, 1978), and a division of labor that often has boys doing home chores that are outdoors, such as mowing lawns. These outside activities have a shaping influence on their interests and values. Jack recalled his adolescence:

> I can remember only one friend that I had from years past. My friend Jim. Just kind of all the fun we had, boyhood fun. We built a fort, lit firecrackers, and all that stuff.

He refers to "boyhood" rather than "childhood" fun, implying that these experiences tended to be shared with other males. Another man recalled:

> The activities shared were a lot of outdoor-type things—fishing, hunting, Tom Sawyer type things. It's a commonality that we both shared that helped bring us together.

Men mentioned activities that ranged from dissecting lizards, riding bikes, and childhood sports to four-wheeling, lifting weights, playing practical jokes on friends, problem solving, and talking about relationships as they reached adulthood. By the time high school graduation arrives, most males have had more experiences and time with men friends than with women friends. Several men commented on this early division in their friendships. For instance, Rick replied:

> Up to the sixth or seventh grade girls are "stay away from the girls!" So during that whole time you only associate with the guys, and you have all these guy friends. And after that you kind of, you know, the first time you go out, you're kind of shy with the girls, and you don't get to know them too well. . . . I really didn't get over being shy with girls until my senior year.

His first contact with girls is in the dating context of "going out," which implies a heterosexual coupling dimension to the relationship in addition to friendship. The segregated contexts of men and women continue into adulthood, and shape the opportunities for expressing intimacy and the expectations of how that intimacy is to be expressed. As a result, men are more familiar with their men friends, and women are more familiar with their women friends.

CONSEQUENCES OF THE SEPARATE WORLDS OF MEN AND WOMEN

Self-Disclosure: Profanity, Sameness, and Group Lingo

As boys move through adolescence surrounded and immersed in friendships with other boys, behavioral differences emerge that distinguish men's and women's friendships. Men develop language patterns that often rely on blunt, crude, and explicitly sexual terms. Bluntness, crudity, and profanity legitimize masculinity by tending to roughen the tone of any statement that a man may make. Swearing serves as a developmental credential in an adolescent boys' maturation process, much as do smoking, drinking, and getting one's driver's license. The "rugged" and "dirty" environment that boys share is translated into a coarser language during adolescence, which is also labeled as "dirty." Men felt that this language was more appropriate around other men, and they often related this sex-specific language to all-male contexts such as military service and sports. Greg responded:

> Well (laugh), not that I cuss a lot, but when I, you know, get around the baseball field and stuff like that. . . . They [women] don't like that. I try to stay away from the crude or harsh humor as much as I can (laugh).

Greg's laughing suggested a tense recognition that men's use of language in "harsh humor" does not easily translate in the company of women.

Men's harsh language and sexual explicitness in joking behavior are censored and muted when interacting with women friends. The censoring of humor to avoid offending women friends testifies to the different meaning and value men and women attribute to the same behaviors. Mike related:

> Around girls you act more of a gentleman. You don't cuss. You watch what you say. Because you don't want to say anything that will offend them.

Men felt more at ease with close men friends, partially from a perception of "sameness." Men assume that male friends will be more empathetic concerning sexual matters since they have similar bodies. Jack related:

> I find it much easier to talk about sexual things with guys, which makes sense.

A majority of the men said it was easier to talk to men about sexual matters than to women. Another man responded to the question, "What are some of the things that would be easier to talk over with a guy?"

> Anything from financial problems to problems with relationships. That's a big thing I really don't like talking to girls about. For some reason I just . . . I don't know . . . I get . . . usually because what I'm saying is from a male's point of view. And I know this is all sounding really sexist. But you know, there are certain things that I view that girls don't necessarily view the same way. And it's just easier talking to guys about that. Well, lately sex is one of those. I mean you can talk about certain sexual things; there are certain things. I had a conversation the other day, and he was talking to me about a sexual act that his girlfriend, his new girlfriend, wanted to do. And he really doesn't care to do it. There's no way I could talk to a girl about what he's talking about.

The men appeared to generalize a common world view to other men that fostered a feeling of comfort. Frank commented:

> I'm more relaxed around guys. You don't have to watch what you say. Around friends like that [men] I wouldn't . . . what could you say? I wouldn't be careful I shouldn't say something like this, or I shouldn't do this. That's because with the guys, they're just like you.

Men friends used the degree of comfort and relaxation experienced with men friends as an indicator of closeness. Matt described this feeling when asked about the meaningful times he has shared with men friends.

> Last week some really good friends of mine in my suite . . . one guy plays the guitar. And so he was just sitting around playing the guitar and we were making up tunes. We were making up songs to this, and that was really a lot of fun. The fun things come to mind. We rented a VCR and some movies and watched those, and just all the laughing together comes to mind as most memorable. As to the most meaningful, those also come pretty close to being the most meaningful, because there was just total relaxation, there. That I felt no need to worry. There's no need to worry about anyone making conversation. The conversation will come. And we can laugh at each other, and you can laugh at yourself, which is handy.

Men were asked to compare their friendships with men to their platonic friendships with women. Generally, men felt more at ease and relaxed when with close men friends than with women friends. John answered the question, "Are there any differences between your friendships with men and your friendships with women?"

> You don't have to worry about the situation you're in. If you have to go to the bathroom, you just run up and go. You don't have to worry about "please excuse me" or anything. And it's a lot more relaxed. A lot more. Like in Jack's house we just go into the kitchen and make ourselves something to eat, you know, part of the family.

The "situation" is comfortable because the men's shared assumptions, cues, and meanings of behavior allow them "not to have to worry" whether they are acting appropriately. The formality associated with women friends is suggestive of Irving Goffman's [1959] concept of "frontstage" behavior, which is more rule-bound and distanced, while the "backstage" behavior with men friends is more intimate because of the lack of censoring and the feeling of informality associated with being "part of the family." The shared history, activities, and perception that other men are "just like you" gave a predictable familiarity to men's interactions. Women also felt that their similarity to each other produced an empathy unique between women friends. A woman responded to a question about the differences between her men and women friends:

> There are some things about a woman's feelings that I don't think a man, having never been in a woman's mind, could ever really understand. Because I think most women are a little more sensitive than men.

Men also developed unique terms with their close friends that expressed their history and connectedness. These terms acknowledged the particular experiences shared between friends and underlined their special relationships. When asked about his most meaningful experiences with his men friends, Tim related:

> The best thing, well the thing is, Rick, Mike, and me kind of have our own lingo. I haven't seen other people use them, like "Bonzo" is one of them. Like "go for Bonzo," and anybody else would just go "Well, whatever."

The "lingo" was derived from activities experienced by the group. The private meaning of the language served as a boundary separating friends from people outside the group.

Doing Versus Talking: The Intimacy of Shared Action

The men were asked, "What was the most meaningful occasion spent with a same-sex friend, and why was it meaningful?" The men mentioned a total of 26 meaningful occasions with men friends, and several men mentioned more than one meaningful experience. We analyzed the responses to clarify the link between sharing an activity and feeling close to a friend. Of those occasions, 20 meaningful times were spent in an activity other than talking. Men related a wide variety of meaningful experiences from "flirting with disaster" in an out-of-control car and winning a court case to being with a close friend the night that the friend found out his sister had committed suicide. Activities such as fishing, playing guitars, diving, backpacking, drinking, and weightlifting were central to men's meaningful experiences.

Nine meaningful experiences directly referenced the sharing of skills and accomplishments. These meaningful times involved the shared enjoyment of learning and mastering skills and accomplishing goals ranging from a sexual experience with a woman to staying up all night on a weekend. The essential ingredients in these experiences seemed to be comfort with a competitive challenge and a sense of shared accomplishment. A man responded to the question of what was the most meaningful experience he had shared with a male friend, and later a group of male friends:

> I've always been extremely shy with women, and one of my friends in between, after high school . . . women were always chasing after him like crazy—and I'm defensive and stuff with women when one time we went to the river. And a couple of girls picked us up and we got laid and everything. And it was kind of, this is going to sound like the standard male thing, but we all kind of went, after, we went and had a few beers and compared notes. You know, and I felt totally accepted because I had just as many good things to say as they did, and I could relate. I knew what they were talking about, because most of my life I've never known what these guys were talking about sexually.

Although this quote might imply sexual exploitation, several aspects should be considered.

First, this man admits he is shy with women, and furthermore he indicates that the women initiated the interaction. He was able to discuss with his men friends a new experience that had been alien to him until this occasion. The argument here is not that the sexual experience was exploitive or intimate; it appears to have been a purely physical encounter between strangers. However, the commonality gained from a shared life experience did provide meaningful interaction among the men.

We further examined the influence of men's active emphasis and women's verbal emphasis on intimate friendship behavior by asking men to compare their friendships with men to their friendships with women. Tim responded to the questions, "What part of you do you share with your men friends and what part of you do you share with your women friends? How would you characterize those two different parts of you?"

> I think that the men characteristics would be the whole thing, would be just the whole thing about being a man. You know, you go out and play sports with your brothers, and have a good time with them. You just . . . you're doing that. And there are some things that you can experience, as far as emotional, [with] your best friends that are men . . . you experience both. And that's what makes it so good is that. With most of the girls you're not going to go out and drink beer and have fun with them. Well, you can, but it's different. I mean it's like a different kind of emotion. It's like with the guys you can have all of it. . . .

Tim says that you can have "the whole thing" with men, suggesting that he can do things and talk about things with his close men friends. With women friends doing things is "different." Tim refers to a "different kind of emotion" and speaks of a "good time" when he is with his men friends. This good feeling may result from the ease and comfort of interacting with close friends who have developed a familiar style of communication from sharing activities.

The value of doing things is apparent in Matt's response, which described his most meaningful times with men friends:

> It was like we were doing a lot of things together. It just seemed like we just grew on each other. Can't think of just one thing that stood out in my mind. It was more like a push-pull type thing. Like I'd pull him through things and he'd pull me through things. It wasn't like there was just one thing I can just think of right now, just a lot of things he did, whatever. Just the things we like to do, we just did them together, and just had a good time.

The closeness is in the "doing"—the sharing of interests and activities. When Matt was asked about his meaningful times with his women friends, he responded:

> It's like things that you'd talk—it's really just like "talk" with them. It's not so much like you'd go out and do something with them, or go out and maybe be with them.

Several men said that with women friends it's "just talking" and referred to interaction with women as "the lighter side of things." For men, it appears that actions speak louder than words and carry greater interpersonal value.

Women were also aware of a difference in men's style of expressing caring. A woman commented on the differences in how her men and women friends let her know that they like her:

> Women talk more about feelings than men do. A man might let me know that he likes me because when he was in New York, he saw a book I'd been looking for and he brought it to me. And so I know that he likes me because he did that. Where a woman might say, twelve days in a row, "I've been looking all over for that book, but I can't find it."

Her male friend expressed his caring through a direct action, while her female friend expressed her caring verbally.

The emphasis on activities in men's friendship shapes their communication of closeness and caring. The significance of the doing/talking emphases in men's and women's styles of intimacy is apparent in the following response of a man to the question, "Why do you think they'd [women friends] be more verbal than your guy friends [in showing that they like you]?"

> I don't know why. I think there's just more ways to . . . I think there's more ways for the males to show me their appreciation that's nonverbal. I don't know why. I just think that if, in the way that they respond to things we do together and stuff like that. There's more ways to show it. Like if we're, I make a good shot in a game or something, just give me a high-five or something like that. You don't have to say anything with the guys. That's just an example, it doesn't have to be just sports. But the same type of things, off the field or whatever, just a thumbs-up type thing from a guy or whatever. There's just more ways to show being around [each other]. Where with the girls, you know, what can they do? You know, run up and just give you a kiss or something, I know girls who would do that in high school. So it—I think their options are just less—so they opt for the verbal type of thing.

He views talking as one option or style of expressing caring. From his activity-oriented perspective, he actually views women as restricted by a lack of alternatives to verbal expressiveness. These expressive alternatives are available to men through cues developed by sharing and understanding common activities. Nonverbal cues, expressed in active settings, contribute to private, covert, and in general, sex-specific styles of intimacy. This suggests that each sex tends to overlook, devalue, and not fully comprehend the other sex's style of expressing care.

Men and women have different styles of intimacy that reflect the often-separate realms in which they express it. The activities and contexts that men share provide a common general experience from which emerge certain values, gestures, and ways of talking about things that show intimacy. Both men and women are restricted in crossing over into each other's realm by early sex segregation, which results in a lack of experience with the meanings and contexts of the other sex. Researchers often underestimate this segregation because of an emphasis on the loosening of sex-related boundaries in the past several decades. Despite such changes, sex segregation still influences men and women, especially during the development of friendships in adolescence.

Covert Intimacy in Sports and Competition

Sports are the primary format for rewarding the attainment and demonstration of physical and emotional skills among adolescent boys. A man stated:

> I would have rather taken my basketball out than I would a girl . . . you know how young men are in the seventh, eighth, and ninth grade . . . if I had the choice I'd play basketball with the guys instead of going out that night.

Researchers have documented detrimental interpersonal consequences that may result from sports participation (Stein and Hoffman, 1978; Coleman, 1976). However, the productive aspects of the sports context have received less attention from researchers. For men, the giving and receiving of help and assistance in a challenge context demonstrates trust and caring in a friendship. Engaging in the risk and drama of performing in a competitive activity provides the glue that secures the men in an intimate process of accomplishing shared goals. Jim responded to the question, "What situations or activities would you choose or would you feel most comfortable in with your close men friends?"

> I'm very comfortable, like playing racquetball. A lot of one-on-one things where you're actually doing something. Playing backgammon. Now being competitive makes it a little easier, because it's like a small battle going on. Not that you're out

to show who's best, but it gives you something more that you two have in common in the situation.

The competition provides a structured context where friends can use their skills to create "something more" than they previously had in common. Each friend brings his own experiences and talents to join the other friend in the common arena of competitive activity. The competition provides an overt and practical meaning; the covert goal, however, is not to "show who's best" but to give "something more that you two have in common."

The sports context provides a common experience whereby men can implicitly demonstrate closeness without directly verbalizing the relationship. Nonverbal communication skills, which are essential for achieving goals in the fast-moving sports context, also provide avenues for communicating intimacy. Greg responded to the question, "What were the most meaningful times that you spent with your men friends?"

> In athletics, the majority of these friends that are close to me were on teams of mine. We played together. We were on the same team, me playing first and him catcher, or at times he played third. You know, first and third looking across the infield at each other. Knowing that we were close friends, and winning the CIF championship. I could just see that it meant a lot to me in terms of friendship too. As soon as that last pitch was made, we just clinched the title, to see the first person that he looked for to give, you know, to hug or congratulate, or whatever, was me. And the same for me to him. That was a big, another emotional thing for the two of us. Because I could just . . . it's just . . . you could just see how close your friends really are, or something like that. When there's twenty-five guys on the team and they're all going crazy, you're just trying to rejoice together, or whatever, for the victory. And the first, the main thing you wanted to do was run across the diamond and get to each other, and just congratulate each other first. And that meant a lot to me emotionally as well as far as friendship is concerned. It was only a split second, because after it was just a mob.

The two friends had grown up playing baseball together. Sharing the accomplishment of winning the championship provided a context where a close friendship could be affirmed and acknowledged nonverbally in "only a split second." Other members of the team, and perhaps even family members, may not have been aware of the intimacy that took place. The nonverbal nature of the glance and the context of excitement in the team's rejoicing after the victory allowed the intimacy to be expressed privately in a covert fashion. Both the intimate style and the context in which the intimacy was expressed contributed to an environment that was relatively safe from ridicule.

How Do I Know You Like Me: Intimacy and Affection in Men's Relationships

When asked, "How do you know that your men friends like you?" only one man responded that his friend tells him directly that he likes him. If men do not tend to self-disclose to each other the closeness of their friendship, how do they evaluate closeness and intimacy with a man friend? In men's friendships with other men, doing something together and choosing a friend and asking him if he wants to participate in an activity demonstrate that they like one another and enjoy being together. These acts have a meaning similar to a boy who asks a girl to a dance; it's assumed that he likes her by the nature of the action. Mike responded to the question, "How do you know or get the idea that they [men friends] like you?"

> When I suggest that we do something I can tell in their voice or the way their actions are that they want to do it. Like hey! they really want to do it. Like, "Anyone want to go to the baseball game?" "Yeah, great! That's exactly what I want to do." That's a good feeling to know that you

> can make some sort of a suggestion that fits. Laughter, the joking, the noise. Knowing that they like to do the things that I like to do and that I like to do the things that they like to do. And it's the same in reverse, and basically I want to do it as well, me agreeing with them. As far as that goes, you'd say, I like it when they show me by asking me, if they want to do it with me.

Men mentioned physical gestures, laughing at jokes, doing one another favors, keeping in touch, "doing stuff," teasing, and just being around friends as ways they know that men friends like each other. The most common responses to the question of how their men friends let them know that they liked them were "doing things together" and "initiating contact." John responded:

> I think it's just something you can sense, that you feel by . . . obviously if you continue to go out and do things with them.

Mike responded:

> Well, they'll call me up and ask me to do stuff, if they have nothing to do, or if they do have something to do and they want me to be a part of it.

Men feel liked by other men as a result of being asked to spend time in activities of common interest. Within such active contexts, reciprocated assistance, physical gestures, language patterns, and joking behaviors all had distinctive meanings that indicated intimacy between male friends.

Reciprocity of Assistance Men mentioned doing favors, which included mailing a letter, fixing a car, loaning money, and talking about problems relating to heterosexual relationships. The men emphasized a reciprocity of assistance and a goal-orientation to both problem solving and situations that involved self-disclosure. This reciprocity demonstrated mutual interest and also was a means to achieve a balanced dependency. Pete responded to the question, "How do your friends let you know that they like you?"

> We help each other out, just like doing favors for someone. Like right now, me and my roommate were going to class, and he was asking me questions because he slept in and didn't study. So I go "what's this—ok, here, just have my notes." Even though I'm going to need them for my thing at three. You know, just little stuff like that.

Matt referred to the assistance given between his closest male friend and himself as a "barter" arrangement.

> Jack and I had a good relationship about this. He's a very good mechanic, and I would ask him. And I would develop something that I do that was rewarding for him. Like I could pull strings and get free boat trips and stuff like that when I was an [diving] instructor. And he would work on my car and I would turn him on to the Islands and dives and stuff. It was sort of a barter situation.

The sharing of skills and access to opportunities fostered interdependency, yet also maintained their independence through a mutual give-and-take.

Physical Gestures Men also reported physically demonstrating affection to each other. However, the physical gestures had a distinctively masculine style that protected them from the fear of an interpretation of a homosexual preference. Men mentioned handshakes, bear hugs, slaps on the back, and an arm on the shoulder as ways that friends demonstrated affection.

Handshakes were the most frequently mentioned. Handshakes offer controlled physical contact between men and are often considered an indicator of strength and manliness. A strong, crisp, and forthright grip is a sign of a "respectable" masculinity while a limp and less robust handshake may be associated with femininity and a homosexual orientation. A bear hug also offers a demonstration of strength, often with one friend lifting another off the ground. Gary described an occasion in his response to the question, "How do your men friends let you know that they like you?"

> I came back from a swim meet in Arkansas last week, and I hadn't seen Mike for two

> weeks. When I came back he came right at me and gave me a big old bear hug, you know, stuff like that. And my mom and dad were in the room, and they're going, "Hey, put my son down!" and we were all laughing.

Men give the affectionate hug a "rugged," nonfeminine veneer by feigning playful aggression through the demonstration of physical strength. The garb and trappings of roughness allow a man to express affection while reducing the risks of making his friend uncomfortable or having his sexual identity ridiculed. A slap on the back is much less risky for a man than a caress on the cheek, although they may have a similar message in the communication of closeness.

Joking Behavior Men developed joking behaviors that communicate closeness and similar ways of viewing the world. Ken responded to the question, "How do you get the idea that they [men friends] are close friends?"

> Laughter is one of them. I'll admit, when I'm around anybody really, not just them, I try to be the world's best comic. Like I said, humor is just important and I love it. I'd rather . . . I just like to laugh. And when they laugh, and they get along with me, and we joke with each other, and not get personal, they don't take it too harshly.

Although Ken says he attempts to be a comic "when I'm around anybody," he goes on to elaborate about the differences in joking when around men or around women:

> For the girls, not so much the laughter because you can't, with the comedian atmosphere, or whatever, you can't tell with the women. . . . Because, you know, if you get together with some girl or someone that likes you a little bit, or whatever—you can tell them that your dog just died, and they'll laugh. You know what I mean, you know how it goes. It's just, they'll laugh at anything, just to . . . I don't know why it is. But you get together with certain girls and they'll just laugh no matter what you say. So it's kind of hard to base it on that. Because the guys, you know, you can judge that with the guys. Because they'll say it's a crappy joke or something like that, or say that was a terrible thing.

Women friends did not respond to his humor in as straightforward and rigorous a manner as did his men friends. This appears to be a result of a covert sexual agenda between the cross-sex friends and a misunderstanding of the cues and nuances of male joking behavior by his women friends. Joking behaviors often are rooted in the contexts of men's shared experience, an experience that women may have little access to. Joking relationships are used by men to show caring and to establish trust in the midst of competitive activities. The following response to the question, "What are the most meaningful occasions that you spent with a male friend?" demonstrated a context where joking behavior expressed intimacy in the midst of competitive action. First, Greg describes the context in which the joking took place.

> The first time I'd been waterskiing was last summer. And among these guys I was really athletic, maybe more so than them even. And he knew how to waterski and I didn't. And we got there, and I tried maybe six or eight times, and couldn't do it, just couldn't do it. I don't know what the deal was because I'm really an athletic person and I figure it wouldn't be that tough, and it was tough. As far as the friendship goes, for Mark, for him to sit there and have the patience to teach me what to do, what was going on, it must have taken an hour or so or more of just intense teaching. Like he was the coach and I was the player, and we got done with that and I did it. And the next time we went I was on one ski, thanks to him. It was that much of an improvement. And to know that we could communicate that well around something that I love, sports, and to know that we could communicate that well in something that we both like a lot, athletics, that meant a lot to our friendship.

Mark provided assistance that altered a potential traumatic experience into a positive success. Specifically, Mark used a joking relationship to reduce the pressure on Greg and allowed him to perform while in a vulnerable position. Mark did not exploit his superior capabilities, but shared them and empowered his friends. Greg explains:

> We were just able to make jokes about it, and we laughed at each other all day. And it finally worked out. I mean it was great for me to be that frustrated and that up-tight about it and know the only thing he was going to do was laugh at me. That may seem bad to some people. They'd have gotten more upset. But for me that was good. . . . It really put things in perspective.

The joking cues expressed acceptance and communicated to Greg that it was okay to fail, and that failing would not jeopardize continuing the lesson. Mark's acceptance of a friend's failures reduced the performance pressure on Greg, and thus released him to concentrate on learning to water-ski.

Joking behavior is important to men because it offers a style of communication that consists of implicit meanings not readily accessible to people outside the group. "In" jokes between friends demand attentiveness to an individual's thinking, emotional states and reactions, and nuances of behavior. They provide a format where a man can be meticulously attentive to the feelings and tastes of another man. An elaborate reciprocation of jokes can be a proxy for more overt forms of caring. Yet, because joking behavior is often used as a distancing gesture and hostile act, joking behavior is not interpreted as an expression of attachment. This adds to the covert nature of the act and further protects men from possible ridicule. The tenuous line between aggression and affection is demonstrated by Tim's response to the question, "Can you think of any other qualities that would be important to a close friendship?"

> Basically that they'll understand you. Like if you do something wrong and they go, "Oh, what a jerk." I mean they can say it, but they'll say it in a different way than some guy who shoots his mouth off, "What a jerk, you fell off your bicycle."

Tim was questioned further, "How would it be different—I know what you mean—but can you describe it?"

> You know, they'll poke fun at you but they'll say it in a friendly way. Where someone else will just laugh, "What an idiot," and they'll mean it. Where your friend will say . . . you know, just make fun of you and stuff. I don't know if I explained it too well.

The same words used by two different men are interpreted and reacted to in very different ways. The tone of voice and social distance between the two men are essential factors in the determination of an understanding friend as opposed to an aggressive enemy. Tim's reactions to both cues reveal the different meanings. The question was asked, "Okay, maybe if I ask another question to get at it, say you fell off your bike, how would you feel when your friend joked about you as opposed to . . . ?"

> I would just start laughing, you know. I mean he'd start laughing at me and I'd just look and he'd go, "You jerk," and I'll start laughing. We know each other and stuff. Some guy off the street—I'll just cuss at him and flip him off, you know. So it's a little different.

Such discriminations are difficult for men to explain and describe. This would suggest that the discriminating task may be even more difficult for women, who have not had the experience in the contexts from which men's friendships have developed. Matt explained how he lets his closest male friend know that he likes him.

> I'll have a tendency to say, "Well, why don't you write?" in a teasing way, and "Okay, when are we going to get together? . . . and this bullshit of you being up there in Stockton."

Coarse language is injected into the teasing to legitimize the implicit meaning that he misses his friend and wishes that they were together. Joking relationships provide men with an implicit form of expressing affection, which is an alternative to explicit forms such as hugging and telling people that they care about them. Joking also may be more personal, since it often relies on a knowledge and sensitivity to a friend's attitudes and tastes, thus recognizing and affirming a unique part of him. The following portion of an interview demonstrates this masculine style. Jim responds to the question, "Why do you think [women are more likely than men to come out and tell you that they like you] that is?"

> Oh, it's just the way you were raised. It's society. You might hug a girl and say, "See you later and good luck on your test tomorrow." Whereas you'll joke around with a guy about it.

"Why would you joke around with a guy?"

> It's just a . . . it's just a different relationship, you know. I think society would accept two girls hugging each other and a guy hugging a girl, but it's a little different when you're two guys. I don't know if you saw the movie *Grease* where there, like Danny and that other guy who's driving the car, they do it, well, like they hug each other, right? After they pull the car out of the shop, it's kind of like that, they stop, they realize what they did. You might even want to, you might wanna say, "Hey, thanks a lot." You do stuff like that. But you don't act silly. You might shake their hand.

Jim was asked if he hugged his closest male friend, to which he responded, "No, I don't do that." He was then asked, "How would it feel if you went up to hug Fred [closest male friend]? How do you think he would react?"

> Well, I can remember a couple of times that we had . . . after a football game when you're real excited and things. It all depends on the situation. If I just did it, you know, out of the clear blue sky, he'd probably look at me and, you know. I could do it jokingly. It might even be pretty funny. I might try that. But I don't think he'd like it. He'd probably think it was a little strange.

Jim was able to hug his friend after a football game, when emotions ran high, and the men's masculinity had just been validated by participating in, and presumably winning, the game. The football context insulated the hugging from being interpreted as unmanly or gay. Jim says, "It depends on the situation." At one point when Jim was asked what it would be like to hug his friend, he interpreted it as a challenge or a dare. "I could do that." However, he translates the act into a joking behavior, "I could do it jokingly," in an effort to stylize the hug as masculine. Men's styles of intimacy attempt to minimize the risks taken when overtly expressing affection. These risks are summed up best by Jim when asked why he would feel strange if he hugged his best male friend. Jim said:

> The guys are more rugged and things, and it wouldn't be rugged to hug another man. That's not a masculine act, where it could be, you know, there's nothing unmasculine about it. But somebody might not see it as masculine and you don't want somebody else to think that you're not, you know—masculine or . . . but you still don't want to be outcast. Nobody I think wants to be outcast.

Thus Jim could not hug his friend "out of the clear blue sky," overtly and without a gender-validating context. The styles of male intimacy attempt to limit these risks. Joking behavior camouflages the hidden agenda of closeness by combining elements of a private awareness of a friend's history and personal nuances with a public tone of aggression and humiliation. A man describes his most meaningful times with men friends.

> The conversation will come and we can kind of laugh at each other. And you can laugh at yourself, which is really handy.

Much as the slap on the back covers an affectionate greeting with an aggressive movement, joking behavior provides a covert avenue in which to express caring and intimacy.

CONCLUSIONS

These findings suggest that microstructural variables, particularly interactional expectations, are powerful explanations for male intimacy styles. Intimacy between men is influenced by their awareness of the restrictive sanctions that are often imposed on men who express certain emotions, such as sadness or fear. Men's intimate verbal style is partially shaped by the fear of sanctions that may be imposed on emotional behaviors deemed culturally unacceptable. Homophobia and the difficulty men have disclosing weaknesses testify to the limitations they experience when attempting to explore certain aspects of their selves. These limitations of male intimacy may distance men from all but their closest men friends, and may also create a premium on privacy and trust in close friendships. Such limitations may be more detrimental later in life where structural settings are less conducive and supportive to maintaining active friendships. A college environment fosters casual access to friendships, and friendships may also be integral and functional for the successful completion of a degree. Thus, the sample in the present study may be experiencing an intimacy that is more difficult for men to maintain in job and career settings.

The interview data show that although constraints in the masculine role limit men in certain situations and in verbal intimacy, men do develop intimate friendship behavior that is based on shared action. Men's intimacy often depends on nonverbal cues that are developed in contexts of active engagement. Men expressing intimacy with close friends by exchanging favors, engaging in competitive action, joking, touching, sharing accomplishments, and including one another in activities. The strengths of men's active style of intimacy involve sharing and empowering each other with the skills necessary for problem solving, and gaining a sense of engagement and control of their lives by sharing resources and accomplishments. Nonverbal cues offered an intimacy based on a private affirmation and exchange of the special history that two men share. This unique form of intimacy cannot be replicated solely by self-disclosure.

In addition to the men's active style of intimacy, they also reported self-disclosure to friends. Contrary to previous research, most men reported that they were more comfortable expressing themselves to a close male friend than to female friends. These men assumed that close male friends would be more understanding because of their shared experiences. Men said that self-disclosure and hugging "depended on the situation," and were more likely to self-disclose in a gender-validating context. Thus, men overcome cultural prohibitions against intimacy with this gender-validating strategy.

There are advantages and disadvantages to both feminine and masculine styles of intimacy. Feminine intimacy is productive for acknowledging fears and weaknesses that comprise a person's vulnerability. Admitting and expressing an emotional problem are enhanced by verbal self-disclosure skills. Masculine styles of intimacy are productive for confronting a fear or weakness with alternative strategies that empower them to creatively deal with a difficulty. Both styles appear necessary for a balanced approach to self-realization and the challenge of integrating that realization into a healthy and productive life.

Although this study focused on generalized sex differences to document a previously unrecognized active style of intimacy, women also demonstrated active styles of intimacy, and men demonstrated verbal styles of intimacy. Thus, although the results are based on generalized tendencies, the data also support the flexibility of gender-based behavior and the ability of men and women to cross over and use both active and verbal styles of intimacy.

The documentation of active styles of intimacy sharpens the understanding of intimate male behavior, and it provides a more accurate and useful interpretation of the "inexpressive male." The deficit model of male expressiveness does not recognize men's active style of intimacy, and stresses men's need to be taught feminine-typed skills to foster intimacy in their relationships. This negation or denial of men's active style of intimacy may alienate and threaten men who then assume that intimacy is a challenge they will fail. An awareness of the strengths in men's covert style of intimacy provides a substantive basis from which to address and augment changes in restrictive and debilitating aspects of masculinity. The finding that gender-validating activities foster male self-disclosure suggests that strategies for developing more intimate capabilities in men would be most successful when accompanied by a gender-validating setting that acknowledges, enhances, and expands the use of the intimate skills that men have previously acquired.

The data suggest the influence that sex-segregated worlds exert on the ways women and men choose, and are most comfortable in expressing, intimacy. The separate adult social worlds that women and men often experience shape the opportunities and forms of intimacy shared between friends. These structural opportunities and the styles of intimacy that become integral to specific opportunities become familiar, expected, and assumed between friends of the same sex, and often are bewildering, inaccessible, and misinterpreted by cross-sex friends or partners.

The implications are clear: men and women will have to be integrated in similar microstructural realms in the private and public spheres if we are to expect men and women to develop fluency in what are now termed "male" and "female" styles of intimacy. If such integration does indeed take place, the reduction of misunderstanding, frustration, and abuse in cross-sex relationships could be profound.

REFERENCES

Allen, J., and Haccoun, D. "Sex Differences in Emotionality: A Multidimensional Approach." *Human Relations* 29 (1976):711–722.

Bakan, D. *The Duality of Human Existence.* Chicago: Rand McNally, 1996.

Balswick, J. "The Inexpressive Male: A Tragedy of American Society." In D. David and R. Brannon (eds.), *The Forty-Nine Percent Majority.* Reading, MA: Addison-Wesley, 1976:55–67.

Bem, S.; Martyna, W.; and Watson, C. "Sex Typing and Androgyny: Further Explorations of the Expressive Domain." *Journal of Personality and Social Psychology* 34 (1976):1016–1023.

Booth, A. "Sex and Social Participation." *American Sociological Review* 37 (1972):183–192.

Caldwell, R., and Peplau, L. "Sex Differences in Same-Sex Friendship." *Sex Roles* 8 (1982):721–732.

Cancian, F. M. "Marital Conflict Over Intimacy." In A. Rossi (ed.), *Gender and the Life Course.* Hawthorne, NY: Aldine, 1985.

———. "The Feminization of Love." *Signs* 11 (1986): 692–709.

Coleman, J. "Athletics in High School." In D. David and R. Brannon (eds.), *The Forty-Nine Percent Majority.* Reading, MA: Addison-Wesley, 1976:264–269.

Cozby, P. "Self-Disclosure: A Literature Review." *Psychological Bulletin* 79 (1973):73–91.

Davidson, J., and Duberman, L. "Same-Sex Friendships: A Gender Comparison of Dyads." *Sex Roles* 8 (1982): 809–822.

Goffman, Irving. *Presentation of Self.* Garden City, NY: Doubleday, 1959.

Hall, J., and Halberstadt, A. "Sex Roles and Nonverbal Communication Skills." *Sex Roles* 7 (1981): 273–287.

Hantover, J. "Boy Scouts and the Validation of Masculinity." *Journal of Social Issues* 34 (1978):184–195.

Henley, N. *Body Politic, Power, Sex and Nonverbal Communication.* Englewood Cliffs, NJ: Prentice-Hall, 1977.

Hoffman, M. "Sex Differences in Empathy and Related Behaviors." *Psychological Bulletin* 84 (1977):712–722.

Komarovsky, M. *Blue-Collar Marriage.* New York: Vintage, 1964.

Lehne, G. "Homophobia Among Men." In D. David and R. Brannon (eds.), *The Forty-Nine Percent Majority.* Reading, MA: Addison-Wesley, 1976:66–68.

Maccoby, E., and Jacklin, C. *The Psychology of Sex Differences.* Stanford University Press, 1974.

Morin, S., and Garfinkle, E. "Male Homophobia." *Journal of Social Issues* 34 (1978):29–47.

Parsons, T., and Bales, R. *Family, Socialization, and Interaction Process.* New York: Free Press, 1955.

Pleck, J. *The Myth of Masculinity.* Cambridge, MA: MIT Press, 1981.

Rubin, L. *Intimate Strangers.* San Francisco, Harper and Row, 1983.

Stein, P., and Hoffman, S. "Sports and Male Role Strain." *Journal of Social Issues* 34 (1978):136–150.

Swain, S. "Male Intimacy in Same-Sex Friendships: The Influence of Gender-Validating Activities." Conference paper presented at the American Sociological Association Annual Meeting, San Antonio, 1984.

Tavris, C., and Wade, C. *The Longest War: Sex Differences in Perspective* (Second Edition). San Diego: Harcourt Brace Jovanovich, 1984.

14

Men Have Friends, Too!

RICHARD NIGHTINGALE

The topic of males' friendships was sparked in my mind in a sociology class called "Gender in American Society." I was hearing stories of how men were jealous of the relationships that female friends shared with one another, especially their abilities to turn to their friends to deal with problems in their lives. Men often describe their friendships with their closest friends as very fulfilling. This may be true to an extent, but if we look at these friendships we see that, more often than not, they are based on shared activities. I feel that the stereotype of men not being able to share intimate relationships with other men, of not being able to share their feelings with one another, is not entirely valid. I do see the relationships that I share with my close friends consisting of many shared activities, but—contrary to what I heard or read in class—in my experiences, trust and open expression of feelings form that sacred friendship bond.

Supposedly, if males do have a friendship in which they share intimate and personal feelings, it tends to be with a woman. Although I have a girlfriend whom I love very much and consider to be one of my best friends, there are times when I can only open up and express my feelings with one of my male friends. It would sadden me to feel that I could only turn to my girlfriend as my sole "release valve." There need to be other intimate relationships for males with other males in their lives.

Over the years, I have built a relationship with an individual that I feel has become intimate, in the sense that the relationship is not based solely on doing things together, but on sharing our feelings as well. We can and do talk about problems, values, or just overall ideas about life. I am proud to say that this relationship does not fit the stereotype of male friendship that I have so often heard about. Sure, on the outside, it might appear that way, but that's fine because I wouldn't want a friendship where all we did was sit and talk about our feelings. Nevertheless, it is very important for men to be able to do this, just as it is for women. The idea that men shouldn't cry is ridiculous to

Previously unpublished. Used by permission of the author.

me. Although I admit I still hold back tears, I know that, if need be, I can open up with my best friend and let them go. It is also important to me to know that he can do this as well.

When I think back and look at my relationship with my friend, not only do I vividly remember the times we played baseball together on the same field, or went to concerts together, but also times when I cried and poured my heart out about a problem that I was facing. These are important memories because they show all of the dimensions of a true intimate relationship with a friend. Having another male friend that you can open up to is important because there are times when the feelings that you need to talk about can only be heard by another male sharing similar problems, or problems that only men face. Although female friends are very helpful, sometimes there are situations in which it only feels comfortable talking to a male ear.

Growing up with my best friend I have seen the problems that he has gone through. His parents are divorced and his father is an alcoholic. His father is a wonderful man, and whenever I saw him he was fun to be around. However, he wasn't around for the last three years because he was jailed for drunk driving. What I saw of my friend's family situation was from outside, so I don't know all the hardships he endured or how he managed. I do know that I would want someone there to be able to talk to if I were facing similar circumstances. Luckily, he did have someone and that someone was me (along with another close friend of ours). He was able to come to me and talk about how difficult it was and the things he was going through. Although we are at separate colleges, which are ten hours apart, it does not affect our relationship, because he is only a phone call away. If our friendship were based only on doing things together, then the relationship would not be possible. Instead, I can call and have him listen to me about the problems that I have. When we talk, we not only listen to each other but we also give each other advice on our problems and feedback on our ideas. Your parents can give you advice or help you with things, but you need to have a same-sex friend to talk to who is in a similar life path and shares the same basic lifestyle as you do.

I will never forget a time when I was going through a serious problem. I felt as if the world was on my shoulders, and I was being bombarded with different emotions. At the time, I didn't feel comfortable talking to my parents about it, though if I needed to they would have been there for me. Since the problem involved my girlfriend, I couldn't go to her. Right then and there what I needed was my best friend. He came over, and we had planned on going somewhere or doing something. However, he knew that could wait. Instead, I poured my emotions out to him and told him all about my problem. He was there for me when I needed him and that was very important to me.

After thinking about this issue and about openly sharing this relationship with you, I called my friend to see what he felt about the subject. Before calling him I wondered about how devastated I would feel if he didn't have the same feelings about our friendship as I did. But when we started talking he told me that he felt the same way I did about our friendship and about male friendships, overall. We talked about how important it is for male friends to feel close enough to open up to each other and, if need be, to cry with each other. Some men may look at a relationship like this and make the stereotypical comment that men don't need to be this way with other men, because we are strong and tough. Well, if you can keep your masculine image and still share an intimate relationship with another male, this shows your "toughness" even more. Men don't need to show this intimate side of their relationship with their best friend to anyone. It's great for those who do that they can. What's important, though, is that it is there, that you have a male friend that you can turn to and share all facets of a true friendship. Hopefully, the more men start to find and build relationships with friends where this is possible, the less men will feel trapped when confronted by feelings that would otherwise tear them up inside.

15

All-American Guys

BERNARD LEFKOWITZ

Within a few weeks of their arrival, the one hundred freshmen had found their places in the social constellation of Glen Ridge High. It wasn't hard to figure out who fit in where.

Every afternoon at about 2 P.M., a chattering, giggling, jostling mass of kids poured out of the exits onto the smoking patio and the lawn of the school. As a group they looked like nothing more than standard-issue suburban teenagers: a wave of Spandex and sweatpants, faded Guess jeans and pastel Benetton blouses. But these youngsters didn't constitute an undifferentiated bloc. The easiest way to distinguish the tribal styles of the kids was to watch how they divided up, the groups they formed naturally, easily on the lawn; who shared cars with whom, who walked together down Ridgewood Avenue.

When the students left the school building, they drifted over to the piece of turf that was claimed by their group. The Freaks, known in Glen Ridge as "Giggers," gathered on the smoking patio. The Greasers, here called "Guidos," sauntered over to the Glen, a grassy area close to the school, where they could consume their stash of beer without adult interference. The "Achievers," except for those involved in after-school clubs, lingered briefly on the lawn and then hurried home. They had homework to do. Meanwhile, the "Jocks" held court in the parking lot. Homework was not their highest priority.

Not many kids drifted from one group to another. They had formed their alliances as early as the elementary and middle grades, and once they found their place on the social map of adolescent Glen Ridge, they stayed there. . . .

The Achievers, Giggers, and Guidos had their friends and admirers, but they were not at the pinnacle of the social pyramid. That rarefied perch was occupied by the Jocks. Otherwise known as "the hunks." With their robust good looks, they partied with the flashiest and, some said, sexiest girls in the school. After a game, especially a winning game, some male teachers fawned over them almost as much as the cheerleaders did.

To those unfamiliar with the high school scene, the word *Jocks* was synonymous with athletes. That wasn't entirely accurate. Not all students who played sports were Jocks with a capital J. Some kids who participated in sports—most Glen Ridge students played on one team or another—gravitated toward the Giggers or the Achievers and, infrequently, the Guidos. But that hard core who considered themselves "real Jocks" did not treat athletes who identified with other social groups as equals. Only certain athletes felt they were truly worthy of the title *Ridger.*

There were two essential criteria for acceptance in the inner circle. First, you had to participate in at least one—or, better still, both—of the two most aggressive and physical sports: football and wrestling. A few baseball players made it into the inner sanctum, but members of the tennis and golf teams were not highly prized by the real Jocks.

The other quality that guaranteed certain kids popularity in Jock circles was an unmistakable style, most dramatically demonstrated by the "Jock

Strut." Going over to the Jock dining table in the cafeteria, or joining their buddies in the parking lot, they didn't amble or shuffle or stroll. They strutted.

The peacock image they projected was not something they had picked up overnight in high school. They had spent years perfecting it. For these young men, the essence of jockdom was a practiced show of contempt for kids and teachers alike. They tried to humiliate any wimpy guy who got in their way, but they reserved their best shots for girls who ignored them or dared to stand up to them.

The only students spared their contempt were the "Jockettes." They were the natural satellites of the Jocks. True Jockettes would not be caught dead at a Gigger party. They basked in the reflected glow of the Jocks. Despite the feminine connotation, the Jockettes consisted of males and females. They were kids who hungered for an invitation to a Jock party, who attended every practice session, cheered them in every game. Jockettes treated their Jocks with an almost maternal loyalty and protectiveness. They were both fan club and palace guard. Over the years, the Jockettes had come to consider the Jocks as not just "the guys." To them, they were always: *Our Guys.*

For a fourteen-year-old freshman surveying the social arrangements in the school, the only cliques that seemed to count were the academic elite and the Jocks. If you didn't fit into one of the two, where were you? Nowhere. The only thing worse was not belonging to any clique at all. Then you got treated like a lost soul, a friendless wanderer. . . .

Were the Jocks comfortable with the roles they were asked to play in the culture of Glen Ridge? A compelling argument can be made that the hyper-masculine style they were asked to assume by parents and brothers and sports enthusiasts was a heavy burden for a kid to carry through adolescence.

Most of these boys did not demonstrate great talent for athletics. They were kids who liked to throw a ball around or wrestle in the mud with their buddies. Yet they were touted by everyone who counted in their world as future stars. More than once, they asked themselves, What if I don't meet my father's expectations? What if I'm not even as good as my older brother—or, worse, my younger brother? When the school upped the ante, when it staked its reputation on their performance on the playing fields, it added significantly to the already outsized image that had been superimposed on a group of young and immature boys.

Then there was the matter of the social role that Glen Ridge conferred on its young athletes. That role decreed that they had to be popular, handsome, desirable to girls. But how many insecure adolescent boys believe, deep down, that they can live up to all that? They must have asked themselves: What if that cheerleader turns me down? What if I'm left out the next time the crowd has a party? They also might wonder whether the love and respect they received from adults and peers was conditioned largely on their athletic performance. If I fail, they might ask, will anyone love me?

The swagger they affected—the Jock swagger—was a handy way to camouflage their doubts. It served three purposes: It compensated for their limitations as athletes by allowing them to pretend they were stars. It was a distorted exaggeration of how they thought dominant men—real men—were supposed to behave. And, conversely, their defiance of authority may well have been a disguised growl of resistance against being programmed to conform to the myths nurtured by the community.

Young men have many social, psychological, and sexual insecurities. The myth of the future jock king added to the list. If they sensed they weren't going to be great athletes, they might try to conceal their insecurity through a show of childish bravado; they might try to lead by intimidation, not merit. If they weren't allowed to grow into their own masculine identity, they could develop their own grotesque version of manhood. . . .

There was nothing about Mary Ryan that would have qualified her as a Jockette. A year older than the students in the class of '89, she was extremely shy and spoke in a whisper that could turn into a whine when someone was mean to her; she dressed conservatively in white blouses and pale-blue and tan cardigan sweaters; she did not wear much makeup, and her brown hair was not permed or streaked or frosted or sprayed into an arabesque monument, as the popular girls did with theirs. Mary Ryan was never known to call attention to herself—except for one time, and that misjudgment would change her life.

In the Jocks' sophomore year, during the first week of February 1987, the shy junior stood up in the school cafeteria and said: "My parents are going to be away next week. I'm going to have a big party. Everybody come." Mary Ryan's secret wish was to be popular, to have people notice her. In Glen Ridge people noticed you if you threw a party when your parents were away. But giving advance notice within the hearing of a hundred or so high school students could be dangerous.

The risk was heightened if the Jocks weren't good friends with the hostess. "First of all, she was a girl none of us liked," Tara Timpanaro recalled. "If someone didn't like you, they're not going to have respect for your home." Charles Figueroa, a wrestler and football player, wasn't her close friend, but he liked Mary. "She smiled a lot and tried to be nice to you, but people wouldn't accept it. She had a kind of weak way about her. She tried not to offend anybody, so people thought they could roll right over her."

There were decorous nonalcoholic parties in Glen Ridge. There were rowdy alcoholic parties. And there were parties that turned violent. Before Tara transferred to Glen Ridge, she "had known for years that Glen Ridge was a major party town. Somebody is always having a party. I can remember my father telling me a guy jumped through a bay window during a party."

Fights broke out at parties. But what got the Jocks really mad was being barred from a party. One of the proudest moments for the '89 Jock clique—a moment that was celebrated in their senior yearbook—was the time they beat up older boys on the lawn of a host's house in Glen Ridge. The reason for the fight: The boys who lived in the house didn't want the Jocks at their party. Indeed, earlier in their sophomore year, on October 11, 1986, Kyle and Kevin were reported to the police when they crashed a party, refused to leave, and "had to be forcibly removed."

Parties could turn ugly when the adolescent partygoers decided they would use the party as a vehicle to hurt, one way or another, the party-giver, who in almost every case was a young woman. These scenes became known among the youth of Glen Ridge as "revenge" parties. The specific reason for the punishment seemed less important than the opportunity to hurt the girl. "If you're a girl and they don't respect you and they don't like you, forget it," said one of Chris Archer's wrestling teammates. This wrestler and other Jocks described what had happened to one of the Jocks' Little Mothers when she drank too much at a party. Like a bag of garbage, the girl was dumped in a closet as the party wound down; the guys locked the closet door and left her confined in the dark to gag on her vomit. Again, the Jocks noted the incident in the yearbook as one of the bright moments of their school year.

Kids who weren't in the cafeteria when Mary Ryan issued her invitation heard about it soon enough, and word traveled swiftly to students in other communities. February was the wrestling season, and high school wrestlers for miles around were told of the impending party. That's how it worked. Wrestlers told wrestlers; cheerleaders told cheerleaders in other towns. Why so much excitement? Parties with parents absent were not uncommon. But this one had the makings of something special. For Glen Ridge kids, the big attraction was that Mary Ryan, a tuition student, lived just across the town border in East Orange. They thought that if the cops busted the party, the guys' parents were less likely to find out. "When you go out of Glen Ridge, you go crazy," one of the athletes recalled. "There are no neighbors to stop you or tell your mother."

The other inducement was Mary's passivity. She was not known as a strong-willed kid, she didn't have many friends to protect her, and her family was not friends with the families of the

Jocks. "If Mary said 'no,' who'd listen to her?" Charlie Figueroa said. "She didn't have anybody who'd fight for her."

Along with everything else, the timing was perfect. The date set for the party was Saturday, February 14, 1987, Valentine's Day. That Saturday also fell in the middle of a three-day weekend, Monday being Washington's Birthday. "It was, like, a party that could go on forever," one Ridger said.

Instead of on Saturday, the party began spontaneously on Friday, February 13. By sundown every parking space for three blocks around the Ryans' house was taken. There were kids from Caldwell, Montclair, Bloomfield, and Verona. From private schools and public schools, from middle school and high school. There were older guys who had graduated three or four years ago. There were even kids from East Side High School in Newark. There were Jocks and Guidos and Giggers, cheerleaders and majorettes, and even a few "band fags." There were girls who looked too young to get into a movie alone and some who seemed old enough to be married and have kids. There were kids who brought bottles, and kids who lugged cases of beer on their shoulders, and some who rolled kegs up the front steps into the kitchen.

They all converged on a narrow three-story white shingle house with a semifinished basement and a small balcony facing a nearby park. The location was perfect for a nonstop party. There were only a few other houses on the block, and they all adjoined the park. You could make as much noise as you wanted with little likelihood of interference.

The kids who got there early made for the upstairs rooms. It was the only place where you could hear yourself talk. The ones who arrived by 10 or 11 o'clock wedged themselves into the kitchen or basement. Sixty or seventy kids jammed together, drinking, smoking, and screaming. Mary Ryan had given up asking who the kids were and where they came from.

Despite all the kids and booze, there was relatively little damage on Friday night. One guy did take out all the crystal glasses and pitchers in the kitchen cabinets, line them up on the table, and fling them, one after the other, against the wall. But that happened at a lot of parties. It was nothing to get excited about.

John Maher, a student who later would be indicted on a charge of conspiracy in the Leslie Faber case, was working on Friday night. "My friends were saying it was a great party, the best," he would say later. "I couldn't believe what was happening. So I made sure I was there Saturday."

Saturday night, February 14, Valentine's Day. More kids. More booze. There were so many bodies in Mary Ryan's house, so many kids jammed into a small smoky space, that they had to open all the windows and the doors. With all the runs to the fridge, the beer couldn't be kept cold. So they gave up on the fridge, cleared everything out, and left the door hanging open.

They started taking the furniture apart. Within an hour the legs had been broken off everything that was standing—coffee tables, kitchen chairs and table, side tables. A couple of guys got the idea of using a leg from the kitchen table as a battering ram. One-two-three—charge. The leg smashed through the plasterboard, leaving a hole the size of a saucer. Back up and start all over. The hole got bigger and bigger, maybe two, three feet in diameter. Okay, let's start on the other wall.

Then some people decided that the amputated remains of the furniture were cluttering up the place. In five minutes every tabletop and chair seat had been heaved into the backyard.

One wall was covered floor to ceiling with a bamboo stand to hold decorative objects. The stand had been attached to the wall. "Betcha can't break that in half," one Jock challenged another. As if he were working out in the school exercise room, the other Jock stood with his back to the bamboo, his arms raised behind his shoulders. A deep breath and pull. The entire bamboo stand, with everything that rested on it, came crashing to the floor. A few minutes later some guys were breaking pieces of the bamboo over their heads and using them as swords in make-believe duels.

One guy stood in front of the fish tank. Thinking. Then he went into the kitchen and returned with a container of Comet detergent. He emptied it into the tank. A half-hour later another kid saw the fish floating dead in the water. He and a friend carried the tank to the door and emptied its contents into the snow.

Mary Ryan would wave her hand in a futile plea to halt the destruction. But one of the girls would take her by the shoulder and guide her out of the room. "We'll help you clean up later," she'd tell her.

Sunday night, February 15, 1987. It seemed as if the whole world under the age of thirty had turned up for the third act at Mary Ryan's house. Among the new notables were the wrestlers from Glen Ridge. They had been forced to miss the Saturday-night festivities because they were competing in a match. Rock-solid and brimming with energy, they could always be counted on to liven up a party.

It was a true gathering of the clan: Richie Corcoran and Kyle and Kevin; Peter Quigley, his companion, Tara, and Peter's older brother, Sean; Chris and Paul Archer. They had no problem finding the house. "You could hear the noise from like a mile away," said one Glen Ridge wrestler. "When you got on the street, it was amazing. It was so cold out and snow was on the ground and there were dozens of kids standing outside. One of the kids I recognized was holding a neon tube over his head and then he smashed it right on his skull. All the lights were on in the house, and you could see people in every window. As I was walking in, part of the frame over the door was hanging down and I almost ran into it.

"There were like a million people in there, all of them drunk. And right away I saw all the wrestlers from the school and I know they can get a little crazy at a party, and I thought, Whew, there's gonna be all sorts of shit tonight. I kept thinking, I'm walking into a movie."

Charlie Terranova, one of the Glen Ridge Giggers, stayed about fifteen minutes. "I went into this place and the things I saw I could not believe. I once worked for a construction company, and there were rooms in this house that looked like a construction crew had gone in there with the crowbars and the pikes and just destroyed the place. I just left. I couldn't stand it."

Some kids may have experienced a letdown when they surveyed the wreckage on the first floor. Really, what was left for them to do? The people who had been there the first two nights seemed to have exhausted all the possibilities. But, on reflection, it was apparent that they hadn't. There were two upstairs floors and a basement, and that left lots of unfinished work.

Chris Archer took the basement. People who were there remembered him rushing down the stairs with a can of spray paint in his hand and spraying every wall with painted graffiti. Another Jock, partygoers recalled, charged upstairs, followed by a pack of football players and wrestlers. First thing they did was dismantle Mary's parents' bed. One kid had the idea of setting the mattress on fire, but another thought that was a stupid idea since it was a waterbed. Let's puncture it, one guy suggested, and start a flood. Some guys began stabbing it with a screwdriver and a kitchen knife.

Other kids carried the bed frame to the top of the landing and, using it as a makeshift toboggan, tried to slide down the stairs. But the frame was too wide to make for a level ride. The kids smashed all the balusters that held up the stair rail. Now there was enough room. They guys sat on the frame, their legs straddling the sides, and slid down the slope.

Mary Ryan had retreated upstairs to her parents' bedroom, where she sat on the floor with another girl and Charlie Figueroa. They heard a roar coming from downstairs and rushed to the door; they saw a bunch of kids charging up the stairs as the boys were sliding down on the bed frame. The boys leaped over the mattress and burst into the Ryans' bedroom. There, they pulled open the dresser, flinging the underwear, blouses, and other clothing on the floor.

Holding their findings over their heads, they marched down the stairs. "Hey, wait a second," Mary shouted, fear in her voice. But who was listening? They put up on the mantlepiece all the personal possessions they had taken from the

Ryans' bedroom dresser. Mary sank down in a corner of the room, her knees up against her chest. "She looked to me like she was getting smaller and smaller, like she wanted to disappear," Charlie said.

Now dozens of kids formed a row, and began snake dancing past the mantel, as though they were performing a religious ritual before an altar. Someone had come up with the perfect description for what was going on. And the snake dancers began to chant it, as they weaved through the room: Ryan's Wreck. Ryan's Wreck. Ryan's Wreck.

It must have got through to Mary that the party was out of control. This had to stop. It wasn't only her life that was being trashed; it was her parents' life, too.

Now she saw a boy pick up her cat by the back of the neck, hanging him high for the crowd to see, and then push him into the microwave. She heard one terrible yowl, smelled burning flesh, burning fur. She screamed: "Stop, you've got to stop." Somebody pulled the cat out, but few people were listening to Mary Ryan.

Mary ran up the stairs, rushed into a room, and flung open the window. She stepped outside onto the balcony. It was not very high, ten feet or so above the ground, but high enough so she could hurt herself if she fell. She leaned against the rail of the balcony and peered down through the darkness at the mob of teenagers who had gathered on the snow-covered incline. "Oh, my God, my house, my house," she screamed. "If you don't stop, I'll jump."

Alarmed, one girl urged her, "Come on, come back in, Mary. Everything's okay. We'll go home." But instantly the sound of her voice was drowned out by dozens of kids chanting: *Jump. Jump. Jump. Jump.*

Charlie Figueroa, standing in Mary's bedroom, decided that all this had to stop. He knew that what he was about to do would break the first rule of Jock solidarity—never squeal. Charlie called the police anyway.

At about 11:15 Officers Chwal and Marinelli, patrolling in East Orange police car number 23, got a radio message: "Loud party in progress. Proceed to the scene." A second police car was also dispatched. When the police got there, kids were still standing beneath the balcony urging Mary Ryan to jump. But they weren't there for long. As soon as they saw the two flashing reds wheeling around the corner, the kids scattered.

The party was over. "Ryan's Wreck" had now passed into the folklore of Glen Ridge High School.

East Orange has a lot more crime than Glen Ridge. The East Orange cops weren't going to spend time chasing a hundred or so kids through the brambles of a park at midnight. But even these experienced cops were impressed by what they found in Mary Ryan's house and recorded in their report: "The reporting officers noticed the front door wide open and the downstairs in shambles. . . . Further investigation showed the entire residence, three floors, in shambles."

The two officers called the crime "malicious mischief" and described the "weapon" used to commit the crime as "physical force." The police detained eleven juveniles, all from Glen Ridge or Glen Ridge High School. These included a boy who would be selected as one of the captains of next year's football team and another football player, Peter Quigley's older brother, Sean, who had already completed his football career at the high school. Also held for questioning was James "Tucker" Litvany, a class of '89 football player, who would be later cited as an unindicted co-conspirator in the Leslie Faber case.

These youngsters were questioned briefly, their parents were informed, and that was the end of that. None of them was charged with a crime. None of them was punished or reprimanded by Glen Ridge High; none of them lost his athletic privileges or eligibility.

Mary never came back to Glen Ridge High. Her parents moved out of their house, and she was reportedly sent to live in another part of the state. A sophomore recalled that one of his teachers heard about the party and briefly discussed it in class: "The kids commented on how drunk people were, how they were breaking things, but how

Mary deserved it. Nobody said they were sorry. Nobody offered to clean up the place. And nobody wanted to pay for the damage."

Two years later the memory of that party remained fresh in the minds of the Jocks and the Jockettes of the class of '89. In a section of their yearbook, where each student listed personal highlights, many of them cited their participation in "Ryan's Wreck" as an outstanding event of the past four years. "It just showed what can happen to a girl when we didn't like her," one Jock would recall. John Maher, another class of '89 football player, would say years later, "That's a party that everybody still talks about."

For that nucleus of sophomore Jocks, this was not a formal initiation rite on the order of their first high school football game. But it was a benchmark experience in their high school years. There had been destructive parties before in Glen Ridge, and there would be others later. But it was under the tutelage of upperclassmen—older, admired football players and wrestlers—that they learned at Mary Ryan's party how much they could get away with. They also learned that the girls who attached themselves to the Jocks could be as pitiless as they were.

The primary lesson was that a bunch of high school kids could raise hell and inflict tremendous pain without being penalized at home or in school. But the party also taught a more advanced lesson. To one father whose daughter was in the class of '89, the boys who participated most enthusiastically at the party behaved as if they were gaining more legitimacy and authority as a group each time they victimized a woman. "If I think back about that period, I can see the group getting stronger, closer, every time they got together and humiliated a girl," he said. "What they enjoyed in common wasn't football. *This* was their *shared* experience. For them, this was what being a man among men was. My daughter would come home with stories—I'd just shake my head and wonder if they thought a girl was human." . . .

4

Men in Families

At some point in their lives, most men marry and father children. Data consistently reveal that around nine out of ten men will someday say "I do," in response to their brides' similar declaration of marital commitment (Coltrane 1998). Despite evidence on the fragility of marriage and the likelihood of divorce (nearly 50 percent of new marriages), men and women continue to march down the aisle. Most husbands and many unmarried men become fathers. The readings in this chapter portray some of what men experience in marriage and parenthood.

Sociologist Jessie Bernard contributed two important ideas that help convey men's experiences in families. The first is the idea of "his" and "her" marriages, meaning that marriage is differently perceived and experienced by men and women. Bringing different expectations, facing different responsibilities, and experiencing different benefits, men and women who marry also reap different levels of happiness or satisfaction. We can stretch Bernard's "two marriages" to note that there seem to be "two parenthoods," "two infidelities," "two divorces," and "two experiences of domestic violence," his and hers.

The second significant contribution Bernard made was her analysis of the male "good provider role," which defined men's major obligation in marriage and fatherhood as breadwinning. As a result of long-term historical trends, men's roles in families came to be defined mostly in terms of their provision of economic support. Whatever else he might be as a husband or father, a man was expected to be a stable and successful provider (Bernard 1981). Before turning to the readings that follow, let's look briefly at some of what we know about both of these important insights.

"HIS MARRIAGE"

Researchers have identified a number of "push" and "pull" factors leading to marriage (Rubin 1976; Nordstrom 1986). Premarital pregnancies, economic hardship, loneliness, and other unwelcome aspects of single living may push people into marriage. Women and men are also pulled into marriage by the companionship, security, sharing, and emotional support that marriage promises. For men these take on greater urgency because they have fewer available alternatives in which to find intimacy (Nordstrom 1986). Gratified to find a depth of emotional support, men more quickly conclude "This must be love." Having more confidants on whom to count, women are slower to deem their relationship "true" love. Thus, men fall in love more quickly and easily than women (Rubin 1980; Lindsey 1997).

As we've seen, men also define and express love somewhat differently than do women. Although a man may believe that he "shows love" instrumentally when he washes his wife's car, repairs something in the house, or brings home a paycheck, such displays may not meet women's more expressive notions of "showing love." Likewise, men more than women believe that they show their love for partners through sex. Women are more likely to believe that "sex is sex" and not a sufficient display of the depth of one's love (Rubin 1976, 1984). Thus, from the above generalizations one can conclude that along with its many splendors, love is a very gendered thing.

What Men Find in Families

Data continue to show the benefits of marriage in men's lives. Mentally and physically, husbands are healthier and more successful than their single male counterparts, especially those who once were married (Horowitz, White, and Howell-White 1996). They live longer, report themselves as happier, and excel more at work. They also suffer less from mental illness and commit less suicide or crime. Men have more difficulty than women adjusting to the end of a marriage. Those who "fly solo" at mid-life are significantly more distressed than their married counterparts (Marks 1996). Fewer men than women initiate divorce, and men remarry more often and more quickly after divorce than do women.

Marriage offers men other benefits, too. Legally sanctioned in the "marriage contract," traditional marriage gave men certain advantages. Although its conditions are unspecified and vary from state to state, traditional legal marriage asserted or assumed that the husband was head of the family (Weitzman 1981). Even as marriage law becomes more gender neutral, remnants of the imbalance of power remain in such conventions as naming after marriage and childbirth, and designation of domicile (which affects eligibility to vote in elections, payment of in-state or out-of-state tuition at public colleges, and so on) (Coltrane 1998).

Within marital relationships, men tend to wield more power, some of which stems from the relative share of household resources (e.g., money) they bring into the relationship. As suggested by Blood and Wolfe's "Resource Theory of Marital Power," the one who brings in the larger share of resources has a legitimate claim on a greater say in decision making (Blood and Wolfe 1960). Men may have more power within marriage, but women's share of power, or role in decision making, increases along with their contribution of resources. In such an equation, the full-time homemaker tends to be relatively powerless (Bernard 1972; Schwartz 1994; Renzetti and Curran 1995).

The resource-power connection pertains most to couples with at least one male (married or cohabiting heterosexuals, gay couples) (Blumstein and Schwartz 1983), but more than just resources are involved. Women do not automatically gain power or dominance in decision making just from wage earning (Pyke 1994; Renzetti and Curran 1995; Eshelman 1997), and men do not automatically become powerless when not bringing in material resources. Even in married couples where women are the sole wage-earners, men retain more decision-making influence than the resource theory would predict (Cohen and Durst, Reading 28). The meanings couples attach to husbands' and wives' employment are as important as the facts surrounding employment and earnings (Eshelman 1997; Renzetti and Curran 1995; Risman and Johnson-Sumerford 1998).

Alexis Walker (Reading 16) shows how even watching television is partly an exercise in gendered power. Decisions about what to watch and for how long (i.e., control of the remote control) reflect the unequal distribution of power in relationships. After reading Reading 16, you may find it harder to laughingly dismiss the familiar image of men "zapping channels."

The politics of marriage are also reflected in the division of housework and childcare. Housework is one of the most complex and contentious dimensions of married life (Hochschild 1989). Although men's share of housework has increased, they still don't do very much of it, whether or not their wives are employed (Coltrane 1998; Hochschild 1989). Women spend many more hours on housework than do men. Among full-time workers, the picture improves, but not completely. According to Hochschild's *The Second Shift,* compared to their husbands, employed women worked the equivalent of an extra month of twenty-four-hour days per year of housework. They also suffered from a measurable leisure gap (Hochschild 1989). Men now do about a quarter of all inside chores (Coltrane 1998).

Childcare is also unequal. Beginning with the transition to parenthood and continuing through their performance of the tasks of parenting, fathers and mothers differ (LaRossa and LaRossa 1981). Generally, men are less involved than their wives in the daily care of children and perceive themselves in supporting rather than primary roles (Simons et al. 1993). Fathers spend between one-third and one-fifth of the time mothers spend in active involvement with their children (Bird 1997). Greater maternal involvement in the daily lives of children extends into how often they think or worry about their children (Walzer, Reading 18) and persists through childhood. In parenting adolescents, there are more "modest" differences in the amount of time mothers and fathers spend in various parent-child activities (Demo 1992). By then, however, children have very different perceptions of their fathers and mothers, seeing fathers as more powerful but mothers as more sympathetic, responsive, and available for them. In fact, regardless of the age of the child, mothers, especially at-home mothers, are typically more invested and involved than fathers in the child's daily life (Thompson and Walker 1989).

Differences persist in what parents do as well as how often they do it. More of the time fathers spend with their children is spent in play activities, whereas mothers do more of the work associated with childcare (bathing, feeding, dressing, changing, and so forth) (LaRossa and LaRossa 1981; Coltrane 1998). More women discover and reap the benefits of early care and interaction with their children. Because fathers do less, they see and "get" less. Instead, many fathers report "doing time" with their kids, "babysitting them," and "helping the mother" (implying that the job is really not his). As parents, mothers display role embracement more than role distance (LaRossa and LaRossa 1981), though both fathers and mothers derive much psychological well-being from the quality of their relationships with their children (Umberson 1989).

Much of what men *don't* do as parents results from their upbringing and social expectations and the fact that women will do it. Some of the literature utilizes a "power" explanation that sees childcare as another responsibility that more powerful actors can avoid by getting less powerful partners to do it (LaRossa 1986). Often, in negotiating responsibilities, men use their higher status and greater power to avoid the unpleasantness associated with housework. The same explanation may or may not hold for caregiving (and may seem offensive when directed there, because it portrays "caregiving and nurturing" as being for "losers" in the power struggle of domestic life).

Of course, there is wide variation in how much housework or childcare is shared between spouses, and in the extent to which they believe in such sharing (Hochschild 1989; Risman and Johnson-Sumerford 1998). Reading 20 describes a variety of paths couples take into "postgender marriages" in which household work and childcare are shared. Having rejected the idea that gender ought to determine what each person does or contributes to the household, such couples have "moved beyond gender" (Risman and Johnson-Sumerford, Reading 20).

Men have the capacity to be attentive, nurturant parents, especially when women are not available to care for children. At-home fathers, shift-rotating fathers, and single, custodial fathers spend a good deal of time caring for children in the absence of mothers. In doing so, they develop sensitivity to children's needs and the abilities to meet those needs that a majority of men never do, because most men don't find themselves in such situations (e.g., Risman 1989; Cohen and Durst 1996). But one shouldn't understate the depth of feeling even "more typical fathers" feel for their children. Mark Neilsen (Reading 25) sensitively expresses his feelings of love and loss as his daughter prepared to leave for college.

Men as Providers

In most contemporary families, men are expected to do more than just provide and are expected to provide more than economic resources (Cohen 1993). They are expected to be intimate partners, companions, and "best friends" with their spouses, and to nurture their children. Still, because they're men they expect and are expected to provide economically. This, then, colors or affects much of what they experience, from the most mundane to the most meaningful matters of family life.

When men become providers, they reap decision-making and power benefits. Additionally, because they provide they avoid (or expect to avoid) housework. Some of what men do and don't do as parents results from the heavy demands of the male economic-provider role. Although work may be something men do for their families, while working they are removed from their families. The opportunity to interact more intensively with one's children is, thus, impeded by the need to support (as sole, primary, or co-provider) those children.

Connections between work and parenting are different for women and men. The burdens and responsibilities associated with childcare weigh more heavily on women, making other, especially occupational, roles more difficult to fulfill freely. In fact, the gap between women's and men's wages (discussed in the next chapter) is partly induced by the ways childrearing differently imposes on men's and women's paid work. Women leave their jobs for childrearing purposes far more often than men do, which can result in lower wages and diminished opportunities to excel, especially if the time off has stretched. Even when they are employed, mothers in two-parent households bear the responsibility for arranging childcare (Bird 1997; Walzer, Reading 18).

Most men escape such consequences because they continue to work through the birth and raising of their children, often increasing their workloads as a way to fulfill their new parental responsibilities and replace reduced or lost incomes from their wives. While avoiding more of the work associated with childcare and escaping the intrusion of parenting into their work lives, men experience the opposite encroachment. Demands of work more consistently impinge on time with and for children (Daly 1996). It is still the case that many men parent around work, while women work around parenting.

Increasingly, many men have shown a desire for more parental involvement (Daly 1993, 1996; Gerson 1993). Thus, the competition and confrontations between jobs and families are every bit as much "men's issues" as they are "women's issues" (see Chapter 5). To maximize men's parental involvement and capitalize on their intentions, structural changes will be necessary. Workplaces will need to be more flexible if more fathers are going to respond to our current celebration of the "new father." Other differences between mothers and fathers are addressed by Susan Walzer (Reading 18).

Research shows that sharing the economic-provider role often facilitates men's involvement in sharing domestic responsibilities (e.g., Coltrane 1996). Among Chicano couples, when wives' earnings approached husband's earnings, husbands' involvement in housework and childcare tended to increase. This outcome resulted from the meanings couples attached to each spouse's work and family work, not just how much wives earned (Coltrane 1996).

Ali Akbar Mahdi (Reading 17) discusses some of the changes in Iranian-American households. Ideological support for the traditional patriarchal structure of Iranian marriage diminishes as wives' educational attainment and their share of wage-earning increase. Husbands then take more active responsibility for household work and wives participate more in some typically masculine domestic tasks. As Mahdi notes, some men are less than enthusiastic about these changes, while a few suffer dire consequences.

THE DOWNSIDE OF MARRIAGE AND FAMILY

Many American households experience a variety of serious problems other than the unequal division of labor or men's monopolization of leisure. Most serious are the trio of troubles consisting of infidelity, divorce, and domestic violence. Unfortunately, given the broader objectives of this volume, we can't fully cover these areas of family life without adding to what is already a very big book. Still, like the aforementioned areas of falling in love, sharing intimacy, and dividing chores, each of these areas is gendered.

More married men than women engage in extramarital sexual relationships, and women and men attach different meanings or bring different motivations to infidelity (Clinard and Meier 1998;

Lindsey 1997). Estimates of infidelity range broadly, but within all studies men's rates exceed women's. Although women are often attracted to such relationships for the emotional support they might get, for men the driving motive is more likely sexual (Lindsey 1997). Overall, men are more permissive in their attitudes about extramarital sex than are women (Clinard and Meier 1998).

The complex phenomena surrounding divorce are clearly gendered (Amato and Rogers 1997). Men and women cite different reasons for seeking a divorce (Amato and Rogers 1997), and their experiences differ through and after a divorce (Furstenberg and Cherlin 1991). Women tend to cite husbands' behavioral or personality problems (e.g., drinking, infidelity, cruelty); men often either externalize the cause (e.g., in-law problems, work commitments) or acknowledge their own shortcomings (e.g., drug use, abusiveness) (Amato and Rogers 1997). Although both genders suffer emotionally from the end of a marriage, men fare much better economically, even experiencing some improvement in their standard of living (Eshelman 1997; Lindsey 1997). Many women and their children are driven downward, even into poverty when a marriage ends because of the combination of child custody and support outcomes (Furstenberg and Cherlin 1991; Eshelman 1997; Lindsey 1997).

More than a million children a year are affected by parental divorce, most of whom (85–90 percent) live with their mothers (Eshelman 1997). Although fathers who fight for custody receive it about half the time, most don't seek it. While children live with their mothers, fathers are expected to support them financially. Unfortunately, many don't. Only about half of non-custodial fathers pay the full amount of court-ordered child support, while about one-quarter of them pay none at all (Eshelman 1997). The U.S. Census Bureau reports that as much as $5 billion in child support is legally owed and unpaid (Sorensen 1997).

Compounding the problem, within two years, nearly half of divorced fathers have ceased all contact with their children (Lindsey 1997). These men, like those who neither marry the mothers of their children nor maintain meaningful father-child relationships, present contemporary fatherhood in an unfavorable light. The selection by Kay Pasley and Carmelle Minton (Reading 21) shows that such male behavior is neither inevitable nor universal. A number of factors promote or obstruct close relationships and frequent contact between divorced fathers and their children.

Finally, consider domestic violence and abuse. Within families, violence occurs with alarming frequency across the range of relationships (e.g., parent on child, child on parent, sibling on sibling, spouse on spouse) and among all races, ethnic groups, and economic classes. However, it clusters at the lower economic levels and, especially, among households where men are unemployed (Lindsey 1997). Neither the extent nor range of violent family relationships can be explained only by reference to gender. Women assault and abuse their children and partners, and lesbian relationships are not immune to violence. Thus, the cultural meanings and structural characteristics that define intimacy and family life contribute to these disturbing phenomena. In the specific instance of spousal violence, however, there are important gender issues.

Although survey research indicates that women are as likely as men to express intimate violence, the context differs. Female violence is more often self-defensive rather than control-oriented. It is less severe, results in fewer injuries or deaths to the male, and is less often a chronic or recurring characteristic of the relationship (Eshelman 1997; Koss et al. 1994). It also lacks the wider cultural acceptance that historically characterized male violence. Many times, women who are abused, like their abusive partners, believe in a man's right to inflict such harm. Although laws have improved, most states allow some forms of spousal exemptions for rape (Renzetti and Curran 1995). Judicial practice further differentiates marital from other forms of rape, with judges being less harsh in their sentencing of husband-rapists (Coltrane 1998).

Data on child abuse show that both fathers and mothers are perpetrators, with mothers actually represented more than fathers. Of course, most primary parenting is done by mothers, thus the "opportunity" for them to be physically abusive toward their children is much greater (Gelles 1998). When one looks only at single custodial fathers and mothers, males are more often physically abusive to their children (Gelles 1998). Additionally, stepfathers and boyfriends of single mothers may abuse children, especially sexually. Some argue that the presence of a father more often "protects" children from excessive exposure to other potential abusers (Popenoe 1996). Nonetheless, David Finkelhor estimates that 1 percent of female children are molested by their fathers (McCaghy and Capron 1997).

Given the multiple factors that account for the wider phenomenon of family violence, and the more gender-specific factors that may account for extremes of wife beating and marital rape, it is important to consider them as distinctive problems. In Reading 22, Michael Johnson suggests that the division of opinion between a "family violence perspective" and a "feminist perspective" results from the fact that they are really dealing with different, though overlapping, phenomena. Once we separate "common couple violence" and more extreme violence against wives, we may need different research strategies, theoretical explanations, and public policies for each of these expressions of male violence.

Men versus Men

Family life varies considerably across cultures and circumstances. Americans conceptualize marriage as a monogamous, freely chosen, love-based relationship in which husbands and wives live in their own household and share cooperatively in making the day-to-day decisions that govern family life, but there is rich, cultural variability to such dimensions of family living. So, too, do men's roles as husbands and fathers vary.

Looking at women and men in Russia, China, India, Japan, Latin America, Israel, Africa, the Arab Middle East, and Scandinavia, men's experiences range from near absolute dominance over wives to legislated equality with women in and outside of marriage (Lindsey 1997). Further complicating the cultural variability are intracultural differences induced by class or regional differences and furthered by economic, religious, and cultural changes. Mahdi (Reading 17) reveals some major changes in Iranian marital gender roles due to immigration and economic mobility.

Across the broad spectrum of cultures, men's involvement with children and in childrearing varies, though it trails mothers' involvement (Lorber 1994; Hoyenga and Hoyenga 1993). For example, among West African fathers there is little direct, nurturant involvement with young children. In urban China, the traditional pattern of fathers as disciplinarians and economic providers but not nurturers is slowly giving way to a more involved version of fathering. Conversely, Aka Pygmy fathers represent the extreme of nurturing, highly involved fathers, displaying a full range of affectionate behaviors, including frequent hugging, kissing, nuzzling, gently playing, and holding (Engle and Breaux 1998). Over the past quarter century in the United States, we greatly enlarged our expectations for fathers to be nurturant and involved (Daly 1993); it remains unclear how closely the "conduct of fatherhood" matches this "culture of fatherhood" (LaRossa 1988). In Reading 19, Gary DeCoker depicts differences between Japanese mothers and fathers. Noting some economic and cultural changes that have occurred in Japan, DeCoker describes the consequences for mothers, fathers, and children.

Within the United States, there are clear racial and class differences in men's experiences of marriage and family. Black men's experiences of unemployment and underemployment are large determinants of family instability and influence their marital and parenting behaviors. Suffering depressed occupational opportunities and facing "higher death rates from violent crime, disease, and poor health care" (Lindsey 1997) make black men less viable husbands and partners. This leads to higher rates of female-headed households from divorce or nonmarital childbearing. Race also shapes men's experiences and coping strategies after divorce (Lawson 1998).

We find more role flexibility and less traditionally divided gender arrangements in middle- and working-class black marriages than among whites or other ethnic and racial groups (Rubin

1994; Lindsey 1997). The socioeconomic instability faced by many black men increases the importance and acceptability of wages earned by black women, making men more open to participating in household routines.

Data on other ethnic groups are more mixed. Rubin, for example, found that Asian and Hispanic men did the least amount of sharing housework of the couples she studied (Rubin 1994). Such patterns have been identified elsewhere as well (see Eshelman 1997), but there are indicators of change, especially among dual-earner couples (Coltrane 1996; Eshelman 1997; Lindsey 1997). Martin Espada's narrative about his son, Clemente (Reading 24), shows how minority status, accompanied as it so often is by distortions and stereotypes about one's cultural background, adds to the weight of what fathers must do to help build positive self-esteem in their children.

White, working-class marriages are more ideologically traditional than middle-class marriages (Rubin 1994, 1976; Eshelman 1997); this traditionalism may diminish depending on spouses' work schedules. Working opposite, nonoverlapping shifts may lead to greater sharing within the household than is found when both spouses work and return home at the same time. More common among the working class, second or third shift schedules create situations in which necessity dictates more male participation in daily household and childcare activities because wives are absent while husbands are present.

Class differences are distorted by the tendencies for working-class men to understate and middle-class men to exaggerate how much they actually do in the household. Because participation may have greater "value" among the middle-class, but be more a matter of practicality or necessity for working-class men, it will be differently reported. Real behavior need not follow accordingly. Regardless of what they say about gendered family roles, there may be more actual sharing among working-class shift workers than among middle-class professionals (Rubin 1994; Ferree 1987, 1990).

Although the bulk of research looks at the working and middle classes, class differences do not end there. Families among the upper class are very gender segregated (Eshelman 1997). Neither husbands nor wives are heavily burdened by domestic responsibilities, because they rely heavily on paid help, but women's roles are nonetheless subordinated to their husbands'. They perform "accommodative and supportive functions" that enable their husbands to function in the economic and political sectors that they come to dominate (Eshelman 1997).

Among lower-class families, men are often absent entirely from day-to-day family life. High divorce rates, increases in the numbers of women bearing children outside of marriage, a persistent gendered wage gap, and a failure of men to provide for their children after divorce, all have "feminized" poverty. Most of those living in poverty are single mothers and their children. Poor men often lack the opportunity, training, and skills to obtain and maintain stable employment. Given the cultural emphasis on the male provider role, this subjects poor men to especially harsh experiences within families. They are less likely to marry and if married to stay married, and they derive fewer of the benefits that more successful men find in marriage.

Gay men's relationships take shape both because they are men, and because relationships between two men are as frowned upon as they are. Gay male couples are both more sexually active and less monogamous than either heterosexual or lesbian relationships because men are more sexually active and less monogamous than women. They display less emotional disclosure because, in general, men are less self-disclosing than women. Role relationships among same-sex couples are more egalitarian and more negotiated than in heterosexual relationships because they cannot fall back on gender differences to allocate or sustain role differences.

There are both similarities and differences in the intimate relationships of gay and heterosexual men. Both heterosexual and homosexual relationships go through similar relationship development. They commence with early infatuation, progress to settling down, and face the same long-term maintenance issues that confront all couples. Both gay and heterosexual couples value attachment and autonomy, and are generally satisfied with their relationships (Renzetti and Curran 1995). They also face the same sorts of conflict areas within their relationships (e.g., money, work, sex). Bob and Rod Jackson-Paris's account of their wedding (Reading 23) demonstrates common relationship issues regardless of sexual orientation. How different are their desires to marry or their feelings for each other from those of other couples entering marriage?

Ultimately, of course, there are major external roadblocks that same-sex couples (gay or lesbian) face that set them apart from heterosexual couples. Stigmatization, homophobia, and a lack of legal rights equivalent to those of heterosexual couples all create additional burdens for gay couples to bear. Thus, although the sentiments expressed in Reading 23 may be achingly familiar, as of the writing of this chapter (January, 2000), nowhere in the United States can two men legally marry. The lack of marriage rights has been challenged in courts in Hawaii and Vermont. In April 2000, Vermont passed a "civil union" law, entitling gay and lesbian coples to over 300 legal supports that provide some marital advantages.

REFERENCES

REFERENCES

Amato, P., and S. Rogers. "A Longitudinal Study of Marital Problems and Subsequent Divorce." *Journal of Marriage and the Family* 59, no. 3 (August 1997): 612–624.

Bernard, J. *The Future of Marriage.* New York: World, 1972.

Bernard, J. "The Good Provider Role: Its Rise and Fall." *American Psychologist* 36 (1981): 1–12.

Bird, C. "Gender Differences in Social and Economic Burdens of Parenting and Psychological Distress." *Journal of Marriage and the Family* 59, no. 4 (1997): 809–823.

Blood, R., and D. Wolfe. *Husbands and Wives: The Dynamics of Married Living.* New York: Free Press, 1960.

Blumstein, P., and P. Schwartz. *American Couples: Money, Work, Sex.* New York: William Morrow, 1983.

Clinard, M., and R. Meier. *Sociology of Deviant Behavior.* 10th ed. Orlando, FL: Harcourt Brace, 1998.

Cohen, T. "What Do Fathers Provide?" In *Men, Work, and Family,* edited by Jane Hood, 1–21. Newbury Park, CA: Sage, 1993.

Cohen, T., and J. Durst. "Daddy's Home: Pathways, Philosophies and Role Attachments of At-Home Fathers in Role-Reversed Couples." Paper presented at the National Council on Family Relations, 58th annual meeting, November 1996.

Coltrane, S. *Family Man: Fatherhood, Housework, and Gender Equity.* New York: Oxford University Press, 1996.

Coltrane, S. *Gender and Families.* Thousand Oaks, CA: Pine Forge Press, 1998.

Daly, K. "Reshaping Fatherhood: Finding the Models." *Journal of Family Issues* 14, no. 4 (1993): 510–530.

Daly, K. "Spending Time with the Kids: Meanings of Family Time for Fathers." *Family Relations* 45, no. 6 (October 1996).

Demo, D. "Parent-Child Relationships: Assessing Recent Changes." *Journal of Marriage and the Family* 54, no. 1 (1992): 104–117.

Engle, P., and C. Breaux. "Fathers' Involvement with Children: Perspectives from Developing Countries." Social Policy Report for the Society for Research in Child Development, XXI, no. 1 (1998): 1–21.

Eshelman, J. *The Family.* 8th ed. Boston: Allyn & Bacon, 1997.

Ferree, M. "Feminism and Family Research." *Journal of Marriage and the Family* 52, no. 4 (1990): 866–884.

Ferree, M. "She Works Hard for a Living." In *Analyzing Gender,* edited by B. Hess and M. Ferree, 322–347. Newbury Park, CA: Sage, 1987.

Furstenberg, F., and A. Cherlin. *Divided Families.* Cambridge, MA: Harvard University Press, 1991.

Gelles, R., and M. Strauss. *Intimate Violence.* New York: Simon & Schuster, 1988.

Gerson, K. *No Man's Land: Men's Changing Commitment to Family and Work.* New York: Basic Books, 1993.

Hochschild, A., with A. Machung. *The Second Shift.* New York: Viking, 1989.

Horowitz, A., H. White, and S. Howell-White. "Becoming Married and Mental Health: A Longitudinal Study of a Cohort of Young Adults." *Journal of Marriage and the Family* 58, no. 4 (1996): 895–907.

Hoyenga, K., and K. Hoyenga. *Gender-Related Differences: Origins and Outcomes.* Boston: Allyn & Bacon, 1993.

Koss, M., et al. *No Safe Haven: Male Violence Against Women at Home, at Work, and in the Community.* Washington, D.C.: American Psychological Association, 1994.

LaRossa, R. *Becoming a Parent.* Beverly Hills, CA: Sage, 1986.

LaRossa, R. "Fatherhood and Social Change." *Family Relations* 37 (1988): 451–457.

LaRossa, R., and M. LaRossa. *Transition to Parenthood.* Beverly Hills, CA: Sage, 1981.

Lawson, E. "Black Men After Divorce: How Do They Cope?" In *The Black Family: Essays and Studies,* edited by R. Staples, 112–127. Belmont, CA: Wadsworth, 1998.

Lindsey, L. *Gender Roles: A Sociological Perspective.* New York: Prentice Hall, 1997.

Lorber, J. *Paradoxes of Gender.* New Haven, CT: Yale University Press, 1994.

Marks, N. "Flying Solo at Midlife: Gender, Marital Status and Psychological Well-Being." *Journal of Marriage and the Family* 58, no. 4 (1996): 917–932.

McCaghy, C., and T. Capron. *Deviant Behavior: Crime, Conflict and Interest Groups.* Needham Heights, MA: Allyn & Bacon, 1997.

Nordstrom, B. "Why Men Get Married: More and Less Traditional Men Compared." In *Men in Families,* edited by R. Lewis and R. Salt. Beverly Hills, CA: Sage, 1986.

Popenoe, D. *Life Without Father.* New York: The Free Press, 1996.

Pyke, K. "Women's Employment as Gift or Burden?" *Gender & Society* 8 (1994): 73–91.

Renzetti, C., and D. Curran. *Women, Men, and Society.* 3rd ed. Boston: Allyn & Bacon, 1995.

Risman, B. "Can Men 'Mother'? Life as a Single Father." In *Gender and Intimate Relationships: A Microstructural Approach,* edited by B. Risman and P. Schwartz. Belmont, CA: Wadsworth, 1989.

Risman, B., and D. Johnson-Sumerford. "Doing It Fairly: A Study of Postgender Marriages." *Journal of Marriage and the Family* 60, no. 1 (1998): 23–40.

Rubin, L. *Families on the Fault Line.* New York: Harper-Collins, 1994.

Rubin, L. *Intimate Strangers: Women and Men Together.* New York: Harper-Perennial, 1984.

Rubin, L. *Worlds of Pain: Life in the Working Class Family.* New York: Basic Books, 1976.

Rubin, Z. "The Love Research." In *Family in Transition,* 3rd ed., edited by A. Skolnick and J. Skolnick. Boston: Little, Brown, 1980.

Schwartz, P. *Peer Marriage.* New York: Free Press, 1994.

Simons, R., J. Beaman, D. Conger, and W. Chao. "Childhood Experience, Conceptions of Parenting, and Attitudes of Spouse as Determinants of Parental Behavior." *Journal of Marriage and the Family* 55, no. 1 (1993): 91–106.

Sorenson, "A National Profile of Non-Resident Fathers and Their Ability to Pay Child Support." *Journal of Marriage and the Family* 59, no. 4 (1997): 785–797.

Thompson, L., and A. Walker. "Women and Men in Marriage, Work, and Parenthood." *Journal of Marriage and the Family* 51, no. 4 (1989): 845–872.

Umberson, D. "Relationships with Children: Explaining Parents' Psychological Well-Being." *Journal of Marriage and the Family* 51 (1989): 999–1012.

Weitzman, L. *The Marriage Contract: A Guide to Living with Lovers and Spouses.* New York: The Free Press, 1981.

FOR ADDITIONAL READING

Given the breadth of topics related to men's experiences in families, there are countless directions one could pursue on *InfoTrac College Edition.* Peter J. Smith and Roderic Beaujot's article, "Men's Orientation Toward Marriage and Family Roles," from the *Journal of Comparative Family Studies* (Summer 1999, vol. 30, no. 3) illustrates the diversity found in men's orientations regarding marriage and family roles, among a sample of Canadian men. Men's attitudes ranged from traditional to liberal, regarding the extent to which women prefer childcare over work. Most men are somewhere in between traditional and modern, and, among them, there is uneven recognition of the various pressures and contradictions that face women more than men.

Those interested more in behavior than beliefs will be interested in two articles that consider variation in men's participation in housework. "Gender Equality and Participation in Housework: A Cross-National Perspective," by Janeen Baxter (*Journal of Comparative Family Studies,* Autumn 1997, vol. 28, no. 3) compares men's and women's participation in housework in the United States, Sweden, Norway, Canada, and Australia. As Baxter reveals, neither the inequality between women and men, nor men's tendencies to report doing more than women credit men with doing, is unique to American families. In "Division of Household Labor Among Black Couples and White Couples" (*Social Forces,* September 1997, vol. 76, no. 1), Terri L. Orbuch and Sandra Eyster look at race differences in who does what in American households. Although somewhat technical in terms of their statistical analysis, Orbuch and Eyster show differences both in terms of what men contribute in households and in what husbands' participation means to wives. David Popenoe's "A World Without Fathers" (The *Wilson Quarterly,* Spring 1996, vol. 20, no. 2) reviews evidence of father absence and considers some of its more problematic consequences for women, children, and the wider society. Popenoe is representative of and influential among social scientists who bemoan men's underinvolvement in family life, and who warn of the disasters resulting from father absence. Finally, "Unfaithfully Yours: A Husband's Tale" (*Men's Fitness,* November 1998, vol. 14, no. 11) by a pseudonymous author, Sean Robinson, illustrates the trauma resulting from infidelity. Although more men than women cheat on their spouses, "Robinson's" tale reveals what cheating feels like by the cheated upon spouse. It is worth considering what aspects of his reaction are uniquely masculine or whether they are more generally "human" reactions.

WADSWORTH VIRTUAL SOCIETY RESOURCE CENTER

http://sociology.wadsworth.com

Visit *Virtual Society* to obtain current updates in the field, surfing tips, career information, and more.

INFOTRAC COLLEGE EDITION: EXERCISE

Go to the *InfoTrac College Edition* web site to find further readings on the topics discussed in this chapter.

16

Couples Watching Television: Gender, Power, and the Remote Control

ALEXIS J. WALKER

Five years ago, my parents bought a second television set because my mother refused to watch television with my father any longer. "I can't stand the way he flips through the channels," she said. Note that my father actually has the use of the new television, and my mother has been relegated to the den with the older model. Nevertheless, mother now has her own set, and conflicts about the remote control device have been reduced considerably.

Several years ago, journalist Ellen Goodman (1993, p. 181) published an essay in which she described the RCD as "the most reactionary implement currently used to undermine equality in modern marriage." Because family scholars rarely study such mundane, everyday life experience, there is little research available to confirm Goodman's sentiments or to assess the prevalence of solutions to television-watching disagreements such as that employed by my parents. RCD use, however, presents a challenging arena in which to examine gender and relationship issues in the experience of daily living.

Over the past 20 years, feminist scholars have shown that ordinary, routine, run-of-the-mill activities that take place inside homes every day bear an uncanny resemblance to the social structure. For example, the distribution of household labor and of child care is gendered in the same way that paid work is gendered: The more boring and less desirable tasks are disproportionately performed by women, and status has a way of reducing men's, but not women's, participation in these tasks. (See Thompson & Walker, 1989, for a review.) Examining television-watching behavior is a way to extend the feminist analysis to couples' leisure.

Despite the fact that television watching is the dominant recreational activity in the United States today (Robinson, 1990), there is little research on this topic in the family studies literature. Indeed, there is little family research on leisure at all (but see Crawford, Geoffrey, & Crouter, 1986; Hill, 1988; Holman & Jacquart, 1988), although scholars often mention that employed wives and mothers have very little of it (e.g., Coverman & Sheley, 1986; Hochschild, 1989; Mederer, 1993). Recently, Firestone and Shelton (1994), using data from the 1981 Study of Time Use, confirmed that married women have less overall leisure time than married men. They also demonstrated gender-divergent patterns in the connection between paid work and both domestic (in-home) and out-of-home leisure time. Women who are employed have less out-of-home leisure time than men do, but men who are employed have less domestic leisure time than employed women. Specifically, they found that employment hours do not affect the amount of leisure time that married women have at home. To explain this surprising finding, Firestone and Shelton speculated that leisure at home appears to be the same for employed and nonemployed wives because leisure is compatible with household chores and with child care. In other words, women combine family work with leisure activities. For example, they iron while watching television.

Although there is little research on leisure in family studies, there is considerable literature on gender and leisure in the field of leisure studies. For example, Henderson (1990), predating Firestone and Shelton (1994), described women's leisure as fragmented because much of it takes place at home where it is mingled with domestic labor. In comparison with men, women say that leisure is less of a priority and that they do not deserve it. Some activities that are defined as leisure pursuits, such as family picnics, are actually occasions for women's work, making leisure a possible source of internal conflict for women (Shank, 1986; Shaw, 1985). In fact, Henderson (1994) called for a deconstruction of leisure because the term embodies contradictions for women, contradictions that may be evident particularly in family leisure (Shaw, 1991).

To develop a way to measure couples' television-watching behavior, I sought guidance from the empirical research on television watching, which also is considerable. Here, studies describe various types of RCD use, sometimes referred to as (a) grazing (sometimes called surfing)—progressing through three or more channels with no more than 5 seconds on any one channel for the purpose of seeing what is available; (b) zapping—switching channels to avoid something, usually a commercial; and (c) zipping—fast-forwarding during a prerecorded program, mostly to avoid commercials (Cornwell et al., 1993; Walker & Bellamy, 1991).

Observational, survey, in-depth interview, and ethnographic data from communications researchers using a wide variety of sampling strategies revealed that, when heterosexual families with children watch television together, fathers dominate in program selection and in the use of the RCD. Sons are active as well, using the RCD more than either their mothers or their sisters. That gender differences are smaller among younger persons, however, suggests a potential for women and men to be more similar in their remote control behavior in the future, when the RCD-using youth of today are adults (Copeland & Schweitzer, 1993; Cornwell et al., 1993; Eastman & Newton, 1995; Heeter & Greenberg, 1985; Krendl, Troiano, Dawson, & Clark, 1993; Lindlof, Shatzer, & Wilkinson, 1988; Morley, 1988; Perse & Ferguson, 1993). Note that the dominance in RCD use by men and boys in a family context is not evident when individuals are observed alone. In an experimental study, women were no less likely to use the RCD than were men (Bryant & Rockwell, 1993). The authors concluded that a social context is necessary to produce such gendered behavior.

Morley (1988) described fathers as using the RCD for unnegotiated channel switching—that is, changing channels when they want to—without explaining their behavior to or consulting other television watchers. Unemployed fathers, by the way, are less likely than employed fathers to use the RCD in this way, suggesting a possible connection between RCD use and the use of legitimate power—that is, power derived from and supported by societal norms and values (e.g., Farrington & Chertok, 1993).

Why so much channel switching? Men say they change channels to avoid commercials, to see if something better is on, to see what they are missing, to watch news reports, because they like variety, to avoid looking up the printed listings, to annoy others, and, my favorite reason, to watch more than one show at the same time (Perse & Ferguson, 1993). By contrast, women say they change channels to watch a specific program. A frightening finding is that the children of heavy RCD users are also heavy users, suggesting that parents pass on this behavior and that we can anticipate more grazing in the future (Heeter & Greenberg, 1985).

Copeland and Schweitzer (1993) concluded: "Men have usually been viewed as the persons who control program selection, and domination of the remote control seems to make visually explicit what may have previously been implicit" (p. 165). This notion of power, clearly stated in the language of the communications researchers, is missing from the family research on leisure. In their studies, however, the communications researchers have focused almost exclusively on parents watching television with their children. Rarely have they studied couples watching television. Furthermore, students of communication rarely have combined data about television watching and RCD use with questions of primary interest to those of us who study close relationships among adults. For example, are there any ways in which watching television with your partner is frustrating? Would you change the way that you watch television with your partner if you could? How do you influence your partner to watch something that *you* want to watch?

These questions address issues of power in relationships described by Aafke Komter (1989). She demonstrated that power is evident not only in the direct, observable resolution of conflict between partners, but also in covert or nonobservable events that reflect structural inequality. Direct expression of power reflects manifest power; covert expression reflects latent power. Examples of latent power are the ability to prevent issues from being raised, the anticipation of the desires of the more powerful partner, and resignation to an undesirable situation due to the fear of a negative reaction from the more powerful partner or worry that change might harm the relationship in some way (see also Huston, 1983; McDonald, 1980). In addition to other domains (e.g., family labor and sexual interaction), Komter included leisure in her study, but she focused only on hobbies and interaction with friends.

I chose to wed the focus of communications researchers on television-watching behavior with Komter's (1989) approach to studying power. I expected that heterosexual couples would "do gender" (West & Zimmerman, 1987) even in such a mundane activity as joint television watching. I anticipated that the creation and affirmation of gender would be evident in the (manifest and latent) power men have over their women partners in the domain of leisure activity. Furthermore, I sought to confirm the importance of gender in partner interaction by examining joint television-watching behavior in lesbian and gay couples as well (see Kollock, Blumstein, & Schwartz, 1985).

METHOD

Participants

The sample for this study was characterized by its diversity. Participants were recruited primarily by students enrolled in an upper-division undergraduate course on gender and family relationships. In recruiting respondent pairs, students worked in groups of four to maximize diversity. Couples were chosen so that each group of four students would select a diverse set of pairs. All respondents were in a romantic (i.e., heterosexual married, heterosexual cohabiting, or cohabiting gay or lesbian) relationship in which both individuals were at least 18 years old. All couples had been living together for at least 1 year and had a television set with an RCD. Within each group, however, participants included (a) couples varying in relationship length, from shorter (1 year) to longer (15 years or more); (b) a lesbian or gay couple; (c) at least one married couple; (d) at least one heterosexual, cohabiting, or unmarried couple; (e) couples with and without children; (f) at least one couple in which at least one partner was Asian American, African American, Latino, or of mixed race; and (g) couples in which both partners were employed and couples in which only one partner was employed. Fourteen percent (n = 5 pairs) of the 36 couples (72 individuals) were gay or lesbian. Here, and for much of this report, I focus attention on the 31 heterosexual pairs.

Women and men in these heterosexual couples did not differ significantly on sociodemographic characteristics. (Table 16.1 shows these characteristics for all couples.) The typical respondent was 34 years old (SD = 12.69) and had completed 2 years of academic work beyond high school. Most (77%, n = 48) were white, although nearly one quarter were either African American, Hispanic, or of mixed race. Nearly three quarters (74%, n = 48) were married; one quarter was cohabiting. On average, their relationships had been in existence for 10 years. Most (77%, n = 48) respondents were employed, and just over 30% (n = 19) were students; only 16% (n = 10) of the

Table 16.1 Characteristics of Respondents in the Sample

	Heterosexual Partners (N = 62)		Lesbian or Gay Partners (N = 10)	
Characteristic	*M* or %	*SD* or *n*	*M* or %	*SD* or *n*
Age	34.1	12.69	36.4	7.52
Education level[a]	3.1	1.70	4.1	1.52
Race (% white)	77.4	48	100.0	10
Relationship status (%)				
Cohabiting	25.8	16	100.0	10
First marriage	58.1	36		
Previous marriage	16.1	10	10.0	1
Years in relationship	10.2	11.22	8.1	3.51
Children at home (%)	32.3	20	0.0	0
Employed (%)	77.4	48	90.0	9
Employment hours[b]	2.8	1.41	3.1	0.38
Income[c]	2.2	1.26	2.4	1.13

Note: Within heterosexual couples, there were no significant differences by gender for any of these variables.

[a]Education measured from 0 (*less than high school*) to 6 (graduate degree).

[b]Employment hours measured from 0 (*0 to 10 hours per week*) to 4 (*more than 40 hours per week*).

[c]Income measured from 1 (*less than $20,000*) to 5 (*more than $80,000*).

sample, however, was nonemployed, nonretired students. Heterosexual respondents represented three income groups. Just over one third earned less than $20,000 annually, one third reported an annual household income between $20,000 and $39,999, and just under one third earned $40,000 or more. One third had children living at home.

Measures

A semistructured interview was administered to each member of the couple. In addition to sociodemographic questions, respondents were asked about the number and location of television sets and videocassette recorders in the home, the frequency with which they and their partners watched television, and other activities they engage in while watching television. They were asked about use of the RCD, in general, while watching with their partner and during the program most recently watched with the partner. They were also asked if their most recent experience was typical of their joint television watching. These questions were quantitative in nature and are similar to the types of questions asked of participants by communications researchers. Additional single-item, quantitative questions were derived from the family studies literature; that is, questions about relationship happiness, happiness with the way things are regarding watching television with the partner, and how much partners enjoy the time they spend together.

Other questions focused on issues of power à la Komter (1989). These questions were open-ended and concerned changed expectations about watching television with the partner over the history of the relationship, how the couple decides on a program to watch together, how partners get each other to watch programs that they want to watch, and their frustrations with watching television with their partner. Respondents were asked if they would like to change anything about the way they watch television together, if they thought they would be successful at making these changes, whether it would be worth it for them to make the changes, and how their partner would react to the changes. In addition, any changes they had already made in their joint television-watching behavior were described.

Procedures

A coin toss was used to determine which partner to interview first. Partners were interviewed separately, usually in their own homes, by trained student interviewers. Interviews were audiotaped and transcribed. SAS was used to analyze the quantitative data, and transcriptions were read and reread for analysis of the open-ended data.

RESULTS

On average, the heterosexual couples had 1.81 television sets ($SD = 0.99$), but some had only 1, and a few had as many as 5. They had 1.30 videocassette recorders ($SD = 0.53$), with a range from 1 to 3. They also had 1.30 RCDs ($SD = 0.68$), with a range from 1 to 3. The typical home had basic cable television (with no extra channels) or a satellite dish.

These individuals watched television quite often—on average almost daily for nearly 3 hours per day ($M = 2.77$, $SD = 1.48$). During the week prior to the interview, they had, on average, watched television together on 4.87 days ($SD = 2.09$). Nearly all, 94% ($n = 29$), of the women and 87% ($n = 27$) of the men reported that, regarding watching television with their partners, they were happy with the way things are. Yet two thirds of the women and three fifths of the men reported that there were things about their joint television watching that were frustrating to them. The interview transcripts were revealing about these frustrations. Women complained about their partners' grazing behavior, both during a show and when they first turned on the television set. One woman in a 3-year cohabiting relationship said:

> I would say that the only thing that's frustrating for me is when we first turn on the TV and he just flips through the channels. It drives me crazy because you can't tell what's on, because he just goes through and goes through and goes through.

Another woman, in the 17th year of a first marriage, reported, "[I get frustrated] only if I get hooked into one show and then he flips it to another one. As soon as I get hooked into something else, he flips it to something else." Such reports from women were common. A married man spontaneously agreed: "We don't watch TV a lot together; I would rather do other activities with my wife. Channel switching wasn't a problem until . . . the remote control." Indeed, many men indicated that their women partners were bothered by this behavior.

In contrast, men reported being frustrated with the quality of the programming or the circumstances of watching, rather than with RCD activity. For example, one husband said, "I wish we had a VCR. . . . I wish we had one of those TVs where you could watch two things on the screen at once." Another said, "It's sort of frustrating when I want to watch something she doesn't, and she goes into the other room and gets sort of pouty about it." A third reported, "No, [nothing is frustrating], but she does talk a little."

I looked specifically at the RCD; for example, where is the RCD usually located? Men were more likely than women to say that they usually hold the RCD or have it near them, $\chi^2(1, n = 62) = 7.38$, $p < .01$, and they were less likely than women to say that their partner usually holds it or has it near them, $\chi^2(1, n = 62) = 14.47$, $p < .001$. In half the couples ($n = 16$), according to both women and men, men have the RCD. In over 80% ($n = 25$) of the couples, according to both the women and the men, the women do not have control of it solely. According to 16% ($n = 5$) of the women and 10% ($n = 3$) of the men, women control the remote. In roughly one third of the couples, the RCD is in a neutral location, or both take turns holding it. The pattern was the same when respondents were asked about the RCD's location during the most recent television show that they watched together. The RCD was more likely to be located near men than near women or near both members of the couple, $\chi^2(2, n = 59)$

= 13.12, $p < .001$. The transcriptions support this general pattern of RCD location, as well. A husband reported, "I usually use the remote because I know how to use it, and it usually sits right in front of me while I am on the couch." A young married woman said, "I had the baby [the RCD] this time. This was a rare occasion." Roger (all names are pseudonyms), a married man, reported:

> I frequently have the remote at my side. I won't change the channel until we are ready to look for something else. If there is someone who wants to change the channel at a commercial, it will be Sally [his wife]. I will hand the remote to her, and she will change it to another favorite show, and then back. And that is very typical.

Sally agreed. The last time they watched television together, the RCD was in "Roger's pocket! Either in his shirt pocket or bathrobe pocket." A young, married man reported:

> I don't hold [the RCD], but I pretty much have control of it, and if I don't care what's on, then I let her have it. Sometimes we fight over it. Not like fight, but, I mean, it's like, "You always have the remote control."

Women were significantly more likely than men to say that RCD use was frustrating to them, $\chi^2(1, n = 62) = 8.42, p < .01$. Only 10% ($n = 3$) of the men, but 42% ($n = 13$) of the women evidenced such frustration. Furthermore, women ($M = 0.61$, $SD = 0.79$) reported that significantly more RCD behaviors were frustrating to them than men reported ($M = 0.15$, $SD = 0.37$) $t(48) = -2.70, p < .01$. Yet 30% ($n = 9$) of the women in the sample and 16% ($n = 5$) of the frustrated men reported that they would like to change how the RCD is used during their joint television watching. This difference was not significant.

What was frustrating about RCD use? Respondents reported being frustrated by the amount of grazing, the speed of grazing, heavy use of the RCD, and the partner taking too long to go back to a channel after switching from it during a commercial. A few respondents actually indicated concern about their own frequent RCD use. Women and men, however, reported similar percentages of other television-watching behaviors that were frustrating (e.g., too much time watching television; bothersome behaviors of the partner, such as making fun of a program); 58% ($n = 18$) of the women and 48% ($n = 15$) of the men were frustrated by these other behaviors.

Thus far, I have shown that men control the RCD more than women and that women are more frustrated by RCD behaviors than men are. I also asked about the other activities engaged in while watching television. Two types of activities were mentioned: family work (e.g., child care, cooking, laundry) and pleasurable activities, such as doing nothing (i.e., relaxing), eating, drinking, playing computer games, and so on. When activities within each type were summed, the findings were revealing. When asked about their most recent joint television-watching episode, men ($M = 1.00$, $SD = 0.52$) responded that they were significantly more likely than women ($M = 0.74$, $SD = 0.44$) to engage in pleasurable behaviors while watching television, $t(62) = 2.11, p < .04$. Women ($M = 0.36$, $SD = 0.71$) were not more likely than men ($M = 0.13$, $SD = 0.34$) to do family work while they watched television, although the data suggested a trend in this regard, $t(62) = -1.60, p < .12$. The small proportion of households with children (32%, $n = 10$) may have contributed to this finding. At least 80% of both women and men described this most recent experience as typical of their joint television watching and of their RCD use. Interestingly, women ($M = 2.84$, $SD = 0.74$) tended more than men ($M = 3.16$, $SD = 0.69$) to describe the particular show they watched as their partner's preference rather than their own, $t(60) = 1.78, p < .08$.

Recall that 30% of the women said they would change the use of the RCD during their joint television watching if they could. Only half as many men would make such a change. The open-ended data support these results. For example, a young married woman described her technique of standing in front of the television to interrupt the signal from the RCD. Another young married woman said that her partner used the RCD to watch more than one program at a time. "I should get him one of those TVs with all the little windows so he can watch them all," she said sarcastically. A middle-aged married woman said that she would like to change their television watching so that she would have "control of the remote for half of our viewing time." Of those who would like to make any changes in their television watching, one in five women expected that they would *not* be successful.

Men typically admitted their heavier RCD use. For example, a middle-aged married man said that he switched channels to avoid commercials. "I'm the guilty party," he said. "My [family members] would leave it there and watch the commercial. I just change it because I'd rather not be insulted by commercials."

One of the most provocative questions asked of respondents was "How do you get your partner to watch a show that you want to watch?" The results were enlightening. A cohabiting woman said, "I tell him that would be a good one to watch, and he says, 'No,' and keeps changing [channels]. I whine, and then usually I don't get [my way]." A middle-aged married woman said:

> Let me think here, when does that occur? [Laughing.] If I really want to watch, I'll say, "I want to watch this one." . . . I'll say, "Come in here and watch this" if he's not in the room, but pretty much we watch the same things a lot, whether or not that's because I let him. He, a lot of the time, turns it on, and I'll come in and join him. But, if it's something I really want to watch, I'll say, "Don't flip the channel; I want to see this."

In contrast, a young married man said that he gets his partner to watch a show he wants to watch in this way:

> I just sneak the remote away from her if she has it, or, if I'm there first, then. . . . I mean, if there's sports on, that's usually what we watch unless there's something else on. I mean, usually if there's . . . some kind of sports game on, we usually watch that, but unless there's . . . . another show on that, you know, she can talk me into, deter my interest, or something. . . .

When asked how his partner gets him to watch something that she wants to watch, he reported:

> Oh, I guess, if there's not anything that I'm . . . real big on watching then I'll let her choose, or if she, you know, she's interested in something. . . . A bunch of times, we watch TV, and it's like, well, we'll go back, and, well, that's kind interesting, we go back and forth.

His wife agreed:

> I usually don't have to beg him. I don't know. [Laughing.] I tie him down, and say, "You're watching this." I don't know. He usually just comes over, and if it's not what he wants, then he'll take the remote and try to find sports.

In other words, this couple watches sports when it is available. If there is no sports program on television and if the husband does not have something else he really wants to watch, then the wife may choose a show, but her husband will be looking for a sports program while her show is on, or at least he will go back and forth between her show and others.

A woman who has been married for 18 years was deliberate in her efforts to watch a particular show:

> I usually start a couple of days ahead of time when I see them advertised, and it is something that I am going to want to watch. I tell him to "get prepared!" I have to be relatively adamant about it. When the time comes up, I have to remind him ahead of time that I told him earlier that I want to watch the program.

When her husband wants to watch a program, however, she said, "He just watches what he wants. He doesn't ask." Finally, a married man reported, "I just say I want to watch something, and if she wants to watch something really bad, I will let her watch what she wants to watch." Ultimately, the authority is his.

The data are much the same when people report on changes they would like to see in the way they watch television together. One man who has been cohabiting with his partner for one year said, "I should probably let her 'drive' sometimes, but [it] would bug me too much not to be able to do it." A woman who has been married for 37 years painted a brighter picture. When asked, "How do you feel about watching TV with your husband? Are you happy with the way things are?" she responded:

> Yes, right now. But see, without the VCR we'd be in trouble because I just tape anything I want to see. Without that, there'd be more conflict. . . . Buying a second TV has changed the way we watch TV. It's made it easier—less stress, less conflict.

A young married man also was more positive. When asked, "Have you changed the way you watch television together?" he replied, "We take turns watching our programs, and I let her hold the remote during her programs."

Earlier, I mentioned that 14% (n = 5) of the couples in this study were gay or lesbian. In these couples, too, one partner usually is more likely than the other to use the RCD. In a gay male couple, one nearly always used the RCD, and the other almost never used it. When the RCD user was asked why they have this typical pattern, he responded: "Why? I don't know. I just like using the remote. I think I'm better at it than he is." In answer to a question regarding whether he used the RCD at all, his partner indicated: "He doesn't let me." In a second gay male couple, one partner again was far more likely to use the remote than the other partner, but both reported using controlling strategies to get their partner to watch a show they wanted to watch. Greg said, "I just tell him I want to watch it, and we do." Rob said, "I just turn it on, and that is what we watch." When asked, "How does your partner get you to watch a program he wants to watch?" Greg replied: "I usually don't watch programs I don't want to watch. If he asked me to watch it with him for a purpose, I would."

In contrast, one partner in a lesbian pair reported, "If we are both here, we try to make sure it's something that we both like." In fact, this couple limited their television viewing to avoid potential problems resulting from their different styles of RCD use. They also made it a practice to talk to each other during the commercials, in part, so that the one partner who tended to do so, would not graze. A second lesbian pair reported similar behavior. When asked, "Think back to the beginning of your relationship with [your partner]. Have your expectations about watching television with her changed over time?" she responded:

> In the beginning, . . . TV watching was something we could do when we didn't know each other very well yet. You know, it was kind of like a sort of neutral or a little bit less personal activity that we could sit and watch TV together as a shared activity. And it's still a shared activity. . . . We don't use it to tune each other out, and if someone wants to talk, we just click the mute button or turn it off.

Becky's partner, Mary, used the RCD much more often than Becky did. According to Becky, however, when Mary grazes, "she's perfectly willing, if I say, 'This looks really good,' she'll stop. She doesn't dominate that way." In fact, when Mary grazes, "she'll just say, 'Is this bothering you?'" Mary agreed that she was the one who usually held the RCD, but that they shared, too. "If Becky has a show she really likes, then I give her the remote so I'm sure I don't play with the TV while she's watching her show." Mary does not "let" her partner hold the RCD; she asks her to hold it to keep her own behavior in check. Indeed, Mary's frustration with their joint television watching comes from her own behavior: "Well, I feel self-conscious about how much I change the channels because I know that she doesn't like to change as often or as fast as I do." Although based on a very modest sample, these findings are intriguing and illustrate how couples successfully develop and maintain egalitarian or peer relationships.

DISCUSSION

These data confirm that for women in heterosexual pairs leisure is a source of conflict—conflict between their own enjoyment and the enjoyment of their partners (Shank, 1986; Shaw, 1985, 1991). The data expose the contradictions between the goal and the reality of leisure for women. Support also comes from the findings that men, more than women, combine other pleasurable pursuits with television watching. Others (Coverman & Sheley, 1986; Firestone & Shelton, 1994; Hochschild, 1989; Mederer, 1993) have shown that women more than men dovetail family labor with their leisure activity.

The data also support previous work suggesting that, when heterosexual couples watch television together, men dominate in program selection and in the use of the RCD (Copeland & Schweitzer, 1993; Cornwell et al., 1993; Eastman & Newton, 1995; Heeter & Greenberg, 1985; Krendl et al., 1993; Lindlof et al., 1988; Morley, 1988; Perse & Ferguson, 1993). Indeed, unnegotiated channel switching by male partners was a frequent occurrence in this sample. Men use the RCD to avoid commercials, to watch more than one show at a time, and to check what else is on (Perse & Ferguson, 1993). And they do so even when their partners are frustrated by these behaviors.

The data reveal that men have power over women in heterosexual relationships (Komter, 1989). Men are more likely than women to watch what they want on television and to do so without considering their partner's wishes. Men control the RCD, which gives them the means to watch what they want, when they want, in the way that they want. Men also persist in RCD use that is frustrating to their women partners. These are examples of manifest power. Men make overt attempts to get their way and are successful at doing so. Men's power is evident, as well, in the lesser power of their women partners. For example, women struggle to get their male partners to watch a program they want to watch and are less able than men to do so. Furthermore, women watch a preferred show on a different television set or videotape it so that they can watch it later. These options do not prevent a husband or male partner from watching a show that he wants to watch when he wants to watch it.

Men's latent power over women is evident as well. Even though women rarely control the RCD, fewer than half report that RCD use is frustrating to them, and only 30% say they would like to change their partner's RCD behavior. According to Komter (1989), resignation to the way things are is evidence of latent power. Another illustration of the effect of latent power is anticipation of a negative reaction. Only four women feel they would be successful if they attempted to change the way their partners use the RCD. In the heterosexual sample, women seem less able than men to raise issues of concern to them, they anticipate the struggles they will encounter when and if their own preferences are made known, and they predict a negative reaction to their wishes from their male partners.

Confirmation of men's latent power over their women partners also was demonstrated by a series of auxiliary analyses. I was unable to explain the dependent variables of respondent's relationship happiness or respondent's enjoyment of the time the couple spends together with independent variables such as frustration with remote control use, dominance of the remote control, or desire to change frustrating remote control use. As Komter (1989) suggested, both lesser power and resignation on the part of women contribute to the appearance of balance in these pairs.

Joint television watching in heterosexual couples is hardly an egalitarian experience. As is true for my mother, some women use a second television or a videocassette recorder to level the playing field (i.e., so that they are able to watch the shows they want to watch). A second television set, however, reduces joint leisure time among those couples who can afford it, and a VCR means that a woman may have to wait to watch her show. Even these solutions to conflict around joint television watching demonstrate that couples watching television are not simple passive couch potatoes. They are doing gender, that is, acting in ways consistent with social structures and helping to create and maintain them at the same time.

Everyday couple interaction is hardly mundane and run-of-the-mill. It is a systematic recreation and reinforcement of social patterns. Couples' leisure behavior is gendered in the same way that household labor is gendered: Social status enhances men's leisure activity relative to women's. Thus, leisure actively has gendered meanings (Ferree, 1990). Through it, women and men are creating and affirming themselves and each other as separate and unequal (Ferree, 1990; Thompson, 1993). In other words, leisure activity is both an occasion for relaxation and an occasion for doing gender (Fenstermaker, West, & Zimmerman, 1991; Shaw, 1991).

As Osmond and Thorne (1993) point out, "Gender relations are basically power relations" (p. 593). Because the power of men in families is legitimate—that is, backed by structural and cultural supports—it constrains the less powerful to act to maintain the social order and the stability of their relationship (Farrington & Chertok, 1993). Few women make demands of their heterosexual partners so that their patterns of television watching change. Instead, they say they are "happy" with their joint television watching. This same pattern is evident when we examine family labor. Most women describe the objectively uneven distribution of household work as fair (Thompson, 1991).

Hochschild (1989) argued that women give up leisure as an indirect strategy to bolster a myth of equality. Rather than resenting her male partner's leisure time, a woman uses the time when he is pursuing his own leisure or interests to engage in what she describes as her interests: housework and child care. Overall, she defines their level of involvement at home as equal, a view that can be sustained only if she ignores her own lack of leisure time, as well as the amount of leisure time her partner has. Hochschild also suggested that a woman sees her male partner's leisure time as more valuable than her own because she feels that more of his identity and time than hers are committed to paid work. She concludes, therefore, that he deserves extra relaxation. In Hochschild's view, and in Komter's (1989), women cannot afford to feel resentment in their close relationships.

In a review of the literature, Szinovacz (1987) wrote that there were few studies of how people in families exercise power. She argued that such studies are needed, as are studies on strategies of resisting power. The data reported here suggest that the exercise of power around couples' television-watching behavior can be overt and relentless. Men's strategies to control the content and style of viewing are ways in which they do gender. Women's resistance strategies (e.g., getting a second television set, using the VCR) are also ways of doing gender. They do little to upset the intracouple power dynamics. Indeed, most women whose male partners are excessive grazers do not describe resistance strategies at all. Instead, they maintain the status quo (Komter, 1989).

Of interest is that in lesbian and gay couples one person was more likely to be the heavier RCD user, as well. Yet these couples had some unique patterns. The behavior of the lesbian couples, in particular, is suggestive for those of us wishing to establish and maintain egalitarian partnerships. One lesbian woman demonstrated a solution to the conflict between partners when one is distressed by the other's RCD behavior. Asked, "Is there anything else you'd like to tell us?" she responded:

> Well, I think that the most important thing for me is to remember to be sensitive to the fact that she doesn't have the same tastes as me, and I try to think about that. And, if she mentions that she likes something, then I ask her before I change the channel if she's done watching it, or if she's not interested, if I could change the channel.

In this act, she elevates the importance of her partner's wishes to the level of her own. She demonstrates the consideration that her partner desires and deserves (Hochschild, 1989; Thompson, 1991). When asked if it would be important for them to make changes in the way they watch television together, her partner expressed insight into her own behavior. Mary, the RCD user, likes to "veg out" and watch TV, but Becky likes to:

> pretend I'm not going to watch, I'm going to get a magazine or the newspaper, . . . or I'll bring some desk paper work over to do. . . . I think, well, I'll just sit there in the living room while Mary watches TV. I'll work on our bills or something. . . . Then, what happens is I'll look around

> and think that looks kind of interesting. Although usually by the time I've looked around, she's changed the channel. . . . I think what happens is that she's more up front about saying, "Hey, I'm going to veg out here and watch some TV," and I pretend I'm going to do more worthwhile things, and I end up just watching TV anyway.

Perhaps these two women, with their honesty to themselves, their sensitivity to each other, and their concern about the ways in which their own behavior is or could be a problem in their relationship, are doing gender, too. They are concerned with the relationship, rather than with getting their own way. This is how women are said to make connection and to demonstrate care, to give what Hochschild (1989) described as a gift of gratitude. Using these strategies, they maximize joint enjoyment of leisure and minimize power imbalances. Rather than reproducing structural hierarchies, they create a bond of equality and provide a different course for the resolution of inherent conflict within couples.

The results from this study are hardly definitive. They are based on a small, volunteer sample, albeit one sufficiently diverse to include different types of close, romantic relationships. Additionally, the very small number of lesbian and gay couples suggests a need to exercise caution when generalizing from these findings. Further study with larger, representative samples will be required to extend these findings beyond the couples interviewed here.

Nevertheless, the patterns I identified are similar to those found in other studies of television watching in families and of the intersection of gender and power in close relationships. Mundane activities are important for understanding the intersection of gender and power in close relationships. Indeed, as Lull (1988) noted:

> Television is not only a technological medium that transmits bits of information from impersonal institutions to anonymous audiences, [but] it is a social medium, too—a means by which audience members communicate and construct strategies to achieve a wide range of personal and social objectives. (p. 258)

Others (Morley, 1988; Spigel, 1992) have pointed out that the way men engage with television programming and women watch more distractedly are illustrations of cultural power. Daytime television programming in the 1950s, for example, was designed to be repetitive and fragmented to facilitate joint housework and television watching for women (Spigel, 1992), thus helping to create and reinforce the view that leisure at home is problematic for women. The availability of the RCD does not change the fact that women's leisure is fragmented.

Recently, I toured an area of Portland, Oregon, billed as "The Street of Dreams," where there are a half dozen homes costing nearly a million dollars apiece. Each year, such homes are opened temporarily to members of a curious public who will never be able to afford them. Inside one was a theater room with three television sets mounted side-by-side along the back wall. A moment after we arrived, a middle-aged heterosexual couple entered the room. The woman smiled and said to the man, "Look, Dan, you could get rid of the remote!" If three television sets in one room is a solution to the problem of being able to watch only one show at a time, gendered struggles inherent in such mundane, everyday activity as watching television are destined to continue. They will do so until women and men are equal in both their microlevel interactions and in the broader social structure. . . .

REFERENCES

Bryant, J., & Rockwell, S. C. (1993). Remote control devices in television program selection: Experimental evidence. In J. R. Walker & R. V. Bellamy, Jr. (Eds.), *The remote control in the new age of television* (pp. 73–85). Westport, CT: Praeger.

Copeland, G. A., & Schweitzer, K. (1993). Domination of the remote control during family viewing. In J. R. Walker & R. V. Bellamy, Jr. (Eds.), *The remote control in the new age of television* (pp. 155–168). Westport, CT: Praeger.

Cornwell, N. C., Everett, S., Everett, S. E., Moriarty, S., Russomanno, J. A., Tracey, M., & Trager, R. (1993). Measuring RCD use: Method matters. In J. R. Walker & R.V. Bellamy, Jr. (Eds.), *The remote control in the new age of television* (pp. 43–55). Westport, CT: Praeger.

Coverman, S., & Sheley, J. F. (1986). Change in men's housework and child-care time, 1965–1975. *Journal of Marriage and the Family, 48,* 413–422.

Crawford, D.W., Geoffrey, G., & Crouter, A. C. (1986). The stability of leisure preferences. *Journal of Leisure Research, 18,* 96–115.

Eastman, S.T., & Newton, G. D. (1995). Delineating grazing: Observations of remote control use. *Journal of Communication, 45,* 77–95.

Farrington, K., & Chertok, E. (1993). Social conflict theories of the family. In P. G. Boss, W. J. Doherty, R. LaRossa, W. R. Schumm, & S. K. Steinmetz (Eds.), *Sourcebook of family theories and methods: A contextual approach* (pp. 357–381). New York: Plenum.

Fenstermaker, S., West, C., & Zimmerman, D. H. (1991). Gender inequality: New conceptual terrain. In R. L. Blumberg (Ed.), *Gender, family, and economy: The triple overlap* (pp. 289–307). Newbury Park, CA: Sage.

Ferree, M. M. (1990). Beyond separate spheres: Feminism and family research. *Journal of Marriage and the Family, 52,* 866–884.

Firestone, J., & Shelton, B. A. (1994). A comparison of women's and men's leisure time: Subtle effects of the double day. *Leisure Sciences, 16,* 45–60.

Goodman, E. (1993). Click. In *Value judgments* (pp. 180–182). New York: Farrar Strauss & Giroux.

Heeter, C., & Greenberg, B. S. (1985). Profiling the zappers. *Journal of Advertising Research, 25,* 15–19.

Henderson, K. A. (1990). The meaning of leisure for women: An integrative review of the research. *Journal of Leisure Research, 22,* 228–243.

Henderson, K. A. (1994). Perspectives on analyzing gender, women, and leisure. *Journal of Leisure Research, 26,* 119–137.

Hill, M. S. (1988). Marital stability and spouses' shared time. *Journal of Family Issues, 9,* 427–451.

Hochschild, A. (with Machung, A.). (1989). *The second shift: Working parents and the revolution at home.* New York: Viking.

Holman, T. B., & Jacquart, M. (1988). Leisure-activity patterns and marital satisfaction: A further test. *Journal of Marriage and the Family, 50,* 69–77.

Huston, T. L. (1983). Power. In H. H. Kelley, E. Berscheid, A. Christensen, J. H. Harvey, T. L. Huston, G. Levinger, E. McClintock, L. A. Peplau, & D. R. Patterson (Eds.), *Close relationships* (pp. 169–219). New York: W. H. Freeman.

Kollock, P., Blumstein, P., & Schwartz, P. (1985). Sex and power in interaction: Conversational privilege and duties. *American Sociological Review, 50,* 34–46.

Komter, A. (1989). Hidden power in marriage. *Gender and Society, 3,* 187–216.

Krendl, K. A., Troiano, C., Dawson, R., & Clark, G. (1993). "OK, where's the remote?" Children, families, and remote control devices. In J. R. Walker & R.V. Bellamy, Jr. (Eds.), *The remote control in the new age of television* (pp. 137–153). Westport, CT: Praeger.

Lindlof, T. R., Shatzer, M. J., & Wilkinson, D. (1988). Accommodation of video and television in the American family. In J. Lull (Ed.), *World families watch television* (pp. 158–192). Newbury Park, CA: Sage.

Lull, J. (1988). Constructing rituals of extension through family television viewing. In J. Lull (Ed.), *World families watch television* (pp. 237–259). Newbury Park, CA: Sage.

McDonald, G.W. (1980). Family power: The assessment of a decade of theory and research, 1970–1979. *Journal of Marriage and the Family, 42,* 841–854.

Mederer, H. (1993). Division of labor in two-earner homes: Task accomplishment versus household management as critical variables in perceptions about family work. *Journal of Marriage and the Family, 55,* 133–145.

Morley, D. (1988). Domestic relations: The framework of family viewing in Great Britain. In J. Lull (Ed.), *World families watch television* (pp. 22–48). Newbury Park, CA: Sage.

Osmond, M.W., & Thorne, B. (1993). Feminist theories: The social construction of gender in families and society. In P. G. Boss, W. J. Doherty, R. LaRossa, W. R. Schumm, & S. K. Steinmetz (Eds.), *Sourcebook of family theories and methods: A contextual approach* (pp. 591–623). New York: Plenum.

Perse, E. M., & Ferguson, D. A. (1993). Gender differences in remote control use. In J. R. Walker & R.V. Bellamy, Jr. (Eds.), *The remote control in the new age of television* (pp. 169–186). Westport, CA: Praeger.

Robinson, J. P. (1990, September). I love my TV. *American Demographics, 12,* 24–27.

Shank, J. W. (1986). An exploration of leisure in the lives of dual career women. *Journal of Leisure Research, 18,* 300–319.

Shaw, S. M. (1985). Gender and leisure: Inequality in the distribution of leisure time. *Journal of Leisure Research, 17,* 266–282.

Shaw, S. M. (1991). Gender, leisure, and constraint: Towards a framework for the analysis of women's leisure. *Journal of Leisure Research, 26,* 8–22.

Spigel, L. (1992). *Make room for TV: television and the family idea in postwar America.* Chicago: University of Chicago.

Szinovacz, M. R. (1987). Family power. In M. B. Sussman & S. K. Steinmetz (Eds.), *Handbook of marriage and the family* (pp. 651–693). New York: Plenum.

Thompson, L. (1991). Family work: Women's sense of fairness. *Journal of Family Issues, 12,* 181–196.

Thompson, L. (1993). Conceptualizing gender in marriage: The case of marital care. *Journal of Marriage and the Family, 55,* 557–569.

Thompson, L., & Walker, A. J. (1989). Gender in families: Women and men in marriage, work, and parenthood. *Journal of Marriage and the Family, 51,* 845–871.

Walker, J. R., & Bellamy, R. V., Jr. (1991). Gratifications of grazing: An exploratory study of remote control use. *Journalism Quarterly, 68,* 422–431.

West, C., & Zimmerman, D. H. (1987). Doing gender. *Gender and Society, 1,* 125–151.

17

Trading Places: Changes in Gender Roles Within the Iranian Immigrant Family

ALI AKBAR MAHDI

Humans are the products of their environment and the socio-cultural forces that shape that environment. Moving away from the safety and stability of life in their homeland, immigrants seek new identities and develop strategies to cope with the demands of these new social and cultural forces. Settling in a host society is never easy. Immigration involves various forms of sacrifice and adjustment. These changes are much harder for those who have left their homeland involuntarily and find the culture of their host society incompatible with that of their native origin. Of the changes an immigrant has to face, changes in the family and gender roles are the most difficult and consequential ones because they involve not only changes in the identity and behavior of individual immigrants but also in relationships with their intimates. The most significant among these changes is the role reversal for members of the family.

Social roles, including family and gender roles, never are fixed. They are negotiated within the new circumstances and are renegotiated if there is any further change in those circumstances. Given the drastic changes in male-female relationships around the world in the past three decades, gender roles have become increasingly a contested site of power, thus open to discursive

Reprinted by permission of the author.

and situational conflicts and negotiations.[1] Although fluid and constructed, the implicit and explicit assumptions regarding one's rights and obligations in a marital relationship serve as a framework for interaction and a basis for construction of one's identity in a partnership. These roles serve as "scripts" defining rules of interactions in the context of marriage and courtship. The definitions given or assumed for these roles often are internalized and enacted as normative expectations that create constraints and possibilities for individual action.

Role changes and reversals in any circumstances and for anyone have important, sometimes, poignant consequences. Under normal circumstances in a homogeneous community, this kind of change proceeds gradually without much difficulty in adjustment. However, in the case of a first generation immigrant, both the pace and the intensity of the change is very high, making it difficult to respond without a great deal of individual hardship and cultural agonies. In the scenario of role reversal in an immigrant family, changes in roles are accompanied with cultural shock, assimilation and counter-assimilation processes, identity deconstruction and reconstruction, and the struggle to establish a new life in the host country. Immigrant studies have been tracing tactics and strategies used by immigrants to establish their new life.[2]

Although the study of immigrant families has gained a great deal of attention in the past two decades, the Iranian immigrant family remains largely unexplored. The story of their identity reconstruction and their attempt to create a space in American social life is a new and interesting chapter in the history of immigration to America. The immigration of Iranian families to the United States is a relatively new phenomenon. Most Iranians who came to the United States during the period of 1950–1980 were students seeking higher education. Because of the 1979 revolution in Iran, many of these students either found it difficult to go back to their homeland or preferred to stay in the United States. In addition to these students, there were also thousands of secular Iranians, especially those of upper and upper-middle classes, who either were forced to leave their country for exile or voluntarily fled the negative consequences of the newly established theocratic state in Iran.[3]

The majority of these Iranian immigrants have college educations and work in high-skilled professions. According to statistics from the 1990 U.S. Census Bureau, Iranian immigrants are among the most highly educated immigrants and their income in some professions exceeds the national average for that profession.[4] According to the statistics reported in the *San Francisco Examiner,* 51 percent of all Iranians in the United States have graduate degrees, compared to approximately 21 percent for the American population as a whole.[5] Most of these Iranians received their education from Western educational institutions or attended Iranian universities modeled after the Western system. Therefore, it is reasonable to say that Iranians living in the United States already have been partially assimilated to the modern, secular, Western values and outlooks. This article attempts to shed light on one aspect of changes in the lives of these Iranians, namely role changes in their families.

THE IMPORTANCE OF FAMILY FOR IRANIAN IMMIGRANTS

Family plays a central role in Iranian society. It is the institution to which the individual relates the most and from which he or she receives his/her identity. This fact that the family is central to the social status of the individual and serves as the foundation of social life in Iran is not lost to the Iranian immigrants who have left their homeland. Family continues to serve as a familiar refuge against the unfamiliar world of the host society. It serves as a buffer to cushion that which challenges against the most cherished and unchangeable normative and behavioral aspects of immigrants' lives and it also serves as an intermediate institution to aid in smoothly adopting new values, norms, and behaviors. Iranian immigrants living in the United States still value family and family relations highly and attempt to protect its integrity and cohesiveness against the forces of disintegration in the host society. Despite such efforts and determination, however, the immigrant family is not immune to the structural pressures of the new culture and society. The lack of family support and a strong social network in the newly adopted country necessitates role adjustments from all family members. The prevalent individualism in American society imposes a higher degree of individual autonomy on family members and rewards the individual achievements of members. This is antithetical to the collectivistic tendencies present in the traditional Iranian family. Iranians' responses to these demands and constraints are often synthetic, combining elements of both cultures and representing continuity and change. As a recent immigrant population, Iranians have begun to define and give a new meaning to their own presence in this country—a synthetic meaning which bears the residues of two lands, two environments, two people, and two different cultures. The composition of these new identities, the meanings attached to them, and the strategies used for survival, provide a ground for the better understanding of group survival in a new environment.

Although this article delineates some of the changes and emerging patterns of interaction and identity within the Iranian immigrant family in the United States, there is something more at work here—something that cannot be detected easily in statistical averages and general trends. It is the experiences of individuals and the diversity of their exposure and entanglement with a large and diverse society like the American one. There can be no definitive definition of the typical Iranian American man or woman immigrant. While each may be subjected to similar structural conditions, the ultimate outcome of their experiences is based on their individual feelings and collective interactions within the communities in which they live.

THEORETICAL OBSERVATIONS

Before discussing changes and adaptations in the Iranian immigrant family in detail, it is important to make some theoretical observations. First, several changes in the immigrant families are generic changes taking place in all urbanized environments and the Iranian immigrant family has been subject to these changes even prior to its arrival in the United States. Industrialization, urbanization, expansion of educational opportunities, greater social and occupational mobility, and a higher standard of living and its associated higher social expectations, all have continued to produce a series of changes affecting men and women throughout the world, especially among those of middle and upper social strata.[6] In the case of women, these changes include participation in the labor force, higher educational achievement, access to previously male-dominated fields and careers, higher levels of political consciousness and participation, higher rates of participation in physical education activities and competitive sports, marrying at older ages and having fewer children, establishing careers and maintaining autonomy, and—in the case of a bad marriage—resorting to divorce with much greater ease.

These changes in women's lives, in turn, have resulted in changes in the status of men within the family and indeed in the whole society. There have been noticeable changes in the roles of men and women at home. Familial responsibilities have been changing for a long time. As noted by several scholars,[7] men's "breadwinner status" has been declining and their responsibilities at home, as well as their identity, have been changing. The new patterns of gender interactions in a host society inevitably leaves its marks on the newcomers. In this case, changes in gender roles occurring in both Iranian and American society over the past five decades have had tremendous effects on these Iranians who have chosen to live in the United States as their adopted country.

Second, researchers have studied "reversed-role" families since the mid-1970s. They have been found not only in the United States[8] but also in Sweden,[9] Australia,[10] and Israel.[11] Even though this type of family, characterized by a full-time employed mother outside of home and a full caretaker father at home, is neither the focus of this article nor a representative of Iranian family in the United States, it is archetypical of the kind of changes in status and relationships of males and females emerging globally. My reference to this type of family is because of the direction of changes taking place in male-female relationships.

Third, not all social scientists agree that the changes experienced by the members of the immigrant/ethnic families are necessarily different from those experienced by the dominant groups in the host society. Acknowledging the historically unique experiences of these immigrants, they believe that the difference between the experiences of these ethnics and those who have been here long enough to be a part of the dominant majority is a matter of degree rather than kind.[12] The first experience of these families in migration is cultural confusion and normlessness. The opportunity to avoid traditional restrictions, combined with restrictions imposed by the new environment, compel these immigrants to develop a comfortable blueprint for their own identity. This identity, however, cannot be very different from the modalities of identities available in the host society. In fact, as Robert Merton argues, in a normless environment a group often develops symmetrical norms that are compatible with the dominant values in that environment.[13] In the case of Iranians, the traditional authoritarian family relationship gives way to a much more egalitarian interpersonal and marital relationship.

Still, the specific norms and values these immigrants bring to American society and the ways in which they synthesize these norms with the dominant values of their host country have crucial impact on the success of these immigrants in assimilating to their new home.[14] The compatibility or incompatibility of immigrants' norms and values with the dominant values of the host society determines the way these immigrants integrate into their host society and forge a new way of life in a society significantly different from that in their country of origin. The attributes of ethnic culture always are mediated through the family. These values concern not only the relationships between family members but also their achievements, lifestyle, and educational and occupational aspirations.

Fourth, changes in the immigrant family structure and relationships are dependent on a host of factors that are not the subject of discussion in this article. These include the immigration laws and reception patterns of the host society, the political atmosphere and relationships between the home and host countries, the time of the immigrants' arrival and the manner of their entry into the host society, the stage of the immigrants' family life cycle, the length of their migration from their homeland, and the age, sex, ethnicity, education, and occupational skills of family members at the time of arrival.[15] Coupled with the political and socio-economic class of the immigrants, these factors have important influences on the kind and degree of changes these immigrants experience in the host society. Therefore, it should be remembered that although the picture of the Iranian immigrant family in the United States illustrated in this article is accurate, it is quite general and does not reflect the variations that might exist among Iranian immigrants of various ethnicities (Arab, Kurdish, Persian, Turkish, etc.), religions (Bahai, Christian, Jewish, Moslem, Zoroastrian, etc.), and socio-economic classes.

Fifth, the United States is a highly industrialized society whose unique cultural and structural features do not leave immigrant families within it unaffected. The new environment within which the immigrant family finds itself is quite different from the one existing in Iran. Many aspects of the traditional Iranian family are being challenged by the socio-economic conditions of life in America. Even those who are determined to maintain their native socio-cultural patterns of relationships find themselves forced to adjust to the structural and cultural imperatives of their host society. Given the normative, expressive, and behavioral differences between Iranian and American families, the task of bridging the gap between these two is not easy. Quite often immigrants deal with these gaps in very unique and creative ways. For instance, arranged marriage—a declining but continuing feature of the Iranian family at home—is almost unheard of among Iranians who have lived long in the United States. However, not being able to find a desirable woman for partnership in America, some Iranian men would arrange to have a suitable bride from home brought over to the United States.[16] This is a new phenomenon on which there is little data and needs investigation.

With these observations in mind, what are some of the specific sociological changes within the Iranian immigrant family? In what follows, I will demonstrate these changes in spousal and parental roles within this family. This will be done by drawing from two field surveys conducted during 1995–97 in Iran and the United States. Early in 1995, a questionnaire comprised of 113 questions was designed and mailed to 821 households in 41 American states. This questionnaire asked a range of questions on decision making in the family, women's views on and attitudes toward female gender roles, and perception of gender roles in the United States and Iran. The results of this survey provided a picture of Iranian women whose views of women's roles inside and outside of home differed from the typical roles ascribed to them by tradition, religion, and society. At that point, it was not clear whether this difference was due to the changes that had taken place as a result of migration to a new country. It was decided to undertake a similar study in Iran in order to determine the significance of the new information about the Iranian women. With that purpose in mind, the same questionnaire was translated, slightly modified to account for sociopolitical and cultural differences between the two environments, and taken to Iran for distribution among Iranian women.

SAMPLES

The respondents in the United States sample included Iranian females drawn from three lists of addresses: the Iranian Cultural Society of Columbus, Ohio (680 addresses, 307 in Ohio, 373 in other states), the Middle East Studies Association, and the Center for Iranian Research and Analysis. Although the latter two databases were biased because their records consisted of a highly educated population, especially interested in Middle Eastern studies, the addresses from the Iranian Cultural Society contained a more diverse population. Drawing information from this database was weighted in order to include more non-academic subjects in the sample. The questionnaire was mailed to the 821 randomly selected addresses. Fifty-seven of these questionnaires were returned because there were no qualified respondents in the household. Of the 743 remaining questionnaires, 158 (21.26 percent) were completed and returned. Nine questionnaires, filled by non-Iranian women married to Iranian men, were excluded, thus leaving the study with 149 completed questionnaires to be used for the analysis.

In the case of the Iranian sample, it became clear that it would be impossible to have a representative random sample without government support and a good amount of time and money, not available to this researcher at the time. Also, the sensitivity of women's issues along with the political and cultural mistrust of a male social scientist by both the government and many women themselves, especially when the researcher is employed by a United States academic institution, made it harder to think of an open research strategy. As a result, it was decided to resort to a macro snowballing strategy, contacting people in six cities in order to distribute the questionnaires in five settings: educational institutions, administrative offices, medical and health care facilities, factories, and neighborhoods. Sites were selected on the basis of availability of a connection, possibility of simultaneous distribution and collection of questionnaires, and security considerations. In each site, questionnaires were given to all available individuals, except those who wished not to participate. Non-participation was reported only in four sites and in each case it did not surpass 15 percent. Reasons for non-participation were lack of trust, fear of persecutions by the government, and possible disapproval by subject's husband. For those who were not able to read and write, the questionnaire was administered by an interviewer.

Since the project had turned into a convenient sampling, efforts were made to maximize the sample by increasing the number of subjects. A total of 1,100 questionnaires were distributed, of which 1,028 were in good order. Given the volume of questionnaires, the security concerns, and the problem of non-randomness of the sample, half of these questionnaires were randomly selected. As a result, the current sample includes 514 subjects randomly selected from 1,028 completed questionnaires.

PROFILE OF RESPONDENTS

Respondents in the U.S. sample are Iranian female immigrants who have lived in the United States an average of 16.05 years (median = 15.98). Close to half (42.9 percent) of the sample are between the ages 31–40 and another 30.6 percent are between the ages 41–50. A majority of them are married (73.2 percent), have children (78.4 percent), work outside the home (77.9 percent), and regard their stay in the United States as permanent (83.9 percent). While the majority are Muslim (72.7 percent), 14.4 percent insist that they do not adhere to any religion. The number of professors, physicians, and businesswomen in the U.S. sample is proportionally very high. Of course, this reflects a bias in the sample. In terms of education, over 50 percent of respondents have graduate degrees, 32.9 percent bachelor degrees, and only 3.4 percent have just a high school diploma. Respondents' husbands were even more educated, 72.6 percent possessing graduate degrees. While respondents had a median family income of $60–75,000.00, their own individual income averaged over $30,000.00 annually. Respondents making less than $10,000.00 comprised 13 percent of the sample, and those having no income also constituted 13 percent of the sample.

Respondents in the Iranian sample are females mostly born in major cities (68.99 percent), as opposed to 25.53 percent born in small towns and 5.49 percent born in villages. Compared to the U.S. sample, respondents in the Iranian sample are much younger. More than half of them are under 30 years old. Half of them have never married (48.48 percent) and work outside the home (54.44 percent), mostly on a full-time basis (68.22 percent). While 92.58 percent are Muslim, 1.65 percent insist that they do not follow any religion. Educationally, over 60 percent of these women have attended college. Like the U.S. sample, their husbands are even more educated, 50 percent having graduate degrees. Occupationally, teachers and clerical workers represent the largest groups in the sample. A third of this sample, even many college educated ones, do not work and are categorized as "housewife." As for income, some 3.17 percent of the sample with an annual household income below 150,000.00 *tomans* are considered poor, 26.19 percent with annual household income between 150,000 and 540,000.00 *tomans* as lower class, and 25.79 percent with an annual household income above 1,200,000.00 *tomans* as upper middle class. The remaining respondents (44.85 percent) may be considered as middle class.

CHANGES IN SPOUSAL ROLES

In the traditional Iranian family, men are the primary wage earners. They provide for their families and demand respect and obedience in return. The general Iranian-Islamic culture also supports men as the authority figure and the source of family livelihood, thus making them responsible for its well-being. Traditional notions of manhood are very much attached to this provision of income for the family and assuming the responsibility for decisions affecting the fate of its members.[17]

Migration, however, has challenged these traditional bases of men's authority. Women have become more educated and are actively participating in the labor market. Their high educational achievement and occupational success have undermined male authority and diminished their roles as sole wage earners in their families. Immigrant men often share the task of providing for their families with their wives and thus have lost the monopoly on decision making in the family. Women's educational and economic gains have provided them with resources that translate to power and status inside and outside the family. This is, of course, a relative phenomenon. The increase in power and status of women is in direct correlation with their level of ambition, education, and social class background. Also, the increase in female autonomy and power at home has not reduced their homemaker's role proportionately. Certainly there is an increase in male

Table 17.1 Mean Percentage of Task-Sharing in the Iranian Immigrant Family as Reported by Wife in the U.S. Sample

	Wife	Husband	Children	Others
Shopping	74.2	25.6	1.2	0.2
Cooking	81.5	15.9	0.6	2.5
Cleaning	63.1	20.8	2.6	12.4
House repairs	21.2	63.5	2.5	14.5
Car maintenance	15.7	63.1	1.1	19.1
Sew/iron/laundry	71.6	17.1	3.4	7.3
Financial planning	42.1	56.1	0.8	0.7
Caring for children	73.9	25.3	0.9	0.6
Chauffeuring kids	68.7	28.5	2.6	0.8
Garden/lawn care	35.2	53.7	3.5	7.7

participation in the household tasks. However, the rate and pace of this change is much slower than other changes. As data from the U.S. survey shows, immigrant women still bear the major responsibility for house chores and child-care (see Table 17.1).

Also, it should be remembered that although the labor participation and professional mobility among Iranian women in the United States have been increasing and have resulted in a decline in the authority of men at home, it is not evident that women's labor participation has acquired a primary significance in the provision of family income. As mentioned by Bozorgmehr, compared to other female immigrants, the rate of female participation in the labor force is low.[18] Men still are the major source of family income, despite the variations in their share in different households. Within my sample the wages and incomes of women were far less than those of men (see Table 17.2). This was particularly true for men engaged in corporate and entrepreneurial activities. Where one sees some comparable income between husbands and wives is in dual career families with professional, medical, and educational occupations.

Table 17.2 Married Respondents' Personal and Family Income in the U.S. Sample

Income Level	Personal Percent	Family (Combined) Percent
None	16.1	0.0
Under 5,000	7.1	0.0
5,001–10,000	7.1	0.0
10,001–15,000	4.0	2.0
15,001–30,000	18.2	4.8
30,001–45,000	18.2	9.8
45,001–60,000	19.2	21.6
60,001–75,000	5.1	15.7
75,001–100,000	2.0	20.6
Above 100,000	3.0	25.5
Total	100.0	100.0

Note: N = 102

Generally speaking, women in the traditional Iranian society were treated like subjects and second class persons rather than decision makers and autonomous individuals. There was a general assumption among traditionalists that, as wives and mothers, women did not need social identities independent of their family identity. This notion still is prevalent among the Islamicists in Iran who have been trying to reduce the independent identities of individual women from movies, and literary and educational materials[19]—a sad reality which finally has prompted President Mohammad Khatami's Minister of Education to take some corrective action against "the depiction of women in high school textbooks."[20]

In the traditional Iranian family, a woman's social and individual identities were often shadowed by a man's and, by and large, defined by her relationship to him as a wife, sister, daughter, or mother. Her happiness was to be measured by her husband's happiness. Her husband's failure was partially hers as well. Her achievements in life were defined more by her husband's and children's achievements than by her own. Whether a wife or a daughter, she was at her best when she was at the service of her husband or father.[21] She was present to help men achieve their emotional, social, occupational, and spiritual goals. So she offered her man any kind of help, whenever and however he needed it. On their part, men took the services and help of their women for granted. They viewed their wives' attachment to them as sources of material and nonmaterial support and they knew they could rely on their women whenever they needed assistance, services, care, and support.

In the new environment of American society, such services and forms of support are not automatically assumed. Such an expectation is neither possible nor granted. Men may still receive a higher degree of support from their wives in their careers and social standing, compared to prevailing norms of male support in the host society, but they cannot take that support for granted as much as they could in Iran. Among professional couples, the support may actually turn into competition. Many men find themselves in competition with their wives for success and social status—a feature I will return to later.

Finding themselves in an environment where most native cultural norms neither apply nor receive any patronage, immigrant men and women are hard pressed to adapt their expectations to the new rules of conduct and to create new modes of relationships and family environment. Given the high educational and occupational achievements of immigrant women, their financial independence from their husbands, acceptance of individualistic values of the American society, competitive nature of the work environment in the United States, and their determination to remove all the traditional limitations on their individual growth, immigrant women have begun to demand a great deal more from their husbands in caring for their children and home affairs. Role differentiation within the family is moving away from a traditional hierarchical division of male-dominated/female-subordinated roles toward a complementary one. Husbands and wives are beginning simultaneously to assume expressive as well as instrumental roles. Data from surveys within the Iranian immigrant family reveal that men are taking a more active role in the household chores, including shopping, cooking, and cleaning (see Table 17.1). Women also are participating actively in the roles traditionally performed by men, such as managing family finances, attending to family business, and even caring for the family car. Iranian women are seeking open equality in doing household chores, in child-rearing, decision-making, ownership of family property, and even in their sexual relationship. To get a sense of these changes, similar statistics from the Iranian sample are presented in Table 17.3.

Table 17.3 shows that in six out of ten activities within the family, men's and women's roles have changed in opposite directions. Where women have become more involved, men have shown less involvement and vice versa. In two areas of house chores (cooking and cleaning) and child care (caring for children and chauffeuring them to their activities) men's involvement has increased dramatically. Interestingly, men's increased participation in the family household is accompanied by a decline in most activities by the children.

While the traditional family did not leave much room for the growth of women's individual self-esteem, the new family in migration has provided an opportunity for women to advance their self-esteem by acquiring many skills and participating in occupations denied to them in their native land.[22] While many find themselves forced to enter the labor market despite many disadvantages and socio-cultural barriers (especially for those with more religious and traditional social outlooks) many others have chosen career paths as a source of identity and livelihood regardless of their husbands' financial and career status. Immigration

Table 17.3 Comparison of Mean Percentage of Task Sharing Within the Iranian Family in Iran and the U.S. as Reported by Wife

		Iran	United States	% Change
Shopping	Wife	53.9	74.2	+ 27.3
	Husband	42.8	25.6	– 40.1
	Children	2.3	1.2	– 47.8
	Others	1.6	0.1	– 93.7
Cooking	Wife	82.2	81.5	– 0.85
	Husband	8.8	15.9	+ 44.6
	Children	0.7	0.6	– 14.2
	Others	1.9	2.5	+ 24.0
Cleaning (floor, rooms, etc.)	Wife	72.7	63.1	– 13.2
	Husband	15.7	20.8	+ 24.5
	Children	2.9	2.6	– 10.3
	Others	7.4	12.4	+ 40.3
House repairs	Wife	14.8	21.2	+ 30.1
	Husband	62.2	63.5	+ 2.0
	Children	4.2	2.5	– 40.4
	Others	16.8	14.5	– 13.6
Car maintenance	Wife	12.8	15.7	+ 18.4
	Husband	70.0	63.1	– 9.8
	Children	3.8	1.1	– 71.0
	Others	13.8	19.1	+ 27.7
Sewing/ironing/laundry	Wife	76.7	71.6	– 6.6
	Husband	14.6	17.1	+ 14.6
	Children	2.0	0.8	– 60.0
	Others	5.2	0.7	– 86.5
Financial planning & management	Wife	36.0	42.1	+ 16.9
	Husband	64.3	56.1	– 12.7
	Children	1.0	0.8	– 20.0
	Others	0.1	0.7	+ 85.7
Caring for children	Wife	81.0	73.9	– 8.7
	Husband	13.4	25.3	+ 47.0
	Children	2.0	0.9	– 55.0
	Others	1.3	0.6	– 53.8
Chauffeuring kids around to their classes/activities	Wife	67.4	68.7	+ 1.8
	Husband	26.0	28.5	+ 8.7
	Children	2.3	2.6	+ 11.5
	Others	1.2	0.8	+ 33.3
Gardening/lawn care	Wife	29.6	35.2	+ 15.9
	Husband	51.8	53.7	+ 3.5
	Children	0.9	3.5	+ 74.2
	Others	15.7	7.7	– 50.9

has had its exactions, but it also has had obvious compensations. It has afforded women many forms of freedom: from the religious restrictions imposed by a theocratic state, from the expectations and constant supervision of their parents and in-laws, from fear of rejection by and isolation from society, from the mores and folkways of their homeland, and from the normative demands of culturally homogeneous Iranian neighborhoods. Although these freedoms have been beneficial to both men and women, women have gained most from them. Since the type and nature of newly acquired roles by women are much more extensive than those acquired by men, the depth and extent of role reversal is greater for women than for men. At the same time, the negative effects of these changes on men have been more severe than on women—a topic discussed later.

New roles for women are accompanied by a high degree of stress but greater satisfaction and realization of their potential. Often being younger and thus more adaptable than their husbands, these women take better advantage of the new educational and professional opportunities in the host society. Their new roles are a challenge for themselves as well as their husbands. However, based on the reports of strains and maladjustments in the Iranian immigrant family, it seems that women demonstrate a higher degree of adjustment to their new roles than men. Of those women who have reported satisfaction with their experiences in the host society, many describe their homes as a battleground in which they confront their husbands' anger and frustrations.[23] Men who fled Iran due to political or religious persecution often left prestigious and lucrative positions behind and came to find themselves unwelcome in the job market in the United States. These men are often in mid-life and find it difficult to cope with the loss of their social status, employment, and authority. Such a loss is accompanied by frustration, alienation, and depression, especially when wives of many of these men are more successful in adapting to American society, achieving educational and occupational success, and enjoying many forms of freedom which they were denied in Iran.

Although acquiring new roles has provided men with a greater realization of their potential as equal partners in marital relationships, the loss of some of the traditional roles and privileges in marital relationships has had devastating effects on some of them, especially on those who have been brought up with a conservative outlook and have a low socio-economic status. These men show less enthusiasm and appreciation for change in their wives' social behaviors and domestic roles.[24] Much of the traditional Iranian/Islamic sensitivities towards and reservations for their women's interaction with male strangers in public often smother by a better understanding of the context of sexual relationships in Western societies, concerns for self-preservation, and values of self-sacrifice and family duty. However, these concerns almost disappear in cases where wives outpace their husbands professionally and come into close contact with other men. As Shahidian argues, there is also a certain degree of ethnic and national pride embedded in these sensitivities as men come to view their women as boundary markers of their national and ethnic identity.[25]

There are numerous reports of high anxiety, depression, suicide, and use of violence among men who have had difficulty adjusting to the changes in their wives' sexual behavior, social roles, and family responsibilities in their newly adopted country.[26] The violent outcomes of many of these cases are referred to as "crimes of honor" (*qatl-e namoosi*). For examples, in February 1996, Jorjik Avanesian, an Iranian immigrant in Glendale, California, killed his wife and six children by setting his house on fire. In his report to the police, he accused his wife of having "illegitimate relationships" with male strangers.[27] In July 1997, Mohammad Talaforush, who was apparently distressed over his wife's request for divorce, put his daughter and son, 8 and 10 years old, respectively, to sleep with an overdose of tranquilizers and then hanged himself.[28] In another incident on November 7, 1996, in response to charges of spouse and child abuse and in anger about a Los Angeles court order that had barred him from further contact with his wife and children, Jafar Derakhti shot his wife to death and later killed himself. Suspicious of his wife's activities, Jafar had barred his wife from using the telephone and the family car.[29] In Sweden, a boy killed his younger sister with the help of his cousin because of her violation of the family honor.[30] Since the incident occurred after a series of similar incidents among the immigrants, the Swedish Parliament organized a hearing where they asked an Iranian sociologist to present his views on the issue.[31]

Since changes in wives' roles require a redefinition of men's roles within the family, it inevitably leads to an identity crisis for men who have to reject their past identity in favor of a new one with less privileges and patriarchal pride. This loss of old privileges often is accompanied by an adaptation to new responsibilities in a more egalitarian marriage where men have to perform tasks in which they never have been trained. In the more successful immigrant families, there is even competition between immigrant husbands and wives in achieving a more independent and secure social status in the host society. In some marriages this competition becomes a source of conflict between the couple and results in divorce. Research reveals that many Iranian families are experiencing a variety of marital and interpersonal problems resulting in a great deal of stress—a stress caused by confusion about interpersonal norms and expectations, domestic arrangements and financial responsibilities, separation from family members in the homeland, a sense of loss of native culture and values, the pressures of adaptation to the new culture and environment, and the changes in gender and family roles, statuses, and identities.[32] The social arrangement within which the couples find themselves is also a contributing factor to such stressful relationships. Lack of family support and a larger kin network reduces the resources available to these couples and increases the amount of time they spend with each other. In situations of conflict and crisis, increased interaction in a highly dense relationship multiplies the number of roles each spouse has to play in relation to the other. In the traditional family in Iran, a husband was "a husband" and a wife "a wife." In the new setting, each spouse not only has to play the role of intimate other but also, in many cases, the role of an absent father, mother, or brother at times of crises. Many of the functions previously solved by interaction with parents and siblings now have to be fulfilled in the relationship with the spouse. Furthermore, dual career immigrants are finding it difficult to balance their time between the heavy demands of their work and their marital responsibilities. Working outside of the family during the day and attending to their children in the evening, many of these couples find themselves too tired to attend to each other and provide emotional support against the hardship of their new lifestyle.

One of the consequences of these conflicts and subsequent stress levels has been a higher rate of divorce. Iranian-Islamic culture views divorce as a tragic solution and a last resort in marital conflict, especially when there is a child in the marriage, but favorable divorce laws in the host society and economic independence of women have made Iranian couples more prone to use this option in the United States. Still, it is worth noting that while divorce in Iran often is followed by a decline in the woman's status, in the United States such a decline of status is not the norm among Iranian female divorcees.[33] Also, it is important to remember that the divorce rate in Iran has also been increasing in the past two decades, especially among the educated and urban population. For instance, in the year 1373 [1994], according to [the] deputy of legal and personal affairs at the National Registry, 32,706 cases of divorce were registered.[34] This marked an increase of 2,394 or 12 percent. This trend continues in later years as well. In the nine months of March-December, 1997, there were 31,424 divorces recorded in Iran, representing an 11 percent increase in the divorce rate from the year before. The rate for Central Province, which is the largest and most urbanized province in the country, was 15.3 percent.[35]

CHANGES IN PARENTAL ROLES

In the context of immigration, the relationship between parents and children is a delicate one. Learned attitudes and relationships between immigrant parents and children are often subject to new rules of applicability, constraints, and demands. The internal dynamics of the Iranian immigrant family is dependent on political, economic, and legal aspects of the host society. In the United States, there are numerous legal measures safeguarding against arbitrary parental treatment, a greater emphasis on children's autonomous growth and personal initiative, and a higher degree of permissiveness in dealing with children and young adults. Iranian families migrating to the United States find themselves compelled to change learned relationships between themselves and their children and adjust to the existing definitions and limitations of parental roles.

The traditional Iranian family was authoritarian and adult-centered. Parental decisions were ultimate and nonnegotiable. Children organized their lives around and according to their parents' decisions, desires, and plans. The new Iranian family is far removed from this model. In the United States, Iranian-American children have become a focal point of the family's status. Children also demand more freedom of movement and expression than they ever have been allowed and often receive a lot more attention than those children raised in Iran. Many Iranian immigrant parents are adjusting their own life choices to the educational needs and life choices of their children and devoting greater amounts of their resources to this end. They want to see their children succeed and some even project their own unfulfilled aspirations onto them. Many parents cite the opportunity for social and educational mobility for their children as the main reason for their immigration to the United States. In the latter case, the economic resources of the family are channeled toward their children and a tremendous amount of pressure is put on children to succeed.

In the homeland, mothers served as role models for their daughters as well as supervisors in most of their affairs, especially those related to the private domain. Conversely, fathers served as role models for their sons as well as supervisors for most of their affairs in the public sphere. In America, these divisions and responsibilities are changing. Men are accepting greater responsibilities in attending to their children, even to their teenage daughters by accompanying them to their athletic activities, entertaining them, supervising their educational programs, monitoring their social activities, and even counseling them in their choice of some of the most private issues traditionally handled by mothers (such as the choice of clothing and makeup). Although these forms of involvement serve as a means of control and as a means of protection for their daughters against undesirable aspects of the host culture, they still represent new definitions of family responsibilities and have the effect of creating new role anxieties and constraints for daughters as well as parents.

Maintaining a family and raising children in the new homeland have advantages and disadvantages. On the positive side, the lack of sociocultural pressures of Iranian society make it easier for parents to allow their children to exercise more autonomy. In Iran, parents are much stricter toward their children, particularly with regard to their daughters' social life. In the United States, in the absence of an extended family and its concomitant obligations, parents spend more time with their children and become more involved in their social lives. The availability of good educational, recreational, and occupational opportunities reduces most parents' anxiety about their children's future, thus making it easier for them to take a more relaxed and liberal approach in their upbringing. Nevertheless, the degree to which this liberal approach is applied to girls is different from that applied to boys. Girls, whose sexuality and lifestyle have a direct bearing on the family honor and dignity, are sometimes subject to closer supervision and stricter home regulations. Though some Iranian parents are more cautious in the social aspects of their young daughters' lives, they rarely let such a sensitivity deter their educational and social mobility. Given the fact that some other Middle Eastern immigrants, especially those of working class background, have demonstrated a propensity for differential treatment of their children, Iranian parents rarely differentiate between girls and boys in terms of educational and career opportunities.[36]

On the negative side, the problems of violence and drugs, coupled with the more liberal sexual standards, increase parents' vigilance and put them on constant alert. The dual career situation of many parents, mixed with a lack of kinship support, increases parental responsibilities, thus causing enormous physical and psychological pressures on these parents and their children. Dual career immigrant families face an acute time crunch. The constant struggle to balance work and family is probably the greatest hurdle for these couples, especially women, who continue to be the primary care givers for the members of their families. Many professional women experience a "superwoman syndrome" by accepting conflicting responsibilities inside and outside of home in addition to the challenges of immigrant life. These women often experience "role overload" and "role conflict"—an experience which at times takes its toll on their children too. These women also experience a lack of emotional support that was accorded to them in Iran when they assumed their traditional roles. Since many of them have changed their gender roles in the new environment, they are deprived of the kind of emotional base necessary for the reproduction of these new identities within the Iranian community. Those women who do not have a healthy relationship or marriage within the Iranian community often leave it for a more supportive relationship with non-Iranian men.

The effects of role changes on men have been more severe. These new marital and parental responsibilities that men carry in the family involve a role shift that they neither are prepared for nor find compatible with their cultural norms and self-image. Given the abrupt and speedy nature of changes taking place in the status and roles of their women, male immigrants are suffering from a cultural lag. Many have a difficult time understanding and/or adjusting to the volume and direction of change in their gender relationships. These changes have resulted in an identity crisis in which men find themselves pulled from all directions: They are separated from their own history, society, family of orientation, and culture. They find themselves confronting pressures and constraints of a new culture which does not have much sympathy for their past culture and identity. They have lost their identity as the breadwinners and the authority in the family. On top of all this, they find themselves estranged even from their most intimate, i.e., their wives.

Neither children of immigrants (born in the United States) nor immigrant children (born outside of the United States but came to this country when they were young) are immune to the negative consequences of migration faced by their parents. The cultural restrictions and ethnic identity imposed by their parents sometimes lead to serious conflicts within the family, resulting in alienation of these children from their parents as well as their friends outside the home. In some cases the outcome of these conflicts has been violence and death.[37] These children also have to endure the comparative disadvantage of not having a wide kin relationship within American society. Often many of these children have a distant memory of their grandparents, uncles, aunts, nieces, cousins, and nephews. The lack of such contacts can generate problems for some teenagers during their formative years in school when they begin to rely on family as a source of identity. Also, in comparison with their counterparts in Iran, most of the younger members of the second generation regularly attend daycare and are denied a long stay in the warm environment of a parental family. The latter experience helps these children to develop a more independent personality though, and approach their future more individualistically than their Iranian parents do.

CONCLUSION

If we establish a curve for the gains and losses of men and women in their gender privileges inside and outside of Iran, we observe two curves that are shaped differently and are traveling in different directions. With the Iranian Revolution of 1979 and the establishment of a theocratic state in Iran, traditional, uneducated Iranian men and women gained higher socio-political statuses, social rewards, and economic benefits. Conversely, most secular, educated, and Westernized Iranians lost a great deal of their social, economic and political statuses. A segment of the latter group either left the country voluntarily or was forced out involuntarily. The gains of the latter group in their newly adopted countries are mixed, varied, and at times contradictory. Most Iranians migrating to the United States, especially the educated and wealthy ones, have succeeded in achieving an economic status equal [to] or higher than the one they had occupied previously in their homeland.

While the economic gains have been common for both men and women, the social gains have not. Men have lost a great deal of the privileges they enjoyed at home. They have lost their authority, the higher level of respect they commanded within home and kin network, and the privileges accorded to them in marriage. Women have had the opposite experience. Not only have they escaped the harsh policies of the Islamic Republic toward women, they have gained a great deal of autonomy, social and educational skill, and a clearer sense of their sexuality, individuality, and identity. It seems among the factors determining the "new identity" of these women, gender has a more prominent influence than other variables. While many of the new roles for men might be accompanied with a sense of anxiety and pain, women's new roles and identities have given them a higher level of satisfaction and self-fulfillment. And of course, both experience the anxieties and problems associated with leaving their country, friends, and loved ones behind and settling in a new land and culture. Both often strive to build new lives in common with rather than separate from one another. How successful they are is a subject for another study.

NOTES

1. See L. L. Lindsey, *Gender Roles: A Sociological Perspective* (Upper Saddle River, NJ: Prentice Hall, 1997); K. Plummer, *Symbolic Interactionism* (Brookfield, VT: Edward Elgar, 1991), Vol. 1; and S. Coltrane, *Gender and Families* (Thousand Oaks, CA: Pine Forge Press, 1998).
2. For studies on challenges faced by the Middle Eastern families in the United States, see Barbara C. Aswad and Barbara Bilge, eds., *Family and Gender Among American Muslims: Issues Facing Middle Eastern Immigrants and Their Descendants* (Philadelphia: Temple University Press, 1996). For examples from other ethnic groups, see G. Buijs, ed., *Migrant Women: Crossing Boundaries and Changing Identities* (Oxford: BERG, 1993); P. Hondagneu-Sotelo, "Overcoming Patriarchal Constraints: The Reconstruction of Gender Relations Among Mexican Immigrant Women and Men," in M. Baca Zine, P. Hondagneu-Sotelo, and M. A. Messner, eds., *Through the Prism of Difference: Readings on Sex and Gender* (Boston: Allyn and Bacon, 1997); C. H. Mindel, R. W. Habenstein, and R. Wright, Jr., eds., *Ethnic Families in America: Patterns and Variations* (Upper Saddle River, NJ: Prentice Hall, 1998); V. Parrillo, "The Immigrant Family: Securing the American Dream," *Journal of Contemporary Family Studies* 12 (1991): 131–45; and several pieces in the following two books: S. J. Ferguson, *Shifting the Center: Understanding Contemporary Families* (Toronto: Mayfield Publishing Co., 1998); V. Demos and M. Texler Segal, eds., *Ethnic Women: A Multiple Status Reality* (Dix Hills, NY: General Hall, Inc., 1994).
3. Abdoulmaboud Ansari, *Iranian Immigrants in the United States: A Case Study of Dual Marginality* (Milwood, NY: Associated Faculty Press, 1992). See also from the same author, *The Making of the Iranian Community in America* (NJ: Pardis Press, 1993).
4. Mehdi Bozorgmehr and George Sabagh, "High-Status Immigrants: A Statistical Profile of Iranians in the United States," *Iranian Studies* 21, no. 2–4 (1988): 5–36.
5. Quoted in *The Iran Times,* August 16, 1996.
6. J. H. Edwards, ed., *The Family and Change* (New York: Alfred A. Knopf, 1960).
7. See K. Gerson, *No Man's Land: Men's Changing Commitments to Family and Work* (New York: Basic Books, 1993); S. Coltrane, *Fatherhood, Housework, and Gender Equity* (New York: Oxford University Press, 1996); and C. West and D. H. Zimmerman, "Doing Gender," *Gender and Society* 1 (1987): 125–51.
8. J. Levine, *Who Will Raise the Children? New Options for Fathers (and Mothers)* (New York: Bantam, 1976).
9. M. E. Lamb, A. M. Frodi, C.-P. Hwang, M. Frodi, and J. Steinberg, "Mother- and Father-Infant Interaction Involving Play and Holding in Traditional and Nontraditional Swedish Families," *Developmental Psychology* 18 (1982): 215–21.
10. J. Harper, *Fathers at Home* (Melbourne: Penguin, 1980).
11. A. Sagi, "Antecedents and Consequences of Various Degrees of Parental Involvement in Childrearing. The Israeli Project," in M. E. Lamb, ed., *Nontraditional Families: Parenting and Child Development* (Hillsdale, NJ: Lawrence Erlbaum, 1982).
12. Thomas Sowell, *Ethnic America: A History* (New York: Basic Books, 1981).

13. Robert Merton, *Social Theory and Social Structure* (Glencoe, IL: Free Press, 1957).
14. Nathan Glazer and Daniel P. Moynihan, *Beyond the Melting Pot* (Cambridge, MA: MIT Press, 1970).
15. See Ali Akbar Mahdi, "The Second Generation Iranians: Questions and Concerns," *The Iranian,* no. 9 (February–March 1997) (an online magazine at www.iranian.com). Also, see Ali Akbar Mahdi, "Ethnic Identity Among Second-Generation Iranians in the United States," *Iranian Studies* 31, no. 1 (1998): 77–95.
16. Parvin Abiyaneh, "Ezdevaaj az Rah-e Dour va 'Aroos-e Vaaredaati" (Long Distance Marriage and Imported Bride), *'Aqaazi Nou* 25 (1993): 14–17; Nahid Shahnavaz and Minoo Gorji, "The New Persian Bride: Tragedy or Triumph," paper presented at the annual meeting of the Center for Iranian Research and Analysis, Atlanta, GA. April, 1997.
17. Behnaz Jalali, "Iranian Families," in M. McGoldrick, J. Giordano, and J. K. Pearce, eds., *Ethnicity and Family Therapy* (New York: The Guilford Press, 1982).
18. Mehdi Bozorgmehr, "Iranians," in David Levison, ed., *Encyclopedia of American Immigrant Cultures* (New York: Macmillan, 1997).
19. A number of studies have documented the efforts by the Islamicists to change the image of Iranian women according to their own perception of what an Islamic woman should look like. See Patricia F. Higgins and Pirouz Shoar-Ghaffari, "Women's Education in the Islamic Republic of Iran," Azar Naficy, "Images of Women in Classical Persian Literature and the Contemporary Iranian Novel," and Hamid Naficy, "Veiled Vision/Powerful Presences: Women in Post-Revolutionary Iranian Cinema," all three in Mahnaz Afkhami and Erika Friedl, eds., *In the Eye of the Storm: Women in Post-Revolutionary Iran* (Syracuse, NY: Syracuse University Press, 1994); and Hammed Shahidian, "The Education of Women in the Islamic Republic of Iran," *Journal of Women's History* 2, no. 3 (1991): 6–38.
20. On 16 Shahrivar, 1378 [7 September, 1999], Hossein Mozaffar, the Minister of Education, acknowledged the lack of understanding in approaching "women's abilities" in the educational system of the IRI. Arguing for a new "presentation of women in high school textbooks," he criticized the current presentation of women in these texts in this way: "The idea that women are handicapped is not desirable because in terms of human values there is no difference between men and women in Islam." *The Iran Times,* September 17, 1999.
21. For a host of religious statements regarding the responsibilities of women to their husbands see Ayatollah Ebrahim Amini, *'Ayin-e Hamsardaari yaa Akhlaaq-e Khaanevaadeh* [The Rules Regarding Care for Spouse or Family Ethics] (Tehran: Enteshaaraat-e Islami, no date); and Zahra Gavahi, *Simaaye Zan dar 'Ayeneh-ye Feq-he Shi'a* [Women as Seen in The Mirror of Shi'i Jurisprudence] (Tehran: The Organization for Islamic Propaganda, 1372 [1993]). For a religiously conservative view, see Seyed Javad Mostafavi, *Behesht-e Khaanevaadeh; Ham'ahangi-ye Aql va Fetrat baa Ketaab va Son-nat dar Masaael-e Zojiyat* [The Heaven of the Family: Coordination of Reason and Nature with the Book and Tradition Regarding Marital Issues] (Mash-had: Hatef Publisher, 1374 [1995]), two volumes. For a religiously liberal view, see Fatemeh Safari, *Olgoo-ye Ejtemaa-i, Paayegaah, va Naqshe Zan-e Mosalmaan dar Jame'eh-ye Eslaami bar Asaas-e Didgaah-haaye Hazrat-e Emaam Khomeini* [The Model, Position, and Role of an Islamic Woman in an Islamic Society, According to Imam Khomeini's Views] (Tehran: The Organization for Islamic Propaganda, 1370 [1991]).
22. See Minoo Moallem, "Ethnic Entrepreneurship and Gender Relations Among Iranians in Montreal, Quebec, Canada," in Asghar Fathi, ed., *Iranian Refugees and Exiles Since Khomeini* (Costa Mesa, CA: Mazda Publishers, 1991); and Arlene Dallalfar, "Iranian Women as Immigrant Entrepreneurs," *Gender and Society* 8, no. 4 (1994): 541–61.
23. Rouhi Shafiee, "The Status of Iranian Female Refugees in England, 1982–92," *Arash,* A Persian Monthly Review of Cultural and Social Affairs, March/April (1994).
24. Hammed Shahidian, "Gender and Sexuality Among Immigrant Iranians in Canada," *Sexualities* 2, no. 2 (1999): 189–222.
25. Ibid.
26. Mehrdad Darvishpour, "The Causes of Violence Against Women in the Immigrant Family," *The Iran Times,* Washington, D.C., February 27, 1998.
27. *The Iran Times,* February 9, 1996.
28. *The Iran Times,* August 1, 1997. The *New York Daily News* characterized this event as a cultural one: "He's traditional; she's Westernized. This was a parent who felt he had no option but to send his kids to paradise."
29. The *Salt Lake Tribune,* November 8, 1996.
30. *Hambastegi,* June–July, 1997.

31. Mehrdad Darvishpour, "Immigrant Women Challenge Men's Role," *Nimrooz* (London), no. 43 (1997).
32. See Shideh Hannasab, "Acculturation and Young Iranian Women: Attitudes Toward Sex Roles and Intimate Relationships," *Journal of Multicultural Counseling and Development* 19, no. 1 (1991): 11–21; Vida Nasehi, "Naqsh-e Zanaan dar Khaanevaadeh-haaye Mohaajer-e Irani [The Role of Women in the Iranian Immigrant Families], *Goft-O-Gu,* no. 9 (Autumn 1374) [1995]: 27–41; Nayereh Tohidi, "Iranian Women and Gender Relations in Los Angeles," in R. Kelley, ed., *Irangeles: Iranians in Los Angeles* (Los Angeles: University of California, 1993); and Mammed Shahidian, "Gender and Sexuality Among Immigrant Iranians in Canada," *Sexualities* 2, no. 2 (1999): 189–222.
33. For reports and analyses of divorces among Iranians abroad see Mehrdad Darvishpour, "The Causes of Family Disintegration Abroad," *Arash,* nos. 14–15, March–April (1992); Mehrdad Hashayekhi, "Causes of Instability in Iranian Families in the United States," *The Iran Times,* nos. 19–21 (July 1990); Vida Nasehi, "Naqsh-e Zanaan dar Khaanevaadeh-haaye Mohaajer-e Irani [the Role of Women in the Iranian Immigrant Families], *Goft-O-Gu,* no. 9 [Autumn 1374] [1995]: 27–41; and Mehrdad Darvishpour, "Divorce Among Iranian Immigrants," *The Iran Times,* vol. XXIV, nos. 14–23 (June 17–August 19, 1994).
34. *Iran News,* 05/04/95. Published by NetIran at www.netiran.com.
35. *The Iran Times,* February 20, 1998.
36. For examples from [the] Arab American community, see Evelyn Shakir, *Bint Arab: Arab and Arab American Women in the United States* (Westport, Connecticut: Praeger, 1997).
37. For an example of these conflicts resulting in a tragedy within the Iranian community abroad, see Asrin Mohammadi, "An interview with Asrin Mohammadi," *Hambastegi,* Paper of the International Federation of Iranian Refugees and Immigrants Councils, June–July, 1997.

18

Thinking About the Baby: Gender and the Division of Infant Care

SUSAN WALZER

The tendency for women and men to become more differentiated from each other in work and family roles upon becoming parents has been documented in longitudinal studies of transitions into parenthood (see summaries in Belsky and Kelly 1994; Cowan and Cowan 1992). New mothers are more apt than new fathers to leave or curtail their employment (Belsky and Kelly 1994). And despite couples' previous intentions (Cowan and Cowan 1992), mothers provide more direct care to babies than fathers do (Belsky and Volling 1987; Berman and Pedersen 1987; Dickie 1987; Thompson and Walker 1989). Fathers tend to act as "helpers" to mothers, who not only spend more time interacting with babies, but planning for them as well (LaRossa 1986).

This pattern of increased gender differentiation following the birth of a baby has been associated with decreases in marital satisfaction, particularly for wives (Belsky, Lang, and Huston 1986; Cowan and Cowan 1988; Harriman 1985; Ruble et al. 1988). A number of researchers have interpreted new mothers' marital dissatisfaction as connected with "violated expectations" of more shared parenting (Belsky 1985; Belsky, Lang, and Huston 1986; Ruble et al. 1988), although some researchers express surprise that wives expect so much in the first place (Ruble et al. 1988). Nevertheless, traditional divisions of household labor have been implicated in marital stress following the birth of a first baby (Belsky, Lang, and Huston 1986; Schuchts and Witkin 1989).

In this paper I focus on the more invisible, mental labor that is involved in taking care of a baby and suggest that gender imbalances in this form of baby care play a particular role in reproducing differentiation between mothers and fathers and stimulating marital tension. My use of the term "mental" labor is meant to distinguish the thinking, feeling, and interpersonal work that accompanies the care of babies from physical tasks, as has been done in recent studies of household labor (see, e.g., Hochschild 1989; DeVault 1991; Mederer 1993).[1] I include in the general category of mental labor what has been referred to as "emotion" work, "thought" work, and "invisible" work (Hochschild 1983; DeVault 1991); that is, I focus on aspects of baby care that involve thinking or feeling, managing thoughts or feelings, and that are not necessarily perceived as work by the person performing it (DeVault 1991).

Using qualitative data from interviews with 50 new mothers and fathers (25 couples), this [reading] describes three categories of mental baby care and suggests that the tendency for mothers to take responsibility for this kind of work is an underrecognized stress on marriages as well as a primary way in which mothering and fathering are reproduced as gendered experiences. While the tendency for mothers to feel ultimately responsible for babies has been identified in other studies (see, e.g., McMahon 1995), this paper describes some of the interactional and institutional contexts within which differences between maternal and paternal responsibilities are reproduced. I suggest that the way that new parents divide the work of thinking about their babies reflects an accountability to socially constructed and institutionalized differentiation between women and men.

DATA AND METHOD

The data grounding this discussion are from a qualitative interview study of 50 mothers and fathers who had become new parents approximately one year before the time of the data collection. The sample was located through birth announcements published in the local newspaper of a small city in upstate New York. This method for locating new parents was an attempt to improve upon the self-selection bias present in many studies of transitions to parenthood in which voluntary samples are generated through obstetrics practices, childbirth classes, or community announcements.

Preliminary letters were sent in stages telling potential respondents about the study and inviting them to be interviewed. These letters were then followed with a phone call to answer any questions and schedule interviews. The response rate for those couples who received letters, were reached by telephone, and fit the sample parameters was 68 percent.

The parents in the sample ranged in age from 21 to 44 years old and the age of the babies ranged from 11 to 18 months. Fourteen of the babies were boys and 11 were girls. Fifteen of the pregnancies were planned while 10 were not. Twenty-three of the couples in the sample were married while two were not. Four of the fathers and three of the mothers had had a previous marriage.

All of the parents in the sample had finished high school or a GED, 23 had a college degree, 4 had master's degrees, and 2 had professional degrees. The median family income range was \$40,000–\$49,999 with 6 families under \$30,000 and five over \$75,000. Two couples reported having received some public assistance. About 40 percent of the sample described growing up in households that could be characterized as poor to working class while 60 percent grew up in middle- to upper middle-class households.

Ten of the mothers were employed full time, 7 were employed part time, and 2 were students. Six of the mothers described themselves as stay-at-home mothers, although two of them provided regular baby-sitting for pay. All except one of the fathers were employed at the time of the interview.

In most cases, wives and husbands were interviewed on the same occasion, first separately and with as much privacy as possible, and then more briefly together upon the completion of their separate interviews. Interview sessions ranged from two to four hours long. In three cases, wives and husbands were interviewed on separate occasions. All of the interviews took place in the couples' homes except for one father who requested an interview in his workplace.

The data used in this paper are part of a larger data set about new parents' transitions into parenthood. The interview protocol was semi-structured and designed to elicit parents' experiences of having become mothers or fathers as well as to discuss possible influences on the nature of their transitions into parenthood. All of the interviews were taped and then transcribed, coded, and analyzed using a constant comparative method (see Glaser and Strauss 1967). This [reading] represents data and analysis that emerged during the course of this grounded theoretical study.

THINKING ABOUT THE BABY AS "WOMEN'S WORK"

Three categories of mental labor associated with taking care of a baby surfaced in my respondents' reports: worrying, processing information, and managing the division of labor (see also Ehrensaft 1987; LaRossa and LaRossa 1989).

Worrying

In this section I contextualize the disproportionate amount of worrying that new mothers do in interactional dynamics between mothers and fathers; that is, I suggest that mothers worry about babies, in part, because fathers do not. In this sense my analysis emphasizes other dynamics surrounding worrying besides the internalization of gendered personality differences (as in Ehrensaft's 1987 account).[2] I also suggest that gender differences in whether and how new parents worry are linked to socially constructed expectations for mothers and fathers to which new parents feel accountable (see West and Fenstermaker 1993 on the role of accountability in reproducing gender).

The "mental" experience of being a new mother—thinking about the baby, worrying "about everything"—is one that many of the women in my sample shared (see also Ehrensaft 1987; Hays 1993):

> I don't walk around like a time bomb ready to explode. I don't want you to think that. It's just that I've got this stuff in the back of my head all the time.
>
> I worry about her getting cavities in teeth that are not even gonna be there for her whole life. Everything is so important to me now. I worry about everything.
>
> It's like now you have this person and you're always responsible for them, the baby. You can have a sitter and go out and have a break, but in the back of your mind, you're still responsible for that person. You're always thinking about that person.

These new mothers described thinking about their babies as something that mothers do: "Mothers worry a lot." Worrying was such an expected part of mothering that the absence of it might challenge one's definition as a good mother. One of my respondents described returning to her job and feeling on her first day back that she should be worrying about her baby. She said that she had to "remind" herself to check on how her baby was doing at the baby-sitter's "or I'd be a bad mother."

Fathers do not necessarily think about their children while they are at work or worry that this reflects on them as parents (Ehrensaft 1987). My respondents did not report feeling like "bad" fathers if they took their minds off of their babies; some even expressed stress when their babies had to have their attention:

> Sitting two hours playing with him, when I first did it was like, this is a waste of my time. I said, "I have more important things to do." And I'm still thinking, "Look at the time I've spent with him. What would I have done otherwise?"

This father's concern with his perceived lack of productivity while spending time with his baby might be a response to the social construction of fathers' roles as primarily economic (see Benson 1968; Thompson and Walker 1989). Another new father in my sample described a sense of loss about time he missed with his baby when he had to travel for work, "but," he said, "it goes back to the idea of being a father . . . I do think in a traditional sense where I'm the father, I'm the husband, it's my job to support the family."

A couple I will call Brendan and Eileen illustrate the relationship between parental worry and social constructions of motherhood and fatherhood. Brendan and Eileen both have professional/managerial careers, each reporting salaries of more than $75,000. When Eileen had to travel for work, Brendan would function as Jimmy's primary caregiver. But when she was home, Eileen wanted to do "the baby stuff." She referred to her caregiving of Jimmy as her "stake" in his life:

> This is going to be hard to say. It's really important to me that Jimmy understands I'm his mother, whatever that means, because I'm probably not a traditional mother by any stretch.

When I asked what it means to her for Jimmy to know that she's his mother, Eileen responded:

> It means that if I come home some night and he's with [his] day care [provider] and he doesn't want to leave her, it'll kill me, is what it means. So I don't know if the rest of this is trying to ensure that doesn't happen. I don't know if the rest is trying to ensure that I have that very special role with him.

For Eileen, anything that she was *not* doing for Jimmy had the potential to damage her "special role" with him. If she was not the most special person to him, she was inadequate as a mother—something she noted that Brendan did not feel. She connected these concerns with what she referred to as "the good mother image": "She's somehow all nurturing and all present and always there." And she added: "Now, I'm not even going to be able to have a shot at it because I'm not a lot of those things."

Brendan said that his behavior with Jimmy was not driven by guilt and anxiety as he perceived Eileen's to be:

> I think she feels that need. She wants to be a good mother . . . Being a father, it's not a guilt thing. It's not like I'm going to do this because I don't want to be a bad daddy.

Brendan noted that his relationship with Jimmy was based on fun while he perceived Eileen's actions to be driven by insecurity. Eileen recognized that her concerns made her less of a good mother in Brendan's eyes. "I think his issue with me as a mother is that I worry a lot. The fact is that I do, but I also think mothers worry a lot." Although Eileen worried that she couldn't match the good mother image she described herself as "absolutely" buying into, worrying itself made her feel like a good mother.

Why is worrying associated with being a mother? I suggest two general reasons, which generate two kinds of worry. The first reason is that worrying is an integral part of taking care of a baby; it evokes, for example, the scheduling of medical appointments, babyproofing, a change in the baby's diet. There appears to be a connection between taking responsibility for physical care and carrying thoughts that reinforce the care (see Coltrane 1989; Ruddick 1983). This kind of worry, which I refer to as "*baby worry,*" is generated by the question: What does the baby need? And babies need a lot. As Luxton (1980:101) points out, women are often anxious because babies are "so totally dependent" and perceived as highly vulnerable to illness and injury.[3]

A second reason that new mothers worry is, as I suggest above, because they are expected to, and because social norms make it difficult for mothers to know whether they are doing the right thing for their babies. I call this "*mother worry,*" and it is generated by the question: Am I being a good mother? While it has been suggested that mothers are more identified with their children than fathers are (Ehrensaft 1987), I suggest that mothers' worrying is induced by external as well as by internal mechanisms. That is, perhaps mothers experience their children as extensions of themselves as Ehrensaft (1987) argues, but mothers are also aware that their children are perceived by others as reflecting on them. Mothers worry, in part, because they are concerned with how others evaluate them as mothers:

> I think that people don't look at you and say, "oh there's a good mother," but they will look at people and say, "oh there's a bad mother."
>
> Being a mother I worry about what everyone else is going to think.

The mother just quoted perceived mothers as uniquely responsible for their children's behavior, and even street violence:

> The behavior of the child reflects the mother's parenting. . . . I mean kids, you have all these things with kids shooting people, and I blame it on . . . mothers not being around.

The association of mothers with worrying provides a source of differentiation between mothers and fathers and presents women with a paradox, often played out in interactions with their male partners. Worrying is associated with irrationality and unnecessary anxiety, and some fathers suggested that their partners worried too much about their babies:

> Sometimes I say, "He's fine, he's fine," but he's not fine enough for her.

However, worrying is perceived as something that "good" mothers do. A number of fathers made an explicit connection between good mothering and their wives' mental vigilance:

> She's a very good mother. She worries a lot.
>
> She's always concerned about how she's doing or she's always worried about if [child's] feelings are hurt or did she say something wrong to her.

This paradoxical message—good mothers worry; worrying too much isn't good—underlies the tendency for mothers to worry and for fathers to express ambivalence about their worrying. One mother described a division of labor in which she was stressed and got things done while her husband's job was to tell her to lighten up:

> I'm the one who stresses out more. He is very laid back. He doesn't worry about things. In fact he procrastinates. And I'm the one, run run run run run. . . . But one of us has to get things done on time and the other one has to keep the other one from totally losing it and make them be more relaxed. So it kind of balances us out.

A father described the care that his wife's worrying ensures for their baby while also suggesting that some of it might be unnecessary:

> She worries a lot. I'm probably too easygoing, but she makes sure he goes to the doctor, makes sure he has fluoride, makes sure he has all of his immunizations. She's hypervigilant to any time he might be acting sick. She's kind of that way herself. I kid her about being a hypochondriac. She makes sure he gets to bed on time, makes sure he's eating enough, whereas I'm a little more lackadaisical on that.

In both of these cases, as with Brendan and Eileen, the respondents described a kind of balance between the mother and father: The mother worries, the father doesn't; his job, in fact, might be to tell her not to worry. This dynamic reinforces a gendered division of mental labor. Although there is a subtext that the mother's worrying is unnecessary and/or neurotic, she does not stop. In fact, the suggestion that the mother "relax" serves to reinforce her worrying because although she does not recognize it as work, she does recognize that worrying gets things done for the baby. If the father offered to share the worrying rather than telling the mother to stop, the outcome might be quite different.

While I would not argue that it is possible for a baby to be cared for without having some assortment of adults performing "baby worry," the "mother worry" I have described here is heightened in our society by assumptions that good parenting is done exclusively and privately by a mother (with perhaps some "help" from a father) and that veering from this model may have severe consequences for children (see Coontz 1992 for a critique of "American standards of childrearing"). Examining another area of mental labor—the work of processing "expert" information about baby care—further reveals the norms attached to the work of thinking about babies.

Processing Information: "What to Expect"

LaRossa and LaRossa (1989) make a direct connection between the fact that wives tend to buy and read how-to books on parenting and their being "in charge" of the baby. Because mothers read the books more thoroughly, they are more informed, and both parents assume that the mother will orchestrate and implement the care: "Her purchase of the books reflects what is generally accepted: Babies are 'women's work'" (LaRossa and LaRossa 1989:144). In this section, I suggest that processing information about baby care is itself part of the work of taking care of a baby. I also argue that the assumption that mothers will do this work is embedded and reinforced in the "information" that mothers get from expert advice (see also Hays 1993).

There are a number of steps that may be involved in the mental labor of processing information about babies:

1. Deciding on the need for advice
2. Locating the advice (often from more than one source)
3. Reading/listening to the advice
4. Involving/instructing one's partner
5. Contemplating and assessing the advice
6. Planning for the implementation of the advice

What I label here as steps 1–3 are carried out by mothers usually (LaRossa and LaRossa 1989; Hays 1993). In my sample, 23 out of 25 of the mothers reported reading parenting literature while 5 of the 25 fathers did.

Step 4 occurs in a number of variations: Mothers tell fathers what to read; mothers tell fathers specifically what they have read; mothers tell fathers what to do based on their own reading. These approaches to disseminating information were apparent in my sample:

> He would say, "Well you're the mother, so what's the answer here?" And I said, "What do you think I have that I would know just because I'm the mother?" But I would do a lot more reading.

> Sarah [wife] has read quite a few and I just pretty much go with her. She hasn't really told me I'm doing anything wrong.

> Every once in a while she might pull something out and show me if she found something she thinks I should read, but I usually don't have time.

Step 5—contemplation and assessment of the advice—is often complicated, since what women find in advice books is ideology as well as information. According to a content analysis performed by Hays (1993; see also Marshall 1991), underlying the advice provided by child care experts is an "ideology of intensive mothering" that, among other things, holds individual mothers primarily responsible for child-rearing and treats mothering as expert-guided, emotionally absorbing, and labor-intensive. Mothers therefore take responsibility for gathering information from sources that reinforce their primary responsibility for the care of babies. As described by one respondent below, mothers have to confront the ideology underlying the advice in order to assess whether they can or want to implement it (Step 6).

The book relied on by a majority of the mothers in my sample was *What to Expect the First Year* (Eisenberg, Murkoff, and Hathaway 1989), a book that one of the authors writes was conceived to address new mothers' "numerous worries." Several of the women in my sample referred to it as their "bible," and, because it is not included in content analyses of expert advice books (see, e.g., Hays 1993; Marshall 1991), I include some excerpts in this paper. These excerpts illustrate how gendered divisions of mental labor are reinforced on an institutional level through "expert" advice for new parents.

The book is divided into two main parts—"The First Year" and "Of Special Concern"—and has a third "Ready Reference" section at the end. "Becoming a Father" is the 25th of 26 chapters in the book and is included in the issues "of special concern." Much of the information given throughout the book is in the form of answers to specific questions that are presented with quotation marks, as if particular mothers had asked them.

One of the mothers in my sample who preferred *What to Expect* over other books nevertheless had questions about its "accuracy." Her statements were sarcastic in response to the book's advice about the effort that mothers should exert to see that their babies eat healthy foods:

> I like to read *What to Expect*. Although I don't think they're too accurate. . . . so your baby should be doing this and the other thing. And never give him any white sugar. Don't give him any cookies. Make sure they're muffins made from fruit juice. Yeah, okay. I'll just pop off in the kitchen and make some muffins.

Following are the comments that open a consideration of when to introduce solid foods to a baby—something that a father can do whether or not the baby is being breast-fed:

> The messages that today's new mother receives about when to start feeding solids are many and confusing. . . . Whom do you listen to? Does mother know best? Or doctor? Or friends? (Eisenberg, Murkoff, and Hathaway 1989:202)

This passage illustrates the mental labor that is expected to accompany the introduction of solid foods: choosing when to do it, consulting with others about the issue, making a decision about whose advice to take. It also presumes that it is the mother who is making the decision in consultation with her mother, doctor, friends, yet her male partner is not mentioned.

The one chapter addressed to fathers begins with the following question from a presumably typical father:

> "I gave up a lot of my favorite foods when my wife was pregnant so I could support her efforts to eat right for our baby. But enough's enough. Now that our son's here, shouldn't I be able to eat what I like?" (Eisenberg, Murkoff, and Hathaway 1989:591)

The tone of the question suggests that the father is getting guff from someone about his diet. Implication: It may not be only babies whose diets new mothers need to worry about. Regardless of who is nagging this father, the question suggests that fathers will not be independently motivated to eat a healthy diet in the interest of their babies and themselves (although given the comments of the mother in my sample, this father may just want some cookies and white sugar in his diet).

As Hays (1993) points out, authors of advice books may not have created gender differentiation in parenting responsibility, but they certainly play a role in reproducing it. Mothers in my sample who already felt that they had the primary responsibility for their babies did not get any disagreement from the advice book they consulted most frequently about "what to expect":

> If your husband, for whatever reason, fails to share the load with you, try to understand why this is so and to communicate clearly where you stand. Don't expect him to change overnight, and don't let your resentment when he doesn't trigger arguments and stress. Instead explain, educate, entice; in time, he'll meet you—partway, if not all the way. (Eisenberg, Murkoff, and Hathaway 1989:545)

This advice directs women to do "emotion work" (Hochschild 1983) to contain their responses to their husbands' lack of participation. Rather than experiencing stress or conflict, new mothers are directed to keep a lid on their feelings and focus on instructing and enticing their husbands into participation (and after all this, not to expect equity).

This kind of "advice" provides reinforcement for new parents' gendered divisions of mental labor, including the tendency for mothers to have the responsibility for getting the advice in the first place. The suggestion from these experts that new mothers should not argue with or expect equity from their husbands may also be a factor in the decreases in marital satisfaction that some experience.

Managing the Division of Labor

In this section I expand the concept of "managing" that has already been applied to infant care in past studies and suggest that it is not only the baby's appointments and supplies that mothers tend to manage (Belsky and Kelly 1994), but their babies' fathers as well (see also Ehrensaft 1987). To use the language from *What to Expect the First Year,* "enticing" fathers into helping out with their babies is another invisible, mental job performed by new mothers, as one respondent said of her husband:

> Peter is very good at helping out if I say, "Peter, I'm tired, I'm sick, you've got to do this for me, you've got to do that," that's fine, he's been more than willing to do that.

Even in situations in which fathers report that they and their partners split tasks equally, mothers often have the extra role of delegating the work, as the following fathers in my sample indicate (Coltrane 1989 and Ehrensaft 1987 also describe "manager-helper" dynamics in couples who "share" the care of children):

> I don't change her [diaper] too often—as much as I can get out of it.

> Then at night either one of us will give him a bath. She'll always give him a bath, or if she can't, she'll tell me to do it because I won't do it unless she tells me, but if she asks me to do it I'll do it.

These quotes from two fathers, who perceived that they split tasks equally with their partners, reflect a division of labor in which the mothers are the ultimate managers. Both of these fathers had described themselves as sharing tasks with their wives—when their wives told them to.

Diaper changes were a particular area in which enticing was evident:

> I mean diapering, that's hard to say. He won't volunteer, but if I say, "Honey, she needs a diaper change, could you do it?" he does it.

> It took me a little while to get him to change the nasty diapers . . . but now he changes 'em all. He's a pro.

Mothers also made decisions about when not to delegate:

> I do diapers. Joel can't handle it well. You know, he does diapers too, but not if there's poop in them.

> I'm pretty much in charge of that, which is fine because it's really not that big of a deal. And she's more, it seems like she's easier for me than she is for him when it comes to diapering 'cause I just all the time do it, you know?

The mother just quoted illustrates how habitual patterns become perceived as making sense—doing becomes a kind of knowing (Daniels 1987; DeVault 1991)—just as being the one to read the book makes the mother the expert. Another woman described how her husband sits and eats while she "knows" what is involved in feeding their baby (and him):

> I know what has to be done. I know that like when we sit down for dinner, she [child] has to have everything cut up, and then you give it to her, you know, where he sits down and he eats his dinner. Then I have to get everything on the table, get her stuff all done. By the time I'm starting to eat, he's almost finished. Then I have to clean up and I also have to get her cleaned up and I know that like she'll always have

> to have a bath, and if she has to have a bath and if I need him to give it to her, "Can you do it?" I have to ask . . . because he just wouldn't do it if I didn't ask him. You know, it's just assumed that he doesn't have to do it.

While on one level it appears that women are "in charge" of the division of labor, the assumption of female responsibility means that, on another level, men are in charge—because it is only with their permission and cooperation that mothers can relinquish their duties. One mother talked about feeling that she had to check with her husband before making plans that did not include their baby, while her husband did not check with her first. She described herself asking her husband, "Can I do this in 3 weeks?" Another woman complained that her partner would leave the house while their child was taking a nap:

> It's always the father that can just say, "Okay, I'm gonna go." Well I obviously can't leave, he's ready for a nap, you know? It's nap time. Mommy seems to always have to stay. I think that fathers have more freedom.

These statements go against suggestions that mothers may not want to relinquish control to their male partners because motherhood is a source of power for women (see, e.g., Kranichfeld 1987). What is powerful, perhaps, is the desire of mothers to be perceived as good mothers, and this may be what they feel they are trading off if they are not taking responsibility for the care of their babies. While mothers may instruct their husbands to do things, the data here suggest that husbands' responses to and compliance with orders are not compulsory (see also DeVault 1991). Fathers who considered themselves equal participants in the division of labor would use the fact that they were "willing" to do diapers as an example:

> We each will do whatever we have to do. It's not like I won't change diapers.

Mothers did not necessarily see any baby task as optional for them:

> It's kind of give and take. As far as diaper changing, I think I do more. . . . It's not one of his favorite tasks.

Women are the "bosses" in the sense that they carry the organizational plan and delegate tasks to their partners, but they manage without the privileges of paid managers. Their ultimate responsibility for baby care may, in fact, disempower them in relation to their husbands, since for many women it means a loss of economic power (see Blumberg and Coleman 1989) and greater dependence on their male partners (LaRossa and LaRossa 1989; Waldron and Routh 1981).

MENTAL BABY CARE AND MARITAL CHANGES

While having a baby may foster greater dependence by women on their husbands, Belsky and Kelly (1994) report that new mothers are often disappointed by the level of emotional support they get from their husbands. I suggest that women's disproportionate responsibility for mental baby care plays an important role in generating women's dissatisfaction. Mothers in my sample were not necessarily appreciated and were even criticized by their partners for worrying; the advice they were in charge of getting told them not to be open with their husbands about their experiences; and their sense of being ultimately responsible for their babies' care affected their access to other sources of validation and power, such as paid work and social networks.

One of the primary ways in which women's sense of responsibility for babies surfaced was in decisions they made about employment. In my sample, many women changed their paid work patterns, quitting jobs or cutting back their hours (see also Cowan and Cowan 1992). These changes had implications for the balance of power in their marriages:

> It's funny now because he is the breadwinner so there have been opportunities where he has interviewed for positions, had opportunities to relocate and get a

> better position and the money was better. You're just put in a position where you have to just follow. Before when we were both working we would talk it out. I'd say, "No, I want to stay here." And now you really can't.

On an institutional level, men's bigger paychecks and women's experiences of low-wage, low-prestige jobs structured some of my respondents' traditional parenting arrangements. But there were also women in my sample, such as Eileen, who made as much money as their husbands and were very satisfied with their jobs, yet felt that they had to answer for their work in ways that their husbands did not. Laura, for example, described her decision to let go of a part of her job that she enjoyed the most because she did not want to see her baby at the sitter for more hours:

> I can't do that, I can't emotionally. I probably could, we'd have to pay more money for the sitter, but I don't want him at the sitter like for 10 hours a day. To me, that's, I'm doing something that I want to do, but in the long run I'm hurting him, you know? In my mind, I think that.

Laura's husband, Stuart, had not cut back on any parts of his job and was struggling with maintaining his performance in extracurricular activities:

> I either want to be involved and do it the right way or I almost don't want to be involved at all. Because I don't want to do a less than good job.

Laura did not mention how the hours required by Stuart's activities influenced the time spent by their child at the baby-sitter's, but she did acknowledge that her marriage was stressed by her resentment of her husband's "freedom." Even though both Laura and Stuart were employed and made similar wages, Laura felt more directly accountable for their baby's care. She could not allow herself to stay at work, she said, because it would hurt their baby: "In my mind, I think that." What Laura resented perhaps is that her husband was free from these kinds of thoughts.

Women's disappointment with their partners may stem from their loneliness in particular with the thinking they do about their babies. One woman in my sample, who did not question her primary responsibility for baby care, was upset by her perception that her husband did not recognize what goes on inside her head (emphases added):

> It really hurts, because he doesn't know how high my intentions or whatever or goals for being a good mom are . . . (crying) *He doesn't know what I think* and when he's at work he doesn't know when she starts screaming and throwing fits, or pulling everything out of the dishwasher when I'm trying to load it, *and I've got all this in the back of my head* that I have to do for school, and the house is a mess, and supper's not cooked and he'll be home in 30 minutes. *He doesn't know that I have to keep telling myself, "Be calm. Love your child."* You know? He doesn't know. So I just get upset when sometimes I really think he would say, "Well she could be a better mom."

Hochschild (1989) notes that when couples experience conflict about housework, it is generally not simply about who does what, but about who should be grateful to whom. This "economy of gratitude," Hochschild suggests, relates to how individuals define what should be expected of them as men and women. Applying this notion to divisions of baby care, if mental labor is defined as an idiosyncrasy of mothers rather than as work, there is nothing for a man to feel grateful for if his wife does it. If fathers are seen as doing mothers a favor when they participate in baby care, fathers will receive more appreciation from their partners than they give back, which may contribute to new mothers' disappointment in the lack of emotional support they receive from their husbands (Belsky and Kelly 1994).

Mothers' sense of responsibility may also keep them from other sources of support, which is another factor that puts stress on their marriages. Several of the mothers in my sample reported that their ability to keep up with social networks was affected by their sense of needing to get home to their babies. Women who lost contact with work or other social networks became more aware of what they did not get from their husbands:

> I need him sometimes to be my girlfriend and he's not. . . . I feel sorry for him because he wasn't ready for that. . . . You don't realize how much you need those other people until you see them less frequently.

Decreases in new mothers' marital satisfaction, I am suggesting, are related both to the lack of recognition and sharing of mental labor by their husbands and to the loss of independence and support from other people that mothers' exclusive mental responsibility generates. If men and women who become parents together shared the mental labor associated with taking care of a baby, neither one would be "free," but perhaps neither one would be unhappy.[4]

INVISIBLE WORK AND DOING GENDER

One prominent explanation for gendered divisions of baby care in general is that the capacity to soothe or respond to a baby's hunger is more innate in mothers than fathers, yet it is not clear why a new mother would feel more worried and driven to read baby-care books if she has more "natural" ability. While the notion that mothers and fathers differ in caretaking competence has been refuted by some research (see, e.g., Parke and Sawin 1976), even those social scientists who argue for the salience of sex differences in caretaking capacity suggest that societies will ascribe more or less meaning to these differences. In Rossi's (1985) "biosocial" approach, biological and cultural factors interact in determining male and female parenting roles. In this discussion I focus on the cultural part of this equation and examine the role of gender in the reproduction of differentiation between new mothers and fathers.

As discussed, new parents experience distinct and different norms attached to motherhood and fatherhood. While fatherhood is equated foremost with economic provision, motherhood is socially constructed as a "constant and exclusive responsibility" (Thompson and Walker 1989:860). These norms have been linked in sociohistorical accounts with Western, dichotomized images of public and private, work and love, that became especially pronounced during nineteenth-century industrialization. As manufacturing took paid work out of households, the "public" sphere of the economy and state became perceived as a male sphere and economic provision the job of fathers, while women were left (at least ideologically if not in reality) to the "private" domain of the household and children (see Glenn 1994; Osmond and Thorne 1993).

This ideology continues to be reflected in pay inequity, occupational segregation, and other gendered workplace processes that both assume and reinforce divisions of labor in which women take primary responsibility for families (Reskin and Padavic 1994). The devaluation of "women's work," both paid and unpaid, has been a particular point of entry for the feminist argument that the ideological separation of public and private, production from reproduction, is a source of exploitation of women (see, e.g., Hartmann 1981).

One question that has puzzled social scientists, however, is why many women do not experience their disproportionate responsibility for household labor as oppressive (Berk 1985; Thompson 1991; Thompson and Walker 1989). The theoretical answers to this question are relevant for understanding divisions of mental baby care. DeVault (1991:11) suggests, for example, that women often do not recognize feeding their families as work because it is perceived as "embedded in family relations," "part of being a parent . . . or of being a wife." The notion that housework is considered to be an integral part of being a wife is reinforced by South and Spitze's (1994) finding in their analysis of housework patterns across marital statuses that married couples have the highest gender gaps in housework.

Studies of divisions of housework provide support for the theoretical notion that imbalances in household labor when men and women live together is a way in which they "do gender"

(West and Zimmerman 1987); that is, they construct their social identities as "women" and "men" through their performance (or not) of "women's work" (Berk 1985; DeVault 1991). The fact that much of the mental work associated with household labor is invisible to the women who do it reinforces the notion that it is simply part of their identities and something for which they are perceived as having a "natural" propensity (Daniels 1987). For women to not do this work might challenge their social definition as women.

Taking responsibility for babies is socially constructed as "women's work," and men and women participate in reproducing this construction through their interactions with each other. One of the women in my sample said of what it means to be "the wife and mother":

> If I hear him [baby] cry during the night, I'm more apt to get right up than Jake. Or if it's time to get up in the morning and I hear him, I'm more apt to get up and go get him. Jake is more apt to stay in bed and see what happens.

According to this respondent, "wives and mothers" do not wait to see if their male partners will take care of the baby. Her husband agreed that although they share their baby's care, his wife does "a little more . . . because she is his mother."

In a hypothetical game of "chicken," in which the winner is the parent who can wait longer for the other parent to take responsibility for a baby's needs, it is difficult for mothers not to lose. There is a much greater threat to their social identities as mothers than there is for fathers if, in any particular moment, they are not taking responsibility for their baby (see also McMahon 1995). One explanation for why new parents reproduce differentiated images of mothering and fathering therefore is because they feel accountable to these already established images, "to normative conceptions regarding the essential womanly nature of child care" (West and Fenstermaker 1993:165).

Perhaps more than any other aspect of gender, Glenn (1994:3) suggests, mothering is perceived as "natural, universal, and unchanging," and in this sense, worrying and knowing about the baby may be constructed simply as part of being a mother in the way that feeding the family is. As with the invisible parts of feeding, men and women becoming parents differentiate themselves as "mothers" and "fathers" by how much they think (or think they should think) about their babies.

I am suggesting here a different spin on new parents' apparent identification with gender-differentiated parental functions. Ehrensaft (1987), for example, suggests that men perceive fathering as something they "do," while women experience mothering as something they "are." Cowan and Cowan (1992) report that new mothers experience a subsuming of themselves into mothering while new fathers become more preoccupied with their abilities as breadwinners. Rather than simply identifying with gender-differentiated images of mothers and fathers, I suggest that new parents feel accountable to these images and reinforce their partner's accountability to these images in order to accomplish parenting and gender at the same time (see Fenstermaker, West, and Zimmerman 1991; West and Fenstermaker 1993).

Whether still employed or not, being in charge of baby care places mothers in a different relationship to paid work than their male partners, whose accountability to the breadwinner image may induce more distance between themselves and their babies. A father can be perceived as a "good" father without thinking about his baby; in fact, his baby may pose a distraction to his doing what he is expected to do. Mothers, on the other hand, are expected to think about their babies. They perform a disproportionate amount of mental baby care not because they are "good" mothers, but in order to be.

CONCLUSION

This discussion has been an attempt to suggest some of the interactional and institutional processes underlying differences in how men and women who become parents together think about babies as well as to highlight the importance of this issue for marital relationships. To the extent that the discussion has emphasized common experiences by gender, this has been intentional, though not necessarily an expected finding. In embarking on the larger study of gender differentiation in transitions into parenthood from which this [reading] is derived, I expected to find variations in the couples in my sample. Although there were indeed variations in how couples approached the care of their babies, the tendency for mothers to be responsible for a variety of forms of mental baby care emerged strikingly in my data as a source of gender differentiation, even in situations of relatively shared physical care.

While this pattern appeared in my sample across employment statuses and family experiences, this study does not claim to be a test of the mediating power of work, family, and other variables, which might be areas for future research. These data are part of a theory-generating study about gender-differentiation in the transition to parenthood and should not be seen as a test of hypotheses. Rather I have tried here to make a theoretical case for the importance of recognizing the mental labor that accompanies physical infant care; I suggest that the way mental labor is divided in male-female couples in transition to parenthood is a way in which women and men recreate motherhood and fatherhood as differentiated social experiences.

Finally, I have suggested that gendered divisions of mental labor may be an underrecognized factor in decreases in marital satisfaction following the birth of a first baby. Women who experience marital dissatisfaction upon becoming new mothers will not necessarily be relieved simply by trading off diaper changes. Only when the work of thinking about the baby is shared can new fathers claim to be truly equal participants and new mothers able to make their economic and other contributions to their babies with less stress and guilt.

NOTES

1. DeVault's (1991) examination of feeding work, for example, elaborates the notion that feeding involves mental labor preceding and beyond the physical act of meal preparation. Feeding the family includes planning, strategizing, juggling various individuals' needs as well as facilitating group interaction.
2. See Risman and Schwartz (1989) for a discussion of individualist versus microstructural approaches to gender; see also Ferree (1991).
3. See Lamb (1978) and LaRossa and LaRossa (1989) for discussions of how babies' dependency contributes to traditionalization in parental roles.
4. Ross, Mirowsky, and Huber (1983), for example, find that wives are less depressed when their husbands help with housework; and the husbands are no more depressed as a result of their contributions. Marshall and Barnett (1995) find that when husbands share supervision of children, both husbands and wives report lower psychological distress.

REFERENCES

Belsky, Jay. 1985. "Exploring individual differences in marital change across the transition to parenthood: The role of violated expectations." Journal of Marriage and the Family, November:1037–1044.

Belsky, Jay, and John Kelly. 1994. The Transition to Parenthood. New York: Delacorte Press.

Belsky, Jay, Mary Lang, and Ted L. Huston. 1986. "Sex typing and division of labor as determinants of marital change across the transition to parenthood." Journal of Personality and Social Psychology 50:517–522.

Belsky, Jay, and Brenda L. Volling. 1987. "Mothering, fathering, and marital interaction in the family triad during infancy: Exploring family systems processes." In Men's Transitions to Parenthood, eds. Phyllis W. Berman and Frank A. Pedersen, 37–63. Hillsdale, N.J.: Lawrence Erlbaum Associates, Inc.

Benson, Leonard. 1968. Fatherhood: A Sociological Perspective. New York: Random House.

Berk, Sarah Fenstermaker. 1985. The Gender Factory. New York: Plenum Press.

Berman, Phyllis W., and Frank W. Pedersen. 1987. "Research on men's transitions to parenthood: An integrative discussion." In Men's Transitions to Parenthood, eds. Phyllis W. Berman and Frank A. Pedersen, 217–242. Hillsdale, N.J.: Lawrence Erlbaum Associates, Inc.

Blumberg, Rae Lesser, and Marion Tolbert Coleman. 1989. "A theoretical look at the gender balance of power in the American couple." Journal of Family Issues 10:225–250.

Coltrane, Scott. 1989. "Household labor and the routine production of gender." Social Problems 36:473–490.

Coontz, Stephanie. 1992. The Way We Never Were: American Families and the Nostalgia Trap. New York: Basic Books.

Cowan, Carolyn Pape, and Philip A. Cowan. 1988. "Who does what when partners become parents: Implications for men, women, and marriage." Marriage and Family Review 12:105–131.

———. 1992. When Partners Become Parents. New York: Basic Books.

Daniels, Arlene Kaplan. 1987. "Invisible work." Social Problems 34:403–414.

DeVault, Marjorie L. 1991. Feeding the Family. Chicago: University of Chicago Press.

Dickie, Jane R. 1987. "Interrelationships within the mother-father-infant triad." In Men's Transitions to Parenthood, eds. Phyllis W. Berman and Frank A. Pedersen, 113–143. Hillsdale, N.J.: Lawrence Erlbaum Associates, Inc.

Ehrensaft, Diane. 1987. Parenting Together. New York: The Free Press.

Eisenberg, Arlene, Heidi E. Murkoff, and Sandee E. Hathaway. 1989. What to Expect the First Year. New York: Workman Publishing.

Fenstermaker, Sarah, Candace West, and Don H. Zimmerman. 1991. "Gender inequality: New conceptual terrain." In Gender, Family, and Economy: The Triple Overlap, ed. Rae Lesser Blumberg, 289–307. Newbury Park, Calif.: Sage Publications, Inc.

Ferree, Myra Marx. 1991. "Feminism and family research." In Contemporary Families, ed. Alan Booth. Minneapolis, Minn.: National Council on Family Relations.

Glaser, Barney G., and Anselm L. Strauss. 1967. The Discovery of Grounded Theory. New York: Aldine De Gruyter.

Glenn, Evelyn Nakano. 1994. "Social constructions of mothering: A thematic overview." In Mothering: Ideology, Experience, and Agency, eds. Evelyn Nakano Glenn, Grace Chang, and Linda Rennie Forcey, 1–29. New York: Routledge.

Harriman, Lynda Cooper. 1985. "Marital adjustment as related to personal and marital changes accompanying parenthood." Family Relations 34:233–239.

Hartmann, Heidi I. 1981. "The family as the locus of gender, class, and political struggle: The example of housework." Signs: Journal of Women in Culture and Society 6:366–394.

Hays, Sharon. 1993. "The cultural contradictions of contemporary motherhood: The social construction and paradoxical persistence of intensive child-rearing." Ph.D. dissertation, University of California, San Diego.

Hochschild, Arlie Russell. 1983. The Managed Heart: Commercialization of Human Feeling. Berkeley: University of California Press.

Hochschild, Arlie, with Anne Machung. 1989. The Second Shift. New York: Viking Press.

Kranichfeld, Marion L. 1987. "Rethinking family power." Journal of Family Issues 8:42–56.

Lamb, Michael E. 1978. "Influence of the child on marital quality and family interaction during the prenatal, perinatal, and infancy periods." In Child Influences on Marital and Family Interaction, eds. Richard M. Lerner and Graham B. Spanier, 137–164. New York: Academic Press.

LaRossa, Ralph. 1986. Becoming a Parent. Beverly Hills, Calif.: Sage Publications.

LaRossa, Ralph, and Maureen Mulligan LaRossa. 1989. "Baby care: Fathers vs. mothers." In Gender in Intimate Relationships: A Microstructural Approach, eds. Barbara J. Risman and Pepper Schwartz, 138–154. Belmont, Calif.: Wadsworth Publishing Company.

Luxton, Meg. 1980. More Than a Labour of Love: Three Generations of Women's Work in the Home. Toronto: Women's Press.

Marshall, Harriette. 1991. "The social construction of motherhood: An analysis of childcare and parenting manuals." In Mothering: Meanings, Practices, and Ideologies, eds. Ann Phoenix, Anne Woollett and Eva Lloyd, 66–85. Newbury Park, Calif.: Sage Publications.

Marshall, Nancy L., and Rosalind C. Barnett. 1995. "Child care, division of labor, and parental emotional well-being among two-earner couples." Paper presented at the 90th Annual Meeting of the American Sociological Association, Washington, D.C.

McMahon, Martha. 1995. Engendering Motherhood: Identity and Self-Transformation in Women's Lives. New York: The Guilford Press.

Mederer, Helen J. 1993. "Division of labor in two-earner homes: Task accomplishment versus household management as critical variables in perceptions about family work." Journal of Marriage and the Family 55:133–145.

Osmond, Marie Withers, and Barrie Thorne. 1993. "Feminist theories: The social construction of gender in families and society." In Sourcebook of Family Theories and Methods: A Contextual Approach, eds. P. G. Boss, W. J. Doherty, R. LaRossa, W. R. Schumm, and S. K. Steinmetz, 591–623. New York: Plenum Press.

Parke, Ross D., and Douglas B. Sawin. 1976. "The father's role in infancy: A re-evaluation." The Family Coordinator 25:365–371.

Reskin, Barbara, and Irene Padavic. 1994. Women and Men at Work. Thousand Oaks, Calif.: Pine Forge Press.

Risman, Barbara J., and Pepper Schwartz. 1989. "Being gendered: A microstructural view of intimate relationships." In Gender in Intimate Relationships: A Microstructural Approach, 1–9. Belmont, Calif.: Wadsworth Publishing Company.

Ross, Catherine E., John Mirowsky, and Joan Huber. 1983. "Dividing work, sharing work, and in-between: Marriage patterns and depression." American Sociological Review 48:809–823.

Rossi, Alice S. 1985. "Gender and parenthood." In Gender and the Life Course, ed. Alice S. Rossi, 161–191. New York: Aldine Publishing Co.

Ruble, Diane N., Alison S. Fleming, Lisa S. Hackel, and Charles Stangor. 1988. "Changes in the marital relationship during the transition to first time motherhood. Effects of violated expectations concerning division of household labor." Journal of Personality and Social Psychology 55:78–87.

Ruddick, Sara. 1983. "Maternal thinking." In Mothering: Essays in Feminist Theory, 213–230. Savage, MD: Rowman & Littlefield Publishers, Inc.

Schuchts, Robert A., and Stanley L. Witkin. 1989. "Assessing marital change during the transition to parenthood." Social Casework: The Journal of Contemporary Social Work, February:67–75.

South, Scott J., and Glenna Spitze. 1994. "Housework in marital and nonmarital households." American Sociological Review 59:327–347.

Thompson, Linda. 1991. "Family work: Women's sense of fairness." Journal of Family Issues 12:181–196.

Thompson, Linda, and Alexis J. Walker. 1989. "Gender in families: Women and men in marriage, work, and parenthood." Journal of Marriage and the Family 51:845–871.

Waldron, Holly, and Donald K. Routh. 1981. "The effect of the first child on the marital relationship." Journal of Marriage and the Family, November:785–788.

West, Candace, and Sarah Fenstermaker. 1993. "Power, inequality, and the accomplishment of gender: An ethnomethodological view." In Theory on Gender/Feminism on Theory, ed. Paula England, 151–174. New York: Aldine de Gruyter.

West, Candace, and Don H. Zimmerman. 1987. "Doing gender." Gender and Society 1:125–151.

19

Japanese Families: The Father's Place in a Changing World

GARY DeCOKER

INTRODUCTION

In the 1930s the anthropologist Malinowski, after years of cross-cultural research on the family, devised what he called "the first law of anthropology." He generalized that in every society each child has a "social father," one male who is the child's father figure. In the past few decades, however, Malinowski's law has not applied in the United States, nor perhaps in Japan. In increasing numbers of American families, mothers raise their children alone. And in Japan a commitment to work often keeps fathers apart from their children.

Even though the family is one of the most basic human entities, most anthropologists have abandoned the search for Malinowski-style "laws" governing family life across cultures. Instead, they strive to understand the institution of the family from within each culture on its own terms. They may use similar concepts to look at the institution of the family in different societies, but because of the way the society and the family interact, they try to put their comparisons into the context of each individual culture. A discussion of fatherhood in Japan, therefore, requires a broad look at Japanese society and a narrow look at specific roles and behaviors within the family. For that reason, we will begin by looking at the way Japan has changed over the last few decades. Then we will explore the way these changes have affected the family.

Since the 1950s, an important decade in postwar Japan and the United States, both societies have experienced significant changes in the birthrate, women's employment rate, and government policies toward the family. The United States has also seen a rise in the divorce rate and number of single-parent families. The poverty rate for children, although it fluctuates, remains around 20 percent—twice that for some minority groups. Sociologists have begun using the term "postmodern" to describe these changes in the American family. If you believe in the theory of convergence, that is, that all societies follow a similar path as they change in response to modernity, you will expect the Japanese family to evolve along the United States model. Most social scientists, however, have come to believe that Japan and the United States are responding differently to the postmodern world. In both societies the institution of the family will continue to change, but in different ways.

In Japan, just as in the United States, the institution of the family has changed. Birthrates have declined. The age of marriage has increased. More women are entering traditionally male professions. But these changes have not dominated the recent history of the Japanese family. Many aspects of family life in Japan have remained the same throughout the postwar years. For instance, the stereotypical father still works long hours and spends little time with his children. And society still defines the ideal mother as a woman who devotes nearly all her time to the education of the children.

THE IMAGE OF THE FAMILY

A culture's image of the family affects the way society adjusts to change. In the United States the image of a "traditional family"—working father, homemaker mother, and children—remains strong. Conservative politicians and religious leaders equate this family type with the strong economy and social stability of the 1950s, and they lament the societal changes that have weakened the traditional family and decreased its number to about one-third of U.S. families with children.[1] On the other hand, sociologists point out that "nontraditional families" have existed throughout the postwar era. Even in the 1950s, the family took on many forms, especially among the poor and in minority groups as they struggled for a place in the economic system. The recent "changes" in the U.S. family took place primarily in the middle class.

Although varying images of the postwar family coexist, the traditional family remains a powerful image in the United States, often becoming the standard measure of a successful family. Within the more recent dual-income family image, for instance, the "supermom," the mother who holds a full-time job while retaining all the previous responsibilities of motherhood, often is judged by her ability to fulfill the family obligations of a traditional mother. And a third and increasingly common family image, the single-parent family, depicts the mother struggling to make ends meet following a divorce or nonmarital birth. The image of the contemporary American family, therefore, now includes the working mother; children at daycare; and a father who may be absent or part of the family.

In Japan the nuclear family remains the dominant image. This image, having grown out of the early postwar "salaryman family," consists of a working father, a full-time housewife, and their children.[2] In this family the father, at work for long hours, leaves all the household responsibilities to his wife. The mother, in turn, devotes all her attention to her children and their schooling, hoping to help them succeed in recreating this family type for themselves as adults. The increasing affluence of the postwar era may have changed the salaryman family activities somewhat over the decades; for example, there is more time for family leisure activities on Sundays. But the traditional family roles of breadwinner and housewife remain intact. Modern household appliances, along with the decreasing size of Japanese families, also may have given the mother more time, but she does not use it to take a full-time job.[3] She keeps her focus on the home.

Unlike the variety of family images in the United States, the image of the salaryman family still dominates in Japan. Not all families in Japan, however, fit the salaryman form. Just as in the United States, the dominant image does not accurately reflect reality. In fact, the postwar salaryman ideal of "pro-fessional housewife" and working father exists in less than 30 percent of Japanese families today (Lock 1996). More women work than stay at home. Still, the traditional salaryman family image remains resistant to change.

Work and the Image of the Family

Images define the family, but social forces create it. Economic and social change constantly recreates the family and affects the experiences of individual members. Family image, however, sometimes mitigates against change. Japanese families, for instance, changed in reaction to postwar economic development when many mothers left home for the work force. But the image of the traditional Japanese family continued its influence. Japan's rapid economic growth during the recovery from the war created an acute shortage of workers. Suddenly more women were needed in the work force. But labor, in deference to the image of the mother in a traditional family, promoted full-time positions only for unmarried women. Married women, when they worked, did so primarily as part-time employees. In addition to labor's position on employment for women, government tax policies influenced women's work choices. Part-time housewives received large tax breaks that

Table 19.1 Percentage of Women in Various Occupations, 1997

Occupation	Japan (percent)	United States (percent)
Clerical	81.1	72.2
Services	55.1	59.4
Agricultural, forestry, livestock farming, hunting and fishing	44.2	19.3
Professional and technical positions	44.1	53.1
Sales	37.2	50.2
Production and transport	27.5	17.2
Managerial positions	9.3	44.3

Source: Japanese Prime Minister's Office web site, http://www.sorifu.go.jp/ danjyo/english/plan2000/1999/p1c201.html.

disappeared when they worked full time. Lastly, a woman's income in Japan still averages less than 60 percent of a man's, thereby further diminishing her desire to seek full-time employment (Uno 1993).

The overall statistics on labor force participation of Japanese women appear at first to differ only slightly from those for U.S. women. Women in both countries work. But further investigation reveals that the overall pattern of work in a woman's life varies greatly in the two countries. The Japanese employment system favors com-panies by ensuring the availability of a pool of low-wage female employees. Until recently the strong economy meant an availability of jobs for women. The 1990s recession, however, clearly revealed the status of women's work. Their jobs were the first to be eliminated, with the assumption that they would be given back when the economy recovered.

The two-track system of Japanese employment also contributes to the marginalization of women employees. The "career track" (*sôgô shoku*) consists primarily of men and the "standard track" (*ippan shoku*) entirely of women. Career-track positions, leading to promotions and in many cases lifetime employment, require long work hours and after-work socializing. Although they are technically open to women, few women apply for these positions. Most women seeking company or government positions instead choose the standard track with the expectation that they will quit upon marriage or the birth of their first child. For this reason, few Japanese women secure administrative/managerial positions in Japanese companies and government offices, as shown in Table 19.1.

Women who quit their standard-track positions, of course, often return to work after their childrearing years. But typically they are part-time employees. Labor-force participation over a Japanese woman's life, therefore, forms an "M-curve" as illustrated in Figure 19.1. In the U.S., by contrast, a similar percentage of women remain employed from about age twenty until retirement, so the employment curve forms an arch that does not dip. The image of the Japanese mother in the postwar salaryman family, therefore, perfectly fits the economic system, offering flexibility and

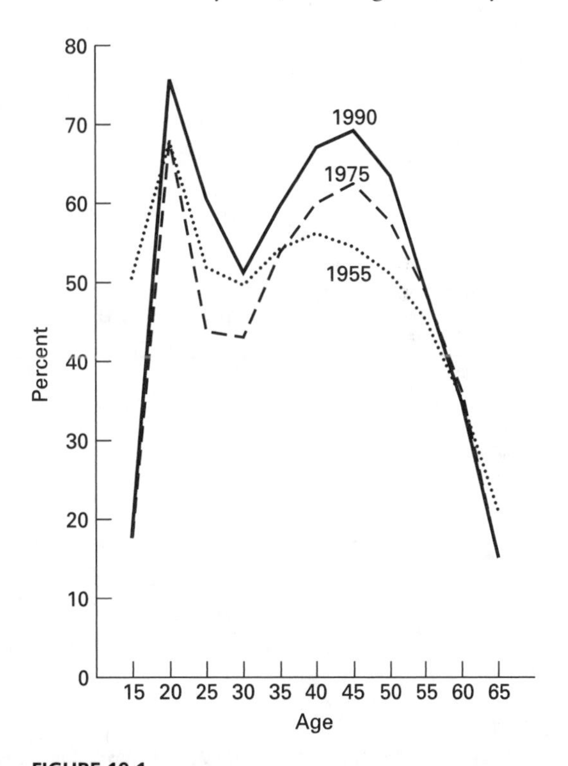

FIGURE 19.1
Percentage of Japanese Women Working, by Age Group, 1955, 1975, 1990.

Prime Minister's Office 1996a, 72.

stability. Women move between family and employment and the salaryman family remains intact.

The difference between the Japanese "M-curve" and the more consistent U.S. pattern of female employment throughout adult life may also result from differences in the economic stability of families in the two countries. In many American families, especially younger and single-parent families, mothers work out of economic necessity. Without their income, the family would plunge into poverty. In Japanese families, on the other hand, the mother's income supplements that of her husband. If she quits during childrearing years or loses her job during an economic recession, the family still can survive.

Attitudes toward work and family roles reflect the difference in women's labor force participation in the two countries. A 1994 Japanese government survey of Japanese and U.S. parents with children younger than age sixteen asked this question: "Which *one* of the choices on this list best describes your opinion as to how involved a father should be in housework and child raising?" In Japan, 54.9 percent of fathers and 41.9 percent of mothers responded that the father should devote all or a majority of his efforts to work. In the United States, 39.4 percent of fathers and 35.4 percent of mothers responded this way. The same survey also asked for opinions about working women. In Japan, 68.6 percent of the respondents indicated that a woman should not work during her childrearing years. In the United States, that figure was only 47.4 percent (Prime Minister's Office 1996b, 65, 67).

Contemporary Japanese attitudes toward work roles, however, look different when compared with Japanese attitudes of earlier decades. Figure 19.2, presenting survey results from the last three decades, clearly illustrates changing Japanese attitudes. These data remind us that comparisons of two countries sometimes make it difficult to notice important changes that are taking place inside the countries. When asked whether women should continue to work after marriage and childbirth, increasing numbers of Japanese respond affirmatively, from about 10 percent in 1972, to nearly 20 percent in 1984, to about 30 percent in 1995.

Table 19.2 also illustrates a gender difference in attitudes toward women working. More Japanese

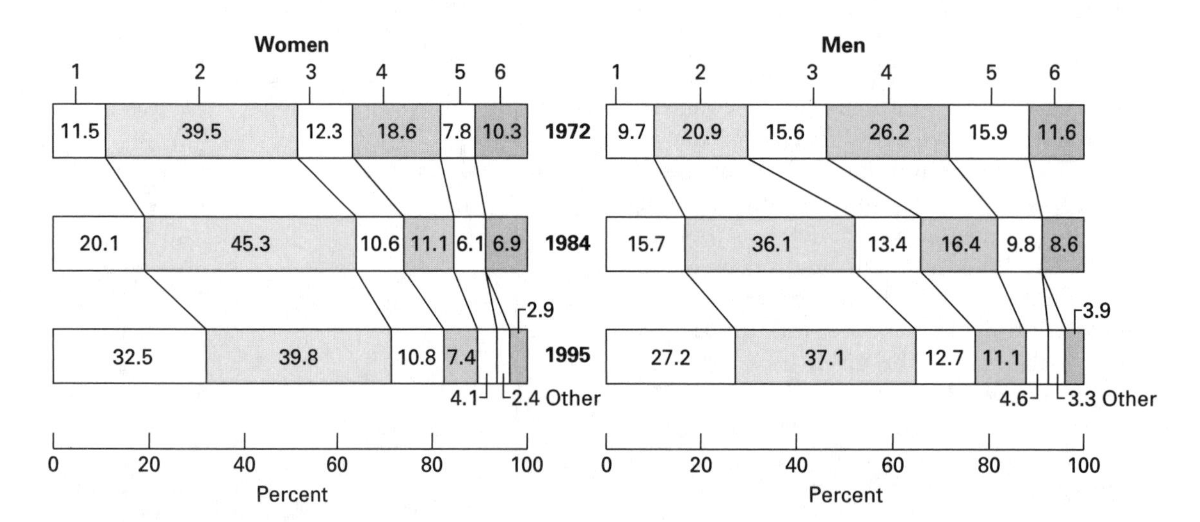

FIGURE 19.2
Response to the Question, "Should Women Work After Giving Birth?" Numbers in chart for women and men correspond to the following responses: 1. Even if they give birth, they should continue to work. 2. After their children grow up, they should go back to work. 3. They should work only until they give birth. 4. They should work only until they are married. 5. Women should not work. 6. Don't know.

Prime Minister's Office 1996a, 88.

Table 19.2 Fertility Rates, 1921 to 1996

	1921	1941	1951	1961	1971	1981	1986	1996
Japan	5.3	4.5	3.2	1.9	2.1	1.7	1.7	1.4
U.S.	3.3	2.3	3.2	3.6	2.3	1.8	1.8	2.0

Source: America's Families, United Nations Populations Division and World Health Organization web page http://www.un.org/Depts/unsd/social/childbr.htm

men than women seem comfortable with traditional images of the family. In fact, some younger Japanese women, stressing the need for emotional fulfillment in marriage, have begun to express a lack of desire to marry at all.[4] A young Japanese woman, working but still living with her parents, can enjoy the freedom her disposable income brings. For these women the Japanese image of the wife at home with the children appears too restrictive. The young working man, on the other hand, still searches for a more traditional woman who will eventually quit work to focus on home life. The 1996 Annual White Paper from the Japanese Ministry of Health and Welfare stated:

> A survey of unmarried persons under 35 years of age regarding what is considered and emphasized when choosing a marriage partner shows that a high percentage of men replied that they emphasize and consider their partner's character and appearance. In contrast, a high percentage of women replied that, in addition to character, they also emphasize and consider their attitudes toward the division of roles between men and women, in which men work to support the family finances and women engage in household work, are reflected in the selection of a marriage partner.[5]

In other words, many Japanese men search for a traditional wife, but many women search for a man who will accept nontraditional family roles. Regardless of the couple's beliefs, a Japanese woman often must balance her roles of mother and employee. She does so, however, by putting these two roles in separate time periods of her life. In a stable society such as Japan, a young mother can take a long-term view of personal development, realizing that the freedom to work exists in the time following childrearing but before retirement, a time referred to as *onna tengoku* or "women's heaven." Within her role of mother, few options exist for the Japanese woman. Anchored in the home with her children, the mother remains there while her husband focuses on his work role. While the children remain at home, she postpones her career and perfects her role as a mother. At this time in her life, stability is more important than taking initiative, and freedom of choice in employment is more of a distraction than an opportunity. Her choices, therefore, come only through stretching the time frame. Delayed entry into marriage, for instance, offers a longer period of freedom while single. Small families and bearing children early in marriage create a similar period of freedom later in a woman's life.

Men, on the other hand, keep their focus on work throughout their lives. One might expect, for instance, that husbands with working wives would devote more time to their families in an effort to compensate for their spouse's absence from the family. A 1991 survey conducted by the Prime Minister's Office, however, noted that the daily schedules of men in dual-income families differed little from those of men in families where the wife did not work.[6] On international surveys of men's involvement in family life, Japanese men show the least involvement.

In contrast to women, the career path for men is much less complicated, unaffected by changes in family status or economic downturns. The stereotypical male professional worker in Japan enters the work force after graduation from college and continues with the same company until retirement at about 60 years of age. For Japanese men in these positions, work leaves little time for family. The practice of lifetime employment, found in many white-collar occupations, creates intense pressure for promotions among age-cohorts in the company. New employees enter the company as a group each April. They experience orientation and company training programs together and over the years move up the hierarchy together. Promotions, however, gradually become more competitive. Early in a man's career a failure to be promoted could result in his being relegated to a nonessential position; toward the end of his career it might lead to forced retirement. These factors exert strong pressure on men to devote themselves to their work.[7]

Extended hours at the office followed by after-work socializing and long commutes make many men appear to be "guests" rather than members of their own families. They often leave for work early and do not return until late at night. A quick bath, dinner, and a few hours' sleep make up the only time many company employees spend at home during the week. Although the newly coined term *karôshi* or "death by overwork" may only apply in extreme cases, it does serve to illustrate the daily expectations of male employees. If there is time for the family, it only exists on Sundays and holidays. Recently, the Japanese government has begun promoting family issues by encouraging companies to decrease work hours and support childcare leaves. Government offices and schools are transitioning to five-day work/school weeks. And some men, freed from financial pressures by their working wives, have begun to leave their company jobs for new ventures.[8] These changes in the working lives of Japanese men may be the beginning of a gradual shift in the male role of the family (Gjerde 1996). But the image of a man's role still places him at work.

CHANGES WITHIN THE FAMILY

Images indeed affect people's decisions about family matters, and in Japan the image of the traditional family serves to hold back change. But change has occurred during Japan's postwar years, just as it has in the United States. These changes show up in data on the age of marriage, birthrates, divorce rates, and the family makeup. Tables 19.2, 19.3, and 19.4 illustrate some of these

Table 19.3 Births to Unmarried Women, by Country: 1980 to 1995

[For U.S. figures, marital status is inferred from a comparison of the child's and parents' surnames on the birth certificate for those states that do not report on marital status. No estimates are included for misstatements on birth records or failures to register births.]

	1980		1990		1994		1995	
Country	Total live births (1,000)	Percent born to unmarried women	Total live births (1,000)	Percent born to unmarried women	Total live births (1,000)	Percent born to unmarried women	Total live births (1,000)	Percent born to unmarried women
United States	**3,612**	**18**	**4,158**	**28**	**3,953**	**33**	**3,900**	**32**
Canada[1]	360	13	398	24	385	25	378	26
Denmark	57	33	63	46	70	47	70	46
France	800	11	762	30	711	36	728[2]	37[2]
Germany[3]	621	8	906	11	770	15	765	16
Italy	640	4	554	6	527	8	521[2]	8[2]
Japan	1,616	1	1,222	1	1,238	1	1,187	1
Netherlands	181	4	198	11	196	14	191	16
Sweden	97	40	124	47	112	52	103	53
United Kingdom	754	12	799	28	751	32	732	34

[1]1980 through 1990 excludes Newfoundland. After 1990, a significant number of births are not allocated according to marital status, resulting in an understatement of the proportion of births to unmarried women. [2]Preliminary. [3]Prior to 1990, data are for former West Germany.

Source: U.S. Department of Commerce. *Statistical Abstract of the United States, 1998: The National Data Book.* Author, 1998, p. 76.

Table 19.4 Marriage and Divorce Rates by Country: 1980 to 1995

	Marriage Rate per 1,000 Population, 15 to 64 Years Old						Divorce Rate per 1,000 Married Women					
Country	**1980**	**1990**	**1992**	**1993**	**1994**	**1995**	**1980**	**1990**	**1992**	**1993**	**1994**	**1995**
Japan	**10**	**8**	**9**	**9**	**9**	**9**	**5**	**5**	**6**	**6**	**6**	**6**
United States[1]	16	15	14	14	14	14	23	21	21	21	20	20
Canada	12	10	9	8	8	8	10	11	11	11	11	11
Denmark	8	9	9	9	10	10	11	13	13	13	13	12
France	10	8	7	7	7	7[2]	6	8	8	8	9	9[2]
Germany[3]	9	9	8	9	8	8	6	8	7	9	9	9
Italy	9	8	8	8	7	7[2]	1	2	2	2	2	2
Netherlands	10	9	9	8	8	8	8	8	9	9	10	10
Sweden	7	7	7	6	6	6	11	12	13	13	14	14
United Kingdom	12	10	9	9	9	8[2]	12[4]	13	12	12	13	13[2]

[1]Includes unlicensed marriages registered in California. [2]Preliminary. [3]Prior to 1991, data are for former West Germany. [4]England and Wales only.

Source: U.S. Department of Commerce. *Statistical Abstract of the United States, 1998: The National Data Book.* Author, 1998, p. 76.

changes and present startling statistics. Many people would point to these data as evidence of America's declining morality and its failure to uphold the traditional family structure. For instance, the divorce rate in the United States, having begun a steep increase in the mid-1960s, now stands at over four times that of Japan. But before we draw any conclusions about the two countries, we first should explore the social changes that lie behind these statistics. This search leads to cultural, social, and governmental policy issues.

Most Americans are surprised to learn that the average age of marriage is higher in Japan than in the United States. But marriage in Japan is confined to a narrow age range. Because the prospects of getting married after age 30 decline greatly, most Japanese feel a great deal of pressure to marry within a certain acceptable age range. Women older than 25 used to be called "Christmas cakes," that is, fresh and desirable at 23, 24, and 25, but stale and undesirable in the years that follow. Because of the rise in the marriage age, this metaphor might not hold, but the pressure to get married still increases for both men and women in their late twenties.

The age range of mothers at the birth of their children also is more compressed in Japan. The societal pressure to give birth within a year or so after marriage quickly puts newlyweds into their roles as parents. At this point, many women quit their jobs and devote their next decade to childrearing. Men, in contrast, increasingly focus their attention on their occupations in their role as breadwinner. In other words, rather than spending the first few years establishing a relationship as a dual-income family, Japanese couples quickly move into their traditional roles as parents.

Traditional roles, however, do not always mean that families are living under the same roof. The Japanese practice of job transfer (*tanshin funin*) sometimes requires businessmen to live apart from their families. Companies often transfer male employees to various subsidiaries and other offices in order to give employees a broad understanding of the organization. Their wives typically remain at home. Americans might wonder why wives do not join their spouses. But the importance of continuity in their children's education, the cost of housing, and caring for elderly parents keep wives rooted to the original home. In 1994 more than 300,000 male workers were living apart from their families.[9]

Even though family roles remain fairly well defined in Japan, not all people would call their marriages fulfilling. In many families, men work long hours and on weekends, and women remain at home with their children or go out to a part-time job or cultural arts classes. Some Japanese refer to an unhappy marriage as a *kateinai rikon* or "home divorce." Although these couples might prefer divorce, they often resist the temptation. Divorced men fear hurting their chances of promotion. Women, aware that Japanese society considers divorce more shameful for women than men, know they will take the brunt of the criticism for the divorce.

Economic circumstances also mitigate against divorce. Women often receive little or no money in divorce settlements, and a majority must support themselves and their children without their husband's help (Axinn 1990). Because most mothers lack full-time employment, they have little hope for a successful life alone. For that reason, those women who choose divorce typically move back to their parents' house. This move gives stability to their children. But the mother's position as *demodori* ("left and come back," a derogatory term for divorced women) leaves her few opportunities for a second marriage. In other words, Japan's low divorce rate, resulting from economic realities and societal images, fails to show the number of families living apart physically and/or psychologically. After living their lives separately, some couples have difficulty adjusting to retirement. As a result, retired males have received the label *nureochiba* (wet, fallen leaves) for the way they stick around the house all day with nothing to do. Other nicknames are even less flattering: *sôdaigomi* (large garbage) and *sangyô haikibutsu* (industrial waste).

The Changing Family and Its Effect on Children

Educators in the United States often note various ways a family's financial situation affects their children's schooling. Children from poor families are more likely to fall behind in school or to drop out altogether. As a group poor children score lower on IQ and achievement tests and are more likely to have learning disabilities and to be placed in special education classes. It follows that these children are also less likely to enroll in postsecondary education. We cannot conclude that poor parents do not want the best for their children. But as poverty increases, parents are under greater pressure and children are at greater risk.

The Japanese postwar family has been much less affected by poverty. In the immediate postwar era, of course, families faced the severe conditions of a country destroyed by war. By the 1970s, however, most Japanese families could expect that their children would graduate from high school. Those young people who chose to pursue postsecondary education had an increasing array of choices. Economic progress paralleled educational progress. The entire country seemed to move in giant steps up the economic and educational ladders during the rapid economic growth of the 1960s and early '70s. Perhaps for this reason, most Japanese continue to describe themselves as middle class. Poverty, as a problem for a distant social class, has not existed in postwar Japan as it does in the United States.

Changes in Japanese families, therefore, have not been driven by poverty nor single parenthood. Instead, the changes that have taken place in the family have evolved around the wife as she moves between the roles of mother and part-time employee. Two images of the mother's role have dominated the postwar era: the *ryôsai kenbo* (good wife, wise mother), which originated in the late nineteenth century, and the more familiar *kyôiku mama* (education mother), which originated in the salaryman family of the 1960s.

The *ryôsai kenbo* image was cultivated in the late Meiji period as the government pulled back from its earlier openness to Western ideas. The Ministry of Education, stressing the differences between men and women, created separate educational agendas for boys and girls. It used the term *ryôsai kenbo* as a label for the kind of woman it hoped to produce. Early in the period, the government stressed a woman's subservient role to her husband and her responsibility as a mother. Later, during the war years, the government focused on a woman's reproductive role in producing more citizens for the state.

After the war, the term *ryôsai kenbo* disappeared from the school curriculum along with the old ethics curriculum of loyalty and filial piety. But the idea of the "good wife, wise mother" continues to influence policy. This ideal makes women low-paid workers in the labor economy and focuses public policy on their reproductive role. Some politicians continue to exhort women to bear more children in order to reverse the declining birthrate. Prime Minister Hashimoto, for instance, when he was the minister of finance in 1990, linked the falling birthrate to the increasing level of higher education of Japanese women. Education, he implied, distracted women from their duties in the home.

In postwar Japan, as women increased their educational levels and continued to work in part-time jobs, a new ideology developed that built on the old *ryôsai kenbo* ideal. The mother in the salaryman family, freed by technology from many of the drudgeries of housework and often freed by urbanization from the imposing presence of her mother-in-law, saw her energy redirected to her children in the ideal of the *kyôiku mama* or "education mother."

Kyôiku mama, used as a term of respect and also of criticism, defines the devotion of the full-time mother to her children. In its positive form she arranges her children's schedules and assists with their schoolwork. In its pathological form she pushes her children so hard that they crack under the pressure, as victims of suicide or as perpetrators of severely antisocial behavior.

Mothers learn their role from the society and from the school. Images of motherhood, many left over from the *ryôsai kenbo* ideal, pervade contemporary Japanese society. For instance, no other job is thought to be better or more suited to women than motherhood, because of the natural bond between mother and child. Mothers are said to provide the best care and education for their young children. These images together create strong societal support for the ideal of the full-time mother (Fujita 1989).[10]

Schools, too, reinforce these images. Young mothers bringing their first-born child to preschool quickly learn that their supporting role is crucial to their child's success. These mothers receive detailed instructions from the schools, which regulate every aspect of the child's appearance—clothing, book bag, *obentô* (homemade box lunch)—and the child's home life—bedtime, weekend, and summer activities (Allison 1996, chap. 4, 5). Of course, preschools vary in the degree to which they stress these regulations. But all schools use them to communicate to the young mother important messages about her role in her child's future schooling.

As her child grows older, the mother takes more of a supporting role by setting up tutoring and cram school lessons, meeting with teachers, and being on call when the child needs help. Home desks with a buzzer for ringing mother in the kitchen may not exist in every home, but they represent the ideal of a devoted *kyôiku mama.* Finally, at the end of her extended tour of duty, the mother accompanies the child to the college entrance examinations and later waits in anticipation for the results. College acceptance signals the end of her role as *kôiku mama.*

Japanese mothers vary greatly in their devotion to the ideal. But no mother can avoid its social pressure. Sometimes this pressure comes from a school's expectation that she needs to offer more to her child. Other times it comes from a mother-in-law seeking a more devoted wife for her husband. The mother even may receive deriding comments from other mothers or acquaintances who criticize her for being too lax or perhaps too intense.

Images of motherhood, of course, convey complementary images of fatherhood: Mother nurtures the children; father provides for them. A 1995 international survey found that, compared with fathers in other countries, Japanese fathers spend the least amount of time with their children (Prime Minister's Office 1996a).[11] In another survey from 194, the most common response to the question of how much time is spent each day with their children was about 30 minutes for Japanese fathers; about one hour for mothers. Most U.S. fathers indicated about 3 hours; mothers, between 6 and 10 hours.[12]

Figure 19.3 illustrates the family dynamics in Japan and the United States. The strongest bond in the United States is between husband and wife. As parents and as individuals, they have a relationship with their children. In Japanese families, however, the strongest bond for the husband lies outside of the family with his work. The wife, left without the strong marital bond with her husband, turns to her children, and her relationship with them typically revolves around their schoolwork. Mothers attend to their children's,

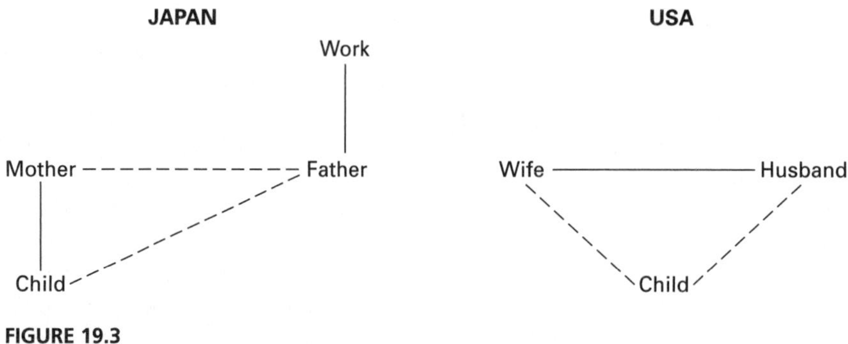

FIGURE 19.3
Family dynamics in Japan and the U.S.A.

From: F. Kumagai, "Filial violence in Japan," *Victimology,* 8:174–194, 1983.

especially their son's, educational success. Fathers attend to their work and their role of provider.

The Japanese male's focus on work leads many people to label Japanese families as "fatherless." And, given the *kyôiku mama* image of the mother, one might conclude that Japanese children feel a paternal void. But many children seem to support the father's traditional role. Indeed, on a 1994 survey of children in eleven countries, 25.6 percent of Japanese children indicated that they preferred a father who put his work ahead of the family. In the United States that number is 3.7 percent. In addition, 30 percent of Japanese children, compared with 14.6 percent in the United States, expressed a preference for a strict father over a friendly one. These images correspond with the attitudes of Japanese adults. In response to the statement, "A father should be stern with his children, while a mother should be affectionate," 55.7 percent of Japanese but only 27.4 percent of Americans agreed (Prime Minister's Office 1996b 38–41). But before concluding that the Japanese father is a "shadowy" tyrant, we should compare the image of the father with the reality of family life.

Although the father might be physically and perhaps emotionally absent from the family, he is psychologically present. Mothers, in their interactions with their children, construct an ideal image of the father and use it in childrearing. When urging her children to comply, for instance, she refers not to her own authority, but to her husband's. In this way she can set boundaries and demands while protecting her nurturing relationship with her children. Because the concept of *amai* (indulgence) defines the Japanese mother-child interaction, mothers seldom exert their authority.[13] Instead, they use the image of the father. This image of the strict father, therefore, actually arises more from the actions of the mother than the father. In reality, "Japanese fathers, when present, are more likely to be playmates than disciplinarians" (Vogel 1996).

CONCLUSION

Although male and female roles in Japanese society are changing, economic and social factors seem to mitigate against that change. More women are working, but opportunities decline in tight economic times. Even though women probably will continue to become more prominent in the work force, most Japanese still view men as providers and women as homemakers. As a result, men's roles in the family seem resistant to change.

Perhaps the place to look when predicting future change is to Japanese young people and the expectations parents have for their sons and daughters. Recent surveys clearly demonstrate the Japanese preference for traditional male and female roles. When asked what level of education they hope their children will reach, 67.5 percent of Japanese parents state university for their boys, but only 39.6 percent aspire to this level for their girls. American parents, as in many European and Asian countries, hold similar educational aspirations for their boys and girls (Prime Minister's Office 1996b, 47). Another survey regarding parental expectations of their child's manners and personal habits illustrates the Japanese tendency to differentiate between boys and girls. Whereas U.S. parents had fairly similar expectations for their male and female children, Japanese parents had higher expectations of their daughters in many of the categories. For their five-year-olds the percentage of parents indicating the importance of table manners was, for boys, 66.7 percent; for girls, 77.9 percent. For cleanliness: boys, 60.3 percent; girls, 70.4 percent. For cleaning up after playing: boys, 57.6 percent; girls, 64.6 percent. For their fifteen-year-olds, regarding cleaning their room: girls, 96.6 percent; boys, 85.3 percent. For cooking for the family: boys, 47.9 percent; girls, 80.7 percent. With the exception of cooking U.S. parents had very similar expectations for their sons and daughters (Prime Minister's Office 1996b, 68).[14]

It is difficult to predict whether future generations of Japanese will follow their parents' values regarding gender roles. But as Merry White points out, "The compartmentalization of both men's and women's lives relies on the particular kind of social and economic stability existing in Japan today; this stability cannot be assumed for the future" (Summer 1987, 162). Japan will surely change. The direction will depend on the way social and economic circumstances interact with the characteristics of Japanese society. Perhaps some indication of the continuation of change in the roles of men and women in Japan can be found in a survey of Japanese women. Adult men and women were asked whether, if they had the opportunity to be reborn, they would choose to be born a man or a woman. In 1950, only 16.4 percent of women chose to be reborn female. In 1987, that number rose to 53.7 percent. Clearly, women find satisfaction in recent societal changes. About 82 percent of men in 1987, on the other hand, indicated a preference for being born a male (Prime Ministers's Office 1996a, 4).[15] Men still appear happier with their gender. This satisfaction may affect the way they adjust to the changes in work and family expectations that societal and economic conditions will continue to bring. Perhaps their feeling of satisfaction with themselves as males will make them more open to change. Or perhaps it will result in their holding firmly to the way things are.

NOTES

1. Some surveys show that "traditional families" make up only about 10 percent of American families, but these do not distinguish between families with children, childless families, and families whose children are no longer at home (Skolnick 1991, 205).
2. See Imamura (1996) for a discussion of changing postwar images of Japanese women.
3. Japan's fertility rate has fallen dramatically since World War II, from 4.5 in 1941 to 1.4 in 1996. See Table 19.3.
4. In 1987, 24.7 percent of Japanese women and 15.7 percent of men agreed with the statement that it was acceptable for people not to marry (Prime Minister's Office 1996a, 32).
5. Japanese Ministry of Health & Welfare web site http://www.mhw.go.jp/english/white_p/book1/p1_c1/c1_sect3.html.
6. Japanese Prime Minister's Office web site http://www.sorifu.go.jp/whitepaper/danjyo/sankaku/1-2-3.html.
7. See John Lorriman and Takashi Kenjo, chap. 8 in *Japan's Winning Margins: Management, Training, and Education* (New York: Oxford University Press, 1994).
8. Fujimura-Fanselow and Kameda 1995, 237–239. Men who have quit their salaryman positions are called *dassara* or "escaped salarymen."
9. Shi'ina, Masae, "Coping Alone: When Work Separates Families," *Japan Quarterly* 34 (Jan.–March 1994): 26–34. In 1994, 321,000 households comprised married men aged 30 to 59 living alone, a 2.3-fold increase since 1975. In the same year, 55.3 percent of Japanese over the age of 65 were living with one of their children. *White Paper: Annual Report on Health and Welfare* (1996), http://www.mhw.go.jp/english/white_p/book1/p1_c1/c1_sect7.html.
10. Fujita explores several images of the ideal mother: "These images may be categorized into three groups: (i) the mother is the best care-taker and educator of children, (ii) the mother-child bonding is the most natural and fundamental one in human relationships (*ningen kankei*); and (iii) no other job is better or more suitable for women than mothering."

11. These averages include working and nonworking parents. Countries surveyed: Japan, South Korea, Thailand, United States, England, Sweden.
12. The difference in the two countries comes from the amount of time Japanese children spend in school. One-third of Japanese parents spend more than 6 hours a day with their newborn to three-year-old children. With their four- to six-year-old children, only 8.3 percent spend that amount of time. Over half spend one hour or less. Many parents in both countries indicate that they spend too little time with their children: Japanese fathers, 59.8 percent; mothers, 38 percent. U.S. fathers, 49.3 percent; mothers, 31.2 percent (Prime Minister's Office 1996b, 30–32).
13. For a discussion of *amai,* a key concept in Japanese socialization, see Merry White, *The Japanese Educational Challenge: A Commitment to Children* (New York: The Free Press, 1987).
14. There also was little gender differentiation in the other countries surveyed: South Korea, Thailand, England, and Sweden.
15. The survey only included women in 1950. The other choices were: men, 59.7 percent; either, 16.3 percent; don't know, 7.6 percent. In the data for 1987, men's choices were: women, 5.8 percent; either, 9.0 percent; don't know, 3.4 percent. Women's choices were: men, 29.7 percent; either, 13.2 percent; don't know, 3.5 percent.

REFERENCES

Allison, Anne. *Permitted and Prohibited Desires: Mothers, Comics, and Censorship in Japan.* Boulder, CO: Westview Press, 1996.

Axinn, June. "Japan: A Special Case." In *The Feminization of Poverty: Only in America?* Edited by Gertrude Schaffner Goldberg and Eleanor Kremen, 91–105. New York: Greenwood Press, 1990.

Fujimura-Fanselow, Kumiko, and Atsuko Kameda, eds. *Japanese Women: New Feminist Perspectives on the Past, Present, and Future.* New York: The Feminist Press, 1995.

Fujita, Mariko. "It's All Mother's Fault: Childcare and the Socialization of Working Mothers in Japan." *Journal of Japanese Studies* 15, no. 1 (Winter 1989): 67–91.

Gjerde, Per F. "Longitudinal Research in a Cultural Context: Reflections, Prospects, Challenges." In *Japanese Childrearing: Two Generations of Scholarship,* edited by David W. Shwalb and Barbara J. Shwalb, 279–299. New York: The Guilford Press, 1996.

Imamura, Anne E. "Introduction." In *Re-Imaging Japanese Women,* edited by Anne E. Imamura, 1–11. Berkeley: University of California Press, 1996.

Kumagai, Fumie. "Research on the Family in Japan." In *The Changing Family in Asia: Bangladesh, India, Japan, Philippines, Thailand,* edited by Unesco Principal Regional Office for Asia and the Pacific. Social and Human Sciences in Asia and the Pacific RUSHSAP Series on Monographs and Occasional Papers 35. Bangkok: Unesco Principal Regional Office for Asia and the Pacific, 1992.

Lock, Margaret. "Centering the Household: The Remaking of Female Maturity in Japan." In *Re-Imaging Japanese Women,* edited by Anne E. Imamura, 73–103. Berkeley: University of California Press, 1996.

Prime Minister's Office. *Chûgakusei no Hahaoya: America no Hahaoya to no Hikaku.* Tokyo: Sôrifu Seishônen Taisaku Honbu, 1991.

———. *Josei no Genjô to Sasaku.* Tokyo: Ôkurashô Insatsukyoku, 1996a.

———. *Kodomo to Katei ni Kan Suru Kokusai Hikaku Chôsa.* Tokyo: Sôrifu Seishônen Taisaku Honbu, 1996b.

U.S. Department of Commerce. *Statistical Abstract of the United States, 1998:* The National Data Book. Author, 1998, p. 76.

Select Committee on Children, Youth, and Families. *America's Families: Conditions, Trends, Hopes and Fears.* Washington, D.C.: U.S. House, 102nd Cong., 2nd sess., February 19, 1992.

Shi'ina, Masae. "Coping Alone: When Work Separates Families." *Japan Quarterly* 41 (Jan.–March, 1994): 26–34.

Skolnick, Arlene. *Embattled Paradise: The American Family in an Age of Uncertainty.* New York: Basic Books, 1991.

Uno, Kathleen S. "The Death of 'Good Wife, Wise Mother'?" In *Postwar Japan as History,* edited by Andrew Gordon, 293–322. Berkeley: University of California Press, 1993.

Vogel, Suzanne Hall. "Urban Middle-Class Japanese Family Life, 1958–1996: A Personal and Evolving Perspective." In *Japanese Childrearing: Two Generations of Scholarship,* edited by David W. Shwalb and Barbara J. Shwalb, 177–200. New York: The Guilford Press, 1996.

White, Merry. "The Virtue of Japanese Mothers: Cultural Definitions of Women's Lives." *Daedalus* 116, no. 3 (Summer 1997): 149–163.

20

Doing It Fairly: A Study of Postgender Marriages

BARBARA J. RISMAN
DANETTE JOHNSON-SUMMERFORD

Little is known about married couples who share the responsibilities of paid and family work without regard for gender prescriptions. Fifteen married couples who divide household work and child care equitably and without regard to gender are interviewed to determine how they arrived at this arrangement and what consequences such a distribution of household labor has on their relationship. Findings suggest that there are four paths to an equitable division of household labor: a dual-career household, a dual-nurturer relationship, a posttraditional relationship, and external forces. An egalitarian arrangement appears to affect both the power and emotional quality of couples' relationships.

The concept of gender itself, with its implied hierarchy in values, symbols, beliefs and statuses, is [a] cornerstone of the edifice of gender inequality.

KOMTER

1989, pp. 213–214

In this article, we explore how both behaviors and meanings in heterosexual marriages change when the spouses reject gender as a basis for organizing marital roles or responsibilities. We work from a feminist theory that defines gender as a system of stratification based on categorization that is created and recreated daily (Bem, 1993; Ferree, 1990; Ferree & Hess, 1987; Lorber, 1994; Risman & Schwartz, 1989; West & Zimmerman, 1987). Historically, the family is a "gender factory," where the polarization of masculine and feminine is created and displayed (Berk, 1985). Perhaps the best explanation for why women do an inequitable share of household labor is because we have defined such work as part of "being a woman" and "doing gender" appropriately (West & Zimmerman, 1987). Although gender is theorized within feminist scholarship as being forged at all levels of social life (Lorber, 1994), it is perhaps most evident in the family and other intimate relationships where gender is still seen even ideologically as a reasonable and legitimate basis for the distribution of rights, power, privilege, and responsibilities. (See review by Thompson & Walker, 1995; and Komter, 1989.) For example, in families in which there is an inequitable division of family work, wives and husbands often compare their household workloads with those of other wives and husbands, rather than with each other, because gender, rather than partner, is considered the appropriate referent (Thompson, 1991).

Much recent research on contemporary families focuses on the division of household labor and child care in American homes. Studies investigating the division of paid work and family work in most families suggest that the revolution of women entering the paid work force in the past decades has not resulted in a consequent shift in the practices of men at home, a fact that Hochschild has termed the "stalled revolution" (Hochschild, 1989). In the majority of American homes, even when wives spend as many hours in the paid labor force as their husbands, they retain primary responsibility for homemaking and childrearing (Berardo, Shehan, & Leslie, 1987; Berk, 1985; Ferree, 1991; Hiller & Philliber, 1986). The power of gender in shaping the household division of labor is apparent in these typical families.

Gender theorists, particularly Connell (1987) and Lorber (1994), suggest that, although structural conditions systematically support male privilege, human beings not only follow social dictates, but create them. The paradox of gender, according to Lorber, is that in order for feminists to eradicate gender, we must first recognize and highlight its ubiquitousness in all social institutions. One project for feminist social scientists is to locate and make visible the power of gender in families and occasionally to highlight when that power begins to diminish, to show that gender is a social institution, and, therefore, social change is possible. The research presented here is a study of heterosexual married couples who have moved beyond hegemonic conceptions of gender as they organize their daily family life. Hegemonic conceptions of gender are essentialist beliefs that are so taken for granted in our society that they are invisible in most families—the view of men as protectors, providers, and sexual aggressors, and women as naturally empathetic, nurturing, and sexually receptive (e.g., traditional mating rituals with girls being courted and then given to their husbands by their fathers in marriage). The couples recruited for this research have moved beyond hegemonic conceptions of gender, of masculinity and femininity, of what it means to "mother" a child, and to "father" one. This is a study of couples who conceptualize their marital relationships in primarily gender-neutral terms. The husband's work is not more central to the family's daily or long-term plans, nor are mothers the primary caretakers of young children. Statistically, such couples may be rare, but theoretically they are very important. If, as Lorber (1994) argues, we can only eradicate inequality by moving beyond gender as an organizing system for all social institutions, then careful scrutiny of those who are moving beyond gender in their daily lives is of utmost importance. We discuss how couples who are making new and postgender marital scripts come to exist—what aspects of gender they must change to form equitable families and how the existence of even a few of these families might affect the cultural meaning of gender itself. The data presented here are based on in-depth interviews with parents in 15 families, as well as field observations of parents and children together in their homes during workday evenings. We describe how these couples came to negotiate their nearly postgender lives and how this affects their relationships.

The language needed to describe such couples is problematic. The couples we study do share roles equitably, but more than that, they have decoupled breadwinning from masculinity and nurturing from notions of femininity. To label them "egalitarian," a word used to describe role-sharing couples, does not necessarily imply a divorcing of gender from family and work responsibilities, nor does the language of equity and fairness. Thompson (1991) shows clearly that perceptions of fairness, and even of justice, are not tied to any objective equality in outcomes in gendered settings. Peer couples, the phrase offered by Schwartz (1994), refers to couples who have moved beyond gendered marital roles, but the word "peer" itself does not imply moving beyond gendered expectations in marital roles because it is logically possible to imagine, even if it is empirically unlikely, peers to be equal in power but to endorse gendered marital roles. For clarity, we call these couples postgender couples. These couples, of course, live in gendered worlds and present themselves as gendered human beings. This language is meant to imply that, in the negotiation of marital roles and responsibilities, they have moved beyond using gender as their guidepost.

INSIGHTS FROM PAST RESEARCH

There have been few studies designed explicitly to examine married couples where husbands and wives have redistributed family work without regard for the typical gendered division of family responsibilities. There is a good reason for this dearth of research—such families are statistically rare. Thus, by necessity, the research in this field is comprised of qualitative studies of recruited volunteer families.

Only 20 years ago, there were few families in which household labor was divided equitably, even when wives were career women. The earliest studies of dual-career families (e.g., Poloma, 1972; Poloma & Garland, 1971; Rapoport & Rapoport, 1972) found no families in which husbands and wives were equally responsible for family work. In these early, dual-career, professional families, middle-class women were grateful to their husbands simply for allowing them the freedom to pursue career goals. Even the wives in these families didn't expect their husbands to do more than help them out with family work. Studies of working-class, dual-worker families (Komarovsky, 1967; Rubin, 1976) in the same era reported similarly traditional gender attitudes and distributions of family labor, even when women had to work for pay.

By the middle of the 1980s, there were some couples who described themselves as sharing equally in the family work. Studies of these couples, however, indicated that this seemed to be more an ideological commitment than a documentable reality (Ehrensaft, 1987; Haas, 1980, 1982; Kimball, 1983). In these studies, husbands and wives were often veterans of the social justice movements of the 1960s and 1970s, ideologically committed to nonsexist, nonracist, nonviolent childrearing. They were professionally educated men and women, often working in human services or education. The sample reporting most egalitarian behaviors was the sample with the fewest parents (Haas, 1980, 1982). Ehrensaft (1987) was particularly adept at identifying and explaining the discrepancies between the ideology and the reality of her respondents' daily lives. The men were committed to doing their share at home and reluctant to assume the total responsibility for economically supporting the family. The wives were enthusiastic about their relatively egalitarian marriages. And yet Ehrensaft reported three ways in which gender emerged as an explanation of differences between fathers and mothers: wardrobe management, worrying, and scheduling. Although in nearly every family such differences were attributed to individual personality traits, the pattern was clear: Women remained responsible for keeping the ship afloat, for seeing what needed to be done, scheduling it, and making sure appearances were appropriate for the outside world. These couples were sharing the labor of family work (although not exactly equally), but not sharing responsibility.

In the late 1980s, two studies showed how some families were beginning to divide labor equally. Coltrane (1989, 1990) studied a sample of households where slightly more than half (12 of 20) of the couples really seemed to be sharing equally both labor and management in their homes. The families in Coltrane's studies were not nearly as ideologically driven as those in the earlier research. Husbands and wives shared the tasks of family work equally because they were committed to quality childrearing. Both parents chose jobs with flexible schedules, and both made children their first priority. By the time of this study, however, there was a culturally available ideology of gender equity for these couples to implicitly—if not explicitly—use to justify their choices. However, ideological commitment alone remains an unstable predictor of egalitarian behavior. In another recent study of 10 couples in which both husbands and wives identified themselves as feminists before their marriage, only two couples actually shared the family work while raising young children (Blaisure & Allen, 1995). By the early part of this decade, research suggests that structural constraints had changed enough to allow equity-oriented men and women to negotiate egalitarian marriages.

The largest and most ambitious study of egalitarian marriages was published recently (Schwartz, 1994). Schwartz returned to the few heterosexual role-sharing couples she had identified a decade before (Blumstein & Schwartz, 1983) and interviewed them again. This study improves previous research because her sample of peer marriages is larger than most (n = 56) and is spread geographically across the U.S. in both urban and rural locations. In addition, she compares the peer couples with 22 "near-peers" (i.e., couples in which both partners were employed but didn't meet her criteria for an equal relationship) and 22 male-dominated couples with a well-defined, gender-based division of labor. Although Schwartz does not specify the criteria for inclusion in the peer marriage sample, there are four commonalities that appeared to be characteristic of peer marriages: (a) the division of household labor stayed within a 40/60 split for labor and responsibility, (b) both spouses believed that each had equal influence over important and disputed decisions, (c) both partners felt each had equal control over family economic resources and equal access to discretionary spending, and (d) each partner's work was given equal weight in the couple's life plans.

Schwartz's study found that the men were usually midranking professionals with incomes modest enough not to encourage them to have unemployed wives or to make their wives' salary unimportant. There were some "power" couples in which both partners had successful and consuming careers, but they were in the minority. Nearly all the couples were between their mid-20s and early 50s, baby-boom generation and younger.

Schwartz reports that there are distinct rewards as well as costs of peer marriages. The major reward (and the reason the men are interested in being peers) is the primacy of the relationship and the importance of the couple's deep friendship. These couples are best friends, committed to each other, and they find each other irreplaceable. The depth and quality of their relationship is the driving force in their lives. However, such intensive commitment to each other and to sharing equally has its costs. Social support is still rare for this kind of marriage, particularly for men. In traditional marriages, women balance work and childrearing, usually by foregoing jobs that are incompatible with parenting. In peer-marriage couples, both partners must make career choices that allow them to co-parent, thus giving up opportunities for advancement that require extensive travel or overtime.

"FAIR FAMILY STUDY" RESEARCH DESIGN

In this research, our theoretical concern is with understanding how marriage operates when both husbands and wives perceive the women as responsible for sharing the breadwinning (the "co-providers" defined by Potuchek, 1992) and the men as responsible for sharing the child care. We focus on the ways that husbands and wives change how gender works within their families.

The research was conducted over 4 years by a team of eight researchers. In 1991, our team spent 3 months writing and revising interview schedules for egalitarian couples and their children. The resulting data from this extensive study are published in their entirety elsewhere. (See Risman, in press.) The findings presented here focus specifically on the married couples' division of family work and paid work. The interviews included open-ended questions about the couple's life histories (e.g., childhood, expectations about adult life), the couple's relationship (e.g., development of their relationship, how the equitable division of labor formed, best and worst aspects of their relationship), their childrearing experiences (e.g., feelings about parenthood, effects of having children, emotional connection to children, responsibilities of child care), and their paid work (e.g., description of current job, work habits, employer's support of family responsibilities).

The interview team for each family consisted of an interviewer for each parent and for each child 4 years old or older and a babysitter if any children under the age of 4 needed to be cared for during the interviews. The number of interviewers matched the number of family members. Interviews, conducted separately with each family member, ranged in length from an hour to just over 3 hours. The interviews were tape-recorded, then transcribed verbatim. Each interviewer also added extensive field notes and an analytic summary to the transcript of the interview. Following the interviews, a member of the research team arranged to conduct an observation of each family in their home. The observer met the family when they arrived home from work and remained in the home until the children were asleep. These observations provided a comprehensive account of all the household activities and behaviors of the members of the family on the evening they were observed. The observer did not tape the evening but provided comprehensive field notes and composed an analytic description of the observations (Lofland & Lofland, 1995). The interviews were coded and analyzed according to the themes that emerged from the data (Glaser & Strauss, 1967). The Ethnograph computer program (Seidel, Kjolseth, & Clark, 1985) was used in the analysis.

We wanted to be certain that we were studying couples who shared housework fairly and that we were not including second-shift families, where women did more than their share. Although husbands' and wives' perceptions of the fairness of their relationship provide fascinating insights into the gendered meanings of equity, our goal in this research was to study couples whose actual division of labor was fair and not subject to such gendered attributions. Therefore, we included measures of actual household labor, as well as perceptions of fairness, as criteria for sample selection. We did not require that couples be legally married, but because only one of the couples in our sample was not legally married, we will refer to the men and women in our sample as husbands and wives for ease of discussion.

Our sampling process involved several strategies. First, we utilized a snowball sampling procedure. We asked everyone we knew for referrals. We also asked respondents in the study to suggest other potential interviewees they knew from their social networks. Second, we advertised. We placed announcements in every PTA newsletter in Wake County, an urban area in North Carolina. We posted signs or mailed advertisements to every day care center in Raleigh, North Carolina, placed announcements in many libraries and fitness centers, and posted ads on grocery store bulletin boards. We advertised in the staff publication of North Carolina State University, as well as with the women's studies programs of Duke University, University of North Carolina at Chapel Hill, and North Carolina State University. We also advertised in local feminist newsletters (NOW and NARAL) and in newsletters of selected local churches and synagogues. We sought respondents primarily in Raleigh, Durham, and Chapel Hill, but also in Greensboro, North Carolina, as well. These recruitment strategies produced an extensive list of potential respondents. We phoned all volunteer respondents and asked them if they "shared equally in the work of earning a living and rearing their children." Those who responded "yes" were sent questionnaires. The recruitment procedure yielded 75 families who volunteered to participate and who appeared eligible after our initial phone conversation.

The questionnaire asked about the division of household labor, such as who was responsible for cooking meals, shopping for groceries, routine housekeeping, making social arrangements, feeding the children, paying bills, as well as who should be responsible for household and child-care tasks, and the extent to which each partner believed their relationship was "fair" in terms of household chores, working for pay, spending money, and child care. Based on this screening, we identified a sample of egalitarian families who met our criteria. Following Schwartz (1994), our standards were not too stringent: The division of labor had to be within a 40/60 split on the short survey checklist, both members of the couple had to agree that they were equally responsible for earning a living and caring for their children, and both members of the couple had to indicate that they believed their relationship was fair on each dimension listed above. . . .

Over half of the parents (eight men and nine women) had PhDs or MDs. Another eight parents had master's degrees (usually in education or business). Three fathers and one mother had only a bachelor's degree, and only one mother had never completed college. These families weren't financially elite; many had given up higher paying jobs to co-parent their children. Others were college professors, a group with notoriously high educational attainments relative to their incomes. All the families were white.

The fact that the sample is educationally elite may reflect, in part, the nature of the population of the study area. The Triangle area of North Carolina is one of the most PhD-dense areas of the U.S. With three major universities and several colleges and the research divisions of multinational corporations in Research Triangle Park, there is a disproportionate number of highly educated persons in our region. We believe that Ferree (1990) is correct when she noted that changes in the structural conditions of women's wage work are a precondition to changes in domestic arrangements. Our explanation for this serious limitation in our sample (that is, we only found highly educated couples) is that women in our society don't have the clout or self-assurance to seek a postgender marriage unless they are highly educated, income-producing professionals. High education and an ability to earn a good living aren't sufficient conditions to create a postgender marriage (as evidenced by their rarity), but these seem to be necessary preconditions. The women in this sample were educated, committed to their work, and had eagerly sought parenthood. Such women tend to marry an equal—their husbands are also educated professionals. The men tended to be family-centered and worked in jobs that were somewhat flexible. The other striking pattern was the number of couples in the same sort of work: three academic couples in the same fields (two even in the same departments) and two couples in similar fields but different workplaces. Another pattern that emerged was that several wives (6 of 15) either earned more or worked in more prestigious occupations than did their husbands. Only two husbands out-ranked their wives. It was more common, however, for the spouses to hold equally prestigious jobs and to earn comparable incomes. Our sample shares this characteristic with samples in previous research (Blaisure & Allen, 1995; Ehrensaft, 1987; Schwartz, 1994), indicating perhaps that couples need this kind of material equality to offset the male privilege embedded in marriage as an institution.

The egalitarian outlooks of these couples are mirrored in the variety of surnames they adopted. In slightly over half the families (8 of 15), the wife assumed her husband's name after marriage in a traditional fashion, and the children bore his name also. In two of the families, the mother and father each carried their own name, and the children's surnames were hyphenated. In three of the families, the mother and father retained their birth names, but the children carried the father's name alone. In one family, the mother alone hyphenated her name (by adding her husband's after her own), but the child used only the father's surname. And finally, in the one family in which the parents were not legally married to each other, both parents retained their own names, and the children carried the mother's name. In the following analyses of these data, however, we follow the simple rule of giving each family a code name, and we use that code name consistently for all members of the family. This rule, although not faithfully representing each family's naming policy, provides the reader with a better understanding of which husbands and wives are in the same family. Just as the family names have been changed to code names, so too identifying characteristics, such as occupation and work site, have been changed to protect the respondents' anonymity.

ANALYTIC FRAMEWORK

In order to analyze how gender continues to influence these families and to what degree they have managed to transform themselves beyond gendered expectations, we make use of conceptual frameworks offered by Connell (1987) in his monograph, *Gender & Power,* and by Komter (1989) in her article on hidden power. Connell argues that an adequate theory of gender requires a focus on external constraint beyond the voluntarism of sex roles, but also recognizes the complexities of individual lives and the freedom that human beings have to recreate their environments, to reflexively reject the worlds they inherit, and to transform them. Connell writes that:

> to describe the structure is to specify what it is in the situation that constrains the play of practice. . . . Since human action involves free invention . . . and human knowledge is reflexive, practice can be turned against what constrains it; so structure can be deliberately the object of practice. (p. 95)

Connell argues, and we agree, that many of the structural analyses of gender in the past have been overly deterministic. Structural conditions constrain possible actions but do not preordain outcomes. Change is historically uneven and is the result of internal contradictions within structural arrangements, as well as the result of conscious struggle of the people involved. In Connell's theory, the gender regime of each social institution (such as family) can be analyzed by attending to three different arenas: labor, control, and cathexis. Labor refers to the division and organization of paid work (including discrimination in wages and sex segregation of jobs), the division of housework and child care, and the relative time and effort devoted to different kinds of work. Control refers to the cultural lines of authority and coercion, the hierarchy of statuses, and the myriad means of social control (from ridicule to the threat of interpersonal violence and prisons). Cathexis refers to the emotional linkages of daily relations. This includes sexuality, but also trust, jealousy, desirability, and the responsibility for and intensity of emotional connections.

We have analyzed our data according to how the sample families deal with labor, power, and cathexis.

FINDINGS

Labor

This sample has little variability in the distribution of household labor and the number of hours each parent devotes to paid work. If labor both inside the home and beyond it were not equally shared, the couple would not have been included in the sample. However, the responses of the couples revealed a number of alternative paths to postgender marriages. Couples' accounts uncovered four different paths to this division of household labor. We have conceptualized the couples according to these routes as dual-career couples, dual-nurturer couples, posttraditionals, and those pushed by external circumstances. In dual-career couples, both partners are interested in their own career growth and success, as well as co-parenting their child or children. Dual-nurturer couples are more child centered than work centered; the two parents organize their work lives almost exclusively around their parental responsibilities. Posttraditional couples have spent at least part of their adult lives in husband-breadwinner and wife-nurturer roles and have consciously adopted a more equitable arrangement. Last, there are couples who are pushed into fair relationships by circumstances beyond their control. In one family, the wife's job was the organizing principle for the family's life because she earned nearly twice as much as her husband. In another family, the wife's chronic illness was at least partly responsible for the husband's domestic labor.

The Dual-Career Couple The most common route of the couples in our sample to postgender marriages is the partnership of two career-oriented professionals in which at least the wife, but often the husband as well, holds egalitarian values. Husbands and wives in this group repeatedly emphasized the ideological importance of husbands and wives sharing the labors of paid and family work. Two thirds of the couples in this study (9 of 15) divided their labor fairly because they were both career and equity oriented. They compare their own labor contributions with those of their spouse, not with those of same-sex peers. Both parents in this group had compromised work goals to balance family and career priorities, but both partners remained committed to their careers, as well as to their childrearing responsibilities. In all but one of these couples, both partners had always studied or worked full-time, shared household responsibilities before parenthood, and never considered anything other than co-parenting. In one couple, the wife had remained home full-time with their infant daughter when her husband held a temporary teaching post for a year. She remembered this time as both atypical and miserable. In these dual-career families, husbands and wives presume that a fulfilling life involves participating in both paid and family work.

A remarkable finding is the near absence of gender expectations used as a basis for organizing labor. Not one of these couples mentioned ever considering that the wife devote herself exclusively to family work. No husbands or wives complained about doing more than their share of housework, nor did the husbands wish their wives were more traditionally domestic. Only two of the nine dual-career couples reported any serious conflict over the division of labor, and usually this conflict was long in the past. Often the couples never really negotiated at all. One mother, a mathematician married to a public policy analyst, explained:

> We consciously divided up tasks. We didn't tend to do the "you'll do half of the time, I'll do half of the time." We tended to divide tasks up, who would do them. So Karl liked to cook. He cooked for himself before we ever married. I didn't cook for myself at all before we were married. I ate out or in the cafeteria where I worked, so we didn't really have to adapt ourselves in that way. If he had wanted to share the cooking because he didn't much like it, I probably would've divided it in some way, half and half, but we tended to sort jobs. He didn't like to do laundry; I like to do laundry. I didn't mind cleaning up, [so] I do the cleaning up . . . the dishes and wiping up counters and stuff. So we found a way that we considered equitable. . . . We did talk about it, and we did consciously do it.

In another household, a similar scenario was revealed, except that the wife loved to cook—she was quite a gourmet—and the husband didn't. This family attempted to divide cooking tasks 50/50, but quickly reverted to sorting tasks so that the wife cooked and the husband cleaned up. The near absence of gender as a criterion for determining labor is evident in another household in which the husband, a philosopher, had considerably more flexibility in his work schedule than the wife, who worked as a corporate manager for a multinational corporation. When asked about how they originally decided about organizing tasks, this husband replied:

> I think within a year after we got married Pam got her degree and started working full-time. Since then, I have taken on the majority of domestic labor. I have never had a grudge. Occasionally I will say, "It would be nice if you did a little more." I have never really felt exploited. If we were both working the same kind of strict hours and one of us did the majority, then I would expect one of us to raise their hand and say this is not fair. My hours are flexible, so it's not a big deal for me to throw a load of laundry in the washer or dryer while I am grading papers.

The one way that gender did manifest itself in the family labor of dual-career couples was that the wives sometimes—but not always—came to the relationship with higher standards for cleanliness. Unlike more traditional homes, however, this difference is not used to justify women's extra burden of housework (Hochschild, 1989). These couples did not consider wives' cleanliness "needs" as justification for their greater contribution to household labor, a rationale many couples employ and one that changes household labor from a necessity for families to women's personal requirements (Thompson, 1991). In the families in our sample, differing standards were perceived as a problem to be worked out equitably, without resorting to gendered justifications. For example, a psychoanalyst married to a commercial real estate salesman explained:

> Stan had been a bachelor for a long time. He was 35 when we got married, so he knew how to do dishes and laundry, and he knew how to cook one thing. He did things like keep his dirty, greasy garbage in a paper bag under the sink, so that the entire sink was so greasy underneath that, after trying to clean it, I had to paint it. I did teach him how to use plastic. So there were a lot of things that he just didn't notice or care about initially, which he has now incorporated into his routine. It's not that he's a man and doesn't do that stuff. It's just that he's absent-minded, and at this point he's a lot more concerned about certain neatness than I am. He'll vacuum periodically; I don't. I don't like the noise. He's the vacuumer, and I tolerate mess more than he does.

Only in one household were there current discussions and negotiations about household labor. A freelance editor married to a psychology professor reported some ongoing negotiations about the standards for household cleanliness.

However, he has reconceptualized the problem in standard feminist language:

> I know the thing that men have the hardest time learning how to do is noticing that there is dust. Men can't see dust. Men don't know what dust is. I still don't see it. I don't know it's there. I know that the nirvana of nonsexist male development is dust. If I get to the dust stage, I'll know that I've really made it.

The Dual-Nurturer Couple We categorize two couples as dual-nurturers, rather than dual-career couples. Dual-nurturers are home, family, and lifestyle oriented, rather than career oriented. They work for pay in order to spend time together and with their families. Only one of the couples in our sample, the Woods, was clearly a dual-nurturer couple. Another couple, the Valts, might be reclassified as a dual-career couple as soon as the most intense years of childrearing are over. (At the time of the interview they had a new baby and a not-quite 4-year-old.) However, we classified the Valts as dual-nurturers because the father had given up a position in a Wall Street trading firm to become a financial planner in order to live with his partner and their two small children. Both parents seemed to spend more time during the work week with their children than at their paid work. In the Woods family, neither parent had consistently worked full-time for years. Neither partner wanted to. The father, an aspiring sculptor, worked half-time as an editor, and the mother was employed as an accountant. They tried to organize their work schedules so that both were not working during the same day, even though their baby was in day care and their other son was in school and attended after-school programs. They explained that evenings were just too hectic if they both spent a full day away at the same time. The work schedules of these parents and the types of jobs they had held varied, based on their family's needs. Neither spouse was strongly attached to the labor force. This couple was particularly focused on the quality of their lives, rather than on material acquisitions or career development.

The Posttraditional Couple The third route into a division of equitable labor is a dissatisfaction with more traditional arrangements. Two of the couples in this study were posttraditional families, in which the couple had shifted from a gender-based to a nongendered division of household labor. In one case, the Germanes, both partners' previous marriages had been organized around traditionally gendered expectations and responsibilities. The wife, in particular, had been clear that she'd left a second marriage (she'd been widowed once before that) in which her husband didn't meet her needs. She was looking for a real "partner" this time around. Ms. Germane had been married to two career diplomats and had moved every 3 years of her adult life. As a bookkeeper in government service, she found it easy to relocate. She had never been unemployed more than 3 months in her entire life because of the "work ethic I grew up with." Paid work was important to this woman's sense of self. When asked about her maternity leave, she answered, "I do feel that 3 months out of work—I don't care who I'm taking care of—is a lot of time." During her daughter's infancy, she had cut back her paid employment to 32 hours per week, but only when her husband's work schedule kept him gone for long days and overnight. Her current husband had returned to school to earn his college degree when their daughter was in the early elementary grades. The wife enthused, "I liked the role reversal when he was the housewife . . . because he walked her to school, and I swear he knew every lady in the school. . . . He became a PTA mom." Both husband and wife in this couple reported always sharing work equally; the wife was aware that they seem to share things "more in the middle" than other people. Even now that the husband was working full-time again, this couple continued to share labor equally. The husband and wife in this posttraditional couple had a particularly strong motivation to keep things fair because they had experienced less satisfying relationships in the past. In contrast, their current relationship was marked by, in Schwartz's (1994) language, a very deep friendship. Both partners wanted to keep things fair in their marriage, with work split "in the middle" to protect this precious friendship.

The other posttraditional couple in our study had renegotiated their gender-based roles when the youngest of their three boys entered kindergarten. It took this family 3–4 years to transform a decade-long pattern of male breadwinner and female homemaker into a postgender relationship. The wife was clearly the instigator and moving force behind this transformation. Ms. Potadman described her resentment at spending Saturdays doing housework after she had returned to her nursing career full-time. Eventually, her family, at her urging, divided up the household tasks using a complicated scheduling system. The wife explained it this way: "I had the idea, let's just list the tasks, and we'll divide 'em up, and whenever he gets his done, he gets 'em done. . . ." This wife still held primary responsibility for planning the scheduling sessions. Each of the three sons, as well as the parents, talked at some length about the complicated charts used for scheduling cooking and cleaning. Each of the sons and the husband cooked one day a week, and the wife twice, with dinner out every Friday. Particular sons were assigned clean-up tasks on days they didn't cook. The assignment of particular days was made according to a formula designed to ensure equity.

The mother seemed to use the work of scheduling as a sort of glue to bind family members together in a common concern. Although there are five members of this family, 50% of the schedule was written in the wife's handwriting. In addition, home observations revealed that, even though the children decided what they would cook on their night as chef, the wife was never far from the kitchen, helping out and advising. Even though all family members maintained the calendar, the wife was able to make them think about their responsibilities by invoking the scheduling ritual. She seemed to spend a great deal of energy maintaining this equitable arrangement. The husband reported that he fully agreed with the equitable arrangement because of his ideological commitment to justice and equity. He simply relied on his wife to determine what his share should be.

External Forces The final route into a postgender relationship is to be pushed there by external forces. In one family in our sample, the Codys, the wife had a better-paying job and a less flexible schedule. In another family, the wife had a chronic illness. These external factors constrained how the division of household labor could be negotiated. However, in both of these families, the ideological apparatus of gender equality was the conceptual framework that families employed to help them make sense of their housework arrangements. In many ways, they perceived these external constraints in gender-neutral terms. For example, in the Cody family, the wife earned two thirds of the household income. The husband had been passed over for a promotion he'd expected, had quit his job, and had changed careers. Mr. Cody was no longer work-focused and was pleased to have the freedom to keep a flexible schedule, to work in his own small business, freed from the economic burden of supporting the family on his income. He explained:

> When we decided that we were going to have children, Marilyn's job situation was such that . . . she worked in a reasonably structured environment, and in that case you can't take time off to bring somebody to gymnastics and, you know, go to the school for plays and if you give parties and do that kind of stuff. So we made a decision that I would do that, and what it does basically is it takes me out of, you know, being the high-powered [careerist] and, you know, it's just . . . I just do something that brings in some money, and it gives me satisfaction that I can do it.

In the Stokes family, where the wife was chronically ill, the husband provided a similar account, using remarkably gender-free language, for why his wife did not do much domestic work. She was unable to drive, so this husband also provided transportation for the daughter's activities and did the grocery shopping for the family. He interpreted their division of housework as just "because of her work schedule and because of her physical limitations. . . . I do almost all of the cleaning because it gets just physically difficult for her to do that."

We cannot know whether the families pushed into equitable arrangements by external circumstances would be postgender families otherwise.

Although the rationales provided suggest that the couples did not employ gender to determine how household tasks should be allocated, what distinguishes these couples is the imposition of contingencies beyond their control. Of course, the families that we have described are ideal types. In casting these families into types, we focused on the themes that emerged from the couples' accounts of their lives, which occasionally reflected more than one theme in their development of an equitable relationship. When this was the case, we compared the couples' accounts with the central features of each type. For example, the Sykes, a dual-career couple by our categorization, might have been labeled dual nurturers. Both Mr. and Ms. Sykes gave up jobs in the private sector to teach high school and took significant pay cuts. Both said they did so for the flexibility to be better parents to their one daughter and to have a higher quality of family life. Yet both adapted to their new environments and were doing well professionally. The had moved into administrative duties and once again were interested in career growth. The Sykes' career interest and commitment is a contrast to those we categorize as dual nurturers. Thus, we have labeled them a dual-career couple, although if we had interviewed them shortly after their decision to leave private sector jobs for teaching, they probably would have appeared to be in the dual-nurturer category. Examples such as the Sykes suggest that there may be a dynamic quality to postgender relationships—that is, the equitable arrangement may occur for different reasons at different points in the family life cycle.

Another caveat about this sample and its gender-free labor is important. In all but two families, paid help was employed for some cleaning tasks, such as vacuuming, dusting, bathrooms, and yard maintenance. Thus, these couples had the luxury of not having to split the most odious homemaking jobs. Most families began hiring help when children were born. Nearly every family had come to the same decision: Weekends were better spent in leisure activities than in housework. These educationally elite families were privileged enough economically to buy themselves out of nearly a day per month of household labor. However, both husbands and wives reported that even before they had housecleaners, they shared the work, and even with the aid of employed help, the majority of household tasks, particularly daily tasks such as cooking, washing the dishes, laundry, routine housekeeping, and child care, were shared by husbands and wives. These upper-middle class couples fall into the relatively privileged sector of society in which most families buy household and child-care services from working-class employees (e.g., Hertz, 1986). However, in the families in this study, both husbands and wives benefited from the use of household help, rather than—as is usually the case—female homemakers being the primary beneficiaries of poorer women's labor (Romero, 1992).

In summary, not one of the 30 parents interviewed suggested that the husband's paid work was more important than the wife's. In fact, four of these families had moved to North Carolina because of the wife's employment, and the husband was the trailing spouse. None thought to mention this as anything unusual. Another four had relocated to North Carolina because of the husband's job. In two of these families, the wife's ability to transfer here was mentioned as a prerequisite for relocation. Another three families relocated because both persons found positions in the area together. Both the men and women in the other four couples lived nearby when they met. Not one of the 30 parents suggested that caring for their children was or should have been more the wife's responsibility. Not one of the 30 parents believed that housework ought to be a wife's responsibility, although half did admit to having conflict about different standards of cleanliness at some point in their relationship. Next, we address how issues of conflict are negotiated and how power is exercised in postgender families.

Control

By "control," Connell (1987) refers to the cultural lines of authority and coercion, the status differential between men and women that underlies the basis of our gender structure. There are countless means, both overt and subtle, by which male privilege is constructed in daily life. For example, fear of ridicule and social disapproval keeps women from girlhood on worried about weight gain and physical attractiveness. The wage gap and sexual harassment in the workplace are other ways that male dominance is reinforced. In families, the continuing expectation that wifehood requires domestic service is yet another way that male privilege and female subordination are recreated in daily life.

Komter's (1989) analysis of power in marriage extends attention even more explicitly beyond overt control to the subtle mechanisms that effectively recreate male privilege at the dyadic level, even without conscious awareness of the individuals involved. Komter articulates how gender norms, gendered selves, and the tacit rules for male-female interaction are hidden forms of power in marriage. Manifest power is the ability to enforce one's will against opposition and can be measured easily. Latent power, however, is more insidious and reflects the force to keep issues from even being raised. Latent power exists when conflict never arises because the needs and wishes of the more powerful are anticipated and met. Last, invisible power extends beyond the harboring of even latent grievances and depends on the systematic differentiation between groups that are so embedded in practice and persons that male privilege is perceived as legitimate, even by women.

Are these complicated and subtle power issues moot in postgender families? The women in our sample of postgender families are both pioneers in and beneficiaries of the women's liberation movement. In order for any family to be included in this study, both the husband and wife had to agree that the division of family work was fair to himself or herself. If any unfairness was reported, it had to be in gender-atypical directions (e.g., the husband did more than half the cleaning). And yet, given the ubiquitous history of male dominance, we would be remiss if we simply accepted a check on a pencil-and-paper survey as an adequate measure of the balance of power and control in these couples.

We explored the issues of power more extensively by asking in interviews how decisions were made about where to live, how to discipline children, how to organize housework, and how to allocate leisure time. In more than half of these couples (9 of 15), we could find no evidence that either member of the couple had more power in the marriage. Neither husband nor wife seemed able to exert control over the other or even seemed to have greater influence than the other partner. What we found was that, although some couples reported manifest power differences (that is, the wife tended to win conflicts), not one spouse reported any latent grievances, and the hegemonic beliefs to support invisible power didn't appear to exist.

In six of these nine power-balanced couples, our probing didn't elicit indications that the couple had any areas of active conflict. Nor did our probes elicit discontent with the allocation of household or child-care tasks in the absence of conflict as an aspect of the hidden power in many marriages (Komter, 1989). When asked about what they would like to change in their marriages, nearly everyone in this group of six couples mentioned the lack of time to enjoy each other's company. This is Ms. Cross's response to a question about the worst thing in her marriage: "I wish I had more time with him. Alan and I don't get to do things together alone, go alone for walks, repaint the house, refinish furniture."

In the other three couples without any power imbalance, either husband or wife, and sometimes both, talked quite openly about areas of conflict, but with no implication that either partner had more influence in resolving such issues. These three couples reported some conflict over differing standards of cleanliness. Yet the tone of discussions appeared to be more about accommodation, rather than about winning or losing the conflict. Mr. Stokes, who did most of the housework, illustrated their arguments in the following way:

> Our conflicts have been mainly over what constitutes an acceptable standard of cleanliness. I'm content to let things get dustier than Rhonda is. Um. She gets much more bothered by the dust and the dirt than I do. I don't think on the full spectrum of humanity that I am very . . . that I'm at an extreme, but I'm more willing to let things go than she is. I care less about messes and dirt than she does.

In another household, the issue never was resolved until 2 years before our interview when the couple hired a housecleaning service to come in once a month.

In the six couples where some power imbalance could be identified, the wives seemed to have more influence. In each of these families, the wives described themselves as out-going, assertive, organized, and ambitious women. They spoke quickly, used declarative sentences, and reported extraordinarily organized and determined work lives. Without exception, the husbands were described by themselves and their wives as introverts, quiet, and laid back. The following excerpts from interviews with the Reluxes show how this couple experienced their personality differences. When asked about how their relationship became marriage oriented, Ms. Relux answered:

> I really liked him, and I intended to finish my training and go back up North because I had done a year of internship in Connecticut, and it was such a relief to get away from very Southern men. When I met Sam, I really liked him, and I wanted to get to know him and probably marry him, and I kept trying to call him.

The husband's interview, conducted in a different room by a different interviewer, confirmed a similar perspective: "Well, I was fortunate to meet Nelly, and for whatever reason, I was immediately the man of her dreams. I'm not sure why."

This wife further reinforced our view that she was the more forceful one in the relationship with her response to questions about their decision to marry and have children:

> Just as I had to drag him into having children, I had to drag and push him into getting married. . . . He's a stick-in-the-mud. He is very happy for everything to be the way that it is always. It's easier.

Her husband told us the same thing: "I've been dragged along through each stage of life."

The Greens also supported the notion that the husbands in this smaller group acquiesce to their wives' needs—in this case, the wife's demand for equity in the responsibility for household labor and not merely in the time spent performing household tasks, an important distinction (Mederer, 1993). Ms. Green told us:

> Yeah, in the beginning, we had this real complex way of keeping track of our finances, which worked well. We divided up halves. In the beginning, it was really hard, and we argued about it all the time. I had to lower my standards, and Stan had to raise his. He had to learn to recognize that certain things had to be done in the house and that he could take that responsibility. He was willing; he was just ignorant.

This woman's husband also told the same story: "It was definitely a process of Roberta having to complain and me having to shape up."

Perhaps Mr. Potadman best articulated how the wives and husbands in this group of six couples differed when he described how their relationship had started:

> She was very vivacious and outgoing, and if I have any faults I . . . I guess those are qualities I feel missing in myself—enthusiasm and things like that. So, I guess, I was drawn to that.

This pattern of self-identified assertive women, who demand equity in their relationships and define it in postgender terms, and self-consciously adaptable men evident in these six power-imbalanced couples differs from the majority of families in this study where the husbands and wives were more power balanced. In no couples in our research did husbands appear to hold more power than wives.

Although these couples live in a world where men, as a class, have more power, property, and prestige than women, they have managed to create marital relationships that honor gender equality. These women have many advantages, but, perhaps most important, they are not economically dependent on their husbands. The work they do for their families, unlike the household work of more traditional women, is not conceptualized as their job. The men benefit from this as well. Their paid work is not the sole source of support for the family. Thus they depend on their wives to shoulder half or more of the breadwinning responsibilities. However, perhaps the major reason that issues of power and control seem so peripheral to the experiences of these couples is that these husbands and wives have made gender an irrelevant basis for organizing their family work and, therefore, irrelevant as a resource for male privilege in marriage. Although a minority of these wives seemed to hold more overt power in times of conflict, most of the couples appeared to be power balanced on this manifest level.

Cathexis

Connell (1987) uses cathexis to refer to the manifestation of the social structure in the gestalt of intimate relationships, in the construction of sexual desire, and in the intensity of emotional attachments. Gender patterns our desire for intimacy—sexual and otherwise. Connell focuses his discussion of cathexis mostly on sexuality, a topic that we chose to omit unless the couple raised the issue. In the short time we had with each family, we felt the need to cover many areas broadly, and we didn't want to alienate our interviewees with sexually explicit questions when they had been recruited for a study on childrearing and family life.

We extend Connell's concern with cathexis by focusing on the emotional connection between the parents and the intensity of emotional bonds between each parent. Although the focus here is on emotion, rather than sexuality, the issues of power are similar. The management of emotion is an area of family work in which women usually do more than their fair share. It is also work that, for the most part, is invisible and thus work for which women do not receive credit in the economy of family labor (DeVault, 1991). And it is the kind of work that cannot easily be reassigned because of ideological preferences. We would be remiss if we didn't explore how these couples who successfully had gone beyond gender expectations in power and labor dealt with the more subtly gendered phenomenon of emotion work.

No single pattern emerged in terms of these couples' relationships except that in nearly every couple both partners mentioned in separate interviews that their spouse was their very best friend, irreplaceable, and precious. The relationships of these couples seem embedded within the framework of a valued, intimate companionship. Apart from this underlying theme of friendship, three different emotional types emerged from the accounts of postgender marriages. First, some couples have a traditionally gendered emotional relationship with one another. These couples agree that the wife is more emotionally sensitive and that this manifests itself in the couple's marriage and parenting. Six families fell into this "mother-as-emotion-expert" category. The second type of family shared the emotional work. In five such households, neither the husband nor the wife believed they were highly emotive, nor that either was more intuitively connected to the child or children. In these families, the work of managing emotions, traditionally done by wives and mothers, was shared by the marital partners. Finally, in four of the families, both husbands and wives reported that they were highly sensitive to emotional nuances, and rather than sharing the emotion work, particularly as parents, they both seemed to do it intensely. These families, in some sense, had twice the emotional management and intensity of the others. In these families there were what we called "parallel emotion workers" (see Table 20.1).

Mothers as Emotion Experts In the families where the mother was the more emotive spouse and parent, the "emotion expert," we observed quite a traditional division of emotional labor. The women tended to keep tabs on the tenor of the marital bond and were considered by both partners more intimately and emotionally connected to the children. In some cases, biological essentialist beliefs were employed as explanations for why the mother had stronger emotional bonds to the children. For example, Mr. Potadman explained his wife's greater emotional connection to his sons when they were young as the result of her pregnancy and breastfeeding. His wife explained her closeness this way:

> I'm more the family organizer. . . . I'm sort of the schedule person; . . . I'm just by nature more prone to do it, so, you

Table 20.1 Postgender Families

Route to a Postgender Family	Emotion Work	Family Name	Ages of Parents	Children
Dual-career couple				
	Mother as emotion expert	Cary	Mother 35 Father 35	5-year-old boy
	Parallel emotion workers	Cross	Mother 39 Father 40	6-year-old girl 4-year-old girl
	Mother as emotion expert	Green	Mother 41 Father 42	6-year-old girl
	Shared emotion workers	Oakley	Mother 39 Father 40	10-year-old boy 4-year-old girl
	Mother as emotion expert	Pretzman	Mother 35 Father 39	8-year-old boy 3-year-old boy
	Mother as emotion expert	Relux	Mother 40 Father 45	6-year-old girl 4-year-old boy
	Shared emotion workers	Staton	Mother 42 Father 47	9-year-old boy 6-year-old boy
	Parallel emotion workers	Sykes	Mother 40 Father 39	9-year-old girl
	Shared emotion workers	Trexler	Mother 31 Father 31	4-year-old boy baby girl
Dual-nurturer couple				
	Parallel emotion workers	Valt	Mother 39 Father 40	3-year-old girl baby girl
	Parallel emotion workers	Woods	Mother 39 Father 41	10-year-old boy baby boy
Posttraditional couple				
	Shared emotion workers	Germane	Mother 45 Father 42	11-year-old girl
	Mother as emotion expert	Potadman	Mother 44 Father 55	17-year-old boy 15-year-old boy
External force				
	Shared emotion workers	Stokes	Mother 44 Father 40	10-year-old girl
	Mother as emotion expert	Cody	Mother 46 Father 43	7-year-old girl 4-year-old boy

know, I do it. Isaac's a little more of a procrastinator. . . . You know, I probably do more of the emotional kinds of stuff. I mean . . . when somebody's unhappy or stuff like that, I go and try to talk to him and try to be available.

In most families, the parents themselves—like Ms. Potadman—simply identified the disparities between their emotional labor as based in personality differences. The Pretzman mother complained that the one problem in their relationship was that her husband didn't have a wide range of emotions and wasn't physically demonstrative. This woman wished for more intensity in the relationship, more emotional highs and lows. "You can't fight with someone who won't get mad," she complained. The lack of emotional intensity led Ms. Pretzman to look for emotional connection elsewhere, an atypical pattern in postgender marriages. She explained. "I don't think he can provide me with the kind of closeness I can get in other friends." Interestingly, there was no indication that Ms. Pretzman was closer emotionally to her children than was her husband. Rather, each parent claimed to have a closer attachment to the child whose personality was seen as more like his or her own. In the Cody family, both parents agreed with Ms. Cody's assessment: "I don't think of myself as an overwhelmingly emotional person, but it would appear that my emotional connections are more emotional than Ronald's." Finally, the Reluxes both believed that the mother was the sensitive, emotionally tuned-in partner. The husband explained how his relationship with the children differs from hers: "My wife pays attention more. . . . She's more sympathetic." Ms. Relux summed up the subjective experience of these postgender families with traditional emotional relationships in her answer to a question about who was most emotionally attached to their daughter during her first year of life:

> I think we were both emotionally connected. I would assume that I was more because I'm the mother. . . . She was the most important thing in my life. . . . Everything else just seemed inconsequential.

It is interesting to note that these men "mothered" their children in every observable or measurable sense—they spent as much time and energy mothering as their wives did. Yet, as Chodorow's (1978) psychoanalytic theory predicts, they didn't mother with as intense emotional connections. Their gendered selves clung, despite their everyday experiences.

Shared Emotion Workers The five fathers in the category of "shared emotion workers" described themselves and were described by their wives as equal partners in managing their own marital relationships and as being intensely connected to their children. What is interesting, however, is that none of the women in this type of family described themselves as tightly emotionally bound to their children as did the majority of mothers in the mother-as-emotional-expert group. Instead, these couples seemed to sense what an "average" amount of attention to emotional issues would be and divided the work accordingly. In the families in which emotion work was shared, the father was as likely as the mother to be the worrier or scheduler. The mother was as likely as the father to be the disciplinarian.

In some of these families, the parents were able to trace how emotional connection changed over time and because of circumstances. In the Oakley family, both parents agreed that they were equally connected to both children, but that the father had been more in tune with the older child in the child's infancy. Ms. Oakley said:

> I felt that James was more emotionally attuned to Peter than I was. James could get him to go to sleep on his shoulder; I couldn't. Stuff like that. I had to nurse him to get him to go to sleep. James could walk him around. James had actually had a lot of experience with kids. His younger brother is 13 years younger than he is, and he took care of him.

Another mother put it this way: "We both just love her to death." In this shared-emotion-work family, that seemed to be an equal opportunity statement.

Parallel Emotion Workers In four families in our study, both parents were self-consciously and carefully in tune with one another. The tenor of the household was unusually charged with warmth and sometimes with jealousy. In one of these homes, the couple consistently employed terms of endearment when speaking about or to their spouse. In another family, we observed that the husband remained with his wife as she washed the dishes, rubbing her back and singing her little love rhymes. In three of the four families in this group, the men are consciously striving to be nurturant human beings. One was a conscientious objector during the Vietnam War. Another was abused as a child and recalled having been uncomfortable in macho settings as long as he could remember. Another is a vegetarian, philosophically opposed to killing even insects that appear inside his home. In contrast, a conscious concern to be gentle and nurturant was not held by the fourth man in the group. He is tall, athletic, and the one conflict related to us by the couple in this marriage concerns how much time the husband spends playing team sports.

The parallel-emotion-worker couples appear to have intense relationships with each other and with their children. In these families, rather than dividing the emotion work, the parents doubled the effort. Accounts of these couples revealed that partners spent a great deal of time and energy processing emotions and attending emotionally to their children. This emotional intensity seems to have both positive and negative consequences for family relationships. The Sykes family appears paradigmatic of this family type. Both parents changed occupations to ones better suited to family life, and the parents reported having closer relationships with each other than with anyone else in the world. Their time spent with their 9-year-old daughter reflected similar connection and attentiveness. For example, the night we observed this home, the daughter's bedtime ritual started with the mother spending 20 minutes brushing her daughter's hair, which extended to the child's thighs. Both parents then went upstairs to her room to read with her about the civil rights movement. The father projected his voice, as if in a play, and used different voices for each character. The parents and their daughter discussed questions that arose from the story (for example, how it must have felt to be a slave). When the reading was over, the lights went out, and the parents squatted at their daughter's bedside, held her hands, and sang folk songs. Although bedtime seemed rushed and hectic in some families, in this home every moment seemed precious and appreciated.

However, there also seems to be a negative consequence to such an intense focus on nurturing. In one family where the father reported "falling head over heels in love" when his baby was born, the father sensed that, as his daughter aged, she was moving from an equal attachment to both parents to a stronger attachment to her mother. This pained him, if only a little. His daughter, the father reported, had a "growing strengthening of her female identification." He felt that his daughter's relationship with her mother was "deepening in ways it isn't with me." This distress when children choose one parent over the other was expressed by parents only in the parallel-emotion-work families in our study. The wife in the Cross family related that her older daughter quickly established herself as a "daddy's girl." The father also described this:

> Pamela was a bit put back at how much Evelyn went for me over her. Evelyn would select me over her, but Pam got her wish. . . . and that's called Nancy. We each got a child.

The wife's interview confirmed that she'd felt rejected but that she felt better now because she believed her elder child was turning back to her for heart-to-heart talks and for cuddling. Ms. Cross wasn't certain about this, though, and she qualified her answer with a wistful "or maybe that is my desire." Mr. Cross believed that his older daughter was still closer to him but admitted, sadly, that things seemed to be slowly changing: "She's unfortunately old enough to see some of these girls-versus-boys things, and so she has girl talks with her mom now. I'm losing out."

This sadness at "losing out" becomes jealousy in some parallel-emotion-work families. For example, when asked how they would feel if the other parent decided to quit paid work and stay home full-time with the children, the immediate response in separate interviews from both parents in the Valt family was "jealous." Ms. Valt continued to explain her unexpected feelings about why she wouldn't want her partner at home full time: "I would be terribly jealous. I would have a problem with it because I would be worried they would love him more, and he would have the edge." She indicated that she had not anticipated this competition with her husband for their children's affection. In retrospect, she said, she had expected, because she was the "mother," to somehow retain an edge of emotional centrality in her children's hearts, and she has not. She explained that she'd assumed "we'd do as much work, but the kid would still love me a little bit more." She found that sharing parenting equally meant sharing the child's deepest affections equally, and this was harder to accept than she anticipated.

Overall, there was not one simple pattern that emerged in terms of how these families handled emotions, the cathexis component of their lives. Instead, six of the families demonstrated traditionally gendered patterns of emotional life, with more sensitive wives and mothers and emotionally reserved men. Clearly, gendered selves can remain even in the absence of gendered behavior and despite changes in institutional and interactional expectations. In the majority of these families, however, we found parents without traditionally gendered emotional selves. In five of the families, the wives and husbands seemed to have moved to a center; the women seemed less emotionally focused than traditional wives and mothers, and the men were more emotionally attuned to relationships and more sensitive than stereotypically masculine men. In four other families, the women seemed as emotionally centered as traditional women, but their husbands were equally emotionally intense. Without longitudinal data, it is impossible for us to offer insights into how couples come to do emotion work in these different ways. One finding is clear, however: There is diversity in how couples accomplish the emotion management and work in these postgender families, and not all have left gender behind.

DISCUSSION AND CONCLUSION

There are several significant findings from this exploratory study of 15 postgendered couples. Supporting the findings of earlier studies, we found that families who appear to be able to move toward a postgender structure are likely to have a family structure in which mothers are highly paid, autonomous professionals (Blaisure & Allen, 1995; Ehrensaft, 1987; Schwartz, 1994). Although income and professional prestige don't necessarily lead to the sharing of family labor by husbands in dual-income homes, without such material status, moving past male privilege appears to be unlikely.

We have found a number of paths that lead couples into postgender family arrangements. Most couples in our sample were dual-career families; both husbands and wives were committed to professional growth and co-parenting their children. In some families, neither parent was career centered, and both parents worked for pay simply to provide for their children. Their focus was more exclusively on the quality of their family time. In other families, there had been a change from more traditional gender arrangements to nongendered divisions of labor. Finally, there were some families who were pushed into postgender arrangements by external circumstances, such as illness or wives' earning power. Whatever the route into these relationships, these women and men established their daily activities and life choices using criteria other than gender. Neither partner appeared to have more bargaining strength or the ability to overcome the other's resistance more often. When power imbalances did seem to exist, the wives, in every case, seemed to be the more powerful partners. The emotional tenor of these families varied widely, although in every family the husband and wife reported the "deep friendship" that earlier research (Schwartz, 1994) has identified.

Although gender exists at the institutional, interactional, and individual levels in our society, the existence of these couples shows us that the consequences of gender are far from deterministic. We have described people who have freed themselves from most of the contemporary constraints of gender inequality in their family lives. The women have managed to command material advantages equal to their husbands and professional status equal to or better than many men. These factors alone, however, do not typically translate into postgender enactments of family. What distinguishes these families from many other dual-income couples is that they have rejected hegemonic notions of gender. To some extent, they have, indeed, moved beyond gender. The women and men in this sample generally have rejected gender as an ideological justification for inequality or even difference in the negotiation of their marital relationships. They do not consider that wifehood involves a script of domestic service or that breadwinning is an aspect of successful masculinity.

The interactions of these couples appear to be guided by rules of fairness and sharing within egalitarian friendships. Husbands and wives compare their contributions to family work with each other, rather than with same-sex peers. This is what differentiates them from most couples. The families in this study are in the process of redefining the cultural rules about being spouses and parents and the cognitive images of spouses and parents. In recreating their families to perform the family work of nurturing, caring, and providing without using gender as a guidepost to direct the tasks, these couples are in the business, whether they know it or not, of recreating our social structure in feminist directions. For that reason alone, analyzing their nontraditional family form is vital.

The limitations of this sample are many. The theoretically driven rules for inclusion—the successful organization of family life without reference to hegemonic, gendered beliefs or behaviors—make this sample so atypical (by race and class, as well as family roles) that it is hard to know whether these findings might be similar in other families. However, it is not the aim of this research to produce a representation of egalitarian families. Rather, we have attempted to address how going beyond gender affects these couples' marital relationships and, in turn, what theoretical implications their relationships might reveal about gender and the possibility for social change. Future research should address how labor, control, and cathexis are negotiated between husbands and wives who share paid and family work but who do not reflect the educational, class, and race privilege of those in our sample. . . .

REFERENCES

Bem, S. (1993). *The lenses of gender: Transforming the debate on sexual inequality.* New Haven, CT: Yale University Press.

Berardo, D., Shehan, C., & Leslie, G. (1987). A residue of tradition: Jobs, careers, and spouse's time in housework. *Journal of Marriage and the Family, 49,* 381–390.

Berk, S. F. (1985). *The gender factory: The apportionment of work in American households.* New York: Plenum Press.

Blaisure, K. R., & Allen, K. R. (1995). Feminists and the ideology and practice of marital equality. *Journal of Marriage and the Family, 57,* 5–19.

Blumstein, P., & Schwartz, P. (1983). *American couples: Money, work, sex.* New York: Basic Books.

Chodorow, N. (1978). *The reproduction of mothering: Psychoanalysis and the sociology of gender.* Berkeley: University of California Press.

Coltrane, S. (1989). Household labor and the routine production of gender. *Social Problems, 36,* 473–490.

Coltrane, S. (1990). Birth timing and the division of labor in dual-earner families: Exploratory findings and suggestions for future research. *Journal of Family Issues, 11,* 157–181.

Connell, R. (1987). *Gender & power: Society, the person, and sexual politics.* Palo Alto, CA: Stanford University Press.

DeVault, M. (1991). *Feeding the family: The social organization of caring as gendered work.* Chicago: University of Chicago Press.

Ehrensaft, D. (1987). *Parenting together: Men and women sharing the care of their children.* New York: Free Press.

Ferree, M. M. (1990). Beyond separate spheres: Feminism and family research. *Journal of Marriage and the Family, 52,* 866–884.

Ferree, M. M. (1991). The gender division of labor in two-earner marriages: Dimensions of variability and change. *Journal of Family Issues, 12,* 158–180.

Ferree, M. M., & Hess, B. (1987). Introduction. In M. M. Ferree & B. Hess (Eds.), *Analyzing gender: A handbook of social science research* (pp. 9–30). Newbury Park, CA: Sage.

Glaser, B., & Strauss, A. (1967). *The discovery of grounded theory: Strategies for qualitative research.* Chicago: Aldine.

Haas, L. (1982). Determinants of role-sharing behavior: A study of egalitarian couples. *Sex Roles, 8,* 747–760.

Haas, L. (1980). Role-sharing couples: a study of egalitarian marriages. *Family Relations, 29,* 289–296.

Hertz, R. (1986). *More equal than others: Women and men in dual-career marriages.* Berkeley: University of California Press.

Hiller, D., & Philliber, W. (1986). The division of labor in contemporary marriage: Expectations, perceptions, and performance. *Social Problems, 33,* 191–201.

Hochschild, A. (1989). *The second shift: Working parents and the revolution at home.* New York: Viking.

Kimball, G. (1983). *50-50 marriage.* Boston: Beacon Press.

Komarovsky, M. (1967). *Blue-collar marriage.* New York: Random House.

Komter, A. (1989). Hidden power in marriage. *Gender & Society, 3,* 2, 187–216.

Lofland, J., & Lofland, L. (1995). *Analyzing social settings: A guide to qualitative observation and analysis.* Belmont, CA: Wadsworth.

Lorber, J. (1994). *Paradoxes of gender.* New Haven, CT: Yale University Press.

Mederer, H. H. (1993). Division of labor in two-earner homes: Task accomplishment versus household management as critical variables in perceptions about family work. *Journal of Marriage and the Family, 55,* 133–145.

Poloma, M. M. (1972). Role conflict and the married professional woman. In C. Safiolios-Rothschild (Ed.), *Toward a sociology of women* (pp. 187–198). Lexington, MA: Xerox College Publishing.

Poloma, M. M., & Garland, T. N. (1971). The married professional woman: A study in the tolerance of domestication. *Journal of Marriage and the Family, 33,* 531–540.

Potuchek, J. L. (1992). Employed wives' orientation to breadwinning: A gender theory analysis. *Journal of Marriage and the Family, 54,* 548–558.

Rapoport, R., & Rapoport, R. N. (1972). The dual-career family: A variant pattern and social change. In C. Safilios-Rothschild (Ed.), *Toward a sociology of women* (pp. 216–244). Lexington, MA: Xerox College Publishing.

Risman, B. J. (in press) *Gender vertigo: Toward a postgender family.* New Haven, CT: Yale University Press.

Risman, B. J., & Schwartz, P. (1989). *Gender in intimate relationships: A microstructural analysis.* Belmont, CA: Wadsworth Press.

Romero, M. (1992). *Maid in the U.S.A.* New York: Routledge.

Rubin, L. (1976). *Worlds of pain: Life in the working-class family.* New York: Basic Books.

Schwartz, P. (1994). *Peer marriage: How love between equals really works.* New York: Free Press.

Seidel, J., Kjolseth, R., & Clark, J. (1985). *The Ethnograph: A user's guide (version 2.0).* Littleton, CO: Qualis Research Associates.

Thompson, L. (1991). Family work: Women's sense of fairness. *Journal of Family Issues, 12,* 181–196.

Thompson, L., & Walker, A. (1995). The place of feminism in family studies. *Journal of Marriage and the Family, 57,* 847–865.

West, C., & Zimmerman, D. (1987). Doing gender. *Gender & Society, 1, 2,* 125–151.

21

Generative Fathering After Divorce and Remarriage: Beyond the "Disappearing Dad"

KAY PASLEY
CARMELLE MINTON

Current estimates suggest that 56% to 62% of all first marriages end in divorce . . . and most divorces include children under the age of 18 years. . . . Most men and women eventually remarry . . . and remarriage occurs fairly quickly after divorce, usually within about 2 years. . . . Because dating is the common pathway to marriage, adults choose a new spouse during the often brief period between marriages. What this means for both adults and children is that there is little time to adjust to the many transitions inherent in terminating one marriage and beginning another. In addition, parenting after divorce and remarriage is complicated.

Unlike most divorced fathers, men who have physical custody of their children often are viewed positively. . . . These fathers are touted as uncommon, often heroic, men who love their children so much that they are willing to assume responsibility for their care. Divorced, nonresident fathers, however, typically are described in pejorative ways that reflect absence, abandonment, or incompetent parenting—they are called disappearing fathers, deadbeat dads, and Disneyland daddies. The negative images of divorced, nonresident fathers as disengaged, uninvolved, and uninterested are pervasive in both the professional and popular literatures.

Here, we focus on the experience of nonresident fathers after divorce from the man's perspective. We address obstacles in marital transitions to a father's continued involvement with his children and offer suggestions for decreasing these obstacles following divorce and remarriage. Several assumptions underlie our thinking. First, we concur with other scholars that fathers are important to children's lives in general. . . . Second, we assume that continued contact with one's father after divorce enhances children's adjustment, and some evidence supports this belief. . . . Third, we believe that continued involvement with one's children is beneficial also to fathers, and some research supports this. . . . Fourth, we assume that most fathers want to be involved with their children and to fulfill their responsibilities as a father . . . ; evidence also supports this assumption. . . . Finally, we assume that good fathering after divorce is hard work that requires especially creative efforts, an important part of fathering. . . . With these assumptions in mind, we turn our attention to fathering and the complications that make it difficult in the face of marital transitions.

GOOD FATHERING AS GOOD PROVIDING

Thompson and Walker . . . suggest that the pervasive societal definition of a good father emphasizes first and foremost being a good provider.

From *Generative Fathering: Beyond Deficit Perspectives,* by A. Hawkins, pp. 118–133. Reprinted by permission of Sage Publications, Inc.

Through assuming responsibility for providing, men maintain status in their families. . . . Loss of job means loss of self, self-respect, and sense of competence to men. Rubin . . . shows that not having a job prompts working-class men to question their value to the family.

It is this latter finding by Rubin . . . that has intrigued us, because it suggests that providing and fathering may not be distinct aspects of a man's self-definition. Much of the literature examining social roles implicitly suggests that roles are unique and distinct from one another. For example, Ameatea, Cross, Clark, and Bobby . . . assess the salience of four social roles (spouse, parent, worker, and home care) as separate entities. They assume that these roles may be associated, but that being a parent is not the same as being a spouse. Our preliminary findings from several focus groups of fathers suggest that men in intact, first marriages may have an integrative view of self. For these men, providing was included in their self-definitions of fathering; being a good father incorporated being a good provider. This is evident in the comments offered by two of our fathers:

> If I am doing my job well, I'm providing a good role model for my children regardless of what I do. . . . My job has taken on a whole new dimension for me as a father.
>
> If you're good in your job and you can make those green stamps, then you can provide more for your son.

Furthermore, our findings show that the inclusiveness of men's self-definitions of fathering incorporated the marital relationship. Several fathers wrote that being a good father meant being responsive to their wife's needs and that this provided their children with a model for their relationships:

> My wife works, too, so I do things my own father would never do, like changing diapers, cooking dinner, or washing clothes. This shows my children (18 months and 3 years old) that it's important to share the work in the family.

We suspect that such overlap in self-definitions may not be the case for women. A woman's self-definitions as a wife and mother are less likely to include providing or spousal behaviors with two exceptions: women in dual-earner families and women in single-parent families, whose self-definitions may reflect overlap in mother and worker. Unlike men, we believe women in general are less likely to evaluate good mothering according to how well they provide for their children financially or what they do in their marital relationship. If the overlap of identities associated with roles that is evident in our work is common, our failure to recognize the inclusive or integrative nature of fathers' self-definitions may cloud our understanding and interpretation of how self-definitions translate into fathering behaviors and how behaviors, in turn, affect self-definitions. Thus, we suggest that men may define themselves as fathers in complex and inclusive ways that reflect overlap of several identities. How this complexity of men's self-definitions is redefined after divorce is unknown, although some scholars . . . have suggested that redefinition occurs.

Because self-definition occurs in a broader context, fathering behavior is culturally influenced. Today, there is more emphasis placed on men's increased involvement in daily family life than in earlier times. . . . Men are being called on to do more with their children as a means of participating in family work. Yet, for many men and women, and for society in general, men's primary responsibility and commitment to the family continues to be judged by their ability and willingness to provide financially. This is evident after divorce as well. Many men are required by the legal system to continue providing for their children by paying child support or maintenance in the full amount and on time. These behaviors become an important criteria by which fathers are evaluated. "Deadbeat dads" are those fathers who fail to fulfill these expectations; such men are seen as poor or bad fathers rather than simply poor or bad workers or providers. Thus, although there is increasing social pressure for men to be involved in child care and household work, fathers are judged on behavior that is more narrowly defined (i.e., providing). As a result, men receive mixed messages about what fathering means and what behaviors constitute good fathering.

If men's self-definitions as fathers are broad and inclusive, but men's performance as fathers is judged primarily on the basis of providing, confusion and ambiguity are likely outcomes. Even among social scientists, there is disagreement about what constitutes fathering. Garbarino . . . suggests that fatherhood is essentially a social invention based on motherhood. Pruett . . . presents an opposing view to fatherhood, stating, "Fathering is not mothering any more than mothering is ever fathering." . . . He suggests that fathers do not need to be mothers to be good parents, but they do need to behave in ways that feel comfortable to them. Men today are being asked to become more involved in the care of their children, but they are also being told to care for them in ways that mothers do. Therefore, they are evaluated on competing criteria (time in child care means time away from providing and vice versa).

The confusion that results from ambiguous and often competing self-definitions is exacerbated following divorce. Fathering behaviors common before divorce that reinforced fathering identity (e.g., getting the child ready for school in the morning) become limited or eliminated altogether. Thus, redefinition of self as father is probable. . . . It is to this ambiguity and the need for redefinition of identity that we now turn our attention as we examine how fathering changes after divorce and remarriage.

A CLOSER LOOK AT NONRESIDENTIAL FATHERING AFTER DIVORCE

Scholars suggest that men are connected to their children through their relationships with their wives . . . and that mothers mediate men's relationships with their children by monitoring, supervising, and delegating certain tasks to fathers. . . . If this is so, then father involvement after divorce should diminish when a man's connection to his wife is severed or dramatically altered. It is too simplistic to assume that the reason fathers disengage from children is because the spousal relationship ends, however; the process of marital dissolution is more complicated.

It is true that many fathers decrease or discontinue their contact with their children after divorce . . . although recent studies show less disengagement than reported earlier. . . . Contact is not the only indicator of father involvement. Payment of child support is another way of assessing father involvement, and surveys show low rates of both child support awards and payments. . . . Many fathers are not court ordered to provide for their children. Of those who are, many fail to do so. The reasons fathers disengage from their children physically and economically and the obstacles that discourage engagement are insufficiently studied . . . but for at least partial answers, we direct our attention to this literature and the results of our research.

What Prompts Fathers to Disengage from Their Children?

Furstenberg . . . summarizes much of the literature on father disengagement in suggesting, "Some fathers are pushed out of the family. . . . Geographic mobility, increased economic demands, and new family responsibilities, which often accompany remarriage, may erode the tenuous bonds between noncustodial fathers and their children." . . . What is clear from the literature is that the decision to limit contact with one's children is neither straightforward nor easy for men to make, although some men experience relief after divorce and welcome freedom from parenting responsibility. If the assumptions we made earlier are accurate (that most fathers want to maintain contact with children and fulfill their paternal responsibility), then most men will experience emotional pain, frustration, anger, and confusion about how to maintain meaningful relationships with their children after divorce. We offer the comments of some of the 92 divorced, nonresident fathers that were part of our earlier study. . . . These two comments emphasize the difficulty of fathering after the end of a marriage:

> The most painful thing was not being able to be with my sons, to tuck them in bed, to say good night to them.
>
> It has not been easy! I have much to learn. I've made many mistakes, but I am praying and trusting God for wisdom. . . . I desperately want to be a faithful father to my children.

Other comments reflect the obstacles to fathering after divorce identified in the literature. We discuss several key obstacles here to demonstrate the increased complexity of fathering following divorce.

The Effects of Legal Decisions Some obstacles to continued father involvement result from court decisions regarding custody, visitation, and child support and the way men perceive these decisions. In most cases (almost 90%), mothers have custody of children after divorce. . . . Although some states give preference to joint legal custody (providing fathers input into decisions affecting their children), how the custody arrangement influences daily life varies. . . . For some, joint legal custody means fathers are equally involved in decisions affecting children's lives (e.g., medical care, education), and children reside part of the week with each parent. For others, joint legal custody does not mean fathers have input into decisions or much access to their children.

We know that decisions about custody and visitation affect the contact between fathers and children, fathers' self-definitions, and the emotional experience of fathers after divorce. For example, joint legal custody is associated with more contact between father and child and the father's feeling closer to the child and more influential in the child's life. . . . Some scholars suggest that joint legal custody implicitly validates fathers' influences in their children's lives, as men adjust better to divorce when they have joint custody. . . . We concur with other scholars . . . that joint legal custody gives fathers the impression that society recognizes their importance to their children. Court orders that fail to recognize the value of fathers, granting only limited access or sole maternal custody, serve as obstacles to father involvement.

The meaning attached to the "visiting status" can be an obstacle to father involvement. Being labeled a visitor results in extensive intrapersonal conflict and emotional turmoil for fathers . . . and maintaining close bonds with children is more difficult because of the arrangements. This means that some men struggle to keep seeing their children, whereas other men do not and then they struggle to feel okay about disengaging from them. Both types of struggles can result in emotional pain; the greater the pain, the less likely the father is to continue visitation. Thus, disengagement may be a coping strategy for managing emotional pain. Many of the divorced fathers in our study expressed their pain and frustration with the legal system and outcomes, as evident in this father's comment:

> I believe that the system [judges, attorneys, etc.] have [sic] little or no consideration for the father. At some point the system creates an environment where the father loses any natural desire to see his children because it becomes so difficult, both financially and emotionally. At that point, he convinces himself that the best thing to do is wait until they are older.

Furthermore, the actual patterns of access and visitation serve as another obstacle to father involvement after divorce. The typical visitation pattern assigned by courts is two weekends a month, or every other weekend. The fathers in our research and those in other studies . . . believed that decisions around visitation relegated them to an insignificant role by denying them liberal access to their children. This was particularly true in situations where fathers saw themselves as highly involved prior to divorce, a finding supported by other scholars. . . . In our research, many divorced fathers wrote about their resentment of visitation limits and the legal system:

> I have been rendered almost impotent/powerless as a father by the legal system.

> I don't get to spend as much time as I would like raising my son. I feel like I got a raw deal in my divorce and would love to have my son with me all the time.
>
> The legal system has essentially invalidated and undermined all of the moral teaching I had spent years to develop and nurture in my children and has made a mockery out of being an upright, involved, and loving father. They have instead showed that their mother's deceitful, immoral lifestyle is rewarded by the state while the honest, etc., father is stripped of everything he holds dear.

The Coparental Relationship Beyond obstacles to father involvement resulting from court decisions around custody and visitation, the way in which former spouses interact may create additional obstacles. Some evidence shows that mothers can and do block and sabotage contact between nonresident fathers and children. . . . "Gatekeeping" by mothers that limits access to children can become a means of getting back at a former spouse. . . . Fathers express anger about such behavior by their former spouses:

> I had problems for the first two years of our divorce with visitations. Going up to three months without being allowed to see my sons. I finally had court-ordered visitation worked out which helped for a while. In the past three years, I've had an average of 20–25 times when my visitation has been denied for no reason. My last two attorneys have given no help in resolving this problem.
>
> She refuses to compromise and it has always been her way or no way. Every situation for which she justified [her behavior] is for the good of the boys, but it will ultimately be for the good of herself.

The level and nature of conflict between former spouses are common obstacles to fathering after divorce, although studies show a decrease in conflict over time. . . . Disagreements over discipline and treatment of children often pervade interaction between former spouses and reduce the frequency of visitation. . . . A primary way former spouses attempt to exercise control over one another is through visitation and the payment of child support, and conflict over these issues is common. Arendell . . . and others . . . show that payment, nonpayment, and delayed or reduced payment are strategies men use to influence former spouses in gaining access to or influence over children. Because society is most concerned over men's failure to meet their financial obligations to children, courts are readier to deal with child support noncompliance than visitation or access issues. In fact, judges often appear to ignore or evade "collateral issues" such as visitation. Evidence suggests that judges feel less comfortable resolving access issues. . . . Men, thus, see themselves as punished by the legal system in this common power struggle, whereas their former wives remain unpunished. . . .

Power struggles evident in the payment of child support may serve as obstacles to father involvement over time. For example, when men perceive the former spouse is squandering the money meant for their children on herself or nonessentials, they are less likely to pay child support in full and on time. They often complain, as this father did, that the system fails to monitor the mother to assure child support is used on the child's behalf:

> It seems as though child support becomes a guise for alimony without any checks and balances. The government places a strangle-hold on those willing to be responsible for child support by an inequitable tax structure, thus placing some negative fault for divorce decree. Trying to comply with judgments on financial matters makes father a dirty and muddled word because of the emphasis placed on an unmonitored amount paid in child support.

Another scenario is that men believe the former spouse obtains more child support than is needed or that a father is not compensated either financially or by simple recognition for the "voluntary" support he provides when the children visit (e.g., they need new shoes for school and he buys them). Under these types of circumstances, he is less likely to pay, especially when he sees the former spouse as making decisions not in the child's best interests.

> My ex-wife often complains about her own finances, but with a $20,000 year job of her own plus about $10,000 a year from me in tax-free money, I have a hard time feeling too sorry for her, especially since she only had a car payment when she left. I resent her for taking my daughter so far away. She did not stop to consider my daughter's needs for two parents who live close.

Yet another scenario is the man who is delinquent in payment, perhaps for good reasons, such as unemployment. Evidence shows unemployment is the best predictor of nonpayment. . . .

In all of these scenarios, the bottom line is that the mother may deny visitation, thinking or saying, "When he pays, he can see the kids." The father may think or say, "When she lets me see the kids, I'll pay." These kinds of impasses place children at risk. Child support and visitation become control strategies for former spouses who fail to consider the effects of nonpayment and lack of contact on the well-being of children. Because the payment of child support and contact or visitation are associated . . . behaviors of former spouses that reduce either the payment of child support or visitation are obstacles to father involvement.

Social Support Being in a social context that is unsupportive of involvement after divorce is another obstacle to fathering. Ihinger-Tallman and her associates . . . argue that, according to identity theory, commitment to an identity increases the chances that the identity is salient to the individual and, thus, reflected in their behavior. Commitment to an identity stems, in part, from a father's social network (e.g., friends, former in-laws, parents). Some evidence shows that support for continued involvement is associated with frequency of visitation, even when the coparental relationship is conflictual. . . .

Preliminary findings from our study also support the relationship between social support and father involvement. We found that among our divorced, nonresident fathers, those who reported receiving more encouragement also reported more involvement in child-related activities ($r = .22$, $p < .05$). Thus, when social support is lacking, as would be more common for divorced fathers, they are less involved.

Other Obstacles Other obstacles that affect the engagement patterns of nonresident fathers after divorce are noted in the literature. Greif and Kritall . . . found that a child may reject the father or that extended kin may interfere and make involvement more difficult. We found these to be themes for our divorced fathers as well:

> The divorce caused my wife to turn my daughter against me, though it took two years to do this after the divorce. Now we [my daughter and I] are trying to turn it around. It is a long, uphill fight, but it is being done, however, slowly.
>
> Her and her family have gone out of their way to keep my son from me. They are using him as a tool to hurt me.

In addition to rejection by one's child or interference by extended kin and former in-laws, geographic distance can impede fathers' connections with their nonresident children, especially visitation. This is often frustrating and emotionally difficult for fathers:

> My daughter and I were extremely close. It was devastating when my ex-wife moved 350 miles away. For several months it was very emotional when I would visit and have to leave. Adjustments have come slowly. I think of her often.
>
> If it were not for the unique opportunities the job I found after divorce affords me, I would be economically deprived of any meaningful involvement and influence

> with my children. [He resides 800 miles away.]

FATHERING AFTER REMARRIAGE

When a remarriage occurs, fathering becomes even more complicated. Men often must redefine themselves to incorporate a new identity as a stepfather or accommodate themselves to their children having a stepfather. In remarriage, relationships between biological parents and children have the longest history and strongest emotional bond. New stepparents are outsiders, lacking knowledge and understanding of preexisting patterns of interaction. As a further complication, the competing developmental tasks of adjusting to life as a newly remarried couple occur simultaneously with the parenting of children.

Under these circumstances, we see at least two factors that make the process of identity redefinition complex. First, scholars . . . have argued that the norms and sanctions around stepfathering are not clear, so stepfathers experience even more confusion about how to parent than do biological fathers. In addition, previous biological parenting experience does not translate well to stepparenting. . . . In other words, parenting behaviors that work in first-marriage families are less effective and sometimes ineffective in stepfamilies. . . . In addition, the acceptable behaviors of supervision and monitoring common among biological fathers are less desirable in stepfathers, especially early on in the marriage. . . .

Second, for many men remarriage following divorce occurs quickly, with 50% of men and 33% of women remarrying within 1 year; . . . adjustment to divorce, however, takes about 3 years. . . . Within this adjustment period, two important changes occur: Fathering is redefined initially, and much of the conflict around coparenting diminishes. When remarriage occurs too quickly, redefinition and coparenting issues are not adequately addressed, so confusion about appropriate behavior as a father or stepfather is likely. Such confusion can undermine the development of cooperation between former spouses, and coparental conflict can exacerbate the confusion around how best to father children and stepchildren. Such role confusion is a reminder that the developmental task of establishing an identity, primarily associated with adolescence, is woven throughout the life cycle . . . and may be especially visible at times when dramatic shifts occur in adult life circumstances.

The legal system provides additional obstacles to fathering after remarriage. In cases where a man becomes a stepfather, the laws governing stepparents fail to clarify his rights and responsibilities. . . . Stepfathering probably is more poorly defined than fathering after divorce, which is already ambiguous. The ambiguity shows up when stepparents attempt to gain access to stepchildren following a divorce or when they attempt to make decisions on behalf of stepchildren in medical emergencies.

It is apparent that the remarriage of either former spouse affects fathering after divorce. The typical pattern is one of decline in contact that may represent a reevaluation of arrangements of residence and custody of children. . . . The least contact is common when only one parent remarries, especially the mother, and more contact when both parents remarry. . . . When one parent remarries, this decline may reflect the need of the new stepfamily to form stronger boundaries so a sense of cohesion develops. If the mother remarries, she may think that one way to do this is to limit contact between the child and his or her father, as evident in the following comment:

> For a period of about three years after my ex-wife remarried, I had very limited visitation privileges with my son. The ex-wife said that she wanted to establish a sense of family with her new husband and I would confuse my son's feelings about this.

Fathers who remarry and become stepfathers often assume some responsibility, financial and emotional, for their stepchildren. Less father involvement with the biological child shows in a decline in payment of child support and a decrease in his availability. . . .

Research also shows that remarriage creates stress in the relationship between the former spouses . . . , which can become an obstacle to fathering. Remarried former spouses report more hostile feelings and negative opinions toward one another than do divorced, nonremarried former spouses. . . . These emotions are associated with fewer child-related discussions between former spouses after remarriage . . . and reduced contact between fathers and children.

Few researchers have examined the effects of the relationship with a former spouse on a father's involvement after remarriage. Buehler and Ryan's . . . findings from 109 divorced fathers are worth noting. First, they found that the relationship with the former spouse was more difficult in stepfamilies when only the father had remarried, but that father involvement was more likely to discontinue when only the mother remarried. In other words, the wife's remarriage but not the husband's served as a barrier to continued father involvement. Buehler and Ryan suggested, "The addition of a new husband (and, possibly, a new father figure) increased the complexity of the binuclear family system in such a way that former husbands found it either unnecessary or too difficult to maintain pre-remarriage levels of involvement with their children." . . . Fathers also may see the stepfather as needing to assume additional responsibility (e.g., financial) for his children, and then they lower their levels of child support. . . . Second, when the mother remarried, higher levels of conflict between the former spouses were associated with *more* father involvement (more frequent and longer visits). This suggests that a nonresident father may react to the remarriage by firmly enacting his commitment to his children and to fathering behaviors. Such an interpretation would be consistent with Snarey's . . . concept of *generativity chill,* that is, feeling a threat to the developmental achievement of learning to care for the next generation. . . . A nonresident father who increases his involvement with his children in the face of his former spouse's remarriage may do so in response to a sensed danger to his generativity. Last, cooperation was strongly associated with child support compliance for remarrieds but was unrelated when neither of the former spouses were remarried. Taken together, these findings show that both remarriage and the coparenting relationship after marriage can serve as obstacles to fathering.

EASING THE TRANSITION

In the face of the multiple changes inherent in divorce and remarriage, maintaining a connection to one's children is difficult. We return to the assumptions we outlined earlier that fathers are important to children's lives, children do best when fathers remain engaged following marital transitions, and fathers want to continue fulfilling their responsibilities. Doing this requires new ways of thinking about self as a father, changes in fathering behaviors, and changes in the broader context to support involvement. In this final section, we discuss some ways to decrease the obstacles and ease the transition to fathering after divorce and remarriage.

We acknowledge that change is difficult for most people, and fathers in marital transition are no exception. All participants in divorce experience discomfort and stress as daily life is altered. Because divorce affects so many aspects of life (e.g., residence, jobs, chores, interaction patterns), distress is common. Recognizing and anticipating changes in the ways nonresident fathers can and are able to function as fathers is essential for them to feel competent and satisfied over time. Fathers who think more flexibly and creatively about how to father best after divorce and remarriage likely experience greater ease in these transitions. This means that one's self-definition must become more individualized to accommodate the ambiguity associated with societal expectations about fathering after divorce. If a nonresident father defines himself and his behaviors by norms that reflect resident fathers (e.g., tucking children in bed nightly), he will evaluate himself negatively after divorce because circumstances may not allow these common behaviors. We believe that divorced, nonresident fathers do best when they define themselves and their behavior in ways that are unique to their situations. And they must reconstruct their relationship with their children on an ethical commitment to the children's well-being. . . . Several examples written by the fathers in our study show such redefinition and reconstruction:

> Basically I talk to my children daily for 15 minutes–1 hour over the phone [bill is $150 a month]. It is surprising how they will open up to you and tell you about their difficulties. I am 300 miles away from my kids and I know more about them than their mother. I also know more about my kids than my father knew about me.
>
> I only see the girls for 10 days at Christmas and 3 months during the summer. I write them weekly, discuss scripture in the letters and how to apply it to their lives. My determination is for them to feel close to me and supported by me even though they live on the other side of the country.
>
> One therapeutic activity for me has been keeping a daily journal of all my thoughts, anger, love, prayers, feelings, etc., that I will give my daughter at a time when she is mature. My ex-wife, I'm sure, is telling her things that are only half true, if true at all. I want her to know my thoughts and that she was loved and thought of daily by me.

Flexibility also means working out a pattern of contact with children that is predictable and stable, yet responsive to changes in the children and life circumstances in general. Research shows that this is good for children and for parents. . . . To do so, former spouses must be able to focus first and foremost on the needs of the children. When this occurs, the quality of the parent-child relationship is enhanced and both children and parents do better. . . . This means that former spouses must overcome the conflict common in the coparenting relationship because conflict serves as a barrier to ongoing engagement.

The trend toward parent education programs and short-term interventions for divorcing parents is encouraging. These programs may help lessen conflict between former spouses and inform participants about support services in their communities, reducing some of the barriers associated with the coparental relationship and social support discussed earlier. A common goal in these programs is conflict management and refocusing parents' attention and commitment to meeting the children's needs. . . . Such programs and additional social support parents access as a result of their participation . . . can help reduce the stress associated with divorce and create less volatile coparenting relationships. . . . These programs will likely be ineffective for the estimated 10% of couples who remain engaged in high conflict after divorce, however. . . . High-conflict couples and those with a history of domestic violence need more intensive and extensive interventions to resolve the issues and promote father involvement. . . .

Last, scholars . . . have called for legal and social reform to reduce some of the obstacles around custody, visitation, child support, and the rights of stepparents. We support their suggestions that the legal system has not gone far enough in validating fathers; rather, for complex reasons it has implicitly and explicitly devalued fathers' and stepfathers' contributions to their children and stepchildren. Courts recognize the financial contributions divorced, nonremarried fathers make to their children, but such contributions are often ignored when they are made by stepfathers. Furthermore, because unemployment and money problems predict child support noncompliance, services that enhance fathers' money management (e.g., debt counseling) and provide job training are warranted. . . . These additional services can help fathers meet their financial responsibilities. Also, because typical child support awards do not adequately meet the actual costs of child rearing . . . , educating fathers about the true costs can diminish their perceptions that mothers squander resources meant for the child.

From the perspective of the divorced, nonresident father, legal reform must do more to enforce access (custody and visitation) agreements. . . . For stepfathers, reform must recognize their contribution to stepchildren, so that if a second divorce occurs, their desire for continued access is not ignored because of their nonbiological status and the law's preference for blood relatives. Men in both cases can feel shut out of children's lives by former spouses and helpless in their attempts to see (step)children. When men have such experiences or feel this way, disengagement increases and children lose. Courts would do well to give equal consideration to issues of access as they do to child support compliance. In this way, the message men receive is that their generative work as fathers extends beyond providing.

Beyond these recommendations, we find it hopeful that many of the fathers in our study wrote sentiments similar to this:

> In the last 4½ years, I have developed an incredibly strong and loving bond with my two sons. I am actively involved in all aspects of their lives. I have even coached their soccer and basketball teams for the last three years. The time I spend with them is very quality time—if anything, the divorce has made me a better and more caring father . . . not to say this would not have happened if my marriage had worked out.

Fathers can and do redefine themselves in ways that allow them to stay engaged and feel good about fathering even in the face of what may seem to be overwhelming obstacles brought about by marital transitions.

22

Patriarchal Terrorism and Common Couple Violence: Two Forms of Violence Against Women

MICHAEL P. JOHNSON

> *You must go through a play of ebb and flow and watch such things as make you sick at heart.*
>
> NGUYEN DU
>
> *(1983)*

We are all too familiar with stories of women who are finally murdered by husbands who have terrorized them for years. In addition, the authors of the 1985 National Family Violence Survey estimate that over six million women are assaulted by their husbands each year in the United States. But are these really the same phenomenon?

This article argues that there are, in fact, two distinct forms of couple violence taking place in American households. Evidence from large-sample survey research and from data gathered from women's shelters and other public agencies suggests that a large number of families suffer from occasional outbursts of violence from either husbands or wives or both, while a significant number of other families are terrorized by systematic male violence enacted in the service of patriarchal control.

SOCIOLOGICAL PERSPECTIVES ON VIOLENCE AGAINST WOMEN IN THE FAMILY

There are two major streams of sociological work on couple violence in families, one that is generally referred to as the *family violence perspective,* and the other of which may be called the *feminist perspective* (Kurz, 1989).

Work in the family violence perspective grew out of family scholars' interest in a variety of family conflict issues, and is generally traced to the early work of Straus (1971) and Gelles (1974). They came together in the early 1970s to develop a research agenda based on the use of interviews to elicit information regarding family violence from large random samples of the adult population of the United States, conducting national surveys in 1975 and 1985. Methodologically, work in this tradition has relied primarily on quantitative analysis of responses to survey questions, utilizing the strengths of random sample surveys in the production of estimates of prevalence, and causal analyses that rely on multivariate statistical techniques. Theoretically, the focus has been largely on commonalities among the various forms of family violence, such as the surprising frequency of violence, the instigating role of stress, and public adherence to norms accepting the use of some violence within the family context.

In contrast, research from the feminist perspective began with a narrower focus on the issue of wife beating, developing a literature that focuses on factors specific to violence perpetrated against women by their male partners (Dobash & Dobash, 1979; Martin, 1981; Roy, 1976; Walker, 1984). Methodologically, feminist analyses have relied heavily upon data collected from battered women, especially those who have come into contact with law enforcement agencies, hospitals, or shelters. Theoretically, the emphasis has been upon historical traditions of the patriarchal family, contemporary constructions of masculinity and femininity, and structural constraints that make escape difficult for women who are systematically beaten.

I do not wish to give the impression that the differences between these two literatures are absolute, although the often-rancorous debates that have gone on between the two groups of scholars seem at times to suggest that there is absolutely no overlap in methodology or theory (e.g., Dobash & Dobash, 1992, pp. 251–284). The truth is that family violence researchers do acknowledge the role of patriarchy in wife abuse (Straus, Gelles, & Steinmetz, 1980, pp. 242–243), and do make use of qualitative data obtained from battered wives (Gelles, 1974). On the other side, many feminist researchers utilize quantitative data (Yllo & Bograd, 1988) and acknowledge the role of factors other than the patriarchal structure of society in precipitating violence against wives (Martin, 1981). As will be seen in the next section, however, family violence researchers and feminist researchers do clearly disagree on some very important issues, and a case can be made that their differences arise from the fact that they are, to a large extent, analyzing different phenomena.

VIOLENCE AGAINST WOMEN IN U.S. FAMILIES: PATRIARCHAL TERRORISM AND COMMON COUPLE VIOLENCE

The findings of the two literatures discussed above lead to strikingly different conclusions regarding a number of the central features of family violence for which they both provide information (gender symmetry/asymmetry, per-couple frequency of violence, escalation of violence, and reciprocity of violence). While these findings suggest to each group of scholars that the other misunderstands the nature of such violence, they suggest to me that these groups are in fact studying two distinctly different phenomena.

The first form of couple violence, which I will call *patriarchal terrorism,* has been the focus of the women's movement and of researchers working in the feminist perspective. Patriarchal terrorism, a product of patriarchal traditions of men's right to control "their" women, is a form of terroristic control of wives by their husbands that involves the systematic use of not only violence, but economic subordination, threats, isolation, and other control tactics.

There are a number of difficult and important terminological issues here. The pattern of violence that I have just described is often referred to with terms such as *wife beating, wife battery,* and *battered women.* I have chosen to avoid these terms for two reasons. I avoid the restrictive term *wife* in order to acknowledge recent literatures that suggest that such a phenomenon may be involved in heterosexual dating relationships (Cate, Henton, Koval, Christopher, & Lloyd, 1982; Stets & Pirog-Good, 1987) and perhaps even in some lesbian relationships (Renzetti, 1992). I have chosen not to switch to a simple nongendered alternative, such as *partner,* because I am convinced that this pattern of violence is rooted in basically patriarchal ideas of male ownership of their female partners.

The terminology of *the battered wife* is also objectionable on the grounds that it shifts the focus to the victim, seeming to imply that the pattern in question adheres to the woman rather than to the man who is in fact behaviorally and morally responsible for the syndrome. The term *patriarchal terrorism* has the advantage of keeping the focus on the perpetrator and of keeping our attention on the systematic, intentional nature of this form of violence. Of course, the term also forces us to attend routinely to the historical and cultural roots of this form of family violence.

The second form of couple violence, which I will call *common couple violence,* is less a product of patriarchy, and more a product of the less-gendered causal processes discussed at length by Straus and his colleagues working in the family violence tradition (Straus & Smith, 1990). The dynamic is one in which conflict occasionally gets "out of hand," leading usually to "minor" forms of violence, and more rarely escalating into serious, sometimes even life-threatening, forms of violence.

Gender Symmetry/Asymmetry

The importance of the distinction between common couple violence and patriarchal terrorism is most forcefully illustrated in the heated debate over the extent to which women are perpetrators of couple violence. One of the surprising findings of Straus and his colleagues' national surveys was that women were evidently as likely to utilize violence in response to couple conflict as were men. One family violence researcher unfortunately chose to refer to these women's use of violence against their partners as "the battered husband syndrome" (Steinmetz, 1978a), suggesting that women's violence against men represented the same sort of phenomenon as the male violence that was being reported to women's shelters across the country. The feminist scholars strongly disagreed (Adams, Jackson, & Lauby, 1988; Berk, Loseke, Berk, & Rauma, 1983; Dobash & Dobash, 1992, pp. 251–284; Dobash, Dobash, Wilson, & Daly, 1992; Fields & Kirchner, 1978; Pleck, Pleck, Grossman, & Bart, 1978; Wardell, Gillespie, & Leffler, 1983). Unfortunately this debate has been structured as an argument about *the* nature of family violence, with both sets of scholars overlooking the possibility that there may be two distinct forms of partner violence, one relatively gender balanced (and tapped by the survey research methodology of the family violence tradition), the other involving men's terroristic attacks on their female partners (and tapped by the research with shelter populations and criminal justice and divorce court data that dominates the work in the feminist tradition).

The Steinmetz (1978) article that introduced the term *battered husband* to the literature relied primarily on data from large-scale survey research to make a case for the position that women are just as violent as men in intimate relationships, and that there was therefore a need for the development of public policy that would address the needs of men who were battered by their wives or lovers. Results from the Conflict Tactics Scale (CTS) used in the National Family Violence Surveys (NFVS)—both the 1975 study upon which Steinmetz relied and the 1985 replication—do indicate almost perfect symmetry in the use of violence by men and women against their partners. (For a thorough methodological critique of the CTS, see Dobash & Dobash, 1992, and Dobash et al., 1992. For earlier responses to many of those criticisms, see Straus, 1990a, 1990b. Although I am in essential agreement with many of the criticisms of the CTS, data presented below indicate that the patterns of violence discovered in shelter samples and national samples differ dramatically even when violence is assessed with the CTS in both settings. This provides strong evidence that the differences are not due merely to the deficiencies of the CTS.) For *any* use of violence, the 1975 national figures for men and women were 12.1% and 11.6%, respectively; in 1985 the comparable figures were 11.3% and 12.1%. For *serious* violence (a subset of the figures for any use of violence, including only acts judged to have a high probability of producing serious injury, such as hitting with a fist), the 1975 figures were 3.8% for men, 4.6% for women; in 1985 the comparable figures were 3.0% and 4.4% (Straus & Gelles, 1990, p. 118). In all cases, the gender differences are less than 2%.

These findings contrast dramatically with those from shelter populations, from hospitals, and from the courts. For example, Gaquin (1978) reported that National Crime Survey data (United States) for the period 1973–75 indicate that 97% of assaults on adults in the family were assaults on wives. Analyses of police files in the U.S. and Britain show similar patterns (Dobash & Dobash, 1992, p. 265; Martin, 1981, pp. 13–14). Kincaid's (1982, p. 91) analysis of family court files in Ontario, Canada, found 17 times as many female as male victims, and Levinger's (1966) study of divorce actions in Cleveland, Ohio, found 12 times more wives than husbands mentioning physical abuse (37% vs. 3%). Fields and Kirchner (1978, p. 218) reported that Crisis Centers in the New York City public hospitals counseled 490 battered wives and only two battered husbands during the last half of 1977.

The most likely explanation for these dramatic differences in the gender patterns of violence in the national surveys and in statistics collected by public agencies is not that one or the other methodology misrepresents the "true" nature of family violence, but that the two information sources deal with nearly nonoverlapping phenomena. The common couple violence that is assessed by the large-scale random survey methodology is in fact gender balanced, and is a product of a violence-prone culture and the privatized setting of most U.S. households. The patriarchal terrorism that is tapped in research with the families encountered by public agencies is a pattern perpetrated almost exclusively by men, and rooted deeply in the patriarchal traditions of the Western family.

Per Couple Frequency

With regard to the frequency of couple violence in "violent" families, we are fortunate to have data using the same data collection instrument (the CTS) with survey samples and shelter samples. According to Straus (1990b), among women who report to NFVS researchers that they have been assaulted by their husbands in the previous year, the average number of such assaults per woman was six ($n = 622$); for those in the sample who had used the services of a shelter, the average was 15.3 ($n = 13$).

In dramatic contrast, Straus cited studies of shelter populations in Maine (Giles-Sims, 1983) and Michigan (Okun, 1986), utilizing the same series of survey questions, that find an average annual number of incidents per woman in the 65 to 68 range! Although Straus argued that the NFVS probably "underrepresents" certain types of violence against women (among the 622 assaulted women in the sample, only four had been assaulted as many as 65 times), he evidently continued to think of this as just another point on a continuum of violence, referring to the missed cases as "cases of extreme violence" (Straus, 1990b, p. 85). Although Straus recognized and discussed the possibilities raised by this "underrepresentation" for resolving differences between the conclusions of shelter research and survey research, and even referred to the possibility of a "qualitatively different experience," he does not seem to have taken the next step, to suggest that perhaps we are dealing with decidedly different phenomena and should adopt a terminology that would mitigate against the mistaken assumption that common couple violence is merely less severe or less frequent than patriarchal terrorism.

Escalation

The two literatures also appear to uncover dramatically different patterns of behavior in terms of escalation. The evidence from the NFVS suggests that so-called minor violence against women does not escalate into more serious forms of violence. Feld and Straus (1990) reported data relevant to this question based on a 1-year follow-up survey of 420 respondents from the 1985 NFVS. My own reanalysis of their published data shows almost no tendency to escalation. For example, among husbands who had perpetrated no acts of minor or severe violence in Year 1 (the year prior to the 1985 interview), 2.6% had moved to severe violence in Year 2. Among those who had committed at least one act of only minor violence, only 5.8% had moved to severe violence; among those who had used severe violence in Year 1, only 30.4% had been that violent in Year 2. Thus, these data indicate that not only is there virtually no tendency to escalation (fully 94% of perpetrators of minor violence do not go on to severe violence), but that in most (70%) of the cases of severe violence there is, in fact, a deescalation. Data on frequency show much the same pattern.

A very different pattern is observed in research with shelter populations. According to Pagelow (1981), "one of the few things about which almost all researchers agree is that the batterings escalate in frequency and intensity over time" (p. 45).

Why does patriarchal terrorism escalate while common couple violence does not? Common couple violence is an intermittent response to the occasional conflicts of everyday life, motivated by a need to control in the specific situation (Milardo & Klein, 1992), but not a more general need to be in charge of the relationship. In contrast, the causal dynamic of patriarchal terrorism is rooted in patriarchal traditions, adopted with a vengeance by men who feel that they must control "their" women by any means necessary. As one husband responded to his wife's protests regarding a violent episode during their honeymoon, "I married you so I own you" (Dobash & Dobash, 1979, p. 94). Escalation in such cases may be prompted by either of two dynamics. First, if his partner resists his control, he may escalate the level of violence until she is subdued. Second, even if she submits, he may be motivated not only by a need to control, but by a need to display that control, yielding a pattern observed by Dobash and Dobash (1979, p. 137), in which no amount of compliance can assure a wife that she will not be beaten:

> For a woman simply to live her daily life she is always in a position in which almost anything she does may be deemed a violation of her wifely duties or a challenge to her husband's authority and thus defined as the cause of the violence she continues to experience. (p. 137)

Reciprocity and Initiation of Violence

On the issue of reciprocity, the NFVS analysts report a pattern in which two-thirds of the families in which the husband has been violent also involve a violent wife, and in which "women initiate violence about as often as men" (Stets & Straus, 1990, p. 161).

Research with shelter populations provides quite a different picture. Pagelow, for example, reported that only 26% of her respondents say they fight back; another 16% indicate that they had once tried, but stopped when it made things worse (Pagelow, 1981, p. 66). She also suggested, although she is not entirely clear (Pagelow, 1981, pp. 65–66), that none of her respondents had initiated the violence in the incidents on which they reported. Giles-Sims's (1983, pp. 49–50) data for a shelter population show dramatic lack of reciprocity in the use of violence, as reported in response to the CTS. The five most severe forms of violence were roughly twice as likely to have been used by the men as the women, and in some cases the differences are even more dramatic (e.g., 84% of the men had beat up their spouse, as compared with 13% of the women), and this in spite of the fact that "the men had almost all abused the women seriously enough to cause injury. In many cases the beatings had been life threatening" (Giles-Sims, 1983, p. 50). The feminist scholars also point out that when women murder they are 7 times more likely than men to have acted in self-defense (Martin, 1981, p. 14). We may sum up the feminist research with testimony to the United States Commission on Civil Rights to the effect that "most women who have been violent towards their husbands have done so only as a last resort, in self-defense against longstanding terror and abuse from their husbands" (United States Commission on Civil Rights, 1978, pp. 450–453, cited in Dobash & Dobash, 1992, p. 257).

Patriarchal Terrorism and Common Couple Violence

The interpersonal dynamic of violence against women uncovered by the researchers working in the feminist tradition is one in which men systematically terrorize their wives, thus the term *patriarchal terrorism*. In these families the beatings occur on average more than once a week, and escalate in seriousness over time. The violence is almost exclusively initiated by the husband, most wives never attempt to fight back, and, among those who do, about one-third quickly desist, leaving only a small minority of cases in which the women respond even with self-defensive violence. These patterns have led researchers in the feminist tradition to conclude that violence against women in the family has its roots in the patriarchal structure of the U.S. family. The central motivating factor behind the violence is a man's desire to exercise *general* control over "his" woman.

It is important not to make the mistake of assuming that this pattern of general control can be indexed simply by high rates of violence. Although the average frequency of violence among cases of patriarchal terrorism may be high, there may well be cases in which the perpetrator does not need to use violence often in order to terrorize his partner. Feminist theorists and shelter activists argue that since patriarchal terrorism has its roots in a motive to exercise general control over one's partner, it is characterized by the use of multiple control tactics (Dobash & Dobash, 1979). The Duluth Domestic Abuse Intervention Project (Pence & Paymar, 1993) has developed a useful graphic representation of this pattern that captures the importance of not becoming overly focused on the violent control tactics that are only part of an overall pattern. . . . The patriarchal terrorist will use any combination of these tactics that will successfully (a) control his partner and (b) satisfy his need to display that control.

Researchers in the family violence perspective describe a dramatically different pattern of violence, one in which the complexities of family life produce conflicts that occasionally get "out of hand" in some families, incidents occurring in those families an average of once every 2 months. The violence is no more likely to be enacted by men than by women, and violent incidents are initiated as often by women as by men. In this common couple violence, there appears to be little likelihood of escalation of the level of violence over time. I would argue that this type of violence is usually not part of a pattern in which one partner is trying to exert general control over his or her partner. Although it is possible that a relatively infrequent, nonescalating use of violence is in some cases part of a generally successful use of other control tactics (the "success" precluding the need to use frequent or extreme violence), I will argue next that it is more likely that the national surveys that uncover this pattern reach only populations in which violence is a relatively isolated reaction to conflict (common couple violence), while studies using data from shelters and other public agencies reach primarily victims of violent, but multifaceted, strategies of control (patriarchal terrorism).

SURVEY SAMPLES AND SHELTER SAMPLES

The debate that has arisen between the feminist researchers and the family violence researchers continues to be framed as a contention over the validity of two radically different descriptions of the nature of couple violence in the United States. The feminists have argued that the description of violence against women that is derived from family violence research is seriously flawed and simply cannot be reconciled with the results of feminist research.

I disagree, arguing that such apparent inconsistencies would be expected if the two literatures are dealing with different phenomena. I propose that the dramatic differences in the patterns of violence described by these two research traditions arise because the sampling decisions of the two traditions have given them access to different, largely nonoverlapping populations, experiencing different forms of violence.

The Sampling Biases of Surveys and Shelters

One of Straus's (1990b) responses to feminist critiques of the NFVS was focused around what he called "the clinical fallacy," which arises "because women whose partners stopped assaulting them are unlikely to seek help from a shelter" (p. 86). The next, perfectly reasonable step in his line of thinking was to call into question generalization from shelter populations to "the general population." The problem is that he did not at this point acknowledge that survey research also misses a significant segment of the general population. Instead, his discussion of the "representative sample fallacy" fell into the trap of assuming not only that the shelter samples are not representative, but that a random sample is. He assumed that random sample surveys provide information regarding "the characteristics and experiences of the total population who manifest a certain problem" (Straus, 1990b, p. 86). I will argue that they do not.

The sampling bias in survey research comes in large part from the fact that even the best designed survey projects are unable to gather information from the total target sample, and nonrespondents may differ in important ways from respondents. For example, men who systematically terrorize their wives would hardly be likely to agree to participate in such a survey, and the women whom they beat would probably be terrified at the possibility that her husband might find out that they had answered such questions. Support for the argument that such families are not represented in the survey data may be found in the fact that among the 182 victims of so-called "wife beating" in the 1985 survey research sample, only four had been assaulted 65 times or more (the average for shelter populations). In contrast, if Straus and his colleagues are correct, and occasional family violence is normative in the sense of being expected and tolerated, if not accepted, then many, if not most, families involved in common couple violence may well agree to participate in a survey on family life.

What about the data sources for most of the feminist research—shelters, hospital emergency rooms, and the criminal and divorce courts? Certainly there are equally serious biases in these sources of data. It is likely, for example, that most families in which couple violence is only intermittent, an unusual response to family conflict, do not need or want such services. The woman or man who is struck or pushed by his or her partner a few times a year will not in most cases report the incident to the police, or go to a shelter, or file for divorce or need to seek medical treatment. Such sources of data are therefore heavily biased in the direction of providing access only to cases of patriarchal terrorism, and, even among those cases, biased in the direction of the most egregious cases.

The biases of shelter samples, although hard to document, are perhaps obvious. The biases of random survey samples, however, may require a bit of documentation.

Do the Survey Numbers Seem to Include Patriarchal Terrorism?

Straus and his colleagues (Gelles & Straus, 1988; Straus & Gelles, 1990) have continued to make use of terminology that implies that their survey data include the phenomenon that is encountered in battered women's shelters, referring to severe violence against wives as "wife beating." However, there are data available from both the NFVS and from women's shelters regarding the number of women who have experienced patriarchal terrorism, data that produce estimates that are so divergent that we must conclude that the NFVS simply does not give access to this phenomenon.

Violence Data Straus (1990b) reported that, in the 1985 NFVS, four women reported a frequency of assaults equal to or above the average for shelter populations. Since this is the number of women above the average for shelter populations, if we assume a symmetrical distribution of frequency of violence for shelter populations, the total population should be projected from double that figure, or eight. The projection to the total U.S. population yields an estimate of about 80,000 women whose beatings fall into a frequency range comparable to that of shelter populations, and who might therefore seriously consider the possibility of moving to shelter housing.

We can compare this figure with an estimate of the number of women actually requesting housing in shelters in the United States. Although the National Coalition on Domestic Violence cannot provide such statistics, this source suggested that I contact the Pennsylvania Coalition on Domestic Violence for the best available statistical data; another source suggested Minnesota. In 1985–86, Pennsylvania shelters housed or turned away 6,262 different women (Pennsylvania Coalition Against Domestic Violence, 1995). Extrapolating that figure to the total U.S. population (i.e., multiplying by 19.20), we get an estimate of 120,230 women who actually tried to use shelter housing, a number roughly 1.5 times the 80,000 that the NFVS suggests might even consider such housing. The Minnesota data provide an even more dramatic contrast. According to the Minnesota Department of Corrections (1987), in 1985, 8,518 women were housed or turned away from shelters. Extrapolating to the U.S. population (i.e., multiplying by 57.72), we get an estimate of 491,659 women to compare with the NFVS estimate of 80,000.

If we take the shelter data as representing the absolute minimum number of women who consider using shelter services each year (they include, after all, only those women who not only considered such action, but took it), we would estimate that the NFVS reaches one-sixth to two-thirds of the victims of patriarchal terrorism in its target sample. However, given the difficulty most women find making the decision to seek help (Kirkwood, 1993), most shelter activists assume that there are at least five terrorized women in the community for every one that seeks shelter, suggesting that the NFVS may collect data from only 1/13 to 1/7 of such couples in its target sample.

Data on Use of Shelter Services There is another potential source of data on patriarchal terrorism in the NFVS. Straus (1990b) reported that there were 13 women in the NFVS who had used shelter services. That figure extrapolates to about 128,600 women nationwide. Unfortunately, the survey wording, referring to the use of the services of a women's shelter, is ambiguous. Most so-called women's shelters in the U.S. actually function as comprehensive resources for women who have been victimized by patriarchal terrorism (many also address issues of sexual assault and child sexual abuse). Most of the women who use the services of such organizations do not actually move into a shelter facility. I will, therefore, compare the figure of 128,600 derived from the NFVS with an estimate from shelters of the number of women who contact them annually regarding domestic violence.

The Pennsylvania Coalition Against Domestic Violence reported 41,425 domestic violence contacts with different women in 1985–86, which extrapolates to 795,360 nationwide. If we assume rough comparability of "domestic violence contacts" in the shelter data and "used the services of a women's shelter" in the NFVS data, we must conclude that the NFVS successfully interviewed about one-sixth of the users of shelter services in its target sample. The Minnesota Department of Corrections reported shelter contacts with 36,189 women in 1985, which extrapolates to about 2.1 million women nationwide, 16 times the extrapolation from the NFVS.

Certainly, there are a great many problems with the statistical manipulations presented above. Pennsylvania and Minnesota are certainly unusual states, having been in the forefront of the shelter movement. In addition, the meaning of "using the services of a shelter" and "shelter contacts" may be different in the two data sources, and there may be hidden problems of distinguishing multiple contacts with the same woman from contacts with different women. The extrapolations from the NFVS and shelter statistics are so divergent, however, that it is unlikely that any of these problems would alter the conclusion that the NFVS simply does not provide valid information regarding the prevalence or nature of patriarchal terrorism.

Response Rates This conclusion implies, of course, that a large number of the cases of patriarchal terrorism in the *target* sample fell into the category of nonrespondents. Were there enough nonrespondents in the NFVS to make this argument tenable? There are two ways to look at the number of nonrespondents in the NFVS. First, we can ask how many of the respondents who were screened as eligible did not complete the interview. According to Gelles and Straus (1991, p. 25), that number is 1,149, representing about 16% of those screened eligible, and providing the basis for the commonly reported 84% response rate in the NFVS. If we assume that roughly half of the nonrespondents are women, we get a figure of 574 female nonrespondents.

However, there is a second way to look at refusals, including as nonrespondents some portion of the 6,166 people whom Gelles and Straus listed as "unable to screen for eligibility." I presume that these are people who refused to answer even the screening questions, since people who were not reached after multiple attempts were listed separately in the discussion of the sampling methodology. Although it therefore makes sense to label these people as refusals, it is likely that only some of them were eligible for the interview in any case, and we probably should not count all of them as refusals. Since 47% of the respondents who *were* screened for eligibility were deemed eligible, it seems reasonable to add 47% of the 6,166—or 2,898—to our pool of nonrespondents, yielding 4,047 nonrespondents and an alternative response rate of 60%, not the 84% usually reported. If we assume that roughly half of the nonrespondents are women, we have 2,024 female nonrespondents.

If we assume the worst, that the eight most severely abused women in the NFVS represent only 1/30 of such women in the target sample, the other 29/30—or 232—women would represent only 40% of the 574 female nonrespondents or 11% of the 2,024 female nonrespondents (the base depending upon your choice of definition of nonresponse). Similarly, if as suggested in the worst case scenario above, the 13 shelter clients who responded represent only 1/16 of such people in the target sample, the other 15/16—or 195—women (shelter clients who presumably refused to participate in the survey) would represent only 34% of the 574 female nonrespondents, or 10% of the 2,024 female nonrespondents. Although these percentages are not small by any means, there are clearly enough nonrespondents in the NFVS to cover even the worst-case estimate of underrepresentation of patriarchal terrorism.

Thus, I would argue that the sampling biases of shelter research and "random" sample research put them in touch with distinct, virtually nonoverlapping populations of violent families. On the one hand, shelter samples include only a small portion of the women who are assaulted at least once by their partner in any particular year. (For 1985, the NFVS estimate is six million such women, while the shelter extrapolations suggest that at most two million women contacted shelters, many fewer seeking services that would make them likely to show up in a shelter research sample.) Of course, this select group is likely to include only women who feel they must enlist help to escape from a man who has entrapped them in a general pattern of violence and control, that is, victims of patriarchal terrorism.

On the other hand, the extrapolations from the NFVS and the Minnesota and Pennsylvania shelter data indicate that survey research reaches only a small fraction of the women who experience severe violence or who make use of the services of shelters. The vast majority of NFVS respondents who experience couple violence have not contacted shelters and have not experienced the level of violence likely to lead them to consider seeking shelter. This select group thus includes only cases in which the women are not generally afraid of their partner—because they have not experienced a general pattern of control—that is, women who are victims of common couple violence.

SUMMARY

Certainly, the case for two forms of violence, one relatively nongendered, the other clearly patriarchal, is not ironclad. However, I am not the first scholar to suggest the possibility that there are multiple forms of couple violence. In fact, at about the time she was developing her case for the "battered husband syndrome," Steinmetz (1978b) published an excellent article making a distinction that is quite similar to the distinction between patriarchal terrorism and common couple violence. More recently, Lloyd and Emery (1994) have emphasized variability in their review of the literature on couple violence. They present nine "tenets" that explicate the interpersonal and contextual dynamics of aggression in intimate relationships. Two of those nine tenets focus on the likelihood of multiple forms of couple violence, and in both cases the authors are able to cite relevant data to support their position (Lloyd & Emery, 1994, pp. 37–40).

Nevertheless, since the heart of the distinction between common couple violence and patriarchal terrorism is one of motivation, the evidence presented above can only be suggestive. What is required is research that can provide insight into motivation. One way to get at motivation would be to gather information concerning a range of conflict and control tactics from each couple. Patriarchal terrorism is presumed to involve acts of violence that are embedded in a larger context of control tactics. Common couple violence is presumed to show a less purposive pattern, erupting as it does from particular conflicts rather than from a general intent to control one's partner. A second approach to motivation is in-depth interviewing of couples who are involved in violence, eliciting interpretations of the psychological and interpersonal causes of specific incidents or patterns of control. The goal is to go beyond the behavioral description of particular acts to develop a narrative of each incident's development, as presented and interpreted by perpetrators and targets of violence. Both of these sorts of data are commonly collected in the work with shelter samples. However, we also need this kind of data from samples that target populations that are more likely to include examples of both forms of violence.

Let me conclude with a partial list of the reasons for my belief that the distinction between common couple violence and patriarchal terrorism is important. The first, and most important, has to do with the role of scientific understanding in the shaping of social policy. The issue is perhaps best illustrated in the debate regarding the gender symmetry/asymmetry of couple violence. The failure to make a distinction between patriarchal terrorism and common couple violence has led some analysts to make the logical error of leaping from (a) the description of a few case studies of terrorism perpetrated against men and (b) frequency estimates of common couple violence against men from survey research to (z) the conclusion that there is a widespread "battered husband" syndrome. This erroneous conclusion may be used in campaigns against funding for women's shelters (Pleck et al., 1978), opponents arguing that shelters should not be funded unless they devote equal resources to male and female victims. Although it is indisputable that *some* men are terrorized by their female partners (I have worked with some at my local shelter), the presentation of survey data that tap only common couple violence as evidence that men are terrorized as frequently as women produces a dangerous distortion of reality.

A similar distortion occurs when stories of patriarchal terrorism against women are used to describe the nature of family violence, while numbers that probably apply only to common couple violence (survey extrapolations) are used to describe its prevalence. If the arguments presented above are correct, random sample surveys cannot produce estimates of the prevalence of patriarchal terrorism. We must develop methods of collecting and extrapolating effectively from shelter, hospital, police, and court data.

A second major problem arises when educational and therapeutic efforts targeted at prevention and intervention are governed by the assumption of one form of couple violence. For example, in women's studies texts and in training manuals at women's centers, one often finds the statement that couple violence always escalates. Unacceptable as any one incident of violence in a relationship may be, if the arguments above are correct, it is certainly not the case that escalation is an inevitable part of male violence, let alone an inevitable part of the violence in lesbian relationships, which is almost certainly more likely to be common couple violence, which does not generally escalate. Thus, advice that is based on a mistaken assumption of impending terrorism may do some women a great disservice. One can also imagine similar scenarios of misinterpretation and misplaced advice in family counseling or other therapeutic relationships. As in most areas of intervention, family practitioners will be most effective if they work with a set of alternative interpretive frameworks rather than with a single-minded assumption that every case of violence fits the same pattern.

The third area in which problems may be created by the conflation of different forms of violence is in theoretical interpretation. If the two forms of violence have different psychological and interpersonal roots, then theory development will either have to proceed along different lines for each, or move in the direction of synergistic theories that explicate the conditions under which particular combinations of the same causal factors might produce qualitatively different patterns of violent behavior. For example, we are beginning to try to develop an understanding of the dynamics of lesbian couple violence, a phenomenon that must seem somewhat mysterious if we assume that all violence within couples follows the pattern found in patriarchal terrorism. If we were to assume a unitary phenomenon, we would develop a theory of lesbian violence that focused heavily on the conditions under which some lesbians might fall into patriarchal family forms. It may be more reasonable to assume that the bulk of violence in lesbian relationships is of the common couple variety and involves causal processes that are very similar to those involved in nonlesbian common couple violence, having little to do with the taking on of patriarchal family values.

Alternatively, using the synergistic approach to theory development, we might note that (a) some, if not all, of the causal factors involved in patriarchal terrorism may also be involved in common couple violence and vice versa, (b) many of these factors are best conceptualized as continuous variables, and (c) although some of them are sex-linked, there is probably considerable overlap in the gender distributions (Taylor, 1993). The following partial list of causal factors may be used to illustrate these three points: (a) motivation to control, (b) normative acceptability of control, (c) inclination to use violence for control, (d) physical strength differences that make violence effective, (e) inclination to expressive violence, (f) victim deference, and (g) structural commitment to the relationship. All could conceivably be involved in the generation of particular cases of either patriarchal terrorism or common couple violence, each can be conceived as a continuous variable, and all are likely to be at least imperfectly linked to gender. The behavior described in this paper as patriarchal terrorism, however, may develop only from the co-occurrence of high values on some particular subset of the causal variables. If all other combinations of the same variables produce either no violence at all or a pattern recognizable as common couple violence, this complex combination of weakly gender-linked, continuous variables would produce a strongly gender-linked pattern of two types of couple violence. Under such conditions, even relatively weak links of the various factors to gender might produce empirical patterns of patriarchal terrorism that occur almost exclusively among men in heterosexual relationships, accompanied by the occasional occurrence of a similar pattern among women—even in lesbian relationships—and in gay male couples.

Finally, we have to ask, "How on earth could two groups of social scientists come to such different conclusions about something as unsubtle as family violence?" We owe it to the families that are the focus of our work not to get so caught up in the defense of our initial positions that we fail to see important insights that can be gained from our disagreements. The social policy, educational, and therapeutic implications of what we do are too important for us to allow our deep moral aversion to violence to blind us to important distinctions. Yes, all family violence is abhorrent, but not all family violence is the same. If there are different patterns that arise from different societal roots and interpersonal dynamics, we must make distinctions in order to maximize our effectiveness in moving toward the goal of peace in our private lives. . . .

REFERENCES

Adams, D., Jackson, J., & Lauby, M. (1988). Family violence research: Aid or obstacle to the battered women's movement. *Response, 11,* 14–16.

Berk, R. A., Loseke, D. R., Berk, S. F., & Rauma, D. (1983). Mutual combat and other family violence myths. In D. Finkelhor, R. Gelles, G. Hotaling, & M. A. Straus (Eds.), *The dark side of families: Current family violence research* (pp. 197–212). Newbury Park, CA: Sage.

Cate, R. M., Henton, J. M., Koval, J., Christopher, F. S., & Lloyd, S. (1982). Premarital abuse: A social psychological perspective. *Journal of Family Issues, 3,* 79–90.

Dobash, R. E., & Dobash, R. P. (1979). *Violence against wives.* New York: Free Press.

Dobash, R. E., & Dobash, R. P. (1992). *Women, violence and social change.* New York: Routledge.

Dobash, R. P., Dobash, R. E., Wilson, M., & Daly, M. (1992). The myth of sexual symmetry in marital violence. *Social Problems, 39,* 71–91.

Feld, S. L., & Straus, M. A. (1990). Escalation and desistance from wife assault in marriage. In M. A. Straus & R. J. Gelles (Eds.), *Physical violence in American families* (pp. 489–505). New Brunswick, NJ: Transaction Publishers.

Fields, M. D., & Kirchner, R. M. (1978). Battered women are still in need: A reply to Steinmetz. *Victimology, 3,* 216–226.

Gaquin, D. A. (1978). Spouse abuse: Data from the National Crime Survey. *Victimology, 2,* 632–643.

Gelles, R. J. (1974). *The violent home: A study of physical aggression between husbands and wives.* Newbury Park, CA: Sage.

Gelles, R. J., & Straus, M. A. (1988). New York: Simon & Schuster.

Gelles, R. J., & Straus, M. A. (1991). Second National Family Violence Survey: Survey methodology (3rd rev.). In R. J. Gelles & M. A. Straus (Eds.), *Physical violence in American families, 1985* (2nd release) [Computer file]. Durham, NH: University of New Hampshire Family Research Laboratory [Producer]. Ann Arbor, MI: Inter-University Consortium for Political and Social Research [Distributor].

Giles-Sims, J. (1983). *Wife-battering: A systems theory approach.* New York: Guilford.

Kincaid, P. J. (1982). *The omitted reality: Husband-wife violence in Ontario and policy implications for education.* Maple, Ontario: Learners Press.

Kirkwood, C. (1993). *Leaving abusive partners.* Newbury Park, CA: Sage.

Kurz, D. (1989). Social science perspectives on wife abuse: Current debates and future directions. *Gender and Society, 3,* 489–505.

Levinger, G. (1966). Sources of marital dissatisfaction among applicants for divorce. *American Journal of Orthopsychiatry, 36,* 803–807.

Lloyd, S. A., & Emery, B. C. (1994). Physically aggressive conflict in romantic relationships. In Dudley Cahn (Ed.), *Conflict in close relationships* (pp. 27–46). Hillsdale, NJ: Lawrence Erlbaum.

Martin, D. (1981). *Battered wives.* Volcano, CA: Volcano Press.

Milardo, R. M., & Klein, R. (1992). *Dominance norms and domestic violence: The justification of aggression in close relationships.* Paper presented at the Pre-Conference Theory Construction and Research Methodology Workshop, National Council on Family Relations annual meeting, Orlando, FL.

Minnesota Department of Corrections. (1987). *Minnesota program for battered women: Advocacy program: Data summary report (through 1986).* St. Paul, MN: Author.

Nguyen Du. (1983). *The tale of Kieu* (Translated and annotated by Huynh Sanh Thong). New Haven: Yale University Press.

Okun, L. (1986). *Woman abuse: Facts replacing myths.* Albany: State University of New York Press.

Pagelow, M. (1981). *Woman-battering: Victims and their experience.* Newbury Park, CA: Sage.

Pence, E., & Paymar, M. (1993). *Education groups for men who batter: The Duluth model.* New York: Springer.

Pennsylvania Coalition Against Domestic Violence. (1995). Personal correspondence.

Pleck, E., Pleck, J. H., Grossman, M., & Bart, P. B. (1978). The battered data syndrome: A comment on Steinmetz' article. *Victimology, 2,* 680–683.

Renzetti, C. M. (1992). *Violent betrayal: Partner abuse in lesbian relationships.* Newbury Park, CA: Sage.

Roy, M. (Ed.). (1976). *Battered women: A psychosocial study of domestic violence.* New York: Van Nostrand Reinhold.

Steinmetz, S. K. (1978a). The battered husband syndrome. *Victimology, 2,* 499–509.

Steinmetz, S. K. (1978b). Wife-beating: A critique and reformulation of existing theory. Bulletin of the *American Academy of Psychiatry and the Law, 6,* 322–334.

Stets, J. E., & Pirog-Good, M. A. (1987). Violence in dating relationships. *Social Psychology Quarterly, 50,* 237–246.

Stets, J. E., & Straus, M. A. (1990). Gender differences in reporting marital violence and its medical and psychological consequences. In M. A. Straus & R. J. Gelles (Eds.), *Physical violence in American families* (pp. 151–180). New Brunswick, NJ: Transaction Publishers.

Straus, M. A. (1971). Some social antecedents of physical punishment: A linkage theory interpretation. *Journal of Marriage and the Family, 33,* 658–663.

Straus, M. A. (1990a). The Conflict Tactics Scales [sic] and its critics: An evaluation and new data on validity and reliability. In M. A. Straus & R. J. Gelles (Eds.), *Physical violence in American families* (pp. 49–73). New Brunswick, NJ: Transaction Publishers.

Straus, M. A. (1990b). Injury and frequency of assault and the "representative sample fallacy" in measuring wife beating and child abuse. In M. A. Straus & R. J. Gelles (Eds.), *Physical violence in American families* (pp. 75–91). New Brunswick, NJ: Transaction Publishers.

Straus, M. A., & Gelles, R. J. (1990). Societal change and change in family violence from 1975–1985 as revealed by two national surveys. In M. A. Straus & R. J. Gelles (Eds.), *Physical violence in American families* (pp. 113–131). New Brunswick, NJ: Transaction Publishers.

Straus, M. A., Gelles, R. J., & Steinmetz, S. K. (1980). *Behind closed doors.* Newbury Park, CA: Sage.

Straus, M. A., & Smith, C. (1990). Family patterns and primary prevention of family violence. In M. A. Straus & R. J. Gelles (Eds.), *Physical violence in American families* (pp. 507–526). New Brunswick, NJ: Transaction Publishers.

Taylor, M. C. (1993). Personal correspondence.

United States Commission on Civil Rights. (1978). *Battered women: Issues of public policy.* Washington, DC: Author.

Walker, L. E. (1984). *The battered woman syndrome.* New York: Springer.

Wardell, L., Gillespie, D. L., & Leffler, A. (1983). Science and violence against wives. In D. Finkelhor, R. J. Gelles, G. T. Hoteling, & M. A. Straus (Eds.), *The dark side of families: Current family violence research* (pp. 69–84). Beverly Hills, CA: Sage.

Yllo, K., & Bograd, M. (Eds.). (1988). *Feminist perspectives on wife abuse.* Newbury Park, CA: Sage.

23

Straight from the Heart

ROD AND BOB JACKSON-PARIS

LET THIS BE OUR DESTINY
TO LOVE, TO LIVE
TO BEGIN EACH NEW DAY TOGETHER
TO SHARE OUR LIVES FOREVER

RODNEY LYNN JACKSON
AND
ROBERT CLARK PARIS
WILL BE JOINED IN MARRIAGE
ON SATURDAY,
THE TWENTY-SECOND OF JULY
NINETEEN HUNDRED AND EIGHTY-NINE
AT ELEVEN O'CLOCK IN THE MORNING
UNITARIAN COMMUNITY CHURCH
1260 EIGHTEENTH STREET
SANTA MONICA, CALIFORNIA
REVEREND ERNEST D. PIPES, JR.

RECEPTION TO FOLLOW

B: Rod and I wanted to be married because in our culture that's the highest level of public and private commitment we could make to one another. We wanted to get married in a church and before God—whatever God is—because we both have deep spiritual beliefs. And we wanted to be married in front of our friends because we wanted them there to witness us making an important statement of commitment.

I was raised a mainline Christian with Southern Baptist overtones and Rod was raised a Lutheran. Neither of us had been active in organized religion for some time because of all the bigotry against gay people. So we knew we wanted to find a minister from a denomination other than the ones we grew up in. We had read about the Unitarian Church in a book called *In Search of Gay America* and learned that the Unitarians have been generally welcoming to gay and lesbian people.

R: Bob was in England on a seminar tour when I first started calling around to find a Unitarian minister who would perform our wedding. The first one I talked to made absolutely clear that he would be doing a "union" ceremony, not a marriage, and that two men couldn't get married. He said, "It's not the same thing. A marriage is for procreating children." Even though I know a church can't legally marry gay people, I felt he was treating me like a second-class citizen, so I told him he sounded like a bigot. He said that he had many gay people in his ministry, but to me that didn't mean anything because he was still setting up a relationship hierarchy in which my relationship to Bob was second-class. So I kept looking.

At the next church I called the secretary said they had many gay members. Then I asked if they were treated like everyone else, and she said, "Of course." She was really open with me and suggested I speak with the minister, Ernie Pipes.

I called the minister and told him our situation, and he said that he wanted to meet with us. He explained that he didn't marry anybody until he had a chance to sit down and talk with them.

B: I got back a week later, and we went in to meet with Reverend Pipes. He had all kinds of awards on the wall of his office for his civil rights work, so we knew we'd come to the right place. We also had the same sense of humor, so we really hit it off with him.

Reverend Pipes told us that he wanted to meet with us three or four times for about an hour each time to talk about why we wanted to get married and how we envisioned our relationship. We explained to him that we wanted to get married because we loved each other and wanted to spend our lives together. And that while we weren't a traditional heterosexual couple, we still wanted to participate in some of the traditions we'd both grown up with.

R: He also asked us about our spiritual beliefs. We told him that while neither of us felt welcome in our denominations because we're gay, we were very spiritual people. We also explained that one of the biggest problems we had with our spirituality was that we couldn't find people who shared our concept of what God is. Reverend Pipes explained to us that the Unitarian Church, a mainline Christian denomination, is humanistically oriented, and they let people form their own beliefs.

B: Reverend Pipes made a point of asking us if we understood that we would invariably face some hard challenges in life. We'd already been together for over three years, so we knew that everything wasn't going to be perfect.

R: So many young couples don't understand that there will inevitably be illness and eventually death to deal with. He said to us, "You can be in love with somebody, but if you don't want to toe the long road, don't bother getting married."

One of the things he asked us was how we resolved our differences. We told him that we fought like cats and dogs, but that we always come to a resolution. He told us that many people say it's best to resolve things before you go to bed, not to let the sun set on your problems. And then he asked if we did that. I said, "No," and explained that there are some problems that are too big for the sun, and that we wait for the sun to go back up again. He laughed and said, "Okay, as long as you both understand that and find that it works for you."

B: Reverend Pipes really emphasized how difficult it was going to be, that being married meant taking care of each other through the good times and the bad. I don't think he was trying to discourage us, but he wanted to be sure that we knew what we were committing to, that marriage wasn't to be taken lightly. We explained to him that this was a very serious decision for us, that we weren't just another number in each other's long list of men, that there was not going to be another true love for either of us. This was it.

By the end of our third meeting, when Reverend Pipes was convinced that this was what we wanted and that we understood what we were doing, we began to discuss the ceremony itself. The reverend explained that the wedding ceremony was basically a piece of theater and that our role was to create a wonderful event for ourselves and the invited guests. The more wonderful the event, the more memorable it would be.

R: We talked about the standard Unitarian wedding ceremony, and then the reverend asked us if we wanted a traditional ceremony or if we wanted to create something on our own. We told him that we wanted a traditional ceremony, including the vows and the part where you say, "You can kiss each other now." We also told him that we wanted to write something to each other that we would read during the wedding, and we told him that we wanted to exchange rings, and that we'd already bought our rings three years earlier, knowing that we would want to get married one day.

B: We were in Lucerne, Switzerland, and stopped to look in the window of a jeweler. These wedding bands caught our eye, so we went in to look at them. They were kind of rough, different from the usual wedding band. They were solid platinum, and one had a band of gold overlay three quarters of the way around, and the other had a gold overlay one quarter of the way around. Together they made two complete circles with two precious metals.

At first the young clerk was kind of confused. I spoke enough German to keep us out of trouble, and she spoke enough English to get herself *in* trouble.

R: She tried to explain to us that these were wedding rings.

B: And I said, "Yes, these are wedding rings, and we want to be fitted for them." At this point, an older woman who was very prim, and very Swiss, came out from the back to see what the problem was. We explained what we wanted, and she said that there was no problem at all and proceeded to take our ring measurements.

R: On the inside of my ring we had inscribed, "Forever, Bob." And on the inside of Bob's ring it says, "Forever, Rod." The young clerk just wrote it all down and never really seemed to understand what was going on and what our relationship was to each other.

B: I think she figured that she had just sold an extraordinary set of friendship rings. So we picked up the rings a few weeks later and put them in a box to save them for our wedding day.

R: With the ceremony arranged, we chose our wedding day, July 22, 1989, and started planning the reception and working on the invitations.

B: We went to the PIP printer on Sunset Boulevard, just above West Hollywood, and we talked to the guy we'd done a lot of business with in the past. He was gay, and when we said that we wanted to look at the books of wedding invitations, he was charmed. We asked to see something that wasn't gaudy, with clean lines, and in the colors of our wedding, which were salmon, silver, and ivory.

We picked out off-ivory stationery, with salmon and silver-lined envelopes. It was pretty standard, with the date and time.

R: We ordered 300 invitations because we decided that we would send them to our entire families, from our grandparents on down to second cousins. I wanted every single person in my family to know I was gay so that the next time they told a faggot joke they'd have to own part of it.

Both of our parents had begged us for years not to come out to our grandparents. But we told them that they were going to find out whether or not we sent them invitations to the wedding. We wanted to do it clean, so we sent them invitations.

B: We also sent invitations to everyone we knew from business, including people like Arnold Schwarzenneger, Maria Shriver, Lou Ferrigno and his wife, Carla, and Joe Weider. We also sent several invitations to people we knew and worked with in the press. This event was for us; we had no idea it would become so political. This wedding invitation left no doubt what our relationship was. And that's how we wanted it, totally up-front.

We included with the invitation the usual things, like the RSVP card and return envelope. We also sent our family photo, which had been published in the *Ironman* article in which I came out. It's the one with Rod sitting in a chair and me standing behind him with Barney, our bird, on my arm. Sam, one of our dogs, was also in the photo. We sent it all out and waited for the returns to come in.

R: Two days after we mailed the invitations, we walked into the gym and there was a buzz. People couldn't believe we were crazy enough or had the courage to do this.

B: I overheard a couple of the old-timers at the gym kind of jokingly nudge each other and say, "Yeah, will you marry me?" We heard all kinds of things, but the one that I think was perfect came from a guy I know who's a bodybuilding writer. He was on the telephone with Joe Weider and asked him what he thought of the "Bob and Rod situation." And Weider said, "They're either very smart or very stupid."

R: Over the next couple of weeks, people got really strange about the whole thing, even some of our friends. People were afraid to show up because they were worried about what people would think if they came to the wedding of two men. We got so fed up that we told several people that we didn't care if they came. We said, "You got an invitation. You're a part of our life. You're welcome to come, but if you have a problem with it, please stay away."

We couldn't believe some of the things people said. One of the bodybuilding judges we invited told us he didn't know what to expect at a gay wedding since he hadn't ever attended one. That was perfectly reasonable, but then he asked if somebody was going to wear a dress. I looked at him and said, "Sure, Art, you can wear a dress. We won't mind."

B: Mostly we didn't hear things directly. People usually sent emissaries to convey their concern. Our friends would tell us that someone who had been invited said blah, blah, blah, and that they told them they were being stupid.

There were some people we knew wouldn't attend, but we sent them invitations anyway, like Arnold and Maria. We never heard anything from them. Given his political affiliations, he couldn't have come even if he wanted to. From what I heard, a few people from the gym who couldn't make up their minds called Arnold to ask him what he thought they should do. I was told he said, "Look, grow up. Do you like these guys? If you do, go."

R: It amazes me that these adults couldn't even make a decision on their own. They had to call Arnold to make sure he wouldn't be mad at them before they made their own decision. It shows me how paralyzed society is when it comes to homosexual issues.

B: Joe Gold, the guy who owned the gym where we worked out, who has been very good to me over the years, gave his invitation to one of the assistants at the gym. He told her to deal with it, somehow. She called me and didn't know what to do. I told her that I really didn't care what Joe did, that if he didn't want to be there, he shouldn't come. I told her that she might point out to him that if I was marrying a woman, he'd be there with bells on. Then she asked if *she* could come with him and I told her she could come with or without him.

This started to happen a lot—people who hadn't been invited to the wedding letting us know that they wanted to come. They wanted to show their support or they were just curious.

R: For example, we got a call from a friend who said, "So and so wants a ticket for himself and his girlfriend." They told him that they wanted to show their support for us by coming to the wedding, but at the same time we knew they wanted to be part of what was quickly becoming the hottest event in town that summer.

B: They wanted "tickets" as if this were a prizefight at the Garden.

R: I heard the conversation Bob was having on the phone with our friend and I went over the top. I was yelling in the background that our wedding wasn't some kind of spectacle. Our friend could hear everything I was saying, so Bob didn't have to do a lot of explaining. Bob just said that we weren't giving out tickets to the event because there were no tickets, just invitations, and we only invited people we knew. It's not that we didn't appreciate their support, but we didn't know these people. The phone rang and rang and rang.

B: Really, it was hilarious that this turned into an event that people felt they had to be a part of.

R: Unfortunately, for the most part our families didn't feel the same way. No one from either side of the family attended. We didn't even get the RSVP cards back from everyone. It was heartbreaking. . . .

R: When Bob and I first planned the wedding, we decided that we wanted to write a poem for each other as part of the ceremony. We'd discussed this with Reverend Pipes, so every time we met with him he asked us how the poems were coming along.

B: Every time he asked, we said, "Coming right along."

R: Nothing like two people who'll lie to a minister.

B: We hadn't written a word.

R: We put it off until the absolute last possible minute. The night before the wedding, with our apartment packed with friends who were staying with us from all over, I got out of the house and went to Zucky's, an all-night deli in Santa Monica. We had to be in the church at seven in the morning, so I didn't have a lot of time.

B: While Rod was gone, I sat in our bedroom and wrote my poem. I knew what I wanted to say, but I wanted to do it in the same meter as "Wood and Stream," the first poem I wrote for Rod, because it was really a continuation of that poem. It turned out to be very complicated to do.

R: I also knew what I wanted to write, but I'm not someone who's organized about it. When I feel it, I sit down and I do it, although I'm not sure what I was feeling at three in the morning, but I had to do it. So I let my heart write for me and finished it in about half an hour.

While I was working on the poem, the waitress who was pouring me cup after cup of coffee asked me what I was doing. She'd seen the two of us there all the time. I'm sure she'd seen us there holding hands. I told her I was writing a poem, and she asked what it was for. I said, "I'm getting married tomorrow at nine." She said, "Oh, I think that's so romantic. It's so nice to hear about couples getting started in their life together." And then she just stood there leaning against the counter caught up in her own thoughts. You could see her thinking back to when she got married. Marriage and love are such universal experiences.

I got home at five in the morning and Bob was still up. He said, "Let me hear it." I said, "Nope, you'll hear it later today." Then he asked if I wanted to hear his, and I said that I'd hear it at the wedding. I wanted to hear it in that moment when we were there together taking our vows. . . .

R: The morning of the wedding I should have been exhausted, but I was more exhilarated than anything else. I was also scared, because I was afraid to get married. I was really afraid of a lifetime commitment and the responsibilities that come with it.

B: We were both scared. Until we took our vows in front of 200 people and God, we could have walked away at any time, no harm, no foul. Of course, there would have been the emotional trauma and it would have been complicated because our lives were so intertwined, but once we were married, we couldn't walk away that easily. We took these vows very seriously.

R: The morning of the wedding was very, very hot, and the church was packed.

B: We came in from the sides and met at the altar. The room was completely silent.

R: The ceremony opened with a prayer from Reverend Pipes, and then our friend Tom sang a beautiful song by Kathy Mattea, "As Long as I Have a Heart": "As long as I have a heart, you have a safe place to dream in, a tree to carve your name in, as long as I have a heart." Right away I was crying. Then our friend Kelliey, from Arizona, read a poem called "What Love Is." After that, I read the poem I wrote for Bob.

I was such a mess while I was reading the poem. I was sweating because of the heat, crying, and the whole time I was trying to keep my hair out of my face.

B: His hair was really long then, and he looked like Cousin It.

R: I heard one of my friends say, "If he doesn't stop playing with his hair, I'm going to cut it all off." I was trying to keep it out of my face so I could read, but I guess I was also nervous.

The whole event was so emotional. There had been all this buildup and the personal turmoil and my fear of, "Oh my God, what if this isn't right?" But it felt right while it was happening, and I couldn't stop crying. I was so happy, and it was such a release. I don't know how anybody understood a word I said because I was snorting and sniffling.

Before I read the poem, I took Bob's hand and held it the whole time. This is what I read to Bob:

If I could promise you
the moon and stars
in all their azure blues
and shooting light
I would
But all that's mine to promise you
is a heart of flesh and blood

If I could promise you
that I could protect you
from all of life's harm
I would
But all that's mine to promise you
is that I will stand proudly next to you
through the days
as they number and cultivate into years
never keeping you from experiencing life
as God has intended in his plans

If I could promise you
the peace of spirit to be
the most honest and compassionate person
you *are* and were meant to be
I would
But all I have to promise you
is a shelter to foster sense of self
a shield of pride
and a space with enough room to grow
to rest when you are weary and to lick
    your wounds

If I could promise you
that I could teach you how
to live life deeply and full
yet lightly and lithe
I would
But all that's mine to promise you
are guiding thoughts
from my sometimes scared and confused
    heart
for I am much more a pupil
than a teacher
and the journey will be equally shared

If I could promise you
that I know *and* would share
the secret to live life true
I would
But all that's mine to promise you
is honesty and valor
to search for a sometimes elusive truth

I could promise you
all the things through time
that all the poets chime
But all that's mine to promise you
when the morning sun casts its first
tendrils of shy, yet persevering light
is the fusion of a maturing heart
    and mind
only me

B: As soon as Rod started reading the poem I started crying, and as I reached for my handkerchief I looked out at the congregation, and with all the handkerchiefs going back and forth it looked like everyone was surrendering.

When Rod finished, I wanted to savor the moment because it was so overwhelming. No one had ever written anything like that to me before. It captured the essence of everything I always felt he couldn't really ever say to me. Yet he said it. But there was no time to pause and savor that moment with Rod because we had to go on with the ceremony. So now it was my turn to read, and I took Rod's hand in mine and said:

Oh my word
I can't believe it.
Here at last
I finally stand.
All dressed up in my tuxedo,
Holding onto Rodney's hand.

With that, everybody burst into laughter. It was such a relief, because there was so much pent-up tension. We both laughed, too, and then I continued:

Look at him.
I see how lucky
I have been
to find the one.
In his eyes
I see the ocean,
stars, the moon
and rising sun.

He's the one
pulled my heart strings
sent me tumbling
head o'er heels.
Picked me up
and through the distance
gave me strength
to drop my shields.

We were walking
through the desert
moonlight bright
nightshadows dance
holding hands
whisper secrets.
God's wondrous gift
this sweet romance.

I want to tell him
I do love him
that I'll never
leave his side.
That he has
to hold forever
my bursting heart
so filled with pride.

Over the miles
we had journeyed
reaching, growing
face to face.
We must remember
oh my darling
it was love that brought us
to this place.

Now it's time
to tell each other
that we'll be
true for life.
The dance
is only just beginning
hold on
through whirlwinds,
strength and strife.

Oh, my word
can you believe it?
Here at last
we finally stand.
All dressed up
in our tuxedos
holding life
right in our hands.

R: I thought it was wonderful. He wrote from the heart. He wasn't trying to be someone else. He was being his own person, my person. It had his sense of humor, his timing. It was wonderful.

After Bob read his poem, we did the "I do" part of the ceremony, and then the minister said, "I now pronounce you spouses for life." He looked at Bob and said, "You may kiss him now." Then he looked at me and said, "You may kiss him now."

B: So we kissed, and everyone burst into applause.

R: We went out of the church first and then everybody followed us in cars to the restaurant for the reception.

24

The Puerto Rican Dummy and the Merciful Son

MARTIN ESPADA

I have a four-year-old son named Clemente. He is not named for Roberto Clemente, the baseball player, as many people are quick to guess, but rather for a Puerto Rican poet. His name, in translation, means "merciful." Like the cheetah, he can reach speeds of up to sixty miles an hour. He is also, demographically speaking, a Latino male, a "macho" for the twenty-first century.

Two years ago, we were watching television together when a ventriloquist appeared with his dummy. The ventriloquist was Anglo; the dummy was a Latino male, Puerto Rican, in fact, like me, like my son. Complete with pencil mustache, greased hair, and jawbreaking Spanish accent, the dummy acted out an Anglo fantasy for an Anglo crowd that roared its approval. My son was transfixed; he did not recognize the character onscreen because he knows no one who fits that description, but he sensed my discomfort. Too late, I changed the channel. The next morning, my son watched Luis and María on *Sesame Street,* but this was inadequate compensation. *Sesame Street* is the only barrio on television, the only neighborhood on television where Latino families live and work, but the comedians are everywhere, with that frat-boy sneer, and so are the crowds.

However, I cannot simply switch off the comedians, or explain them (how do you explain to a preschooler that a crowd of strangers is angrily laughing at the idea of *him*?). We live in western Massachusetts, not far from Springfield and Holyoke, hardscrabble small cities that, in the last generation, have witnessed a huge influx of Puerto Ricans, now constituting some of the poorest Puerto Rican communities in the country. The evening news from Springfield features what I call "the Puerto Rican minute." This is the one minute of the newscast where we see the faces of Puerto Rican men, the mug shot or the arraignment in court or witnesses pointing to the bloodstained sidewalk, while the newscaster solemnly intones the mantra of gangs, drugs, jail. The notion of spending the Puerto Rican minute on a teacher or a health care worker or an artist in the community never occurs to the television journalists who produce this programming.

The Latino male is the bogeyman of the Pioneer Valley, which includes the area where we live. Recently, there was a rumor circulating in the atmosphere that Latino gangs would be prowling the streets on Halloween, shooting anyone in costume. My wife, Katherine, reports that one Anglo gentleman at the local swimming pool took responsibility for warning everyone, a veritable Paul Revere in swim trunks wailing that "The Latinos are going to kill kids on Halloween!" Note how (1) Latino gangs became "Latinos" and (2) Latinos and "kids" became mutually exclusive categories. My wife wondered if this warning contemplated the Latino males in her life, if this racially paranoid imagination included visions of her professor husband and his toddling offspring as

From *Muy Macho* by R. Gonzalez (ed.) ©1996 Anchor Books.

gunslingers in full macho swagger, hunting for "gringos" in Halloween costumes. The rumor, needless to say, was unfounded.

Then there is the national political climate. In 1995, we saw the spectacle of a politician, California Governor Pete Wilson, being seriously considered for the presidency on the strength of his support for Proposition 187, the most blatantly anti-Latino initiative in recent memory. There is no guarantee, as my son grows older, that this political pendulum will swing back to the left; if anything, the pendulum may well swing farther to the right. That means more fear and fury and bitter laughter.

Into this world enters Clemente, which raises certain questions: How do I think of my son as a Latino male? How do I teach him to disappoint and disorient the bigots everywhere around him, all of whom have bought tickets to see the macho pantomime? At the same time, how do I teach him to inoculate himself against the very real diseases of violence and sexism and homophobia infecting our community? How do I teach Clemente to be Clemente?

My son's identity as a Puerto Rican male has already been reinforced by a number of experiences I did not have at so early an age. At age four, he has already spent time in Puerto Rico, whereas I did not visit the island until I was ten years old. From the time he was a few months old, he has witnessed his Puerto Rican father engaged in the decidedly nonstereotypical business of giving poetry readings. We savor new Spanish words together the same way we devour mangos together, knowing the same tartness and succulence.

And yet, that same identity will be shaped by negative as well as positive experiences. The ventriloquist and his Puerto Rican dummy offered Clemente a glimpse of his inevitable future: Not only bigotry, but his growing awareness of that bigotry, his realization that some people have contempt for him because he is Puerto Rican. Here his sense of maleness will come into play, because he must learn to deal with his own rage, his inability to extinguish the source of his torment.

My father has good reason for rage. A brown-skinned man, he learned rage when he was arrested in Biloxi, Mississippi, in 1950, and spent a week in jail for refusing to go to the back of the bus. He learned rage when he was denied a college education and instead struggled for years working for an electrical contractor, hating his work and yearning for so much more. He learned rage as the political triumphs of the 1960s he helped to achieve were attacked from without and betrayed from within. My father externalized his rage. He raged at his enemies and he raged at us. A tremendous ethical and cultural influence for us nonetheless, he must have considered himself a failure by the male career-obsessed standards of the decade into which I was born: the 1950s.

By adolescence, I had learned to internalize my rage. I learned to do this, not so much in response to my father, but more in response to my own growing awareness of bigotry. Having left my Brooklyn birthplace for the town of Valley Stream, Long Island, I was dubbed a spic in an endless torrent of taunting, bullying, and brawling. To defend myself against a few people would have been feasible; to defend myself against dozens and dozens of people deeply in love with their own racism was a practical impossibility. So I told no one, no parent or counselor or teacher or friend, about the constant racial hostility. Instead, I punished a lamp, not once but twice, and watched the blood ooze between my knuckles as if somehow I could leech the poison from my body. My evolving manhood was defined by how well I could take punishment, and paradoxically I punished myself for not being man enough to end my own humiliation. Later in life, I would emulate my father and rage openly. Rarely, however, was the real enemy within earshot, or even visible.

Someday, my son will be called a spic for the first time; this is as much a part of the Puerto Rican experience as the music he gleefully dances to. I hope he will tell me. I hope that I can help him handle the glowing toxic waste of his rage. I hope that I can explain clearly why there are those waiting for him to explode, to confirm their stereotypes of the hot-blooded, bad-tempered Latino male who has, without provocation, injured the Anglo innocents. His anger—and that anger must come—has to be controlled, directed, creatively channeled, articulated—but not all-consuming, neither destructive nor self-destructive. I keep it between the covers of the books I write.

The anger will continue to manifest itself as he matures and discovers the utter resourcefulness of bigotry, the ability of racism to change shape and survive all attempts to snuff it out. "Spic" is a crude expression of certain sentiments that become subtle and sophisticated and insidious at other levels. Speaking of crudity, I am reminded of a group organized by white ethnics in New York during the 1960s under the acronym of SPONGE. The Society for the Prevention of the Niggers Getting Everything. When affirmative action is criticized today by Anglo politicians and pundits with exquisite diction and erudite vocabulary, that is still SPONGE. When and if my son is admitted to school or obtains a job by way of affirmative action, and is resented for it by his colleagues, that will be SPONGE, too.

Violence is the first cousin to rage. If learning to confront rage is an important element of developing Latino manhood, then the question of violence must be addressed with equal urgency. Violence is terribly seductive; all of us, especially males, are trained to gaze upon violence until it becomes beautiful. Beautiful violence is not only the way to victory for armies and football teams; this becomes the solution to everyday problems as well. For many characters on the movie or television screen, problems are solved by *shooting* them. This is certainly the most emphatic way to win an argument.

Katherine and I try to minimize the seductiveness of violence for Clemente. No guns, no soldiers, and so on. But his dinosaurs still eat each other with great relish. His trains still crash, to their delight. He is experimenting with power and control, with action and reaction, which brings him to an imitation of violence. Needless to say, there is a vast difference between stegosaurus and Desert Storm.

Again, all I can do is call upon my own experience as an example. I not only found violence seductive; at some point, I found myself enjoying it. I remember one brawl in Valley Stream when I snatched a chain away from an assailant, knocked him down, and needlessly lashed the chain across his knees as he lay sobbing in the street. That I was now the assailant with the chain did not occur to me.

I also remember the day I stopped enjoying the act of fistfighting. I was working as a bouncer in a bar, and found myself struggling with a man who was so drunk that he appeared numb to the blows bouncing off his cranium. Suddenly, I heard my echo: *thok*. I was sickened by the sound. Later, I learned that I had broken my right ring finger with that punch, but all I could recall was the headache I must have caused him. I never had a fistfight again. Parenthetically, that job ended another romance: the one with alcohol. Too much of my job consisted of ministering to people who had passed out at the bar, finding their hats and coats, calling a cab, dragging them in their stupor down the stairs. Years later, I channeled those instincts cultivated as a bouncer into my work as a legal services lawyer, representing Latino tenants, finding landlords who forgot to heat buildings in winter or exterminate rats to be more deserving targets of my wrath. Eventually, I even left the law.

Will I urge my son to be a pacifist, thereby gutting one of the foundations of traditional manhood, the pleasure taken in violence and the power derived from it? That is an ideal state. I hope that he lives a life that permits him pacifism. I hope that the world around him evolves in such a way that pacifism is a viable choice. Still, I would not deny him the option of physical self-defense. I would not deny him, on philosophical grounds, the right to resistance in any form that resistance must take to be effective. Nor would I have him deny that right to others, with the luxury of distance. Too many people in this world still need a revolution.

When he is old enough, Clemente and I will talk about matters of justification, which must be carefully and narrowly defined. He must understand that abstractions like "respect" and "honor" are not reasons to fight in the street, and abstractions like "patriotism" and "country" are not reasons to fight on the battlefield. He must understand that violence against women is not acceptable, a message which will have to be somehow repeated every time another movie trailer blazes the art of misogyny across his subconscious mind. Rather than sloganizing, however, the best way I can communicate that message is by the way I treat his mother. How else will he know that jealousy is not love, that a lover is not property? . . .

The behavior we collectively refer to as "macho" has deep historical roots, but the trigger is often a profound insecurity, a sense of being threatened. Clemente will be as secure as possible, and that security will stem in large part from self-knowledge. He will know the meaning of his name.

Clemente Soto Vélez was a great Puerto Rican poet, a fighter for the independence of Puerto Rico who spent years in prison as a result. He was also our good friend. The two Clementes met once, when the elder Clemente was eighty-seven years old and the younger Clemente was nine months. Fittingly, it was Columbus Day, 1992, the five-hundredth anniversary of the conquest. We passed the day with a man who devoted his life and his art to battling the very colonialism personified by Columbus. The two Clementes traced the topography of one another's faces. Even from his sickbed, the elder Clemente was gentle and generous. We took photographs, signed books. Clemente Soto Vélez died the following spring, and eventually my family and I visited the grave in the mountains of Puerto Rico. We found the grave unmarked but for a stick with a number and letter, so we bought a gravestone and gave the poet his name back. My son still asks about the man with the long white hair who gave him *his* name. This will be family legend, family ritual, the origins of the name explained in greater and greater detail as the years pass, a source of knowledge and power as meaningful as the Book of Genesis.

Thankfully, Clemente also has a literal meaning: "merciful." Every time my son asks about his name, an opportunity presents itself to teach the power of mercy, the power of compassion. When Clemente, in later years, consciously acts out these qualities, he does so knowing that he is doing what his very name expects of him. His name gives him the beginnings of a moral code, a goal to which he can aspire. "Merciful": Not the first word scrawled on the mental blackboard next to the phrase "Puerto Rican male." Yet how appropriate, given that, for Katherine and me, the act of mercy has become an expression of gratitude for Clemente's existence.

BECAUSE CLEMENTE MEANS MERCIFUL

—for Clemente Gilbert-Espada
February 1992

At three AM, we watched
the emergency room doctor
press a thumb against your cheekbone
to bleach your eye with light.
The spinal fluid was clear, drained
from the hole in your back,
but the X ray film
grew a stain on the lung,
explained the seizing cough,
the wailing heat of fever:
pneumonia at the age
of six weeks, a bedside vigil.
Your mother slept beside you,
the stitches of birth still burning.
When I asked, "Will he be OK?"
no one would answer: "Yes."
I closed my eyes and dreamed
my father dead, naked on a steel table
as I turned away. In the dream,
when I looked again,
my father had become my son.

So the hospital kept us: the oxygen mask,
a frayed wire taped to your toe
for reading the blood,
the medication forgotten from shift to shift,
a doctor bickering with radiology over
the film,
the bald girl with a cancerous rib removed,
the pediatrician who never called, the
yawning intern,
the hospital roommate's father
from Guatemala, ignored by the doctors
as if he had picked their morning coffee,
the checkmarks and initials at five AM,
the pages of forms flipping like a deck
of cards,

recordkeeping for the records office,
the lawyers and the morgue.

One day, while the laundry
in the basement hissed white sheets,
and sheets of paper documented dwindling
breath,
you spat mucus, gulped air, and lived.
We listened to the bassoon of your lungs,
the cadenza of the next century, resonate.
The Guatemalan father
did not need a stethoscope to hear
the breathing, and he grinned.
I grinned too, and because Clemente
means merciful, stood beside the
Guatemalteco,
repeating in Spanish everything
that was not said to him.

I know someday you'll stand beside
the Guatemalan fathers,
speak in the tongue
of all the shunned faces,
breathe in a music
we have never heard, and live
by the meaning of your name

. . . While Latino male behavior is, indeed, all too often sexist and violent, Latino males in this country are in fact no worse in that regard than their Anglo counterparts. Arguably, European and European-American males have set the world standard for violence in the twentieth century, from the Holocaust to Hiroshima to Vietnam.

Yet, any assertiveness on the part of Latino males, especially any form of resistance to Anglo authority, is labeled macho and instantly discredited. I can recall one occasion, working for an "alternative" radio station in Wisconsin, when I became involved in a protest over the station's refusal to air a Spanish-language program for the local Chicano community. When a meeting was held to debate the issue, the protesters, myself included, became frustrated and staged a walkout. The meeting went on without us, and we later learned that we were *defended,* ironically enough, by someone who saw us as acting macho. "It's their culture," this person explained apologetically to the gathered liberal intelligentsia. We got the program on the air.

I return, ultimately, to that ventriloquist and his Puerto Rican dummy, and I return, too, to the simple fact that my example as a father will have much to do with whether Clemente frustrates the worshipers of stereotype. To begin with, my very *presence*—as an attentive father and husband—contradicts the stereotype. However, too many times in my life, I have been that Puerto Rican dummy, with someone else's voice coming out of my mouth, someone else's hand in my back making me flail my arms. I have read aloud a script of cruelty or rage, and swung wildly at imagined or distant enemies. I have satisfied audiences who expected the macho brute, who were thrilled when my shouting verified all their anthropological theories about my species. I served the purposes of those who would see the Puerto Rican species self-destruct, become as rare as the parrots of our own rain forest.

But, in recent years, I have betrayed my puppeteers and disappointed the crowd. When my new sister-in-law met me, she pouted that I did not look Puerto Rican. I was not as "scary" as she expected me to be; I did not roar and flail. When a teacher at a suburban school invited me to read there, and openly expressed the usual unspoken expectations, the following incident occurred, proving that sometimes a belly laugh is infinitely more revolutionary than the howl of outrage that would have left me pegged, yet again, as a snarling, stubborn, macho.

MY NATIVE COSTUME

When you come to visit,
said a teacher
from the suburban school,

don't forget to wear
your native costume.

But I'm a lawyer,
I said.
My native costume
is a pinstriped suit.

You know, the teacher said,
a Puerto Rican costume.

Like a guayabera?
The shirt? I said.
But it's February.

The children want to see
a native costume,
the teacher said.

So I went
to the suburban school,
embroidered guayabera
short sleeved shirt
over a turtleneck,
and said, Look kids,
cultural adaptation.

The Puerto Rican dummy brought his own poems to read today. *Claro que sí.* His son is always watching.

25

Stardust Memories

MARK NEILSEN

It was 11:15 P.M. I raised my head off the pillow, trying to emerge from sleep in my darkened bedroom. What had I just heard? The murmur of an intruder? Some disturbance on the street below? My wife slept on next to me, the boys had gone to bed an hour ago, and even the dog remained peacefully curled on the rug next to the bed. Straining, I listened again and gradually it became clear. Muffled by the rush of water from the bathroom came a carefree and unselfconscious rendition of "Stardust," my daughter singing to herself in a late-night shower. So charmingly did her young womanhood wrap around those plaintive lines that I caught my breath, instantly aware what a privilege it was to hear such an unintended concert—and just as instantly aware that my time for such privileges was running out. Quite fully awake now, I almost wished it had been only an intruder.

In a scant few months, my daughter would be headed for college, where, if she ever sang in the shower, I wouldn't know about it. In fact, I would know nothing at all of her life unless she chose to tell me. For she would be on her own, and whatever that might mean for her, it meant for me the end of one era of being a father.

Certainly, all fathers have faced the inevitable moment when little girls grow up, hugs become less frequent and idolizing gazes vanish altogether. Driving a car, dating, maturing physically—all these things transform our daughters, and leave us fathers reeling just a bit. We remember all too well what it is like to be a young man, and even if we trust our daughter's judgment, we nonetheless remember enough to trust no man fully.

I extol the joys of being a father every chance I get. Having changed diapers, given baths, fed, clothed and cuddled, I can say with authority that

From *America,* June 19, 1999, vol. 180, i, 21, p. 18.

it is indeed a wonderful experience. But I also know that letting go, not only of the child but of the consolations of a particular stage of fathering, is not entirely joyful, no matter how necessary it might be.

In a cold instant, all of this has shaken me out of unconsciousness, and I head down to the kitchen for a midnight cup of tea and a crossword puzzle.

For months now I had foreseen the pain of parting—the last good-byes, the car emptied of her belongings, the final embraces. I had seen it coming even while I was encouraging her to go, traveling with her months ago to make her college visits. Just the two of us drove along the browning corn and soybean fields, she fingering the radio dial incessantly, browsing the local radio stations for her music. A young woman who has always kept her own counsel, my daughter had lately grown monosyllabic in response to my inquiries about her college plans.

I had accepted this, aware that my relationship with her was far more important than how much I would have to say about where she went to college or why. So the miles rolled on more or less in silence, the silence of a truce: amicable, but a truce nonetheless.

The result was good. She picked a wonderful school that seemed a good fit to her personality and talents. As we sat on a bench during a break from touring the campus and planned our afternoon, she said she would like to visit a class in the English department, her proposed major.

"Okay," I said, "I'll meet you here after class is over."

"Well . . . aren't you coming, too?" she asked with just a hint of trepidation. My paternal radar having suddenly detected the sense of being needed, I said, "Okay, I'll come along."

However little I relished the prospect of spending another hour in a college classroom, I was pleased to be able to do something for my daughter and charmed by the prospect of still providing some sort of emotional ballast for this increasingly independent young woman, so deeply involved in making "an important decision." It felt good to be needed.

We found the classroom easily, milled about with the students and took two empty seats. The students welcomed my daughter and gave her copies of relevant class documents. The professor entered the room, and just as my eyes prepared to glaze over, my daughter turned to me and asked, once more with a touch of trepidation, "Are you going to stay?"

"Ah, no, I'll meet you outside after class is over." Easy come ballast, and easy go. It seemed she would be able to manage without me after all.

To some extent, her very self-assurance is a bit unsettling to me. I now feel at times almost as nervous about the future as I did when I went away to college myself. Back then, knowing little about endings and beginnings, and so, no big deal, I went through it smoothly enough. Now, those same pivotal events are a bigger deal, for they are touching reminders of time's passage. No more able at 50 than at 18 to prevent change from happening, I am now more aware—of the losses as well as the gains. Sometimes, the awareness makes it difficult to sleep.

Going off to college is a touchstone in the evolving parent-child relationship. No longer will she be under our roof, with all the predictability that implies. No longer will I be aware—if not from moment to moment, at least day to day—how she is living. And that concerns me. Actually, I remind myself, as I sip my tea, she has been going her own way for some time now, growing and developing her own responses to the challenges of life—and not necessarily according to parental precepts. Growing up is easily one of the most difficult times of one's life. It is also one of the most exciting.

Letting go of a daughter is not done all that easily. Last Christmas, I gave her a calendar featuring the original Beatrix Potter illustrations of Peter Rabbit, which I thought very sweet. In response to her evident bewilderment, I explained that, well, maybe this was a last-ditch effort to hang on to my little girl. "I know," she said, and agreed to my offer of exchanging the calendar for another. Georgia O'Keeffe, it seems, was more what she had in mind.

In the evolutionary scheme of things, paternal protection has been important for offspring. Sadly, we live in an age when it must be said that not all fathers abuse their children and not all protectiveness is pathological. Still, all healthy fledglings leave the nest; and from the very first, that is what I wanted for my children. The hard part is letting this benign separation happen. The time has come for me to shed a role, however cherished, as a snake sheds its skin.

Nor is that all bad. Shedding a role is an opportunity to return to who one really is. When a certain kind of parenting is no longer required, we are reminded that we are more than our roles. Just as children go through phases, so do parents, although we tend not to be as aware of that. But there is a big difference between bundling up a child in a snowsuit, buckling her into a car seat to travel to Grandma's and putting her on a 747 to fly to Paris. We need different resources for each task.

When I was younger, my wife and I strapped the family on my back—sometimes literally—and off we went. If something needed doing, one of us simply did it. Gradually, the children have needed to do more for themselves, and we consequently have had to let them. I say we have had to let them because often it would have been easier to grab the reins from them and accomplish what needed accomplishing.

Aware of all these things, I drink my tea and finish my crossword puzzle, knowing, too, that it is time to let her go, that in many ways we have already let her go. I think again of the exquisite beauty of the song and of her young life. I think of the unrestrained joy of singing, of being on your own, of beginning to write your own song. I swell with a father's pride: She is able to be on her own, able to stand for what she believes, able to sing for it. I will miss her mightily, but it bodes well for her future that she can go.

And it bodes well for us that we can let her go, though the sweet sorrow of parting grabs us on occasion. For sorrow is not the only thing her late-night concert has stirred in me. Suddenly, I am looking at my life anew. I will spend less energy caring for this child's welfare as she begins to spend more. Where might my new-found bonus of energy now go? This is a question for another time, but as I make my way back to my bedroom through the darkened house, I realize that I have a couple of more songs to sing myself.

5

Men and Employment

No other area of social life has so dominated our thinking about men's lives as work and wage-earning. It has been argued that work is *what men do,* and that their occupations define *who men are* (Rubin 1976, 1994). Though somewhat true, these arguments may occasionally be overstated. Men's lives are more multi-dimensional. On top of working, most men are sons, brothers, friends, husbands, and/or fathers. Certainly, though, work is an important activity in the lives of most men, and, as we have seen, interacts with many of their other roles and relationships. The following pages address some of the meanings of work in men's lives and how it is experienced by men of varying backgrounds or social circumstances.

MEN AND WORK

Most men are employed in some capacity. Data from 1997 showed that three out of four men age 16 and older were in the paid labor force. Men's participation rate has slightly and gradually slid since the mid-1970s, when it peaked at 78 percent. Most social scientists point to the more remarkable changes that have occurred in women's labor force participation (currently at approximately 60 percent), especially over the past forty years. There have been a number of major social changes accompanying that increase, many of which continue to be felt in homes and workplaces today.

Work and Masculinity

Although both men's and women's primary motivation for work is financial, work has particular meanings in men's lives. With no rites of passage to unambiguously mark a male's entrance into manhood, work may fill much of that void. So argued many mid-century sociologists (e.g., Parsons 1942; Zelditch 1974), who suggested that "normal" adult men have jobs, and that men's jobs define their status as men.

Work is intertwined with masculinity in other powerful ways. Men's success at work denotes their success at being men; in Robert Gould's terms, the "bigger the paycheck," the "bigger the man" (Gould 1974). This monetary emphasis should not be seen as total; like women, men seek more than money from the work they do. Feelings generated at work can also affect how men see themselves and whether they consider themselves successful.

At the opposite end, failure at work often symbolizes failure as a man, while falling short of success in the workplace may result in a damaged male identity. There are numerous good sociological examples, stretching over many years, that revealed what happened to men whose experiences of work were filled with failure (e.g., Liebow 1967; Rubin 1976, 1994; Newman 1988). For some, failure at work and wage-earning seeped into their other roles, soiling the friendships they formed, the marriages they entered and left behind, and their styles of fathering (Liebow 1967; Newman 1988). Lillian Rubin's exemplary studies of working-class families demonstrated how failing to succeed at work, though not as traumatic as complete failure, altered men's behavior at home (Rubin 1976; 1994). Needing to feel important or productive and wanting to be in charge somewhere led men to seek those feelings at home in their relationships with their wives.

Rubin provides a poignant illustration of the potent connection between men's experiences at work or as providers and their sense of themselves as men (Reading 27). As she observes, "Going to work isn't just what he does, it's deeply linked to who he is." As one of her informants remarked, "It's not just the income; you lose a lot more than that. . . . It's terrible; something goes out of you." Thus, even if women's and men's labor force participation rates continue toward convergence, there are considerable differences in the symbolic meanings of work in men's and women's lives.

Cohen and Durst's study of twenty nontraditional couples (Reading 28) reveals that the meanings attached to not working affect its consequences in men's lives. Although men tended to suffer some negative consequences from their at-home, nonwork status, those outcomes were joined by positive consequences in their family relationships, especially relationships with their children.

Men's Jobs and Men's Families

As we've seen, work is especially intertwined with men's experiences of family life. The literature on families long portrayed men's roles as husbands and fathers as extensions of their work roles (Bernard 1981). Men's responsibilities consisted of earning the family income or the bulk of it. One can overstate such a characterization, but in the century from industrialization through the 1960s, most families were supported financially by men in their provider roles.

Simultaneously, men's families can be instruments of their economic success and victims of their ongoing efforts to succeed. Papanek's notion of the "two-person career" illustrated how important wives were to the professional success experienced by their husbands (Papanek 1973). Based obviously on middle-class ideas of careers, Papanek's notion applies almost as well to men of lower social standing. Their jobs could be designed in such ways that they were expected to "give their all," while someone else—at-home wives—would care for the home and raise the kids.

Still largely structured on such a male model of work, most jobs demand that workers fit their family lives around their jobs, while the jobs "act as though" workers have no family responsibilities to concern themselves with (Kanter 1977; Margolis 1979; Hewlett 1991; Hochschild 1997). This is illustrated in Arlie Hochschild's description of Amerco (the name she gave the company she studied), where employees had unexpected responses to the more common work-family conflicts (Reading 26). Many workers preferred being at work to being at home, and most of the flexible options the company offered went unused. As she shows, some of that response stems from a climate in which time spent at work is perceived to reflect one's commitment to work. The claims jobs make in men's lives are often extreme.

Although work-family conflicts affect most working women and men, efforts to create greater fit between jobs and families are commonly considered "women's issues." Such designations suggest that men are unconcerned about the competition and conflict between the demands of their jobs and the needs of their families. Recent research suggests that this is untrue, or at least not true for all men (Pleck 1993; Gerson 1993; Hochschild 1997). Continued labeling of such policy initiatives in gender-specific ways only adds to the ambivalence men feel about requesting or utilizing whatever family supports are available at work.

EXPERIENCING EMPLOYMENT

Men versus Women

The work world is hardly gender-blind. Job opportunities and experiences are segregated; wages are unequal; autonomy, authority, and power are unevenly distributed; and, overall, women's work experiences relative to men's receive less recognition, value, support, or compensation. Men reap certain benefits from their work experience that women don't (Reskin and Padavic 1994).

Examinations of occupations, salaries, mobility, authority, or autonomy paint consistent pictures of a male-dominated workplace. Where women are largely "ghettoized" in pink-collar occupations that place them on the underside of the wage gap, men are more prosperous. While women are trapped on "sticky floors" beneath "glass ceilings" that bar their access to higher status, higher paying positions, many men ride "glass escalators" upward (Reskin and Padavic 1994). These gaps have been narrowing in recent decades, but remain substantial. Women's wages, for example, continue to lag behind men's wages (women earn approximately 75 percent of what men earn), and men continue to monopolize positions of greater authority and autonomy.

Many factors account for such sex differences in work experience. Males and females are differently socialized about the meaning of work and what constitutes appropriate aspirations. They are then differently reinforced and tracked educationally, differently recruited or assisted in pursuing select occupations, and face different obstacles in finding jobs and moving upward (Reskin and Padavic 1994; Renzetti and Curran 1995). As a result, women's economic status suffers in comparison to men's.

Men, too, suffer certain costly consequences as a result of their work. The social dominance of work in men's lives often comes at the expense of other meaningful sources of identity or involvement in their families. Additionally, many men face risks to their health and longevity as a result of their experiences at work. More than 90 percent of occupational deaths occur to men (Farrell 1993). Recognizing this reality of men's work lives helps balance an otherwise misleading one-dimensional understanding in which men "only benefit."

One work-related area in which women clearly suffer more than men do is in the experience of sexual harassment. Harassment refers to "unwanted or repeated verbal or physical sexual advances, or remarks or behavior that are offensive to" and cause discomfort to the recipient, interfering with job performance (Reskin and Padavic 1994). It includes demands for sex in exchange for the right to keep one's job or receive certain benefits, and "hostile work environments," in which sexualized language or setting makes the employee uncomfortable enough that job performance is disturbed. Some estimate that as many as half of working women are harassed during their working lives. Only a small proportion of victimized women file formal complaints (Koss et al. 1994).

Women most at risk of harassment are women in traditional male occupations. Harassment victims suffer in a variety of ways, from quitting or losing their jobs, through frequent absenteeism and lower morale, to damaging health and mental health outcomes (Koss et al. 1994). Men who harass tend to be hard to characterize or distinguish because they lack any particular demographic or social profile. The characteristics that mark the typical harasser (older, white, married man) also describe the typical male worker (Koss et al. 1994).

Men versus Men

Simplistic generalizations about men's work experiences or economic privileges ignore other critical and complicating factors like class, race, age, and sexuality. Most men are not "movers and shakers" but rather employees in jobs that deny them the opportunity to fully realize the American dream of success. Despite fairly widespread belief in the "achievement ideology" (MacLeod 1995), where hard work and ability yield success, most men fall short of their ultimate goals. The numbers who make it to the top are few, as are the top positions themselves. Thus, male domination of the workplace realistically pertains to only select categories of men.

Black and Hispanic men earn less and experience more occupational instability than their white counterparts. Black men who are employed full-time, year-round, earn less than three-quarters of white men's earnings; Hispanic men earn less than two-thirds of what white men earn (Statistical Abstract of the United States 1998). Over the last two decades African American and Hispanic men have been losing ground compared to their white counterparts (Reskin and Padavic 1994). Men of all racial and ethnic backgrounds find that their chances of "climbing to the top" are much greater the higher up the ladder from which they start to climb. Those who start at or near the bottom typically stay there or climb only a few rungs higher.

In Reading 29, William Julius Wilson looks at the economic plight of low-income, inner-city black men. Specifically, Wilson illustrates some of the attitudes held by prospective employers that make them reluctant to hire inner-city blacks, especially black men. Such biases are not limited to white employers, and can be heard in the comments of both black employers and inner-city residents. These prejudices worsen a situation created by the restructuring of the economy, which itself reduced the number of good jobs available to those who live in poor urban areas.

As James Woods shows, gay men's experience of the workplace sets them apart from their heterosexual peers (Reading 30). Discrimination in hiring curtails the occupational chances of openly gay men. The fear of potential discrimination pushes other gay men into an "office closet," where they sharply segregate their personal lives from the workplace and workmates. Using a variety of concealment strategies, many gay men cope only by rigidly guarding their actions and monitoring their conversations. Their daily work life becomes a performance accomplished at great expense with careful planning and concerted effort. As both Wilson and Woods show in their discussions of race and sexual orientation, ignoring the many substantial differences that define men's experiences of work renders any discussion of work and masculinity virtually meaningless.

MEN WITHOUT WORK

Looking at which men don't work reveals some interesting patterns in which the important effects of race and ethnicity are especially visible. Unemployment rates for Hispanic men are nearly 75 percent higher than those found among white males, and black men's unemployment rates were more than double the rate among white men (Eshelman 1997). Since wages for both groups of men are much lower than incomes earned by white men, when unemployment is not a problem, underemployment often is. Economic realities such as these spill over into other areas of social life. Without jobs (or without good jobs), men don't make particularly attractive husbands or especially successful fathers (Wilson 1987, 1996). As Kenneth Brown's experience illustrates (Read-ing 31), they also suffer more intangibly, growing less self-confident and losing some of their sense of manhood. These effects, too, may be more acute depending on one's race.

The centrality of work as men's most visible social activity assures that as men age they will have to deal with the eventual surrender of their social status. Retirement may be met with mixed emotions or even welcomed with a sense of accomplishment and anticipation, depending on such factors as marital status, race, occupational status, and financial stability. When it is mandatory, not well planned for, and followed by economic hardship, its consequences are more negative. Such factors deepen the blow that can accompany the rolelessness of retirement, when men confront the lack of other meaningful sources of male identity. Postretirement depression is a consequence for many men. Bob Greene's comments about his father's retirement (Reading 32) illustrate the dominance of work in how men see themselves and are seen by others, and how all that is shaken by the end of one's working days.

Finally, it is worth mentioning, again, the experiences of men who willingly withdraw from paid work in order to raise their children while their spouses provide the needed financial support. While aware of how and what they contribute to their families' stability and success, they also know they stand apart from most other men. They are seen as "heroes" to some, novelties or even failures to others. Significantly, despite being at home with children, responsible for housework, and economically dependent on their spouses, their experiences are different from the experiences of similarly situated women, because they are men in a society in which most men work.

REFERENCES

Bernard, J. "The Good Provider: Its Rise and Fall." *American Psychologist* 36 (1981): 1–12.

Eshelman, J. *The Family.* 8th ed. Needham Heights, MA: Allyn & Bacon, 1997.

Farrell, W. *The Myth of Male Power.* New York: Simon & Schuster, 1993.

Gerson, K. *No Man's Land: Men's Changing Commitment to Family and Work.* New York: Basic Books, 1993.

Gould, R. "Measuring Masculinity by the Size of a Paycheck." In *Men and Masculinity,* edited by J. Pleck and J. Sawyer, 96–100. Englewood Cliffs, NJ: Prentice Hall, 1974.

Hewlett, S. *When the Bough Breaks: The Cost of Neglecting Our Children.* New York: Basic Books, 1991.

Hochschild, A. *The Time Bind: When Work Becomes Home and Home Becomes Work.* New York: Henry Holt, 1997.

Kanter, R. *Work and Family in the United States: A Critical Review and Agenda for Research and Policy.* New York: Russell Sage, 1977.

Koss, M., L. Goodman, L. Fitzgerald, N. Russo, G. Keita, and A. Browne. *No Safe Haven: Male Violence Against Women at Home, at Work, and in the Community.* Washington, D.C.: American Psychological Association, 1994.

Liebow, E. *Tally's Corner.* Boston: Little, Brown, 1967.

MacLeod, J. *Ain't No Makin' It.* 2nd ed. Boulder, CO: Westview Press, 1995.

Margolis, D. *The Managers: Corporate Life in America.* New York: William Morrow, 1979.

Newman, K. *Falling from Grace: The Experience of Downward Mobility in the American Middle Class.* New York: Random House, 1988.

Papanek, H. "Men, Women, and Work: Reflections on the Two-Person Career." In *Changing Women in a Changing Society,* edited by J. Huber. Chicago: University of Chicago Press, 1973.

Parsons, T. "Age and Sex in the Social Structure." *American Sociological Review* 7 (October 1942): 604–616.

Pleck, J. "Are 'Family-Supportive' Employer Policies Relevant to Men?" In *Men, Work, and Family,* edited by J. Hood. Thousand Oaks, CA: Sage, 1993.

Renzetti, C., and D. Curran. *Women, Men, and Society.* 3rd ed. Boston: Allyn & Bacon, 1995.

Reskin, B., and I. Padavic. *Women and Men at Work.* Thousand Oaks, CA: Pine Forge Press, 1994.

Rubin, L. *Families on the Fault Line.* New York: HarperCollins, 1994.

Rubin, L. *Worlds of Pain: Life in the Working Class Family.* New York: Basic Books, 1976.

U.S. Bureau of the Census. *Statistical Abstract of the United States: 1998.* Bureau of the Census. Washington, D.C., 1998.

Wilson, W. J. *The Truly Disadvantaged: The Inner City, the Underclass, and Public Policy.* Chicago: University of Chicago Press, 1987.

Woods, J. *The Corporate Closet: The Professional Lives of Gay Men in America.* New York: The Free Press, 1993.

Wilson, W. J. *When Work Disappears: The World of the New Urban Poor.* New York: Knopf, 1996.

Zelditch, M. "Role Differentiation in the Nuclear Family." In *The Family: Its Structure and Functions,* 2nd ed., edited by R. L. Coser, 256–258. New York: St. Martin's Press, 1974.

FOR ADDITIONAL READING

There are a variety of ways to use *InfoTrac* to follow up on issues raised in the readings in this chapter. For further consideration of the ways in which meanings attached to work, work assignments and workplaces are "gendered," read Mats Alvesson's ethnography of a Swedish advertising agency, "Gender Relations and Identity at Work: a Case Study of Masculinities and Femininities in an Advertising Agency" (*Human Relations,* August 1998, vol. 51, no. 8). Alvesson shows how when work becomes "feminized," male workers often act in ways to restore what masculine feelings they expect to derive from their jobs. Kevin Yelvington's "Flirting in the Factory" (*Journal of the Royal Anthropological Institute,* June 1996, vol. 2, no. 2) is an ethnographic study of a factory in Trinidad in which sexual themes emerge in the "flirting" behaviors between female line workers and male co-workers and supervisors. Although both the men and women portray flirting as an expression of sexual attraction, Yelvington also analyzes the power dynamics embedded in such behavior.

Finally, in keeping with discussion of variations and inequalities in men's economic experiences, especially by race and class, read "The Declining Social and Economic Fortunes of African American Males: A Critical Assessment of Four Perspectives," from *The Review of Black Political Economy* (Spring 1998, vol. 25, no. 4) by James Johnson, Walter Farrell Jr., and Jennifer Stoloff. After detailing indicators of the social and economic status of black males, the authors compare how the *spatial isolation, cultural capital, "search and destroy,"* and *social capital* theses each attempt to explain the experiences of African American men.

WADSWORTH VIRTUAL SOCIETY RESOURCE CENTER

http://sociology.wadsworth.com

Visit *Virtual Society* to obtain current updates in the field, surfing tips, career information, and more.

INFOTRAC COLLEGE EDITION: EXERCISE

Go to the *InfoTrac College Edition* web site to find further readings on the topics discussed in this chapter.

26

Giving at the Office

ARLIE RUSSELL HOCHSCHILD

Career: 1. Progress or general course of action of a person through life . . . 2. An occupation or profession, especially one requiring social training, followed as one's life work . . . 3. A course, especially a swift one . . . 4. Speed, especially full speed, "the horse stumbled in full career . . ."

Random House Dictionary of the English Language

Entering the company cafeteria to pick up ham and cheese sandwiches for our noon interview, Bill Denton, a senior manager who oversees all personnel issues at Amerco including the Work-Life Balance program, banters with the cashier. "Is the pickle in my sandwich free?" She smiles. On our way out, he nudges a young man hunched over a spreadsheet who is munching on a sandwich, "We'll expect that report in half an hour." They both laugh. In the elevator, he chats amiably with a secretary as we rise to the ninth floor and the offices of Amerco's most powerful executives. Inside his office, he motions me to a chair, leans forward in his, and says, "I've set aside an hour for you."

After thirty years with the company, Bill Denton, fifty-two, exudes vigor, warmth, and a powerful sense of direction. He is a sturdy man about five feet ten inches tall. He has neatly trimmed brown hair and a rapid, confident way of speaking that leads one to assume he is right about what he's saying. His four children, posed in framed pictures behind his desk, look remarkably like him. How, I ask, seeking a neutral place to begin, did he get started at Amerco? He answers as if I had asked him to describe the principle according to which he'd risen to the top, and this brings him immediately to the matter of time:

> Time has a way of sorting out people at this company. A lot of people that don't make it to the top work long hours. But all the people I know who *do* make it work long hours, some more than others. The members of the Management Committee of this company aren't the smartest people in this company, we're the hardest working. We work like dogs. We out-work the others. We out-practice them. We out-train them. By the time people get within three or four levels of the Management Committee, they're all very good, or else they wouldn't be there. So from that point on, what counts is work and commitment. People don't say, "He works like a dog." You just start to see performance differences created by a willingness to work all the time.
>
> Then there's a final elimination. Some people flame out, get weird because they work all the time, or they're no fun, so they don't get promoted. The people at the top are very smart, work like crazy, and don't flame out. They're still able to maintain a good mental set, and keep their family life together. *They* win the race.

Curiously unrushed, this sixty-hour-a-week manager of three hundred employees is a winner of that race—all those hours and he's still a nice guy. "We hire very good people with a strong work ethic to start with," he observes.

> People look around and see that. So then they work hard to try to keep up, and I don't think we can do anything about that. . . . It's going to be a long time before somebody becomes the CEO of a company saying, "I'm going to be a wonderfully balanced person"—because there are just too many others who aren't. The environment here is very competitive.

Bill himself averaged ten hours a day, and given his handsome salary, his love of his work, and a willing wife, he was happy to do so. The twelve top managers I interviewed all worked between fifty- and seventy-hour weeks. One described himself as a "twelve-hour player," another as a "controlled workaholic." A third said, "They tell us to get the job done—but not to spend too much time on it. But the job takes time." Most executives came in weekends and all of them took work home. Interestingly, though, Bill estimated that only a third of the employees he considered workaholic "made a real difference" to the company while two-thirds of them did not. Managers often started or ended meetings with workaholic jokes. A colleague quipped at an 8 A.M. meeting, "How's the weather in Tokyo, Jim?" to a colleague who had arrived directly from the airport. "When I get home from a trip, I never know if I'm kissing my wife hello or goodbye," chimed in another to a round of rueful laughter.

Parking lots told a similar story about a workaholic company culture. The executive lot began filling around 7 A.M. and thinned out slowly after 5 P.M. Even on the Fourth of July, the one day of the year other than Christmas when one might most expect Amerco employees to be off duty, there was a sprinkling of cars and vans in the parking lots around the central administration and engineering buildings, their windows rolled part-way down, as if their owners were saying, "I won't be long." A man in shorts, a row of pens in his shirt pocket, walked in rapid, long strides toward his office. Another was just leaving, briefcase in hand. A third had a child in tow.

To Bill, long hours did not seem imposed from on high. Instead, in his view, the corporation simply attracted people ready to attune themselves to company needs. "No one tells us to work long hours," Bill explained matter-of-factly. "You don't get the 'leaving *early* again?' We impose it on ourselves. We're our own worst enemy." Like the Protestants in Max Weber's classic study *The Protestant Ethic and the Spirit of Capitalism,* they seemed to respond not to God's wagging finger, but to some internal urgency that pressed them to extend each workday. "You hear stories of managers who drag workers into conversation just as they're packing up to leave, or who make 5 P.M. bunker checks to see who's still there," one man told me, "but that's the exception." Of his three hundred employees, Bill noted,

> I don't decide how much work they do. *They* decide. If they could talk their coworkers into working less, then they could probably work less themselves.

If Bill's hours were long, they were also hours in a privileged zone, well protected from unwanted interruptions. At home, his wife screened his calls and greeted visitors at the door; his secretary did the same for him at work. Together, these two women took much of the uncertainty out of his workday. Like other top executives Bill told none of those stories so commonly heard from employees farther down the Amerco hierarchy—about disappearing cats, suddenly feverish children, emergency calls from elderly relatives, or missing babysitters. In a polite way, Bill's wife and secretary patrolled Bill's time, keeping a vigilant eye out for time-thieves or unauthorized time-squatters. Bill's secretary was his clock: she sorted out his schedule and his daily priorities for him, telling him when he had to do what. This allowed him to respond "spontaneously" to tasks as they presented themselves to him, and it generally left him available to concentrate on any one of those tasks until he got it done.[1] As he described it,

> I immerse myself. I love my job. I really enjoy it. If I don't like it, I don't do it at all. I have a bad habit that way. I'm *undisciplined.*[2]

Another manager commented: "I get in at 7:30 A.M., get myself a cup of coffee, and look over my schedule. Then I'm off and running. I don't look up till about 4 P.M."

In telling his story, Bill Denton frequently spoke of "players" on a "football team" or of winners on a "playing field." This image of work at Amerco as a football game came up regularly in conversations with those at the top but was absent from discussions with those at the bottom. When women in the boardroom, who seldom if ever watched professional football, spoke of their careers, they often relied on this vision of players on a football field. When men on the assembly line who did watch pro football described their jobs, they didn't use this image. Chess, poker, Monopoly, hunting—any of these might have been more apt—but football was the prevalent imagery in the executive offices of Amerco.

Metaphors guide how we feel. The image of football focuses attention on an engrossing, competitive enterprise that calls for exquisitely close coordination among all members of a team. One is doing something right if speed feels exciting instead of silly or frightening. If work is a football game, it imparts vitality, urgency, the potential thrill of victory to the often mundane tasks at hand. More important, to think in football terms is to set aside the parts of life that exist "off the field." The relationship between work and family life naturally disappears as an issue.

Bill could be a "player" because of a prior understanding with his wife, who lived entirely off the field. As he described their relationship,

> We made a bargain. If I was going to be as successful as we both wanted, I was going to have to spend tremendous amounts of time at it. Her end of the bargain was that she wouldn't go out to work. So I was able to take the good stuff and she did the hard work—the car pools, dinner, gymnastic lessons. In those days, it was easier to do. All her friends were in the same boat. Today, I don't know what somebody would do if they chose to do what Emily did. There might not be any other homemakers to share this life with.
>
> Emily left Oakmont College after two years when we got married. After we had our son, she decided she preferred to manage the home. Later, she finished her degree at Lawrence College, but even then, she felt the best use of her time was managing me and the family.
>
> I really had it made. I worked very long hours and Emily just managed things. I never had to worry about getting the laundry, figuring out how to get the kids here and there. Emily made that her life's work. The kids did eventually go away to boarding school, but before that I arranged to go to school plays in the middle of the afternoon and sporting events that started at 4 P.M. When the kids wanted me to be there, I was there.

It was the ultimate privilege, Bill felt, to live with a wife who wasn't counting the coin of sacrifice. Emily, too, was "in management." But these days, Emily's "business," the home, was being marginalized. Women more than a decade younger than his wife were becoming the junior executives Bill managed. But Bill presented Emily as the Good Sport Housewife. She didn't feel left behind or mind, he said, being dependent on him. She had enjoyed raising their four children before they set off for boarding school, and she still declared herself happy rooting for Bill from the stands. For his part, Bill made it clear that he felt lucky compared to men and women in two-career families. He had more time.

Given this situation, Bill wasn't escaping home for work, or work for home. He wasn't seeking a "haven" for he had two safe worlds, though one of them overwhelmingly dominated the other. . . .

It was these top male executives, living protected lives, adequate husbands and fathers by their own lights, who were expected to implement Amerco's new family-friendly policies. They were to be the first line of defense for the harried, time-starved employees below them, desperately juggling commitments at home and at work. Some of these men were members of the Amerco Corrective Action Team, which dealt with the allotment of family-friendly benefits.

They were to understand a mass of employees whose concerns were so different from theirs that they might have been living on another planet.

Half of Amerco workers cared for children thirteen or under, for elderly relatives, or for both, or said on a survey that they expected to provide such care in the near future; and most workers who cared for the young and old had working spouses who were as pressed for time as they were. Was there a way to help such workers balance their lives and benefit Amerco too? This was the question the CEO had asked Bill Denton to answer. Bill was skilled at taking on missions that required him to understand circumstances different from his own. He was by no means a self-centered or inflexible man. He knew that his subordinates might feel different pressures and need work schedules different from his.

Bill Denton was unusual among managers in his willingness to throw out the old rules. If a woman was the best "man" for the job, so be it. Nor did Bill avoid the problem of family needs by shying away from female employees, who he felt were the ones most likely to press the issue. Bill believed that new talent was often to be found among women with families and among men with working wives. The company that took advantage of this workforce and adapted to its needs would have a leg up in the battle for a share of the global market. Bill also knew that the issue of work-family balance was coming to his company, his division, his workplace. He did not think, as some top managers did, that work-family balance was a problem for only "5 percent." He didn't minimize the need for such policies, at least not on the surface.

But two things kept Bill from acting on his understanding of the problem: his sympathy for the family circumstances of Amerco's workers had to compete with other urgent company concerns, such as meeting production goals; and he lived in a social bubble among men who also worked very long hours, had (house)wives at home, and assumed the normality of this arrangement. These two factors may have made him impatient with the issue of work-family balance. At one point he blurted out, "I'm *tired* of dealing with it. I wish we could just be done with it."

DADDY IS AT WORK

Bill faced a dilemma. He was supposed to help create a more family-friendly workplace. But circumstances undermined his motivation to act forcefully to do so. He did implement some changes, though, in line with the new emphasis on family friendliness: he became a "good daddy"—at work.

Like many of his peers, Bill Denton had practically grown up at Amerco. It was in his thirty years with the company that many of his most significant rites of passage had taken place. It was there that he felt most secure. So he naturally came to view the office as the proper place to express family feeling.

As Bill saw it, home was ideally a branch office of Amerco, but in addition, Amerco should now feel like a home. Part of a manager's job, Bill felt, was to make work an emotionally comfortable environment in which to be efficient. His job, as he and his colleagues intuitively understood it, was to give work a homelike feel by taking on the daddy role there.[3] Bill himself made an effort to be continuously available to those he managed and to "motivate" them to work well and hard. But for some employees the meaning of his concern was more personal.

Men who were curtailing their own roles as fathers at home spent long hours with "fathers" like Bill at work. Among both women and men who had lost fathers to workaholism or divorce, many were now rediscovering as working adults what it was like to have a dad who taught them to do something, a dad who scolded, who demanded, who coaxed, who cared. At last, after all these years, they could actually catch his attention.

Bill was a better father at work than he had been at home—first of all in the sense that he was nearly always there. At the office, Bill found he could handle his employees' mistakes with an equanimity he had rarely been able to maintain with his own children. At the office, his "sons"—and more recently "daughters"—were generally grateful and eager to learn. At home, his own son had no desire to learn Bill's trade and now was asking where his father had been all those years of childhood.

In truth, it was hard for most of these executive father figures to imagine pulling themselves away from work. It was simply more satisfying being Dad here than anywhere else. As one of them put it,

> I know I shouldn't be seen here as many hours as I am, because this is creating an unwritten message about work hours. But I'm here because I like it, and I have a lot to do on top of that. That's a real dilemma for me. Sometimes I think I ought to work in my library at home. But then I wouldn't be here, and that diminishes the amount of time I'm available to people. A lot of my job is to be available to people, to give encouragement, and once in a while, a good word. People like to know I'm here.

After the company introduced the Total Quality philosophy of management, Bill found himself calling one meeting after another to "monitor the process." In these sessions, Bill led by encouraging one person, praising another, teasing a third; while employees from different divisions got to see each other, chat, joke, and develop camaraderie. Periodically, he would become exasperated with the time spent this way—up to a third of many workdays—and would urge managers in his division to reduce the number and length of meetings they held. In one such mood he noted with disgust but also a barely disguised hint of satisfaction,

> Partly, people come to meetings to get hugs. Partly, they come for blessings: "Yes, my child, it will be all right." Partly, they come to be seen. Partly, they come to share risks. Partly, they come to get information. But why are we group-solving our problems? That's what individuals are paid to do. I'm not sure a great idea has ever come out of a meeting. We can downsize meetings. It's madness.

But after blowing off steam, Bill called another meeting. Sure, they ate up time and extended the day, but people wanted to attend them and, well, why not?

As with the Spotted Deer Childcare Center, Bill's workplace often seemed to outshine the family for which it was supposed to be a poor substitute, offering satisfactions that its employees had been brought up to believe only the family should give. It was there, not home, that many of his office children felt themselves most appreciated.

Still, there were limits. "Office fathers" had to take orders from an office father yet higher up, and sometimes they were ordered to do things that were "bad for the children." They had to get more work out of people. They had to demote and occasionally fire people. This was a company, after all, and there was work to be done. When layoffs were to be made, Amerco typically turned to outside consultants to "wear the black hat," perhaps because to have done otherwise would have undermined the bond between workers and managers.

Being a good father at work came easily to Bill Denton, given his traditional idea of fatherhood and given his relationship to his own first boss at Amerco. As he explained,

> When I started with this company, my supervisor's first speech was this: "You've got young children at home. There are going to be plays, ballet recitals, soccer games. I expect you're going to need to go to those, and you should find the time to make that happen—but I'll still hold you accountable for the job getting done." I can't believe that these days managers don't say the same thing to their employees. I don't know what the problem is.

"Fatherhood" brought to Bill's mind an image of a child's performance—the part of childhood that comes closest to being a career. Like a business meeting, each concert or soccer game is a slot in time scheduled in advance by someone else. A concert recital might start at 2 P.M. and end at 3:30 P.M. A soccer game might go from 4 to 6 P.M., a play from 8 to 10 P.M. When Bill spoke of being a good father or of balancing work and family life, what came to mind were these well-bounded events in the "careers" of his children, extracurricular equivalents to the events of his day.

In addition to such performances, medical emergencies also seemed legitimate reasons for a good father to take time off from his job. A car accident, a football injury, a sudden illness might draw him out of even the most important meeting. Work demands stopped at the hospital door. Focused on performances and emergencies—the best times and the worst, he said—he knew little about those times when his children were offstage, unable to get started on something, discouraged, or confused. True, Bill lived in a town where work was near home. But something in Bill's ideas about fatherhood and time made the everyday part of home life seem very far away.

Sons and daughters now grown, many managers did look back on their fathering years with a kind of mild regret that they had spent so little time with their children. Perhaps this simply reflected the influence of new ideas about fatherhood, even on these "ten-hour players" in their fifties. "Did you have enough time with your boys when they were growing up?" I asked one manager reputed to have blocked a paternity leave for a gifted young engineer in his plant. He replied,

> No. No. Well, the youngest one, yes. But I didn't bond well with my oldest child. Being the ambitious person I was, I worked incredibly long hours when I first started. We have a good relationship now, but I look back and I didn't get wise soon enough. I didn't take vacations for the first six years of my oldest child's life.

"If you had it to do over again," I asked, "would you do anything differently?" Here, he hesitated,

> I don't know. I can't answer that. Probably not. I'll tell you why. I was the youngest of six kids, the only one to go to college. My father was a machinist on the railroad in Peoria, Illinois. When I was in high school, he was laid off from work. It was a crushing experience for him, and for me. I wanted to do it differently.

His own success at work, his stilling of the memory of his father's failure, these were more important to him than being a good father at home, even in retrospect. As for Bill Denton, he avoided the question by putting it this way: "I'm pleased with how my kids came out."

It was not these top executives but their wives who spoke ruefully of their husbands' absences from family life. One whose husband had been fired from an unprofitable division came close to a breakdown in Amerco's Office of Career Transitions. Sitting next to her husband, she exploded at the young outplacement officer:

> My husband missed our children's *birthdays*! He missed their *games*! He missed the father-daughter *banquets*! Didn't the company get *enough* of his time? Because we saw *nothing* of him!

When top male managers themselves spoke of regrets for that lost time with children at home, it often took the form of a report on what their wives thought. One noted,

> My wife would tell you she missed me playing the role of father. She would tell you she was both mother and father. She's there to see firsthand what the costs are for our kids, so her insights would be better.

Another recalled,

> Only after twenty years did my wife tell me that she was hurt when I told her my job was my number one priority and my family was number two. I'm not sure I even remember having said it, but I'm sure I did.

Some of these executives were beginning to find the post–child-raising years unsettling at home, but this only reinforced their commitment to long hours at work. To be sure there were tantrums, rivalries, and strains at work, but for many of them, these were minor matters compared to the problems flaring up at home. Difficult as work could be, the exhausted executive's life there was more predictable and more protected from bad feelings than his life at home. It was in family life that the most troubling questions arose—Am I really worthy of love? Do I really love? These were not questions executives generally had to face straight on at Amerco. One's deepest motivations were more vulnerable to critical scrutiny at home than at the office, with its comforting, built-in limits. As one executive confided,

> I told my wife I work long hours at the office for her, but she doesn't believe it. She says I just work for myself, because I put in long hours when I know she doesn't want me to.

Some wives believed their husband used work as a mistress, and they often blamed the mistress, not the man. When one executive suddenly dropped dead of a heart attack, his grieving widow invited no one from the company to speak at the funeral. "Why should I?" she exclaimed to a friend. "It was the *company* that killed him!" . . .

Given this work and family culture among top executives at Amerco, Bill took up the question of company family-friendly policies with a curiously split consciousness. Part of the time, he spoke as if there was hardly any problem at all. "In a small town like this," he explained, "you're a minute away from your house. Just get in the car and go. I did. I made my son's soccer games."

But when pressed, he slipped almost imperceptibly into a different line of reasoning. As all top managers did, Bill began talking about time as if it were hardwired not just to workers' skills but to their career aspirations. He refused to accept, he declared, the meritocratic principle suggested by some, "Judge the work, not the face time." His belief in flexibility stood in direct contradiction to one unbending principle: The time a worker works, in and of itself, has to count as much as the results accomplished within that time. Time is a symbol of commitment.

Whether time mattered more than results was a key point in contention. But it became buried in the company's rhetoric. In an hour and a half meeting of the Corrective Action Team that I attended, Bill assembled a group of employees drawn from different sectors of the company to help implement and monitor Amerco's family-friendly policies. Bill sat with five other team members intently reviewing possible expressions of "Amerco's philosophy" with regard to work-family balance. Should it be:

> Consistent with our valuing the individual, we believe it essential that our people lead balanced lives.

Or

> We recognize the legitimacy of the demands and pressures of our employees' lives outside work.

Or

> We believe that a solid work ethic is an important foundation block upon which a career can be built. However, hard work is not an end in itself, and alone, is not valued in and of itself.

Or

> Tired, overworked people, worried about their children, parents, and other nonwork issues do not give their best effort to the job, nor can they be expected to.

By the end of the meeting, the committee had settled on the first statement. But the basic question—Did the company evaluate employees based on their output or their work schedules?—remained only half-addressed. When Amy Truett asked Bill, "How would you define 'commitment'?" he answered immediately:

> I don't think we can get commitment with less than fifty or sixty hours a week. That's what other corporations are doing. To be competitive, that's what we need to do. In my gut, I can't believe we can do it very differently.

A chorus of voices arose in protest. But no one quite dared to ask the underlying questions. What is a "balanced life" then? With this kind of commitment, what room is left for family?

Top executives in another division of Amerco responded to Amerco's statement on work-family balance with less sympathy than Bill had when it appeared in an office memo a week later. Generally, they saw balance as strictly a woman manager's problem and so a limited one: a job share here, a part-time position there, and everything would be fine.

One thoughtful thirty-year-old male junior manager described to me a meeting of top executives who responded angrily, though confidentially, to the memo. His own childhood family had disintegrated as his parents spent less and less time at home, and finally divorced. "Now my mother lives in a small apartment in Orange County, California. My father married again and lives in Texas. I don't know where the crib I slept in as a baby is. I don't know where any of my old toys are. They must have given them away." Disguised in a seventy-hour-a-week company uniform, he spoke as a potential defector from an uncaring system:

> I was reporting on the results of a climate survey in my division. I said that the people I worked with wanted a better balance between work and family. I got it right between the eyes. Dave blew up at me: "Don't *ever* bring up 'balance' again! I don't want to hear about it! Period! Everyone in this company has to work hard. *We* work hard. *They* have to work hard. That's the way it is. Just because a few women are concerned about balance doesn't mean we change the rules. If they chose this career, they're going to have to pay for it in hours, just like the rest of us." . . .

NOTES

1. See also Helga Nowotny, "Time Structuring and Time Measurement: On the Interrelation between Timekeepers and Social Time," in J. Fraser et al., eds., *The Study of Time II* (Amherst: University of Massachusetts Press, 1975).
2. In his book, *Hidden Rhythms,* sociologist Eviatar Zerubavel describes another social world in which time was tightly scheduled and controlled—the medieval Benedictine monastery. Instructions about how to keep busy all day appeared in the Rule of Saint Benedict (Eviatar Zerubavel, *Hidden Rhythms: Schedules and Calendars in Social Life* [Chicago: University of Chicago Press, 1981], pp. 34, 35, 39). In the contemporary company, too, time is extremely ordered, regulated, rationalized. When executives at Amerco explained why they "loved" their work, they often mentioned the satisfaction of solving challenging problems, being part of the community of employees, and, of course, receiving money and prestige. But I wondered if some of that "love" didn't arise from the comforting regularity of their tasks.
3. See Roma Hanks and Marvin B. Sussman, eds., "Where Does Family End and Corporation Begin: The Consequences of Rapid Transformation," in *Corporations, Businesses and Families* (New York: Haworth Press, 1990), p. 6.

27

"When You Get Laid Off, It's Like You Lose a Part of Yourself"

LILLIAN B. RUBIN

For Larry Meecham, "downsizing" is more than a trendy word on the pages of the *Wall Street Journal* or the business section of the *New York Times*. "I was with the same company for over twelve years; I had good seniority. Then all of a sudden they laid off almost half the people who worked there, closed down whole departments, including mine," he says, his troubled brown eyes fixed on some distant point as he speaks. "One day you got a job; the next day you're out of work, just like that," he concludes, shaking his head as if he still can't believe it.

Nearly 15 percent of the men in the families I interviewed were jobless when I met them.[1] Another 20 percent had suffered episodic bouts of unemployment—sometimes related to the recession of the early 1990s, sometimes simply because job security is fragile in the blue-collar world, especially among the younger, less experienced workers. With the latest recession, however, age and experience don't count for much; every man feels at risk.[2]

Tenuous as the situation is for white men, it's worse for men of color, especially African-Americans. The last hired, they're likely to be the first fired. And when the axe falls, they have even fewer resources than whites to help them through the tough times. "After kicking around doing shit work for a long time, I finally got a job that paid decent," explains twenty-nine-year-old George Faucett, a black father of two who lost his factory job when the company was restructured—another word that came into vogue during the economic upheaval of the 1990s. "I worked there for two years, but I didn't have seniority, so when they started to lay guys off, I was it. We never really had a chance to catch up on all the bills before it was all over," he concludes dispiritedly.

I speak of men here partly because they're usually the biggest wage earners in intact families. Therefore, when father loses his job, it's likely to be a crushing blow to the family economy. And partly, also, it's because the issues unemployment raises are different for men and for women. For most women, identity is multi-faceted, which means that the loss of a job isn't equivalent to the loss of self. No matter how invested a woman may be in her work, no matter how much her sense of self and competence are connected to it, work remains only one part of identity—a central part perhaps, especially for a professional woman, but still only a part. She's mother, wife, friend, daughter, sister—all valued facets of the self, none wholly obscuring the others. For the working-class woman in this study, therefore, even those who were divorced or single mothers responsible for the support of young children, the loss of a job may have been met with pain, fear, and anxiety, but it didn't call their identity into question.

For a man, however, work is likely to be connected to the core of self. Going to work isn't just what he does, it's deeply linked to who he is. Obviously, a man is also father, husband, friend, son, brother. But these are likely to be roles he assumes, not without depth and meaning, to be sure, but not self-defining in the same way as he experiences work. Ask a man for a statement of his identity, and he'll almost always respond by telling you first what he does for a living. The same question asked of a woman brings forth a less predictable, more varied response, one that's embedded in the web of relationships that are central to her life.[3]

Some researchers studying the impact of male unemployment have observed a sequenced series of psychological responses.[4] The first, they say, is shock, followed by denial and a sense of optimism, a belief that this is temporary, a holiday, like a hiatus between jobs rather than joblessness. This period is marked by heightened activity at home, a burst of do-it-yourself projects that had been long neglected for lack of time. But soon the novelty is gone and the projects wear thin, ushering in the second phase, a time of increasing distress, when inertia trades places with activity and anxiety succeeds denial. Now a jobless man awakens every day to the reality of unemployment. And, lest he forget, the weekly trip to the unemployment office is an unpleasant reminder. In the third phase, inertia deepens into depression, fed by feelings of identity loss, inadequacy, hopelessness, a lack of self-confidence, and a general failure of self-esteem. He's tense, irritable, and feels increasingly alienated and isolated from both social and personal relationships.

This may be an apt description of what happens in normal times. But in periods of economic crisis, when losing a job isn't a singular and essentially lonely event, the predictable pattern breaks down.[5] During the years I was interviewing families for this book, millions of jobs disappeared almost overnight. Nearly everyone I met, therefore, knew someone—a family member, a neighbor, a friend—who was out of work. "My brother's been out of a job for a long time; now my brother-in-law just got laid off. It seems like every time I turn around, somebody's losing his job. I've been lucky so far, but it makes you wonder how long it'll last."

At such times, nothing cushions the reality of losing a job. When the unbelievable becomes commonplace and the unexpected is part of the mosaic of the times, denial is difficult and optimism impossible. Instead, any layoff, even if it's defined as temporary, is experienced immediately and viscerally as a potentially devastating, cataclysmic event.

It's always a shock when a person loses a job, of course. But disbelief? Denial? Not for those who have been living under a cloud of anxiety—those who leave work each night grateful for another day of safety, who wonder as they set off the next morning whether this is the day the axe will fall on them. "I tell my wife not to worry because she gets panicked about the bills. But the truth is, I stew about it plenty. The economy's gone to hell; guys are out of work all around me. I'd be nuts if I wasn't worried."

It's true that when a working-class man finds himself without a job he'll try to keep busy with projects around the house. But these aren't undertaken in the kind of holiday spirit earlier researchers describe.[6] Rather, building a fence, cleaning the garage, painting the family room, or the dozens of other tasks that might occupy him are a way of coping with his anxiety, of distracting himself from the fears that threaten to overwhelm him, of warding off the depression that lurks just below the surface of his activity. Each thrust of the saw, each blow of the hammer helps to keep the demons at bay. "Since he lost his job, he's been out there hammering away at one thing or another like a maniac," says Janet Kovacs, a white thirty-four-year-old waitress. "First it was the fence; he built the whole thing in a few days. Then it was fixing the siding on the garage. Now he's up on the roof. He didn't even stop to watch the football game last Sunday."

Her husband, Mike, a cement finisher, explains it this way: "If I don't keep busy, I feel like I'll go nuts. It's funny," he says with a caustic, ironic laugh, "before I got laid off my wife was always complaining about me watching the ball games; now she keeps nagging me to watch. What do you make of that, huh? I guess she's trying to make me feel better."

"Why didn't you watch the game last Sunday?" I ask.

"I don't know, maybe I'm kind of scared if I sit down there in front of that TV, I won't want to get up again," he replies, his shoulders hunched, his fingers raking his hair. "Besides, when I was working, I figured I had a right."

His words startled me, and I kept turning them over in my mind long after he spoke them: "When I was working, I figured I had a right." It's a sentence any of the unemployed men I met might have uttered. For it's in getting up and going to work every day that they feel they've earned the right to their manhood, to their place in the world, to the respect of their family, even the right to relax with a sporting event on TV.

It isn't that there are no gratifying moments, that getting laid off has no positive side at all. When unemployment first hits, family members usually gather around to offer support, to buoy a man's spirits and their own. Even in families where conflict is high, people tend to come together, at least at the beginning. "Considering that we weren't getting along so well before, my wife was really good about it when I got laid off," says Joe Phillips, an unemployed black truck driver. "She gave me a lot of support at first, and I appreciated it."

"You said 'at first.' Has that changed?" I ask.

"Hell, yes. It didn't last long. But maybe I can't blame it all on her. I've been no picnic to live with since I got canned."

In families with young children, there may be a period of relief—for the parents, the relief of not having to send small children off to child care every day, of knowing that one of them is there to welcome the children when they come home from school; for the children, the exhilarating novelty of having a parent, especially daddy, at home all day. "The one good thing about him not working is that there's someone home with the kids now," says twenty-five-year-old Gloria Lewis, a black hairdresser whose husband has been unemployed for just a few weeks. "That part's been a godsend. But I don't know what we'll do if he doesn't find work soon. We can't make it this way."

Teenagers, too, sometimes speak about the excitement of having father around at first. "It was great having my dad home when he first got laid off," says Kevin Sollars, a white fourteen-year-old. "We got to do things together after school sometimes. He likes to build ship models—old sailing ships. I don't know why, but he never wanted to teach me how to do it. He didn't even like it when I just wanted to watch; he'd say, 'Haven't you got something else to do?' But when he first got laid off, it was different. When I'd come home from school and he was working on a ship, he'd let me help him."

But the good times usually don't last long. "After a little while, he got really grumpy and mean, jumped on everybody over nothing," Kevin continues. "My mom used to say we had to be patient because he was so worried about money and all that. Boy, was I glad when he went back to work."

Fathers may also tell of the pleasure in getting to spend time with their children, in being a part of their daily life in ways unknown before. "There's a silver lining in every cloud, I guess. I got to know my kids like I never did before," says Kevin's father, who felt the sting of unemployment for seven months before he finally found another job. "It's just that being out of work gets old pretty fast. I ran out of stuff to do around the house; we were running out of money; and there I was sitting on my keister and stewing all day long while my wife was out working. I couldn't even enjoy building my little ships."

Once in a while, especially for a younger man, getting laid off or fired actually opens up the possibility of a new beginning. "I figured, what the hell, if I'm here, I might as well learn how to cook," says twenty-eight-year-old Darnell Jones, a black father of two who, until he was laid off, had worked steadily but always at relatively menial, low-paying jobs in which he had little interest or satisfaction. "Turned out I liked to cook, got to be real good at it, too, better than my wife," he grins proudly. "So then we talked about it and decided there was no sense in sitting around waiting for something to happen when there were no good jobs out there, especially for a black man, and we figured I should go to cooking school and learn how to do it professionally. Now I've got this job as a cook; it's only part-time, right now, but the pay's pretty good, and I think maybe I'll go full-time soon. If I could get regular work, maybe we could even save some money and I could open my own restaurant someday. That's what I really want to do."

But this outcome is rare, made possible by the fact that Darnell's wife has a middle-management position in a large corporation that pays her $38,000 a year. His willingness to try something new was a factor, of course. But that, too, was grounded in what was possible. In most young working-class families of any color or ethnic group, debts are high, savings are nonexistent, and women don't earn nearly enough to bail the family out while the men go into a training program to learn new skills. A situation that doesn't offer much encouragement for a man to dream, let alone to believe his dream could be realized.[7]

As I have already indicated, the struggles around the division of labor shift somewhat when father loses his job. The man who's home all day while his wife goes off to work can't easily justify maintaining the traditional household gender roles. Therefore, many of the unemployed men pick up tasks that were formerly left to their wives alone. "I figure if she's working and I'm not, I ought to take up some of the slack around here. So I keep the place up, run the kids around if they need it, things like that," says twenty-nine-year-old Jim Anderson, a white unemployed electrician.

As wives feel their household burdens eased, the strains that are almost always a part of life in a two-job family are somewhat relieved. "Maybe it sounds crazy to you, but my life's so much easier since he's out of work, I wish it could stay this way," says Jim's wife, Loreen, a twenty-nine-year-old accounting clerk. "If only I could make enough money, I'd be happy for him to stay home and play Mr. Mom."

But it's only a fantasy—first because she can't make enough money; second, and equally important, because while she likes the relief from household responsibilities, she's also uneasy about such a dramatic shift in family roles. So in the next breath, Loreen says, "I worry about him, though. He doesn't feel so good about himself being unemployed and playing house."

"Is it only him you worry about? Or is there something that's hard for you, too?" I ask.

She's quiet for a moment, then acknowledges that her feelings are complicated: "I'm not sure what I think anymore. I mean, I don't think it's fair that men always have to be the support for the family; it's too hard for them sometimes. And I don't mind working; I really don't. In fact, I like it a lot better than being home with the house and the kids all the time. But I guess deep down I still have that old-fashioned idea that it's a man's job to support his family. So, yeah, then I begin to feel—I don't know how to say it—uncomfortable, right here inside me," she says, pointing to her midsection, "like maybe I won't respect him so much if he can't do that. I mean, it's okay for now," she hastens to reassure me, perhaps herself as well. "But if it goes on for a real long time like with some men, then I think I'll feel different."

Men know their wives feel this way, even when the words are never spoken, which only heightens their own anxieties about being unemployed. "Don't get me wrong; I'm glad she has her job. I don't know what we'd do if she wasn't working," says Jim. "It's just that . . . ," he hesitates, trying to frame his thoughts clearly. "I know this is going to sound pretty male, but it's my job to take care of this family. I mean, it's great that she can help out, but the responsibility is mine, not hers. She won't say so, but I know she feels the same way, and I don't blame her."

It seems, then, that no matter what the family's initial response is, whatever the good moments may be, the economic and psychological strains that attend unemployment soon overwhelm the good intentions on all sides. "It's not just the income; you lose a lot more than that," says Marvin Reed, a forty-year-old white machinist, out of work for nearly eight months. He pauses, reflects on his words, then continues. "When you get laid off, it's like you lose a part of yourself. It's terrible; something goes out of you. Then, on top of that, by staying home and not going to work and associating with people of your own level, you begin to lose the sharpness you developed at work. Everything gets slower; you move slower; your mind works slower.

"It's a real shocker to realize that about yourself, to feel like you're all slowed down . . . ," he hesitates again, this time to find the words. "I don't know how to explain it exactly, maybe like your mind's pushing a load of mud around all the time," he concludes, his graying head bowed so as not to meet my eyes.

"Everything gets slower"—a sign of the depression that's so often the unwelcome companion of unemployment. As days turn into weeks and weeks into months, it gets harder and harder to believe in the future. "I've been working since I was fourteen," says Marvin, "and I was never out of work for more than a week or two before. Now I don't know; I don't know when I'll get work again. The jobs are gone. How do you find a job when there's none out there anymore?"

The men I talked with try to remind themselves that it's not their fault, that the layoffs at the plant have nothing to do with them or their competence, that it's all part of the economic problems of the nation. But it's hard not to doubt themselves, not to wonder whether there's something else they could have done, something they might have foreseen and planned for. "I don't know; I keep thinking I could have done something different," says Lou Coltrane, a black twenty-eight-year-old auto worker, as he looks away to hide his pain. "I know it's crazy; they closed most of the plant. But, you know, you can't help thinking, maybe this, maybe that. It keeps going round and round in my head: Maybe I should have done this; maybe I should have done that. Know what I mean?"

But even when they can accept the reality that they had no control over the situation, there's little surcease in the understanding. Instead, such thoughts increase their feelings of vulnerability and helplessness—feelings no one accepts easily. "I worked nineteen years for this damned company and how do they pay me back?" asks Eric Hueng, a forty-four-year-old unemployed Asian factory worker, as he leaves his chair and paces the room in a vain attempt to escape his torment. "They move the plant down to some godforsaken place in South America where people work for peanuts and you got no choice but to sit there and watch it happen. Even the government doesn't do a damn thing about it. They just sit back and let it happen, so how could I do anything?" he concludes, his words etched in bitterness.

For American men—men who have been nurtured and nourished in the belief that they're masters of their fate—it's almost impossible to bear such feelings of helplessness. So they find themselves in a cruel double bind. If they convince themselves that their situation is beyond their control, there's nothing left but resignation and despair. To fight their way out of the hopelessness that follows, they begin to blame themselves. But this only leaves them, as one man said, "kicking myself around the block"—kicks that, paradoxically, allow them to feel less helpless and out of control, while they also send them deeper into depression, since now it's no one's fault but their own.

"I can't believe what a fool I was," says Paul Santos, a forty-six-year-old Latino tool and die maker, his fingers drumming the table nervously as he speaks. "I was with this one company for over fifteen years, then this other job came along a couple of years ago. It seemed like a good outfit, solid, and it was a better job, more money and all. I don't know what happened; I guess they got overextended. All I know is they laid off 30 percent of the company without a day's notice. Now I feel stupid; if I had stayed where I was, I'd still be working."

Shame, too, makes an appearance, adding to the self-blame, to Paul's feeling that he did something wrong, something stupid—that if he'd somehow been better, smarter, more prescient, the outcome would be different. And the depression deepens. "I've been working all my life. Now it's like I've got nothing left," Paul explains, his eyes downcast, his voice choked with emotion. "When you work, you associate with a group of people you respect. Now you're not part of the group anymore; you don't belong anywhere. Except," he adds with disgust, "on the unemployment line.

"Now that's a sad sight, all these guys shuffling around, nobody looking at anybody else. Every time I go there, I think, *Hey, what the hell am I doing here? I don't belong here, not with these people. They're deadbeats.* Then I think, *Yeah, well you're here, so it looks like you're no better than them, doesn't it?*"

Like so many other men, Paul hasn't just lost a job; he's lost a life. For his job meant more than a living wage. It meant knowing he had an identity and a place in the world—a place where his competence was affirmed, where he had friends who respected and admired him, men with whom he could share both the frustrations and satisfactions of life on the job.

It's not just for men that the job site is a mirror in which they see themselves reflected, a mirror that reflects back an image that reassures them that they're valued contributors to the social world in which they live. It functions this way for all of us. But it's particularly important for men because when the job disappears, all this goes too, including the friendships that were so important in the validation of the self.[8]

In my earlier research on friendship, the men I interviewed spoke repeatedly about how, once they left the job, they lost contact with the friends they had made at work.[9] Sometimes these men acknowledged that it was "out of sight, out of mind." But others insisted that, even though they might never see each other again, these friendships represented lasting bonds. "Maybe they don't continue to see each other once the activity [or job] doesn't keep them together," said one man I interviewed then, "but that doesn't mean they don't share very deep and lasting bonds, does it?"[10] Perhaps. But the bonds, if they exist, can't replace the face-to-face interactions that are so important to the maintenance of the self.

"I don't see anybody anymore," mourns Bill Costas, a thirty-four-year-old unemployed white meat packer who had worked in the same plant for nine years. "The guys I worked with were my buddies; after all those years of working together, they were my friends. We'd go out after work and have a beer and shoot the bull. Now I don't even know what they're doing anymore."

For wives and children, it's both disturbing and frightening to watch husband and father sink ever deeper into despair. "Being out of work is real hard on him; it's hard to see him like this, so sad and jumpy all the time," laments Bill's wife, Eunice, a part-time bank teller who's anxiously looking for full-time work. "He's always been a good provider, never out of work hardly a day since we got married. Then all of a sudden this happens. It's like he lost his self-respect when he lost that job."

His self-respect and also the family's medical benefits, since Eunice doesn't qualify for benefits in her part-time job. "The scariest part about Bill being out of a job is we don't have any medical insurance anymore. My daughter got pneumonia real bad last winter and I had to borrow money from my sister for the doctor bill and her medicine. Just the medicine was almost $100. The doctor wanted to put her in the hospital, but we couldn't because we don't have any health insurance."

Her husband recalls his daughter's illness, in a voice clogged with rage and grief. "Do you know what it's like listening to your kid when she can't breathe and you can't send her to the hospital because you lost your benefits when you got laid off?"

In such circumstances, some men just sit, silent, turned inward, enveloped in the gray fog of depression from which they can't rouse themselves. "I leave to go to work in the morning and he's sitting there doing nothing, and when I come home at night, it's the same thing. It's like he didn't move the whole day," worries thirty-four-year-old Deidre Limage, the wife of a black factory worker who has been jobless for over a year.

Other men defend against feeling the pain, fear, and sadness, covering them over with a flurry of activity, with angry, defensive, often irrational outbursts at wife and children—or with some combination of the two. As the financial strain of unemployment becomes crushing, everyone's fears escalate. Wives, unable to keep silent, give voice to their concerns. Their husbands, unable to tolerate what they hear as criticism and blame—spoken or not—lash out. "It seems like the more you try to pull yourself up, the more you get pushed back down," sighs Beverly Coleride, a white twenty-five-year-old cashier with two children, whose husband has worked at a variety of odd jobs in their seven-year marriage. "No matter how hard we try, we can't seem to set everything right. I don't know what we're going to do now; we don't have next month's rent. If Kenny doesn't get something steady real quick, we could be on the street."

"We could be on the street"—a fear that clutches at the hearts and gnaws at the souls of the families in this study, not only those who are unemployed. Nothing exemplifies the change in the twenty years since I last studied working families than the fear of being "on the street." Then, homelessness was something that happened somewhere else, in India or some other far-off and alien land. Then, we wept when we read about the poor people who lived on the streets in those other places. *What kind of society doesn't provide this most basic of life's needs?* we asked ourselves. Now, the steadily increasing numbers of homeless in our own land have become an ever-present and frightening reminder of just how precarious life in this society can be. Now, they're in our face, on our streets, an accepted category of American social life—"the homeless." . . .

For Beverly Coleride, as for the other women and men I met, sustaining the denial has become increasingly difficult. No matter how much they want to obliterate the images of the homeless from consciousness, the specter haunts them, a frightening reminder of what's possible if they trip and fall. Perhaps it's because there's so much at stake now, because the unthinkable has become a reality, that anxieties escalate so quickly. So as Beverly contemplates the terror of being "on the street," she begins to blame her husband. "I keep telling myself it's not his fault, but it's real hard not to let it get you down. So then I think, well, maybe he's not trying hard enough, and I get on his case, and he gets mad, and well, I guess you know the rest," she concludes with a harsh laugh that sounds more like a cry of pain.

She doesn't *want* to hurt her husband, but she can't tolerate feeling so helpless and out of control. If it's his fault rather than the workings of some impersonal force, then he can do something about it. For her husband, it's an impossible bind. "I keep trying, looking for something, but there's nothing out there, leastwise not for me. I don't know what to do anymore; I've tried everything, every place I know," he says disconsolately.

But he, too, can't live easily with such feelings of helplessness. His sense of his manhood, already under threat because he can't support his family, is eroded further by his wife's complaints. So he turns on her in anger: "It's hard enough being out of work, but then my wife gets on my case, yakking all the time about how we're going to be on the street if I don't get off my butt, like it's my fault or something that there's no work out there. When she starts up like that, I swear I want to hit her, anything just to shut her mouth," he says, his shoulders tensed, his fists clenched in an unconscious expression of his rage.

"And do you?" I ask.

The tension breaks; he laughs. "No, not yet. I don't know; I don't want to," he says, his hand brushing across his face. "But I get mad enough so I could. Jesus, doesn't she know I feel bad enough? Does she have to make it worse by getting on me like that? Maybe you could clue her, would you?"

"Maybe you could clue her"—a desperate plea for someone to intervene, to save him from his own rageful impulses. For Kenny Coleride isn't a violent man. But the stress and conflict in families where father loses his job can give rise to the kind of interaction described here, a dynamic that all too frequently ends in physical assaults against women and children.

Some kind of violence—sometimes against children only, more often against both women and children—is the admitted reality of life in about 14 percent of the families in this study.[11] I say "admitted reality" because this remains one of the most closely guarded secrets in family life. So it's reasonable to assume that the proportion of families victimized by violence could be substantially higher.

Sometimes my questions about domestic violence were met with evasion: "I don't really know anything about that."

Sometimes there was outright denial, even when I could see the evidence with my own eyes: "I was visiting my sister the other day, and I tripped and fell down the steps in front of her house."

And sometimes teenage children, anguished about what they see around them, refused to participate in the cover-up. "I bet they didn't tell you that he beats my mother up, did they? Nobody's allowed to talk about it; we're supposed to pretend like it doesn't happen. I hate him; I could kill him when he does that to her. My mom, she says he can't help it; it's because he's so upset since he got fired. But that's just her excuse now. I mean, yeah, maybe it's worse than it was before, but he did it before, too. I don't understand. Why does she let him do it to her?"

"Why does she let him do it to her?" A question the children in these families are not alone in asking, one to which there are few satisfactory answers. But one thing is clear: The depression men suffer and their struggle against it significantly increase the probability of alcohol abuse, which in turn makes these kinds of eruptions more likely to occur.[12]

"My father's really changed since he got laid off," complains Buddy Truelman, the fifteen-year-old son of an unemployed white steel worker. "It's like he's always mad about something, you know, ready to bite your head off over nothing. I mean, he's never been an at-ease guy, but now nothing you do is okay with him; he's always got something to say, like he butts in where it's none of his business, and if you don't jump to, he gets mad as hell, carries on like a crazy man." He pauses, shifts nervously in his chair, then continues angrily, "He and my mom are always fighting, too. It's a real pain. I don't hang around here any more than I have to." . . .

Many of the unemployed men admit turning to alcohol to relieve the anxiety, loneliness, and fear they experience as they wait day after day, week after week for, as one man put it, "something to happen." "You begin to feel as if you're going nuts, so you drink a few beers to take the edge off," explains thirty-seven-year-old Bill Anstett, a white unemployed construction worker.

It seems so easy. A few beers and he gets a respite from his unwanted feelings—fleeting, perhaps, but effective in affording some relief from the suffering they inflict. But a few beers often turn out to be enough to allow him to throw normal constraints to the wind. For getting drunk can be a way of absenting the conscious self so that it can't be held responsible for actions undertaken. Indeed, this may be as much his unconscious purpose as the need to rid himself of his discomfort. "I admit it, sometimes it's more than a few and I fall over the edge," Bill grants. "My wife, she tells me it's like I turn into somebody else, but I don't know about that because I never remember."

With enough alcohol, inhibitions can be put on hold; conscience can go underground. "It's the liquor talking," we say when we want to exempt someone from responsibility for word or deed. The responsibility for untoward behavior falls to the effects of the alcohol. The self is in the clear, absolved of any wrongdoing. So it is with domestic violence and alcohol. When a man gets drunk, the inner voice that speaks his failure and shame is momentarily stilled. Most men just relax gratefully into the relief of the internal quiet. But the man who becomes violent needs someone to blame, someone onto whom he can project the feelings that cause him such misery. Alcohol helps. It gives him license to find a target. With enough of it, the doubts and recriminations that plague him are no longer his but theirs—his wife's, his children's; "them" out there, whoever they may be. With enough of it, there's nothing to stay his hand when his helpless rage boils over. "I don't know what happens. It's like something I can't control comes over me. Then afterward I feel terrible," Peter DiAngelo, an unemployed thirty-two-year-old truck driver, says remorsefully.

One-fifth of the men in this study have a problem with alcohol, not all of them unemployed. Nor is domestic violence perfectly correlated with either alcohol abuse or unemployment. But the combination is a potentially deadly one that exponentially increases the likelihood that a man will act out his anger on the bodies of his wife and children. "My husband drinks a lot more now; I mean, he always drank some, but not like now," says Inez Reynoso, a twenty-eight-year-old Latina nurse's aide and mother of three children who is disturbed about her husband's mistreatment of their youngest child, a three-year-old boy. "I guess he tries to drink away his troubles, but it only makes more trouble. I tell him, but he doesn't listen. He has a fiery temper, always has. But since he lost his job, it's real bad, and his drinking doesn't help it none.

"I worry about it; he treats my little boy so terrible. He's always had a little trouble with the boy because he's not one of those big, strong kids. He's not like my older kids; he's a timid one, still wakes up scared and crying a lot in the night. Before he got fired, my husband just didn't pay him much attention. But now he's always picking on him; it's like he can't stand having him around. So he makes fun of him something terrible, or he punches him around." . . .

"Does he hit you, too?" I ask Inez.

She squirms in her chair; her fingers pick agitatedly at her jeans. I wait quietly, watching as she shakes her head no. But when she speaks, the words say something else. "He did a couple of times lately, but only when he had too many beers. He didn't mean it. It's just that he's so upset about being out of work, so then when he thinks I protect the boy too much he gets real mad."

When unemployment strikes, sex also becomes an increasingly difficult issue between wives and husbands. A recent study in Great Britain found that the number of couples seeking counseling for sexual problems increased in direct proportion to the rise in the unemployment rate.[13] Anxiety, fear, anger, depression—all emotions that commonly accompany unemployment—are not generators of sexual desire. Sometimes it's the woman whose ardor cools because she's frightened about the future: "I'm so scared all the time, I can't think about sex." Or because she's angry with her husband: "He's supposed to be supporting us and look where we are." More often it's the men who lose their libido along with their jobs—a double whammy for them since male identity rests so heavily in their sexual competence as well as in their work.[14]

This was the one thing the men in this study couldn't talk about. I say "couldn't" because it seemed so clearly more than just "wouldn't." Psychologically, it was nearly impossible for them to formulate the words and say them aloud. They had no trouble complaining about their wives' lack of sexual appetite. But when it was they who lost interest or who become impotent, it was another matter. Then, their tongues were stilled by overwhelming feelings of shame, by the terrible threat their impotence posed to the very foundation of their masculinity.

Their wives, knowing this, are alarmed about their flagging sex lives, trying to understand what happened, wondering what they can do to be helpful. "Sex used to be a big thing for him, but since he's been out of work, he's hardly interested anymore," Dale Meecham, a white thirty-five-year-old waitress, says, her anxiety palpable in the room. "Sometimes when we try to do it, he can't, and then he acts like it's the end of the world—depressed and moody, and I can't get near him. It's scary. He won't talk about it, but I can see it's eating at him. So I worry a lot about it. But I don't know what to do, because if I try to, you know, seduce him and it doesn't work, then it only makes things worse."

The financial and emotional turmoil that engulfs families when a man loses his job all too frequently pushes marriages that were already fragile over the brink.[15] Among the families in this study, 10 percent attributed their ruptured marriages directly to the strains that accompanied unemployment. "I don't know, maybe we could have made it if he hadn't lost his job," Maryanne Wallace, a twenty-eight-year-old white welfare mother, says sadly. "I mean, we had problems before, but we were managing. Then he got laid off, and he couldn't find another job, and, I don't know, it was like he went crazy. He was drinking; he hit me; he was mean to the kids. There was no talking to him, so I left, took the kids and went home to my mom's. I thought maybe I'd just give him a scare, you know, be gone for a few days. But when I came back, he was gone, just gone. Nobody's seen him for nearly a year," she says, her voice limping to a halt as if she still can't believe her own story.

Economic issues alone aren't responsible for divorce, of course, as is evident when we look at the 1930s. Then, despite the economic devastation wrought by the Great Depression, the divorce rate didn't rise. Indeed, it was probably the economic privations of that period that helped to keep marriages intact. Since it was so difficult to maintain one household, few people could consider the possibility of having to support two.

But these economic considerations exist today as well, yet recent research shows that when family income drops 25 percent, divorce rises by more than 10 percent.[16] Culture and the institutions of our times make a difference. Then, divorce was a stigma. Now, it's part of the sociology and psychology of the age, an acceptable remedy for the disappointment of our dreams.

Then, too, one-fourth of the work force was unemployed—an economic disaster that engulfed the whole nation. In such cataclysmic moments, the events outside the family tend to overtake and supersede the discontents inside. Now, unemployment is spottier, located largely in the working class, and people feel less like they're in the middle of a social catastrophe than a personal one. Under such circumstances, it's easier to act out their anger against each other.

And finally, the social safety net that came into being after the Great Depression—social security, unemployment benefits, public aid programs targeted specifically to single-parent families—combined with the increasing numbers of women in the work force to make divorce more feasible economically.

Are there no families, then, that stick together and get through the crisis of unemployment without all this trauma? The answer? Of course there are. But they're rare. And they manage it relatively well only if the layoff is short and their resources are long.

Almost always, these are older families where the men have a long and stable work history and where there are fewer debts, some savings, perhaps a home they can refinance. But even among these relatively privileged ones, the pressures soon begin to take their toll. "We did okay for a while, but the longer it lasts, the harder it gets," says forty-six-year-old Karen Brownstone, a white hotel desk clerk whose husband, Dan, lost his welding job nearly six months ago. "After the kids were grown, we finally managed to put some money by. Dan even did some investments, and we made some money. But we're using it up very fast, and I get real scared. What are we going to do when his unemployment runs out?"[17] . . .

When I talk with Karen's husband, Dan, he leans forward in his chair and says angrily, "I can't go out and get one of those damn flunky jobs like my wife wants me to. I've been working all my life, making a decent living, too, and I got pride in what I do. I try to tell her, but she won't listen." He stops, sighs, puts his head in his hands and speaks more softly: "I'm the only one in my whole family who was doing all right; I even helped my son go to college. I was proud of that; we all were. Now what do I do? It's like I have to go back to where I started. How can you do that at my age?"

He pauses again, looks around the room with an appraising eye, and asks: "What's going to happen to us? I know my wife's scared; that's why she's on my case so much. I worry, too, but what can I do if there's no work? Even she doesn't think I should go sling hamburgers at McDonald's for some goddamn minimum wage."

"There's something between minimum wage jobs and the kind you had before you were laid off, isn't there?" I remark.

"Yeah, I know; you sound like her now," he says, his features softening into a small smile. "But I can't, not yet. I feel like I've got to be ready in case something comes up. Meanwhile, it's not like I'm just sitting around doing nothing. I've hustled up some odd jobs, building things for people, so I pick up a little extra change on the side every now and then. It's not a big deal, but it helps, especially since it doesn't get reported. I don't know, I suppose if things get bad enough, I'll have to do something else. But," he adds, his anger rising again, "dammit, why should I? The kind of jobs you're talking about pay half what I was making. How are we supposed to live on that, tell me that, will you?"

Eventually, men like Dan Brownstone who once held high-paying skilled jobs have no choice but to pocket their pride and take a step down to another kind of work, to one of the service jobs that usually pay a fraction of their former earnings—that is, if they're lucky enough to find one. It's never easy in our youth-oriented society for a man past forty to move to another job or another line of work. But it becomes doubly difficult in times of economic distress when the pool of younger workers is so large and so eager. "Either you're overqualified or you're over the hill," Ed Kruetsman, a forty-nine-year-old unemployed white factory worker, observes in a tired voice.

But young or old, when a man is forced into lower-paying, less skilled work, the move comes with heavy costs—both economic and psychological. Economically, it means a drastic reduction in the family's way of life. "Things were going great. We worked hard, but we finally got enough together so we could buy a house that had enough room for all of us," says thirty-six-year-old Nadine Materie, a white data processor in a bank clearing center. "Tina, my oldest girl, even had her own room; she was so happy about it. Then my husband lost his job, and the only thing he could find was one that pays a lot less, *a lot less.* On his salary now we just couldn't make the payments. We had no choice; we had to sell out and move. Now look at this place!" she commands, with a dismissive sweep of her hand. Then, as we survey the dark, cramped quarters into which this family of five is not jammed, she concludes tearfully, "I hate it, every damn inch of it; I hate it." . . .

Psychologically, the loss of status can be almost as difficult to bear as the financial strain. "I used to drive a long-distance rig, but the company I worked for went broke," explains Greg Northsen, a thirty-four-year-old white man whose wife is an office worker. "I was out of work for eleven and a half months. Want to know how many days that is? Maybe how many hours? I counted every damn one," he quips acidly.

"After all that time, I was ready to take whatever I could get. So now I work as an orderly in a nursing home. Instead of cargo, I'm hauling old people around. The pay's shit and it's damn dirty work. They don't treat those old people good. Everybody's always impatient with them, ordering them around, screaming at them, talking to them like they're dumb kids or something. But with three kids to feed, I've got no choice."

He stops talking, stares wordlessly at some spot on the opposite wall for a few moments, then, his eyes clouded with unshed tears, he rakes his fingers through his hair and says hoarsely, "It's goddamn hard. This is no kind of a job for a guy like me. It's not just the money; it's . . ." He hesitates, searching for the words, then, "It's like I got chopped off at the knees, like . . . aw, hell, I don't know how to say it." Finally, with a hopeless shrug, he concludes, "What's the use? It's no use talking about it. It makes no damn difference; nothing's going to make a difference. I don't understand it. What the hell's happening to this country when there's no decent jobs for men who want to work?"

Companies go bankrupt; they merge; they downsize; they restructure; they move—all reported as part of the economic indicators, the cold statistics that tell us how the economy is doing. But each such move means more loss, more suffering, more families falling victim to the despair that comes when father loses his job, more people shouting rage and torment: "What the hell's happening to this country?"

NOTES

1. It's not possible to compare the rate of unemployment in these families with those I interviewed two decades ago because the previous sample was made up of men who were employed. But comparing the unemployment rates in 1970 and 1991 is instructive. Among white men with less than four years in high school, 4.5 percent were unemployed in 1970, 10.3 percent in 1991. The figures for high-school graduates are 2.7 percent and 5.4 percent, respectively. For blacks with less than four years in high school, the 1970 unemployment rate stood at 5.2 percent, compared to 14.7 percent in 1991. For black high-school graduates, the rates are 5.2 and 9.9, respectively (*Statistical Abstract* [U.S. Bureau of the Census, 1992, Table 637, p. 400]). The number of food stamp recipients, which typically rises as the unemployment rate climbs, jumped to an all-time high in 1993, when one in ten Americans were in the food stamp program.
2. Barbara Ehrenreich, *Fear of Falling* (New York: Pantheon Books, 1989), and Katherine S. Newman, *Falling from Grace* (New York: Free Press, 1988), write compellingly about middle-class fears of what Newman calls "falling from grace." But these fears probably are more prevalent among working-class families, and with good reason, since job security

is still so much more tenuous there than in the middle class.

3. Cf. Rubin, *Worlds of Pain,* and Lillian B. Rubin, *Women of a Certain Age: The Midlife Search for Self* (New York: Harper Perennial, 1986).

4. John Hill, "The Psychological Impact of Unemployment," *New Society* 43 (1978): 118–120; and Linford W. Rees, "Medical Aspects of Unemployment," *British Medical Journal* 6307 (1981): 1630–1631.

5. See Newman, *Falling from Grace,* pp. 174–201, for an excellent analysis of what happened when the Singer Sewing Machine plant in Elizabeth, New Jersey, closed and downward mobility inundated a whole community.

6. Hill, "The Psychological Impact of Unemployment"; and Rees, "Medical Aspects of Unemployment."

7. Barry Glassner, *Career Crash* (New York: Simon & Schuster, 1994) studied career crashes among baby boomer managers and professionals and provides an interesting counterpoint to the people I'm writing about here. Unlike the men and women of the working class, Glassner found that the people he studied have a range of options and a variety of resources to help cushion the blow of unemployment.

8. Women's friendships on and off the job are very different from those men form. Especially among working-class men, friendships on the job are likely to be compartmentalized and segregated from the rest of their lives. For women, however, these friendships tend to become an integral part of their social lives, therefore usually are sustained by both face-to-face and telephone interactions after they leave the job. See Lillian B. Rubin, *Just Friends: The Role of Friendship in Our Lives* (New York: Harper Perennial, 1986).

9. Rubin, *Just Friends,* p. 73. For similar findings about the fragility of male friendships, see Robert Brain, *Friends and Lovers* (New York: Basic Books, 1976); Sarah A. Haley, "Some of My Best Friends Are Dead," in William E. Kelley, ed., *Post-Traumatic Stress Disorder and the War Veteran Patient* (New York: Brunner/Mazel, 1986); Stuart Miller, *Men and Friendship* (Boston: Houghton Mifflin, 1983); and John M. Reisman, *Anatomy of Friendship* (New York: Irvington Publishers, 1979).

10. Rubin, *Just Friends,* p. 73.

11. A few researchers argue that, since the majority of men who batter their wives are gainfully employed, unemployment is of little value in explaining battering (H. Saville et al., "Sex Roles, Inequality and Spouse Abuse," *Australian and New Zealand Journal of Sociology* 17 [1981]: 83–88; and Martin D. Schwartz, "Work Status, Resource Equality, Injury and Wife Battery," *Creative Sociology* 18 [1990]: 57–61). But the evidence is much stronger in the direction of a relationship between unemployment and family violence; see Frances J. Fitch and Andre Papantonio, "Men Who Batter," *Journal of Nervous and Mental Disease* 171 (1983): 190–191; Richard J. Gelles and Murray A. Straus, "Violence in the American Family," *Journal of Social Issues* 35 (1979): 15–39; New York State Task Force on Domestic Violence, *Domestic Violence: Report to the Governor and Legislature: Families and Change* (New York: Praeger Publishers, 1984); and Suzanne K. Steinmetz, "Violence-prone Families," *Annals of the New York Academy of Sciences* 347 (1980): 251–265.

12. John A. Byles, "Violence, Alcohol Problems and Other Problems in Disintegrating Families," *Journal of Studies on Alcohol* 39 (1978): 551–553; Ronald W. Fagan, Ola W. Barnett, and John B. Patton, "Reasons for Alcohol Use in Maritally Violent Men," *Journal of Drug and Alcohol Abuse* 14 (1988): 371–392; Fitch and Papantonio, "Men Who Batter"; Kenneth E. Leonard et al., "Patterns of Alcohol Use and Physically Aggressive Behavior in Men," *Journal of Studies on Alcohol* 46 (1985): 279–282; Larry R. Livingston, "Measuring Domestic Violence in an Alcoholic Population," *Journal of Sociology and Social Welfare* 13 (1986): 934–951; Albert R. Roberts, "Substance Abuse Among Men Who Batter Their Mates," *Journal of Substance Abuse Treatment* 5 (1988): 83–87; J. M. Schuerger and N. Reigle, "Personality and Biographic Data That Characterize Men Who Abuse Their Wives," *Journal of Clinical Psychology* 44 (1988): 75–81; and Steinmetz, "Violence-prone Families."

13. Reported in the *San Francisco Chronicle,* February 14, 1992. The study found that in the same year that unemployment rose from 6.5 to 9.2 percent, there was a 30 percent increase in the number of couples seeking advice from marriage counselors about their waning sex lives.

14. Ethel Spector Person, "Sexuality as the Mainstay of Identity," *Signs* 5 (1980): 605–630.

15. An article in the *San Francisco Chronicle,* October 19, 1992, surveyed several recent studies of divorce, one of which found that when income drops 25 percent, divorce rises by more than 10 percent; another predicted ten thousand divorces for every 1 percent rise in unemployment.

16. Cited in the *San Francisco Chronicle,* October 19, 1992.
17. Unemployment benefits vary from state to state. In California, a state where benefits are among the most generous, the range is $40–230 a week for a maximum of twenty-six weeks. How much a person actually collects depends upon how long she worked and how much she earned. Even at the highest benefit level, available only to workers who have worked steadily at one of the relatively well-paid blue-collar jobs, the income loss is staggering. For workers in the lower-level jobs, for those who worked intermittently through no fault of their own, or for those who depended on the underground economy to supplement their meager wages, benefits can be so small as to be relatively meaningless.

28

Leaving Work and Staying Home: The Impact on Men of Terminating the Male Economic-Provider Role

THEODORE F. COHEN
JOHN DURST

INTRODUCTION: THE MALE ECONOMIC-PROVIDER ROLE AND AMERICAN FAMILIES

Throughout much of the history of the industrialized United States, men's adult roles have centered around their performance as workers (Bernard 1983; Kimmel 1996). In the absence of any formalized rites of passage, the transition to manhood was defined largely by taking and keeping a job. To be a man meant to earn a steady wage, the size of which helped determine one's masculine worth (Gould 1974). Without gainful employment, adult men were seen as failures in the ultimate demonstration of masculinity and manhood (Parsons 1942; Bernard 1983).

The association of masculinity with wage earning was especially evident in men's roles in their families. To be a husband or father one had to work, and one's work amounted to much more than wage-earning. Men were "breadwinners" or "providers." According to Bernard, if men were steadily employed, nearly anything else they did with or for their spouses and children was above and beyond what was expected of them. If, however, they failed at successful providing, they also failed as husbands or fathers. There was little they could do to compensate for this failure (Bernard 1983). The cultural connection between men's work and families had a structural counterpart. Paid work was designed on the assumptions that most workers were men, and that men's contributions to their families consisted of financial support earned away from families (Reskin and Padavic 1994). As a consequence, most jobs and almost all professions were structured to demand much of their (male) employees, who would then dedicate themselves to these jobs as a means of demonstrating love for their wives and children and as a method of caring for their family's instrumental needs.

There was the further assumption that men *could* so dedicate themselves because of their receipt of expressive support and domestic labor from wives. Although it sounds as if it refers mostly to middle-class professionals, Hannah Papanek's idea of "the two-person career" is a reminder of the gender interdependence implied in the male economic-provider role (Papanek 1975). To be a "really good provider" one needed freedom from responsibility or concern for the daily care of one's children and household. Such freedom came in the form of at-home wives who became supports upon which husbands' jobs were built.

According to both popular and scholarly analysis, at-home wives lived difficult and unfulfilling lives. Being at home presented women with "problems with no names" (Friedan 1963) and led energetic, talented young women to "dwindle into wives" (Bernard 1972). Because they were dependent on their spouses for economic support, women were subjected to male power and authority, derived from men's provision of the economic resources on which families lived (Blood and Wolfe 1960).

More recent scholarship continues to cast a negative light on the predicament of contemporary at-home women who are economically dependent on their husbands (Schwartz 1994). Reportedly, they bear most to all of the domestic burden, enjoy little to no control over the financial resources, and are often overlooked in the making of "big financial decisions." Additionally, they are compelled to concede to their husbands in order to maintain the relationships on which they are financially dependent. Ultimately, this dependence, like the disregard that many traditional husbands have for their wives' sacrifices, may become a source of resentment on the part of many at-home wives, leaving them less than gratified with their marriages (Schwartz 1994). In this study we look at whether *gender* spares at-home men from these experiences of their female counterparts.

Table 28.1 Labor Force of Married Women with Children 6 Years Old or Younger

1960	1970	1980	1990	1997
19%	30%	45%	59%	64%

Source: U.S. Census Bureau, 1998.

Of course, most families today do not consist of breadwinner-housewife households. Over the last four decades we have seen the erosion of the foundation upon which both the "good provider" and his at-home wife rested, and the rejection of this model as the "ideal marriage." Women's labor force participation has steadily and extensively increased to the point where a majority of married women, with spouses present and young children at home, are employed full-time away from the home (see Table 28-1). Furthermore, while estimates vary, few put the percentage of breadwinner-homemaker families at much more than 10 or 15 percent of family households in the United States (Coontz 1997; Strong, DeVault, and Sayad 1998).

With the growth in dual-earner households, there have been changes in the culture of "providing." "Co-provider families" are more common and accepted as either economic necessities or preferred household arrangements (Hood 1986; Coltrane 1997). As they became the typical arrangement within which children were raised, there followed an enlarged cultural emphasis on the need for men to expand what they do *in* (as opposed to merely *for*) their families. Ideological support for greater male involvement in maintaining households and raising children has been provided by the mass media and by many in the child development professions. As a result, whatever their levels of real participation in family life, men are urged to be fuller participants (Hochschild 1989; Gerson 1993; Rubin 1994; Coltrane 1998).

Despite such shifts in the realities and ideologies of work and family roles, culturally and structurally, the male economic provider role still lives in the way many men think about their lives. Many men continue to identify their occupations as central to their self-identities. They see their work as among the major contributions they make to their families and as one way they show love and care (Newman 1988; Rubin 1994). This comes across especially poignantly in data on men who lose their jobs or suffer significant downward mobility and find themselves grasping for some lost sense of self (Rubin 1994). Such men witness the erosion of their status in their families due to their inability to maintain their jobs and their families' accustomed standards of living (Newman 1988).

On the institutional level, many workplaces remain relatively inflexible, seemingly trapped in the notion that people who work outside the home have people at home to care for their households and families. Thus, the United States has been among the slowest countries to implement supports like flexible workplaces, parental leave, and childcare (Hewlett 1991). Informal "cultures of work" deter even those workers who find themselves in workplaces that officially offer family supports from using them (Hochschild 1997). By measuring "commitment" and productivity in terms of hours at work, such workplace cultures leave workers no better off than if they had no family supports at all (Hochschild 1997).

"ROLE REVERSERS" AND THE MALE-PROVIDER ROLE: CONCEPTS UNDER CONSTRUCTION

We are interested here in the experiences of men whose lives depart from both the "male as economic provider" and the dual-earning, "co-providing" models. Our data are drawn from interviews with a sample of couples in which men have left behind all to most of the provider role.[1] We specifically explore the range of impacts men experience as a consequence of "not providing."

The twenty families who are our focus here are supported financially by the wages of wives, while children are cared for and households are tended by at-home fathers. Thus, our sample couples represent gender reversals of the breadwinner/homemaker pattern. They are a lot like their Australian counterparts studied by Graeme Russell in the early 1980s (Russell 1987). Like Russell, we call them "role reversed," though for us this is as much a reflection of a lack of alternative terminology as it is an accurate depiction of their lifestyles and ideologies. In fact, we consistently call attention to differences between what our male informants experienced and what at-home mothers experience, to show that something other than a complete "reversal" occurs in the lives of our sample couples.

We explore the consequences for at-home men of both leaving work and staying home. In assessing such experiences we examine the following areas of impact: changes in men's economic statuses and opportunities, their relationships with their spouses and children, their social relationships outside their families, and their identities as men. Ultimately we consider whether and why men who are at-home and economically dependent upon their wives suffer or are spared the same sorts of consequences that women in traditional male-provider households suffer.

METHODOLOGY

We conducted separate but simultaneous, semi-structured interviews with the at-home husbands and wage-earning wives in twenty couples. Couples came through referrals from contacts of ours, via ads in local newspapers, or from postings at medical offices, childcare centers, schools, or large regional employers. We are well aware of the inherent limitations in this sort of sampling, but given the relative uniqueness of the populations that interest us, we deemed any sort of probability sampling too impractical. Of course, given our sampling techniques, we make no attempt to infer from such a sample to the wider population of such households.

In building our sample we used one consistent guideline to assess a couple's "fit" with our research, namely, whether the couple would be seen as relatively traditional if the husband and wife were in the other's place. By this standard, like Russell's research on Australian role-reversed couples, we included couples in which the husband contributed to the household income by generating some income from a home-based business. Unlike Russell, we had only one case in which the husband "went to work," in this case driving a school bus part-time (Russell 1987). In no case did we include the couples in which either the hours men worked or the wages they earned made them equal co-providers.

Additionally, although 80 percent of our couples arrived at their domestic arrangement by at least some choice,[2] we intentionally included five couples for whom the husband's move homeward was "reluctant," triggered by unemployment or disability. We are aware of some significant differences between couples who choose to embark on such a novel lifestyle and those who wind up there because of circumstances imposed upon them. They typically confront different issues and may have different reactions to what they face. Still, what they experience daily, and how they are seen by outsiders, makes them very much alike. At present, we are more concerned with describing the outcomes for men of *being* there than we are with the processes of *getting* there.

In other ways, our informants fit the following profile. Sample husbands were, on average, 38 years old; wives were 37. They were married an average of 11.4 years (ranging from 3 to 21 years), and most were in their first marriages. Five individuals in 4 couples were in a second or subsequent marriage. Couples had an average of 2 children, ranging in age from 3 months to 19 years old. Because this lifestyle is often selected for some childcare purpose, it is worth noting that the average age of the youngest children in these households was 3.9 years old (range = 3 months to 17) with a median of 3. Six couples had only 1 child.

Economically, couples' incomes ranged from a low of under $10,000 to a high in excess of $100,000. The median income was $54,395. On average, wives were the sources of 75 percent of their households' incomes. The only times husbands were the sources of more than half of their household's financial resources, their contributions resulted from inheritance or disability payments, not wages. At the other extreme, in eight of these households women were the sole sources of income.

In allocating wage earning, housework, and childcare, our sample couples were unusual. While the women were the sole or dominant providers for their families, the men either shared or shouldered the domestic work and childcare. Thus, their households were unlike either traditional male-provider/female-homemaker or more typical dual-earning couples. Although we look more thoroughly at their divisions of housework and childcare in later sections, note that in the majority of our couples men did most to all of the cooking, dishes, laundry, cleaning, and repairs. Of these tasks, only repairs fit what men do in more traditional divisions of labor. Even among couples where the men retained some economic activity based in the home, both spouses perceived the man's role as consisting of housework and the care and supervision of children.

Interviewing

We devised our interview strategy to ensure that couples would neither collaborate nor contaminate each other's accounts by having prior knowledge of what the other reported. Since we each asked the husbands and wives the same questions, we were able to compare what each of us had heard and locate any substantial differences in account, perception, or assessment that surfaced. We also intentionally avoided "specializing" in interviewing "only husbands" or "only wives."

We constructed and utilized an interview guide built around these six key questions: What backgrounds did these couples come from? How did they get into their household arrangements? How do they divide responsibilities? How do they feel about their arrangements? What reactions have they received from others regarding their domestic arrangements: Finally, what do they identify as the major areas of impact of their role reversals in their lives? Although the sixth question was the most directly related to our aim here, in pursuing some other concerns (e.g., how others have reacted to them) we were able to more completely identify the impact of role reversals in their lives.

Interviews ranged from around 90 minutes to over three hours; most were approximately two hours long. They consisted of a minimum of 60 questions organized into the sections and order depicted above. All interviews were tape recorded and transcribed. We compared the interviews for each spouse in a couple, and compared each couple to the other couples already interviewed. Sensitive to the possibility of obtaining conflicting and incompatible accounts, we systematically compared accounts immediately following each interview, offering each other our major impressions of key points. We identified dominant patterns and critical dimensions of variation across our sample as well as areas that were variations of what one would predict, based on the literature on work and family roles. We will identify the most significant response patterns to our questions about the impacts on men of leaving work and staying home.

THE MULTIPLE IMPACTS OF LEAVING WORK AND STAYING HOME

Economic and Career Impact: Male Sacrifice and Female Mobility

The most immediate effect of men's leaving work was an economic one. By this we mean much more than an income effect, though clearly there were financial consequences. Most of our sample couples went from being two-earner, dual-income households to either female-breadwinner or female major-provider households. However, the actual monetary impact was both less than we anticipated and difficult to quantify because of the nature of men's work experiences and the fact that their lost wages were greatly offset by the disappearance of childcare costs and other expenses couples would incur if both spouses were employed outside the home.

Few of our male informants left (by choice or circumstance) high-paying, high-mobility, "fast-track" occupations. This made their departure from paid work more manageable financially. While they were not "failures" as workers, neither were they glaring successes. On the other hand, a half dozen (30 percent) of the women were very high achievers, with their prospects now further aided by their husbands staying home. Thus, over a period of a few years, the financial effect of men's move homeward might be completely absorbed in women's increased upward mobility, which might not have happened had the men not stayed home.

As many couples noted, their arrangement saved them whatever monies they would have spent on childcare expenses. For some, this was a primary motive for their unusual arrangements, usually joined by other claims of benefits of having children home with parents as opposed to being in outside care. Among our more circumstance-driven couples, the income effect was further reduced by men receiving either some disability or unemployment payments associated with their forced move home.

The economic impact went well beyond finances. Leaving work required men to step back and often "sacrifice" their own chances for mobility. Among our choice-based reversers we found a mixture of responses. A few men perceived themselves to be taking a risk, possibly even sacrificing their potential occupational futures for the familial present. They knew that in leaving paid work for home they were suspending, perhaps forever, their own chances for upward mobility in exchange for the familial arrangement that they came to value more heavily. For example, this comment from a 39-year-old father of three suggests that he worries about his eventual ability to return to work when his children are older.

> I'm not using my time to train myself, become computer literate or anything. There is enough time in the day [so that] I probably could do something. That's what

> I've thought about lately. I need to start doing something looking three or four years down the road.

As Russell (1987) found among his sample of role-reversed fathers, men's "economic reactions" were mixed. Some men welcomed the opportunity to leave unpleasant or dangerous jobs, vacate the work role, or at least take a kind of "time out" during which they could "recharge" and "refocus." Others managed to develop new career directions or initiate cottage industries from their homes (Russell 1987).

For four of our five circumstance-based, at-home fathers, leaving work was unanticipated and unwelcome. It forced them to question and possibly surrender their provider identities. Typically, some rationalization followed, even if only a questioning, "what choice do I have?" sentiment. Such thinking was an essential ingredient in making these circumstance-induced arrangements run harmoniously. Nevertheless, even in these cases, a recognition of the positive impact of being at home and the value of being with their children helped them further neutralize their sense of having "lost" the provider role.

By staying home, men were in the equivalent economic positions of full-time housewives and were largely dependent on the wages earned by their spouses. A few men reacted to this situation in ways similar to what is reported about housewives (Friedan 1963; Bernard 1972; Schwartz 1994). For example, a 44-year-old father of three made the following comment:

> Probably one of the most difficult things for me is that I don't feel that I have my own income. . . . You know, all of our money is . . . in joint accounts. . . . I don't feel like I can make, with certain rare exceptions, unilateral decisions about spending money. You know, because it's not "my money." When I was working I generally didn't make those decisions when it came to important purchases but I could, you know . . . I felt that I could.

More typically, though, such dependency did not have the same consequences for our male informants as it reportedly has on women at home (Schwartz 1994). Unlike economically dependent housewives, most of the men were able to retain their status within the household as at least equal partners in decision making. When we inquired about decision making or areas of conflict, neither husbands nor wives described either themselves or their spouses as less than equally involved or as consistent "winners" or "losers" in marital conflicts. Furthermore, men felt neither unappreciated nor vulnerable, as housewives often feel. In Pepper Schwartz's terms, they and their spouses seemed to have "reconceptualized the goods to be provided," seeing their domestic and nurturant contributions as ways of providing, especially for their children (Schwartz 1994).

There were two economic factors that help explain why our male informants were differently affected than are housewives by their at-home statuses. First, most knew that they had economic opportunities outside the household should this arrangement fail. Thus, while they were currently dependent economically on their wives, they knew that this did not have to be the case. Second, many men were contributing something to the financial base upon which their families depended. Whether by home-based businesses (e.g., photography, computer refurbishing, writing), part-time work (e.g., as a school bus driver), or through inheritance, disability, or unemployment payments, slightly more than half of our sample generated some of the income on which their families depended.[3] Such economic contributions, though typically less than one-fourth of their household incomes, probably helped keep some men from feeling the consequences attributed to economically dependent housewives.

The Social Impact: Isolation, Stigmatization, and Support

Being at home had significant impact on men's wider social lives. The most notable effects were the social isolation and stigmatizing that men suffered as a result of their unusual lifestyles. Their isolation was more extreme than that experienced by comparably situated women, both because women draw upon more of a community of other mothers (shrinking though it may be) and because men typically derive much of their social lives and daily interaction from work. Having left work and now absent a workplace, men were without the structured, daily contact around which men socialize, and where, as one 43-year-old father of two said, "these sorts of pure relationships develop." For example:

> I miss shootin' the breeze. I miss . . . being in an NCAA [basketball] pool . . . that I kinda miss . . . I mean, I love sports. . . . And, too, . . . guys [at work] will be sittin' around on break time talkin' about "I went to see this movie, I went to see that movie," I mean obviously it [his situation] is limiting socially. (39-year-old father of one)

Additionally, being at home and responsible for children meant that when opportunities arose for interacting with other parents of young children they usually consisted of being the only father among a group of mothers. For some fathers this was so uncomfortable that they avoided such outside contact, like play-groups or parenting co-operatives. But even if men felt comfortable mixing with neighborhood mothers and their children, they sometimes felt as if their presence bothered the women with whom they mixed.

> The first time—Yeah, I felt a little funny. I was just self-conscious, but not only about being the only dad but . . . with a beard and longer hair. . . . I felt like there was a certain moment of . . . for a little while . . . they all had to get used to me. (39-year-old father of two)

It is impossible to tell with certainty whether men exaggerated the discomfort of others or projected their own sense of uneasiness onto women who actually may have been more welcoming of their involvement. In fact, it doesn't matter, because they acted on what they perceived to be the likely discomfort that others might feel toward them. It is also the case that a few men thought that they were "better received" because they were men than their wives would be in similar circumstances.

The at-home lifestyle imposed other restrictions on facets of men's social lives and time for peer involvement. Although some men commented that being at home gave them more time, often they found that their time did not easily correspond to either the amount of time friends had for social relationships or the specific hours they and their friends were each available. Couple friendships were not abundant either. Some of this was due to the lifestyles they had created for themselves, since those lifestyles put them into a fundamentally different gendered situation than those experienced by most couples they knew. This was well illustrated in the following comment from a 32-year-old sales representative and mother of two:

> There are friends [from his old job] that don't understand it and they're distancing themselves, especially since he quit. . . . And I don't know if it's because he's not in that clique anymore . . . or if it's the situation. . . . It's never really brought up. It's really strange, but when . . . we get together . . . and all hang out . . . the guys [are] talking sports and all the girls [are] talking about how, "Oh, I went clothes shopping for Taylor the other day," and I've never gotten into that. So I get really lost and just start talking sports, and I inevitably find my husband coming over and talking recipes with the women. And he's so natural about it! . . . He's very comfortable with it, and I'm comfortable with it. . . . I just wonder, though, I mean they really don't call us as much anymore.

Compounding this, couples become greedy of their weekend and evening time. There was a slight compensatory nature to this as women sought to maximize their family time, and couples attempted to spend quality couple time whenever they could be together. Raising children is a labor-intensive, time-depleting activity. Even among those in more conventional lifestyles, there is often a withdrawal from active socializing as time and energy grow scarce and priorities are reshuffled. But our couples were not conventional. In choosing these arrangements, they were emphasizing the value they attach to raising their children. Thus, the impact of their lifestyles on their social ties was greater than what their life stage alone would otherwise create.

Scrutinizing and Stigmatizing All of our sample couples told stories about some reactions that their lifestyles generated. These reactions ranged from positive to negative, from slight to strong, and consisted of everything from curiosity, to insensitivity, to criticism. Such reactions came from a variety of sources, including family, friends, co-workers, former co-workers, neighbors, and service people. Informants indicated that people struggled to "figure them out" and understand their unusual lifestyles. The following two examples illustrate the kinds of commentary. Describing his in-laws, a 44-year-old father of three stated:

> I think they can understand it to a point and I think they accept it. . . . They're not gonna be congratulatory about the decisions that we made . . . but I'm not expecting any accolades here. . . . On the other hand, I'm not expecting any criticism. . . . I don't know, maybe I'm imagining this, but I sense a certain amount of disapproval. . . . And it's been very subtle. . . . I mean almost every time we see them it's, "Well, have you been looking for work? . . . Are you planning on going back to work?" . . . Nobody ever says, "Well, what's your problem?" But you kind of sense maybe they are thinking of asking you that question. (44-year-old father of three)

Similarly, a 32-year-old woman told this story about an encounter her husband had at a party.

> [He] was just talking to a bunch of people and they were all talking about what they did, and they said, "Yeah, he [husband] used to work at—." And this ditz says to him, "What do you do now?" And he says—oh, what does he call himself now?—I think he said, "I'm a full-time househusband." And she went, "Oh . . . how nice!" And then, later on, she came back to him and said, "So what *do* you do? What exactly do you do?" . . . And he says, "I stay home full-time and raise my kids." And she just didn't get it. . . . People don't. . . . My mom is really still confused about this whole thing.

As these examples show, it was the men who drew the attention and received most of the reactions from curious or even somewhat critical others. After all, our female informants were not visibly different in either what they did or where they did it than other employed women with children. The same was obviously not true for the men, who were very apparent in their departures from convention.

Support and Pedestalization For the most part, the stigmatizing remained relatively mild and was counterbalanced by numerous stories of support. Parents and friends were the most likely others to be described as "supportive." Parents and parents-in-law, even when ambivalent about some aspects of their children's lifestyles, were repeatedly described as "appreciating" the care that their grandchildren were receiving from their stay-at-home fathers. Furthermore, whereas friends' support was restricted to acceptance, parental support often included assistance with daily family life.

A 34-year-old father of three described his parents and in-laws as particularly supportive, especially when compared with outsiders.

> They . . . say that they never imagined me doing what I'm doing . . . they never thought that I would just give it all up and stay home with the kids. My mother-in-law, she watches me with the kids and she says, "I've never seen a man with the finesse of doing—if I didn't know any better, I would think you were like a woman." My mother tells me the same thing. She's very supportive, my father's very supportive . . . my sister was supportive. . . . But the outside world—I get looked at like I'm a divorced parent and it's my

weekend with the children . . . like I'm very inexperienced. . . . And in the male community—"Aren't you gonna get a job? Don't ya think you should get a job?"

Some of the social support couples described was clearly "gendered." Numerous wives told us that they had heard some version of the following kind of comment, especially from co-workers: "He's so wonderful; you are *so* lucky." Seeing their husbands "put on pedestals," mostly by other women, for doing what women traditionally did made some wives ambivalent. They were proud of their husbands and they appreciated the sacrifices and risks their husbands were undertaking, but their lifestyles were built on both his and her contributions. Greater external recognition of that would have lessened the women's ambivalence (Russell 1987).

Compared to reactions received by housewives, our male informants confronted some obvious differences. At-home mothers are rarely the recipients of glorification. Taken for granted, even dismissed or condescended to, they may come to feel defensive about their role. At-home men, who may feel defensive even without direct cause, become the recipients of praise, while their wives can become the target of envy from female co-workers. Furthermore, while staying home often renders full-time housewives invisible, it drew consistent attention to the similarly situated men. To us, this suggests the importance of gender in determining the effects of lifestyles that seem to disregard or dismiss gender as a role-determining factor.

The Impact on Marital Relationships

As noted earlier, most of our male informants fell outside conventional patterns of housework and childcare (Baxter 1997). On top of the more "masculine" chores (e.g., home repairs) they did most of the daily cooking and participated in much of the cleaning, the shopping, and the laundry. To determine "who does what" in terms of household labor, we asked informants to consider seven specific household tasks: cooking, washing dishes, shopping, doing laundry, cleaning the house or apartment, paying bills, and home maintenance and repair. We asked whether they or their spouses did "all or most" of that task or whether responsibility was "shared." In Table 28.2, we show the percentage of husbands and wives who place responsibility for each task into one of the following categories: "all or most done by husband," "shared," and "all or most done by wife." These couples do not divide these chores in the expected and more common manner.

In a majority of sample couples, men assumed most to all responsibility for the cooking, laundry, cleaning, and repairs. Furthermore, adding those who "shared" to those couples in which men did "all to most" of the specific task reveals that a majority of men at least shared in the daily domestic tasks of cooking, dishwashing, shopping, laundry, and cleaning. This is in stark contrast to recent survey data on men's participation in housework among dual-earner couples, which shows men in the United States contributing between 20 to 30 percent of the daily cooking, cleaning up after meals, laundry, housekeeping, and grocery shopping (Baxter 1997). Men in dual-earner couples in the United States, Sweden, Norway, Canada, and Australia contribute between 25 to 28 percent (Baxter 1997).

It is important to point out that the pathways couples took to their unconventional lifestyle were associated with their resulting divisions of household labor. Whether a man was home by choice or circumstance shaped both how much he did in the home and what he thought about being home (Russell 1987). In general, the more reluctant at-home fathers tended to do less than those men home by choice. Their lower levels of participation skew the above data in such a way as to understate the consistently high levels of male participation within the other fifteen couples. If one looks only at the fifteen men who were home by choice, a majority of husbands were credited with responsibility for "all or most" of each domestic task by both the men and women we interviewed.

As impressed as we were with the patterns depicted above, men's experiences staying home are not adequately conveyed by frequencies or percentages. The narrative data tell a stronger story about the breadth and depth of men's involvement

Table 28.2 Percent of Informants Reporting on Responsibility for Domestic Tasks

	Cooking		Dishes		Shopping		Laundry		Cleaning		Repairs		Bills	
	Husbands	Wives	H	W	H	W	H	W	H	W	H	W	H	W
All/most done by husband	75%	75%	50	40	45	50	55	50	55	50	85	45	30	30
Shared	5%	15%	40	30	25	20	5	5	20	20	10	20	15	10
All/most done by wife	20%	10%	5	30	30	30	40	45	25	30	5	0*	55	60

*A number of women said that "neither" they nor their husbands was responsible for repairs.

in day-to-day domestic life. Such stories are vividly represented by this comment from a 34-year-old at-home father. Describing the domestic and childcare responsibilities that make up his day, he reported:

> I've got my schedule where I start at that corner where the coffee pot is and I work my way around. If it gets dirty again, it's there until tomorrow. My dogs have a shedding problem. . . . So vacuuming might be two times a day, all right? [In] the morning . . . I feed [my two daughters] then [my son] gets up. . . . I change him, I get him down here—I feed [him]. . . . I kiss [my older daughter] good-bye and [she's off] to the bus stop. . . . [My other daughter] sometimes doesn't like to eat with the rest of them . . . so . . . I feed her whatever she wants. I kind of run my house like a restaurant, meaning you don't have to have exactly what I'm making. . . . Then there's play time for [son], there's play time for [other daughter], then there's her school time where she's learning how to trace her letters and doing that—or . . . she does the computer. Then I go through my routine of cleaning and straightening. . . . There's everyday cleaning and then there's the cleaning that you don't do every day, like taking the curtains down and cleaning the curtains. People look at me, when I say these things, and they're like, "Are you an ordinary guy?" I'm like, "Hey, it's gotta be done—someone's gotta do it." . . . My wife thinks I'm crazy for taking on so much responsibility. In one day you can hear, "Daddy, I'm thirsty." Then my youngest [starts] crying. Then one of the dogs starts barking because she wants to go outside, so she'll bark all day long, wanting to play. . . . And then the birds will run out of food and . . . they'll start screaming, [but] I don't hear it as screaming or barking or crying, I hear it all as "Daddy . . . Daddy, Daddy, *Daddy*!"

We were also struck by the consensus between husbands and wives over the division of labor. Based on a long and consistent history of evidence depicting the gender-duality of marital experience, we expected to receive two substantially different accounts from each of our role-reversed couples about how their arrangements affected their marital relationships. We most anticipated hearing divergent depictions of the division of labor. We didn't. Men and women both spoke of husbands doing the vast bulk of the household chores. Using the same categories as reported above ("All or most husband," "shared," and "all or most wife"), there was surprisingly widespread agreement within each couple as to who bore the responsibility for the various domestic tasks. Looking only at the five most clearly "domestic tasks" (see Table 28.3), levels of within-couple agreement were highest regarding shopping and the laundry. Only one husband and wife reported different answers. Levels of agreement remained high across the range of domestic tasks.

The division of labor was only one example among many of strong consistency between husbands' and wives' versions of familial reality. It is noteworthy because it is often an area where disparate accounts are obtained and is used to indicate the phenomenon of "his" and "her"

Table 28.3

Task	Number of Couples in Which Both Spouses Gave Same Answer	Percent of Sample with Within-Couple Consensus
Shopping	19	95%
Laundry	19	95%
Cleaning	18	90%
Cooking	16	80%
Dishes	14	70%

marriages. Thus, these couples run counter to the larger trend, both in their perception of having created a more equitable arrangement of paid work, housework, and childcare, and in their construction of a shared marital account. In telling us their stories, recounting their histories, identifying significant turning points, or assessing what impact the role reversal had on them, our sample couples consistently revealed surprising consensus. We believe that this consensus resulted from both the ways our couples live their lives and the ways we collected our data.

This lifestyle is sufficiently novel that it requires substantial amounts of contemplation and conversation. Choice-based reversers, especially, need to make sure that they can afford to live on only the wife's income, and that each can live comfortably with their somewhat unusual role responsibilities. This not only forces them to talk things over before they "switch places" but engenders much checking as to how each is experiencing the arrangement. Of course, our interviewing strategy also contributed to the widely overlapping accounts, because it kept each spouse from engaging in excessive distortion. Each knew that as she or he was answering our questions, the spouse was answering them in a nearby room.

There is more, though, than just agreement reflected in couples' accounts. Spouses displayed considerable empathy for each other's sacrifices and responsibilities. We believe that such empathic understanding arose from the fact that each has lived the other's life, enabling them to respect each other's sacrifices and struggles. Breadwinning wives not only appreciate the risks their husbands take by being home, but they appreciate what being home means in a way that breadwinning men likely never do. Likewise, at-home men know about workday pressures and the burdens of breadwinning because they've done it. Thus, each spouse possessed a sensitivity to the daily life of the other that allowed them to share with us the nuances of each other's experiences. Such insight might otherwise be lost to spouses who lack the experiences that breed empathy. The marital impact among our couples was more consistently positive than reported elsewhere in the scant literature on such household types (Russell 1987).

Significantly, our role-reversed couples were not just traditional marriages in reverse. Although the literature is filled with some stark accounts of marriage for at-home women, we did not witness such outcomes for at-home men. Thus, something more than the burdens of housework and childcare, or other than the privileges that follow provision of economic resources, was at work. Once again, we believe this "something more" is best identified as gender. Both women and men in our sample couples were aware that the man's masculine identity was threatened by his at-home status and needed to be protected and preserved in other meaningful ways. Wives actively buffered their husbands from criticism and bolstered their male identities. They went out of their way to acknowledge what their husbands were doing and express appreciation. Such considerations prevented our at-home fathers from experiencing the costs that otherwise occur for housewives and at-home mothers.

Having husbands at home, however, allowed women to experience some of the benefits that more typically go to male providers. With husbands tending to many of the daily needs of the household and children, women had more time and energy to devote to their own occupational success. Moreover, they had greater legitimacy in doing the things that accompany upward mobility and which make prosperity more likely. They could work late, travel for business, and count on returning home to households where they would not be buried beneath the burdensome "second shift." Thus, these arrangements mimicked much of the "two-person career" (Papanek 1975; Hunt

Table 28.4 Percent of Informants Reporting on Responsibility for Childcare Tasks

	Bathing		Bedtime		Reading		Play		Transporting		School		Social	
	Husbands	Wives	H	W	H	W	H	W	H	W	H	W	H	W
All/most done by husband	19%	23%	17	18	7	6	74	57	86	65	18	40	27	38
Shared	44%	46%	44	35	57	56	26	36	7	9	55	30	45	25
All/most done by wife	37%	31%	39	47	32	38	0	7	7	6	27	30	27	38

and Hunt 1982) so often essential to men's occupational success and so rare in women's experiences of work and family.

Although many women enjoyed the freedom and status that their work and wage earning gave them, these did not fully displace their sense of themselves as mothers and wives. Most women continued to make substantial efforts to involve themselves with their children and to express their support and gratitude to their husbands. Women's intentional involvement at home and expressed appreciation are often lacking from males who are sole breadwinners. Again, this indicates their continued awareness and the persistent relevance of gender.

Men as At-Home Fathers: The Impact on Relationships with Children

Perhaps the single biggest impact of men's at-home status was in their relationships with their children. Uniformly, whether home by choice or circumstance, men perceived their relationships with their children as deeper and stronger than what they would have been had they been co-providers. This sentiment is not necessarily reflected in the data we collected on responsibility for children, but that is more a reflection of the mix of ages of children than it is a lack of father's responsibility.

As with household tasks above, we inquired about seven specific childcare activities or tasks: bathing children, monitoring and handling children's bedtime routines, reading with and playing with children, chauffeuring kids, participating in school activities, and overseeing children's social lives. Unfortunately, because of the ages of the children our childcare data are less complete than we had hoped. A number of the aforementioned activities are age-specific and, therefore, are not relevant for every sample couple (e.g., giving baths or putting children to bed). Thus, none of the categories in Table 28.4 contains data from more than sixteen couples.

In comparing our childcare and housework data, a couple of noteworthy differences surfaced. First, most husbands and wives reported childcare activities as shared activities (Russell 1987). For the tasks of bathing, reading, and involvement in school and social activities, the most common response category among both husbands and wives was "shared." Second, for most of those same items, the second most common response category was "all or mostly wives" (in the case of putting children to bed, these two outcomes were reversed).

These two patterns reveal that women retained greater independent involvement in childcare tasks than they did in household tasks. As we explored the narrative accounts, we found that this resulted more from women wanting to remain connected to their children than from men's failure or unwillingness to care for children. In fact, looking at the amount of time children spend at home with only one parent, we estimate that our sample fathers were responsible for 75 percent of the sole parenting. Although the specific aforementioned tasks were more likely shared or performed by mothers, most of the time children were home they were under the care and supervision of their fathers (Russell 1987).

Once prompted to think about specific (lifestyle) effects on their children, couples emphasized three clearly beneficial outcomes. All informants pointed to a substantially increased closeness between fathers and children that would be impossible without dad at home (Russell 1987).

They further thought that their children benefited by being home with a parent as opposed to spending much of their time in outside childcare. Finally, most couples also acknowledged the likely role model effects that might come from having a nurturing, domestically competent, at-home father and an employed mother.

Men talked often about how much closer their children were to them than they had been to their fathers. This is in keeping with the observation that men are seeking to be more nurturant than the fathers they themselves had (Daly 1993; Hawkins et al. 1993). As in this comment from a 42-year-old father of two, these men came to recognize that their relationships with their children were deeper and closer than the typical father-child relationship.

> I'm a heck of a lot closer to my kids. I've said many times to many people that I think every male should have to take some time, a three- to five-year period, saying, "This is your job now—you stay at home, you cook, you clean, you take care of the kids, you know, you do all that." . . . I think the kids and I, I know we're definitely much closer. . . . And I think that's good.

Even the most reluctant, circumstance-induced reversals left men aware and appreciative of the subsequent closeness they found in their relationships with their children. This comment from a 39-year-old father of two, forced out of work by a disabling injury, reflects that well.

> I think the most positive thing about this is spending time with the kids. I . . . say that because I look back and see what I missed with [my oldest]. I mean, I didn't hardly ever see him . . . and I can see everything that I missed in it. . . . If I could do things over again . . . if I could choose to stay home, I would have. . . . I didn't know what I was missing.

Informants expressed the additional perception that children are better off not being cared for by "strangers," not being in day care, and having consistent parental care at home. These are all identified as benefits. For example:

> I personally think me staying home is much more positive than them being with a babysitter. . . . I mean, we're not communicating with the kids through a babysitter or a day care. . . . We know what they do during the day. . . . I'm not saying that you can't be close with your kids and, you know, still be [a] two-job family and all that, but it is different. . . . I guess it's just much more difficult when both parents are working full-time . . . to know what [your kids] are always doing, and to let them know there is always someone there . . . that they can talk to. (39-year-old father of two)

In most of the more choice-based reversals, this was the dominating motivation behind men's decisions to be at home. A 34-year-old father of three recalls:

> [To] this day, I don't believe in day care. . . . I used to go to work around 2 P.M. [and] I would drop my daughter off [at a day care center] before then. And what really broke my heart, I would say what created me [as an at-home father]—she was in the room, on the other side of the glass, and I was watching her do things. And I watched her open her arms and go to this complete stranger, and I said, "Is this why I had children? Is this why my wife had children? All these people with children putting them in day care, are they missing something here?" And that's when I started going to work at 4:00, and then 5:00, and then . . . I stopped going to work.

Couples also saw the possible role model effects of the combination of at-home, caregiving fathers and successful, wage-earning mothers as beneficial, even if unrealistic. Parents of daughters noted, for example, that they might be "spoiling their girls" by "raising their expectations too high" in terms of what they could expect in their own marriages and families later. Parents of sons recognized that their sons saw a different kind of fathering, one that involved men in nurturing positions.

> I think I am nurturing. I'm there for the kids when they need me, and I think it's important for them to see me in a nontraditional role and to know that they can also achieve in a fashion that is not necessarily—that they're not restricted in terms of what they can do, or what they choose not to do, for that matter. (44-year-old at-home father of three)

These various parental outcomes were frequently the source of men's willingness to enter these domestic arrangements, their commitment to remain in them, or their acceptance of having to be at home. Even when it was the only bright spot in an otherwise gloomy reaction for those most reluctant "forced reversers," all the men we interviewed identified positive consequences in their relationships with their children.

Being a father at home also led the men to alter their perceptions of being fathers and men. They clearly define the essence of "providing" much more broadly than economic provision. They provide care, consistency, and a role model for their children, and they provide support for their wives. At the same time, they have embraced a more balanced and shared notion of gender. To them, being a man includes caring and nurturing qualities. Although they express some uncertainty about the impact of their being at home on their children, they maintain a belief in its ultimate benefits.

The Impact on Self: Reshuffling Priorities and Restructuring Identities

Being at-home fathers also affected how men saw themselves. After all, in a society where most men work, they don't. Whereas most other men derived their self-identities in large part from their jobs, they can't. Finally, there are few men that they know who are like them and are home with children. Thus, they needed to reconstruct a sense of self that was different both from how they saw themselves before they left work and from what they thought most men were like. Finally, they had to reconfigure the priorities in their lives, putting their families, especially their children, at the center of who they were, what they were doing, and why they were doing it. The shift in priorities is nicely suggested in the following comment from a 37-year-old father of four:

> This is for all those guys out there who are working too doggone much, okay? Kids are unimpressed with your office life, okay? They don't care. They want your time and they want your attention. . . . Your kids do not look at you and say, "Oh, here's the guy that brings the shoes and clothes and feeds us." That is not what they see. . . . Even though I knew that I was working for my family [before], and that was the main thing, it has really brought home to me that they *are* more important than anything else.

Repeatedly, men commented that being home with the children had changed *them*. In acknowledging such personal consequences, men included changes in how they look at life and at what's important. A 39-year-old father of two reported:

> I'm better at recognizing things that need to be done . . . at seeing things and acting on it. I think it's a huge exercise in patience and compromise and everything else. . . . Just being able to handle more, emotionally and physically, than I was used to . . . I don't know, it just made me better. I appreciate things more.

In the same vein, others commented on becoming "more understanding," "feeling more confident" in their nurturing abilities, and "finding a balance" since leaving work and staying home. Thus, not only their priorities but their personalities were perceived to be affected by their move homeward.[4]

Being at home also meant that men had to terminate their provider identities much as they had ceased their providing activities, and replace them with an increased emphasis on their identities as parents. In becoming highly involved fathers they reexamined some of the earlier ideas they had of men, work, and family.

> I see myself in a more positive light than I saw myself ten years ago. Ten years ago I was worried about what was I gonna do as a "man." What was I gonna do when I grew up, and you have all those pressures on you—you have to have a career, you have to do things. And somewhere along the line I made a change where it doesn't matter what car you drive, it doesn't matter [if you rent]. . . . I've owned a house, I've done that and we want to do that again. But now every move, in my mind, that my wife and I make—it's "How does this affect [the children]? And that's all that matters. (34-year-old father of three)

There were a few men, especially but not exclusively those driven into role reversals by circumstances like disability or unemployment, who expressed ambivalence about their status. No longer primary breadwinners, they had neither easily surrendered the idea that they should provide, nor completely redefined themselves around their roles as nurturers. One reluctant at-home father put it bluntly when he reported, "Let me be perfectly clear. The only reason I am home, the *only* reason, is because I hurt my back." There were also a few men who, though clearly home by choice, nonetheless expressed a desire conveyed through one man's comment that,

> I do feel like I . . . would like to be able to support everybody and be the traditional "Big Daddy," but I also know that I'd be losing doing some things that I wouldn't want to lose. . . . I may have my time down the road anyway, so I don't know. . . . I'd like to have my shot at being the big financial provider. (39-year-old father of two)

Comments such as these indicate that if the male as economic-provider role is dead in how these couples structure their lives, its specter occasionally haunts men who remain aware of its longstanding cultural dominance and wrestle with its mandates. It must be stressed, however, that comments such as these surfaced in some fashion in fewer than a third of our interviews with at-home fathers.

Of the five areas of impact, the personal impact seems most related to and influenced by the others. For example, the consequences of being home for a man's sense of self result from his withdrawal from work, the role he then takes in his marriage, the social reactions he receives, and the relationship he develops with his children. Although there are interconnections, too, among the other four impact areas, no other one is as much the product of all of the other four.

CONCLUSION

Four key points stand out. First, leaving work to be at home full-time brings men a variety of consequences, some negative and others quite positive. Men sacrifice their own economic identities and career success, suffer reduced contact with friends, and encounter a variety of social reactions that make them stand out or open them to the commentary, scrutiny, and stigmatizing of others. So why do it? In addition to these less than positive impacts are the relationship consequences that informants unambiguously identified as positive. This is especially true regarding their relationships with their children, but their marital relationships also become enriched by the high levels of empathy and support that each spouse provides the other. Thus, they make a trade-off between sacrifices and gains. Although the sacrifices may be more numerous, the couples we spoke with attach greater value to the rewards associated with their domestic arrangements.

Second, it is possible to sort out the effects of being at home on men in terms of whether they result more from leaving work or from staying home. Of course, this imposes somewhat arbitrary and artificial dimensions on our attempt to understand men's experiences. Since men are only home because they've left work, it may be impossible to ascertain which step wrought which outcome. To us, it seemed analytically worthwhile to suggest what we believed to be the greater source of what these men and women reported.

We believe that leaving work was responsible for the economic impact discussed above as well as much of the social impact, especially the sense of isolation and some of the stigmatizing. Surrendering or losing one's work and the wages it generated puts one's economic future in some doubt and separates men from many of their former social contacts. It also leads outsiders to question or scrutinize what these men were doing and/or why they were doing it that way. Some of what resulted within their relationships with their wives also stemmed from having left work and being dependent upon their spouses. However, unlike what others (e.g., Schwartz 1994) report of the deleterious consequences for housewives in households structured around an economic provider–homemaker division of labor, outcomes for our sample were much more benign. In fact, some of the empathy otherwise associated with peer marriages grew within these relationships because men were home. They knew and appreciated what their wives had taken on, and women recognized and protected their husbands from the gender judgments associated with losing or surrendering their provider roles.

Being home and taking responsibility for households and childcare had other specific consequences. The most notable are the enlarged father-child relationships, the high level of male domestic involvement, and some of the social consequences, including being scrutinized and stigmatized, but also supported and pedestalized. Furthermore, being home caused men to reshuffle their priorities and develop a new sense of identity.

Third, behind the impact of both leaving work and staying home is a restructuring of the temporal dimension of men's lives (Daly 1996). By leaving the highly structured time of the workplace for the more flexibly structured relational time of the household, men have a time structure that allows them to further reorganize their social lives. This made possible many of the more specific consequences depicted above. Being at home while their spouses shouldered most of the economic provision gave men the time to alter or leave the provider role, develop greater support and rapport within their marriages, and be with their children, where they could develop a keener sense of what children are like or what they need. Leaving work and staying home also gave men more time to reconstruct their sense of identity and reconfigure their priorities and use of time (Daly 1996). How men used their time at home varied, but clearly there was a greater temporal opportunity in being at home than is available to the more typical male co-provider or the traditional male economic provider.

Fourth, as we repeatedly noted, this work illustrates the ongoing influence and relative importance of gender, even within households that appear to defy gender considerations. In comparing the situations of at-home fathers to that of at-home mothers, some clear similarities and differences surface. The similarities indicate the importance of being at home, while the differences highlight the significance of being *men* who happen to be at home. Said differently, the similarities between at-home men and housewives indicate areas where gender matters less, while the differences between them suggest areas in which gender matters more.

Both at-home mothers and fathers experience some curtailment of their outside social and economic lives, receive some dismissive or judgmental reactions, and take on most of the unceasing responsibility for daily housework and childcare. However, the differences in their experiences are more significant. In particular, their statuses within their marriages and households and in the eyes of outsiders are quite different, despite the fact that their lives are structurally similar. This has important implications, especially in what it says about the relative importance of situations and of gender.

As we saw from our interviews, men's lives can be remade by the situations in which they find themselves. By being at home, by choice or circumstance, men developed deeper and more intimate relationships within their families and a greater capacity for nurturance than they claimed otherwise would have been possible. These outcomes resulted despite the absence of much earlier socialization toward such lifestyles and in the absence of any real role models to emulate. Like other research (e.g., Risman 1989; Gerson 1993), this suggests that the situations we encounter have the possibility of redirecting our lives in ways that our gender socialization did not anticipate. Thus, along with whatever personal qualities men possessed that might predispose them to "moving homeward," men developed certain qualities because they are at home.

However, the fact that their experiences are so different from those experienced by full-time housewives suggests that neither their socialization nor their situations are sufficient to account for the experiences they have. Of equal importance is their gender. The pedestalizing and scrutinizing that outsiders do of these at-home men, the appreciation displayed by wives for their husbands' sacrifices and contributions, the extent wives go to involve husbands in financial and other decision making and to protect their male identities, and the roles men continue to play in the running of their households, result because they are men. Thus, although in some ways they live their lives "outside" of or "beyond" gender, they live inside a very gendered world, retain recognition of that fact, and are affected by how they and others "do gender" (West and Zimmerman 1987).

NOTES

1. The data here are drawn from a larger study of forty couples. Our larger sample includes the twenty role-reversed couples and an additional twenty couples in which husbands and wives are employed outside the home in opposite, mostly nonoverlapping shifts.
2. By this we mean that five of twenty couples were pressed into their arrangement by circumstances like job loss or disability that left them no other choice. All other couples either chose freely to switch places, or saw such a switch as the optimum alternative given circumstances that confronted them (e.g., a child with chronic medical problems).
3. In eight of these households men contributed none of the income.
4. Personality characteristics operate in the other direction, too. Although we attempted no specific formal personality assessments, our narrative data suggest that being at home and out of the more public environment at work fits well with the kinds of social and personal characteristics many of these men reported of themselves. As a group they were not highly sociable, depicting themselves as having "a few friends" or "not too many," and as not needing to be around people all the time.

REFERENCES

Baxter, J. "Gender Equality and Participation in Housework: A Cross-National Perspective." *Journal of Comparative Family Studies* 28 (Autumn 1997): 220–248.

Bernard, J. *The Future of Marriage.* New York: Bantam, 1972.

Blood, R., and D. Wolfe. *Husbands and Wives.* New York: The Free Press, 1960.

Bernard, J. "The Good Provider Role: Its Rise and Fall." In *Family in Transition,* edited by A. and J. Skolnick, 125–144. Boston: Little, Brown, 1983.

Coltrane, S. *Family Man: Fatherhood, Housework, and Gender Equality.* New York: Oxford University Press, 1997.

Coltrane, S. *Gender and Families.* Thousand Oaks, CA: Pine Forge Press, 1998.

Coontz, S. *The Way We Really Are: Coming to Terms with America's Changing Families.* New York: Basic Books, 1997.

Daly, K. *Families and Time: Keeping Pace in a Hurried Culture.* Thousand Oaks, CA: Sage, 1996.

Daly, K. "Reshaping Fatherhood: Finding the Models." *Journal of Family Issues* 14, no. 4 (1993): 510–530.

Friedan, B. *The Feminine Mystique.* New York: W. W. Norton, 1963.

Gerson, K. *No Man's Land: Men's Changing Commitment to Family and Work.* New York: Basic Books, 1993.

Gould, R. "Measuring Masculinity by the Size of the Paycheck." In *Men and Masculinity,* edited by J. H. Pleck and J. Sawyer, 96–100. Englewood Cliffs, NJ: Prentice Hall, 1974.

Hawkins, A., and S. Christiansen, K. Sargent, and E. Jeffrey Hill, "Rethinking Fathers' Involvement in Childcare: A Developmental Perspective." *Journal of Family Issues* 14, no. 4 (1993): 531–539.

Hewlett, S. *When the Bough Breaks: The Cost of Neglecting Our Children*. New York: Basic Books, 1991.

Hochschild, A. *The Second Shift*. New York: Viking, 1989.

Hochschild, A. *The Time Bind*. New York: Metropolitan Books, 1997.

Hood, J. "The Provider Role: Its Meaning and Measurement." *Journal of Marriage and the Family* 48 (May 1986): 349–359.

Hunt, J., and L. Hunt. "The Dualities of Careers and Families: New Integrations or New Polarizations?" *Social Problems* 29, no. 5 (June 1982): 499–510.

Kimmel, M. *Manhood in America: A Cultural History*. New York: The Free Press, 1996.

Newman, K. *Falling from Grace*. New York: The Free Press, 1988.

Papanek, H. "Men, Women, and Work: Reflections on the Two-Person Career." *American Journal of Sociology* 78 (January 1975): 853–872.

Parsons, T. "Age and Sex in the Social Structure." *American Sociological Review* 7 (October 1942): 604–616.

Risman, B. "Can Men Mother? Life as a Single Father." In *Gender in Intimate Relationships: A Microstructural Approach,* edited by B. Risman and P. Schwartz, 155–164. Belmont, CA: Wadsworth, 1989.

Rubin, L. *Families on the Fault Line: America's Working Class Speaks About the Family, Economy, and Ethnicity.* New York: HarperCollins, 1994.

Russell, G. "Problems in Role Reversed Families." In *Reassessing Fatherhood,* edited by C. Lewis and M. O'Brien. London: Sage, 1987.

Schwartz, P. *Peer Marriage.* New York: The Free Press, 1994.

Strong, B., C. DeVault, and B. Sayad. *The Marriage and Family Experience.* (7th ed.) Belmont, CA: Wadsworth, 1998.

U.S. Bureau of the Census. *Statistical Abstract of the United States,* Table #654. Washington, D.C.: 1998.

West, C., and D. Zimmerman. "Doing Gender." *Gender & Society* 1 (1987): 125–151.

29

The Meaning and Significance of Race: Employers and Inner-City Workers

WILLIAM JULIUS WILSON

Blacks reside in neighborhoods and are engaged in social networks and households that are less conducive to employment than those of other ethnic and racial groups in the inner city. In the eyes of employers in metropolitan Chicago, these differences render inner-city blacks less desirable as workers, and therefore many are reluctant to hire them. The degree to which this perception is based on racial bias or represents an objective assessment of worker qualifications is not easy to determine. Although empirical studies on race and employer attitudes are limited, the available research does suggest that African-Americans, more than any other major racial or ethnic group, face negative employer perceptions about their qualifications and their work ethic.

The Urban Poverty and Family Life Study's survey of a representative sample of Chicago-area employers . . . indicates that many consider inner-city workers—especially young black males—to be uneducated, unstable, uncooperative, and dishonest. Furthermore, racial stereotyping is greater among employers with lower proportions of blacks in their workforce—especially blue-collar employers, who tend to stress the importance of qualities, such as work attitudes, that are difficult to measure in a job interview.

The survey featured face-to-face interviews with employers representing 179 firms in the city of Chicago and in surrounding Cook County that provided entry-level jobs. The sample is representative of the distribution of employment by industry and firm size in the county. The survey included a number of open-ended questions concerning employer perceptions of inner-city workers that yielded views concerning job skills, basic skills, work ethic, dependability, attitudes, and interpersonal skills. Of the 170 employers who provided comments on one or more of these traits, 126 (or 74 percent) expressed views of inner-city blacks that were coded as "negative"—that is, they expressed views (whether in terms of environmental or neighborhood influences, family influences, or personal characteristics) asserting that inner-city black workers—especially black males—bring to the workplace traits, including level of training and education, that negatively affect their job performance. . . .

The personnel manager of a suburban bakery stated: "We have some problems with blacks. . . . I find that the blacks aren't as hard workers as the Hispanics and—or the Italian or whatever. Their ethic is much different where they have more of the pride. The black kind of has a, you-owe-me kind of an attitude." . . .

A manufacturer introduced a class distinction with this blunt assertion about the work ethic among blacks: "The black work ethic. There's no work ethic. At least at the unskilled. I'm sure with the skilled . . . as you go up, it's a lot different." A more nuanced discussion of the work ethic among blacks in terms of economic class status was provided by the director of an inner-city human resources firm:

> The question is, Is there a difference in the work ethic? . . . I see a tremendous amount of difference in the work ethics of the individuals who come out of different income groups . . . and that's where the difference is and if a black individual we're talking about comes out of an income group of, uh . . . you know, middle class, successful situation, I think the work ethic is probably exactly what it is for a white person coming out of the same kind of background. The reality, of course, is that there are many, many, many more black persons that come out of the other kind of milieu, but I don't know whether I'm begging the question or not. I really don't think it's a racial thing.

If some employers view the work ethic of inner-city poor blacks as problematic, many also express concerns about their honesty, cultural attitudes, and dependability—traits that are frequently associated with the neighborhoods in which they live. A suburban retail drugstore manager expressed his reluctance to hire someone from a poor inner-city neighborhood. "You'd be afraid they're going to steal from you," he stated. "They grow up that way [laughs]. They grow up dishonest and I guess you'd feel like, geez, how are they going to be honest here?" . . .

Questions about the employment woes of blacks sometimes involved assumptions about cultural and family influences in the inner city. One employer asserted that "part of the culture that is dented in their minds [is] that the best thing to do is just go on to welfare and have the state support them"; another argued that blacks in the poorer neighborhoods are "culturally not prepared to work"; and a vice president of an inner-city health service firm related the high jobless rate in the inner city to the disproportionate number of families that have weak employment histories:

> I think it depends on previous generations . . . as to whether a family member has . . . worked, what kind of jobs they've held, how successfully they've held jobs and the like. I think that's where the difference is. . . . I think statistics will show that in your black and Hispanic areas there . . . are greater numbers who have not worked and therefore the work ethic of future generations is less.

The most common belief among the employers was that the social dislocations in the inner city are mainly a function of the environment in which blacks live. As one employer, a retail caterer, put it, "You and I can grow up going to school, coming home, looking forward to coming home or looking forward to going to a movie. In the inner-city neighborhood, they don't look forward to going to school because the streets are unsafe, the schools are unsafe. They don't look forward to coming home because home for them is on the street basically. And it becomes a way of life." A plant manager at an inner-city firm added:

> The neighborhood itself that they live in is a real tough place to work and there's a lot of outside pressure that causes these people not to come to work and I believe a lot of employers look at past history and it seems like nowadays you can't really get a true background from an employee.

. . . An employer from a computer software firm in Chicago expressed the view that "in many businesses the ability to meet the public is paramount and you do not talk street talk to the buying public. Almost all your black welfare people talk street talk. And who's going to sit them down and change their speech patterns?" . . .

Finally, an inner-city banker claimed that many blacks in the ghetto "simply cannot read. When you're talking our type of business, that disqualifies them immediately; we don't have a job here that doesn't require that somebody have minimum reading and writing skills."

Although many of the employers' negative comments reflected general criticisms on inner-city blacks, when specific reference was made to gender, black males bore the brunt of their criticisms.

Table 29.1 Observations as to Why Inner-City Black Males Cannot Find or Retain Jobs Easily

	FREQUENCY (%) OF RESPONSES BY EMPLOYERS' PROFESSION				
Rationale	**Customer Service**	**Clerical**	**Craft**	**Blue-Collar**	**All Employers**
Lack of job skills	9.0	7.1	12.5	17.6	11.7
Lack of basic skills	44.5	37.5	37.5	36.8	38.5
Lack of work ethic	25.0	48.0	25.0	52.9	36.9
Lack of dependability	13.6	14.3	12.5	22.0	16.8
Bad attitude	15.9	16.1	25.0	19.1	17.3
Lack of interpersonal skills	18.2	10.7	0	3.0	8.9
Racial discrimination	15.9	14.3	0	13.2	13.4
Unweighted *N*	44	56	8	68	179

Source: Data from the 1988 employers' survey conducted as a part of the Urban Poverty and Family Life Study, Chicago.

Indeed, as we shall soon see, employers expressed a clear preference for black females over black males. A significant number of the employers stated that previous experiences had soured their opinion of inner-city black male workers. As seen in Table 29.1, a substantial percentage of the employers in each occupational category feel that a lack of basic skills and a lack of work ethic are the two main reasons why inner-city black males have difficulty finding and retaining employment.

The following explanation for the inner-city black male's employment woes was offered by an employer at a two-year suburban college.

> If they get the job, in the first couple of weeks or so, everything seems to be fine, or maybe even the first 90 days but somehow when they get past that, you see a definite, a marked difference. . . . They tend to laziness or there's something there. I've seen this pattern over and over again, you know. I think people are willing to give them a chance and then they get the chance and then it's like they really don't want to work.

The vice president of a Chicago offset-printing firm stated:

> Well, I worked with them in the military, and the first chance they get, they'll slack off, they don't want to do the job, they feel like they don't have to, they're a minority. They want to take the credit and shift the blame. It's like this guy who runs the elevator [a young black man operated the elevator in the buildings], he's like that. They procrastinate. Some of them try. The ones that have higher education are better than that, but a lot of them don't get an education.

A suburban employer drew upon previous experience to offer the following reasons why inner-city black men cannot find jobs:

> It's not every case but the experiences that I've had is the fact that they're not willing to set themselves straight, put 100% effort into their job and try to develop and build within a company. The experiences that I've run into with it is that they develop bad habits, I guess is the best way to put it. Not showing up to work on time. Not showing up to work. Somewhere down the road they didn't develop good work habits.

Employers at inner-city firms tended to be the most critical of inner-city black male workers. One stated: "I just personally, I've had problems with them in the past. . . . They seem to have a lot of associated problems going on at the same time, personal problems, marital problems, that made it difficult for them to get to work every day on time. . . . It's mostly a problem of just getting to work." An inner-city manufacturer at a tool, die, and metal-stamping plant cited past interviews of job seekers when discussing the reasons for his unfavorable opinion of black male workers:

> Ah, let's see, I just went through spot welder and I interviewed over 30 of them, the majority of them were black, pretty bad. Yes, I would say that the majority of them have an unstable history. And you can tell attitude just by talking to a person, you know. It's subjective, but it's me talking with 30 years experience, but yeah, I can stake my claim, my reputation, and I can do it. I can interview, and somebody comes in with cut-off shorts on and looking for a job, I just send them away.

A hotel employer in Chicago indicated that he had had some good success with black male workers but that one of the reasons so many of them do not find employment is that their applications reveal high job turnover. He pointed out that when asked why they had left their previous job,

> they'll, on the application itself, just say something like "didn't get along with the supervisor" and then the next job, reason for leaving, "didn't get along with supervisor," next job reason for leaving, "didn't like it," and they'll have gone through three or four jobs in a matter of six or eight months and then they don't understand why they don't get hired here.

A suburban employer added: "They don't know how to dress when they come to an interview. They bring fourteen other people with them."

Tardiness and absenteeism was a concern expressed by several employers. "We've hired black guys before and they don't show up and they call in sick," stated the general manager of an inner-city restaurant. The chairman of a car transport service voiced a similar complaint. When asked why inner-city black men cannot find jobs, he stated:

> Number 1 . . . they're not dependable. They have never been taught that when you have a job you have to be there at a certain time and you're to stay there until the time is finished. They may not show up on time. They just disappear for an hour or two at a time. They'll call you up and say, "Ahhh, I'm not coming in today" and they don't even call you up.
>
> *Interviewer:* So they're undependable, that's one.
>
> *Respondent:* And the second thing is theft.

Another employer expressed his misgivings about inner-city black males in the course of relating his experiences with one of his previous workers. Agreeing that discrimination probably plays a role among most employers in the hiring of blacks, he went on to explain why:

> I think one of the reasons in all honesty is because we've had bad experience in that sector . . . and believe me I've tried. And as I say, if I find . . . whether he's black or white, if he's good and you know we'll hire him. We are not shutting out any black specifically. But I will say that our experience factor has been bad. We've had more bad black employees over the years than we had good. . . . We hire a young black [as a stockman] and he just absolutely hated anybody telling him anything. I mean if you criticized him or if you gave him an order or an instruction he absolutely resented it. And after a while he started fighting with the other employees. . . . One of the women says, look, I need a case of this or this or this and he doesn't do it, doesn't get it for them and they are waiting to fill an order and he ignores them and then when they complained he would get mad and start swearing at them. You know so it's things like this that you know you can't, can't tolerate it. And he was one of the few that we have let go. And believe me I can count on one hand over the years how many employees we've actually fired. We don't do it indiscriminately. . . .
>
> *Interviewer:* So do you think because of experiences like that, do you think that you are a little bit more leery when a black man comes in here than a white man?

> *Respondent:* Yes, in all honesty I probably am, but . . . as I say, I hired this other one, he was a gem.

Other employers expressed reservations about inner-city black men in terms of work-related skills. They "just don't have the language skills," stated a suburban employer. The president of an inner-city advertising agency highlighted the problem of spelling.

> I needed a temporary a couple months ago, and they sent me a black man. And I dictated a letter to him, he took shorthand, which was good. Something like "Dear Mr. So and So, I am writing to ask about how your business is doing." And then he typed the letter, and a good while later, now not because he was black, I don't know why it took so long. But I read the letter, and it's "I am writing to *ax* about your business." Now you hear about them speaking a different language and all that, and they say *ax* for ask. Well, I don't care about that, but I didn't say *ax,* I said ask.

Several of the employers of blue-collar workers drew a connection between the problems of inner-city black male joblessness and the failure of some applicants to pass drug-screening tests. For example, the manager of a suburban glass-container firm pointed out:

> We've got the unfortunate situation of, through our drug screening, disqualifying roughly 30 percent of those people that get through the screening process and get to the physical exam. We're losing about 30 percent of them through the drug screening process. I think it's a shame, I think it's a sin, it's a disgrace to our society, and as far as I'm concerned it's one of the number one things that we've got to attack.

The president of an inner-city trucking firm likewise stated: "You're going to find a lot of them coming through that comes in there—we've found—we drug test them as part of the physical—and there's a lot of them on drugs. We used to—we were a customs bonded carrier and we used to polygraph them all and we would find that a lot of them are thieves."

Many employers often develop negative opinions of black male workers in the absence of previous firsthand experience. A manufacturer explained that nobody wanted to hire the inner-city black male because of the stereotype that

> they don't want to work, they don't want to do anything. I think that's a big part of it. I don't think anybody wants to admit it but I think that's primarily it. . . . They're ignorant, they don't work, they don't want to work . . . they got a real bad rap, and uh . . . nobody, I don't think anybody will come out and admit it, but I think that's the first thing they consider in a black applicant.

The UPFLS employer survey clearly suggests that although black women also suffer as a consequence of the negative attitudes held by employers, nonetheless, in an overwhelming majority of cases in which inner-city black males and females are compared, the employers preferred black women. When asked how the situation of inner-city black males compares with that of black females, almost one-half of the employers stated that there is a gender difference in inner-city workers' success in finding and retaining employment. Only 14 percent indicated that there was no difference between the employment experiences of inner-city black males and those of black females. A large proportion (43 percent) had no opinion, however, mainly because they had not had any direct employment-related experiences with blacks in general or with black men or black women in particular.

As revealed in Table 29.2, of those respondents who felt that the employment situation of inner-city black men and that of black women differs, almost 78 percent felt that black women have a better chance of finding and retaining employment because they are either more responsible and determined or have better attitudes and a better work ethic.

"I think that probably they [inner-city black females] are much more responsible in what we've found," stated the general manager of a suburban luggage retail store. "So many single-family

Table 29.2 Employers' Observations About Gender Differences in Inner-City Blacks' Abilities to Find and Retain a Job

	FREQUENCY (%) OF RESPONSES BY EMPLOYERS' PROFESSION				
Observations on Blacks' Chances for Employment	**Customer Service**	**Clerical**	**Craft**	**Blue-Collar**	**All Employers**
Positive toward women	61.1	93.1	100	71.4	77.9
Negative toward women*	5.6	3.4	0	28.6	13.0
Unweighted *N*	18	29	2	2	77

*Employers with negative feelings about black women's chances for employment expressed concerns about child care and other family responsibilities.

Source: Data from the 1988 employers' survey conducted as a part of the Urban Poverty and Family Life Study, Chicago.

homes right now you'll find that they're working two jobs trying to support a family. You see it all the time. In many, many cases they're the ones that are supporting their four kids and the husband's whereabouts are unknown." . . .

A small number of the respondents remarked that black males were less successful than black females because they were more threatening to employers. According to one respondent who hires clerical workers:

> Black men present a particularly menacing demeanor to white men. I think they are frightened by them. I think they do not speak the same language. I don't think they use the same codes, I think the whole communication process is a very threatening one. And to the black male, whose need to assert himself is so crucial, because he feels so totally battered in his environment, his sense of manhood is very turned off or intimidated or feels the need to rail against any efforts on the part of the white male establishment to in any way emasculate him further. And I think, there's a tremendous communications block that they come to because both are in some ways frightened and intimidated by the other. Therefore I think many times, when companies hire black males, they hire the most complacent, the least aggressive, the most eunuchish type they can get because they don't want to have some crazy, who's going to become some kind of warmonger, running around the company and spatting. They hire the ones that are most acceptable, and sometimes they're not necessarily the brightest or the most capable.

The executive director of an inner-city charity had similar views:

> People are afraid of black men. If it is a choice of a black man or a woman to do the job I think that an employer would take a black woman. But, if I were going to hire for a job I would take a black woman over a black man because of this situation and then also I would be less afraid of a black woman. I would say, well, maybe he's got a criminal record, but, he's—or I would just be a little bit more apprehensive.

. . . Regardless of whether there has been a shift in attitudes among employers, race is obviously a factor in many of their current decisions; however, the issues are complex and cannot be reduced to the simple notion of employer racism. . . .

If discrimination is a significant factor in the employment woes of inner-city blacks, it is not recognized as such by a substantial majority of the employers in this survey. When asked the reason for the high levels of unemployment in Chicago's inner-city neighborhoods, only 4 percent of the 179 employers mentioned discrimination. Indeed, employers tend to dismiss the charge of discrimination, even though some of their statements indicate that it does exist.

When asked about the problem of discrimination in connection with the employment experiences of inner-city blacks, the director of personnel at a Chicago department store said, "I have a different view on it because I think it's used as an excuse many times. I've found that to be true in the experience here that it is a real convenient excuse for a minority to use and it's frustrating to me as an employer." . . . The views of the director of personnel at a Chicago hospital perhaps best capture the sentiments of many employers: "People are much more aware . . . of their rights and their—a lot of them abuse it too. A lot of frivolous claims. There's a lot of unjustified filing of charges. . . . It's a big problem." . . .

Conclusions about the role of prejudice and discrimination in the labor market are usually based on analyses of the interaction between white employers and minority employees. Many readers will interpret the negative comments of the employers as indicative of the larger problem of racism and racial discrimination in American society. It is therefore instructive to consider, for separate analysis, the perceptions of the African-American employers who were interviewed in our survey. Their responses suggest that it would be a mistake to characterize the overall comments of the employers in our survey as racist, even though some clearly contain racist sentiments. Indeed, it is significant to note that of the fifteen African-American employers in our survey, twelve expressed views about inner-city black workers, in response to our open-ended questions, that were coded as negative. Only two of the black employers offered comments that could be described as positive, and one expressed views that were coded as neutral. Thus, whereas 74 percent of all the white employers who responded to the open-ended questions expressed negative views of the job-related traits of inner-city blacks, 80 percent of the black employers did so as well.

The black president and CEO of an inner-city wholesale firm described what he saw as the effects of living in a highly concentrated poverty area:

> So, you put . . . a bunch of poor people together, [rushed and emphatic] I don't give a damn whether they're white, green or grizzly, you got a bad deal. You're going to create crime and everything else that's under the sun, dope. Anytime you put all like people together—and particularly if they're on a low level—you destroy them. They not, how you going to expect . . . one's going to stand up like a flower? He don't see no reason to stand up. When he gets up in the morning he sees people laying around doing nothing. He goes to bed at night, the same damn thing. All they think of, do I get to eat and sleep?
>
> *Interviewer:* So, you understand this wariness of some employers?
>
> *Respondent:* Sure.

. . . When asked why black men cannot find jobs, the black personnel manager of a retail food and drug store stated: "I think that's really the culture. I think that a lot of the black males look at some of the jobs that they feel are beneath their dignity to work and rather than accept any job, they'd rather stay out of employment rolls." A black employer in a Chicago insurance company argued that

> there is a perception that most of your kids are black and they don't have the proper skills . . . they don't know how to write. They don't know how to speak. They don't act in a business fashion or dress in a business manner . . . in a way that the business community would like. And they just don't feel that they're getting a quality employee. . . .
>
> *Interviewer:* Do you think—is that all a false perception or is something there or—?
>
> *Respondent:* I think there's some truth to it.

In answer to the same question, another black employer gave this negative response:

> Attitude. Poor attitude. I'm very vocal on that. They lazy, a lot of them. You know, when you trapped, you realize you're trapped, but if you don't try to do something about it yourself, then you'll always be trapped. If you get into a welfare mode then you becomes a slave. And if that's

> what you want to be, so be it as an individual but I don't want to be a slave. I'm going to work. It's an attitude problem, that's all I can tell you and I've known, I been around them. I know what's happening.

Finally, the black personnel manager of a security services firm dismissed the problem of job discrimination: "It's a myth to me and it's something that has been another part where our society constantly has pushed it in the minds of a lot of people and they lean on this as a crutch. And when you lean on this as a crutch well, then how will you ever pick up yourself or apply yourself to anything? So, no. Discrimination can be thrown out the window."

Although the criticisms of employers were directed at both inner-city black males and females, the harshest and most frequent criticisms were aimed at the black male. Employer comments about inner-city black males revealed a wide range of complaints, including assertions that they procrastinate, are lazy, belligerent, and dangerous, have high rates of tardiness and absenteeism, carry employment histories with many job turnovers, and frequently fail to pass drug screening tests.

Many of their comments clearly suggest a direct link between their assessment of the quality of the black inner-city workforce and their hiring strategies. There are many ways in which employers can deny inner-city workers employment or access to employment if they are reluctant or do not want to hire them. Direct, overt discrimination in refusing to consider any black applicants at all or in not seriously reviewing their applications for employment is the most obvious, but only a handful of employers admitted to such practices. Others consider black job applicants but frequently screen them out because their applications are not considered strong, or they do not make a good impression during the interview process, or they fail to pass an employment or skills tests.

The job interview provides job applicants with an opportunity to challenge employer stereotypes. However, as Kathryn Neckerman and Joleen Kirschenman, two of the six researchers who conducted the interviews for the UPFLS employer survey, pointed out:

> Inner-city black job seekers with limited work experience and little familiarity with the white, middle-class world are also likely to have difficulty in the typical job interview. A spotty work record will have to be justified; misunderstanding and suspicion may undermine rapport and hamper communication. However qualified they are for the job, inner-city black applicants are more likely to fail subjective "tests" of productivity during the interview.

Skills tests are less biased than the subjective assessments used in the typical interview. The data reveal that city employers—that is, those with firms within the city of Chicago—who apply skills tests have a higher average proportion of black workers in entry-level jobs than do those who do not use these tests, even when one takes into consideration the size of the firm, the occupation, and the percentage of blacks in the neighborhood.

However, many Chicago employers engage in recruitment practices that automatically eliminate or significantly reduce the number of inner-city applicants who could apply for entry-level jobs in their firms. Selective recruitment—that is, limiting the search for job candidates in various ways—is widely practiced by the Chicago employers in this survey. Although some of the employers justified this strategy in terms of practicality and efficiency, noting that they save time and money by relying on the referrals of employees instead of screening large numbers of applicants from newspaper ads, most indicated that it yields a higher-quality applicant.

Although the formal criteria of applicant quality were based on neither race nor class, the recruitment strategies designed to attract high-quality applicants were. Inner-city populations were often overlooked if employers limited their recruitment efforts to certain neighborhoods or institutions and placed ads only or mainly in ethnic or neighborhood newspapers. Indeed, the recruitment practices of the employers reflect their perceptions of the quality of the inner-city workers. . . .

In light of what this employers' survey tells us, the selective recruitment practiced by many city employers results in the systematic exclusion of numerous inner-city blacks from jobs in Chicago. And given the negative view that employers have of inner-city workers, it is reasonable to conclude that although these practices are characterized by employers as necessary to recruit a higher-quality worker, they in fact are deliberately designed to exclude inner-city blacks from the employment applicant pool.

How should we interpret the negative attitudes and actions of employers? To what extent do they represent an aversion to blacks per se and to what degree do they reflect judgments based on the job-related skills and training of inner-city blacks in a changing labor market? As pointed out earlier, the statements made by the African-American employers concerning the qualifications of inner-city black workers do not differ significantly from those of the white employers. This raises a question about the meaning and significance of race in certain situations—in other words, how race intersects with other factors. A key hypothesis is that given the recent shifts in the economy, employers are looking for workers with a broad range of abilities: "hard" skills (literacy, numeracy, basic mechanical ability, and other testable attributes) and "soft" skills (personalities suitable to the work environment, good grooming, group-oriented work behaviors, etc.). While hard skills are the product of education and training—benefits that are apparently in short supply in inner-city schools—soft skills are strongly tied to culture, and are therefore shaped by the harsh environment of the inner-city ghetto. If employers are indeed reacting to the difference in skills between white and black applicants, it becomes increasingly difficult to discern the motives of employers: are they rejecting inner-city black applicants out of overt racial discrimination or on the basis of qualifications? In this connection, one study conducted in Los Angeles found that even after education, income, family background, and place of residence were taken into account, dark-skinned black men were 52 percent less likely to be working than light-skinned black men. Although this finding strongly suggests that racial discrimination plays a significant role in the jobless rate of black men, the study did not pursue the extent to which employers associate darkness of skin color with the social and cultural environment of the inner-city ghetto.

Nonetheless, many of the selective recruitment practices do represent what economists call statistical discrimination: employers make assumptions about the inner-city black workers *in general* and reach decisions based on those assumptions before they have had a chance to review systematically the qualifications of an individual applicant. The net effect is that many black inner-city applicants are never given the chance to prove their qualifications on an individual level because they are systematically screened out by the selective recruitment process. Statistical discrimination, although representing elements of class bias against poor workers in the inner city, is clearly a matter of race. The selective recruitment patterns effectively screen out far more black workers from the inner city than Hispanic or white workers from the same types of backgrounds. But race is also a factor, even in those decisions to deny employment to inner-city black workers on the basis of objective and thorough evaluations of their qualifications. The hard and soft skills among inner-city blacks that do not match the current needs of the labor market are products of racially segregated communities, communities that have historically featured widespread social constraints and restricted opportunities. . . .

Many inner-city residents have a strong sense of the negative attitudes which employers tend to have toward them. A 33-year-old employed janitor from a poor South Side neighborhood had this observation: "I went to a coupla jobs where a couple of the receptionists told me in confidence: 'You know what they do with these applications from blacks as soon as the day is over?' Say 'we rip them and throw 'em in the garbage.'" In addition to concerns about being rejected because of race, the fears that some inner-city residents have of being denied employment simply because of their inner-city address or neighborhood are not unfounded. . . .

A 34-year-old single and unemployed black man put it this way: "If you're from a nice neighborhood I believe it's easier for you to get a job and stuff. I have been on jobs and such and gotten looks from folks and such, 'I wonder if he is the type who do those things that happen in that neighborhood.'"

Although the employers' perceptions of inner-city workers make it difficult for low-income blacks to find or retain employment, it is interesting to note that there is one area where the views of employers and those of many inner-city residents converge—namely, in their attitudes toward inner-city black males. Inner-city residents are aware of the problems of male joblessness in their neighborhoods. For example, more than half the black UPFLS survey respondents from neighborhoods with poverty rates of at least 40 percent felt that very few or none of the men in their neighborhood were working steadily. More than one-third of the respondents from neighborhoods with poverty rates of at least 30 percent expressed that view as well. Forty percent of the black respondents in all neighborhoods in the UPFLS felt that the number of men with jobs has steadily decreased over the past ten years. However, responses to the open-ended questions in our Social Opportunity Survey and data from our ethnographic field interviews reveal a consistent pattern of negative views among the respondents concerning inner-city black males, especially young black males.

Some provided explanations in which they acknowledged the constraints that black men face. An employed 25-year-old unmarried father of one child from North Lawndale stated:

> I know a lot of guys that's my age, that don't work and I know some that works temporary, but wanna work, they just can't get the jobs. You know, they got a high school diploma and that . . . but the thing is, these jobs always say: Not enough experience. How can you get some experience if you never had a chance to get any experience?

Others, however, expressed views that echoed those of the employers. For example, a 30-year-old married father of three children who lives in North Lawndale and works the night shift in a factory stated:

> I say about 65 percent—of black males, I say, don't wanna work, and when I say don't wanna work I say don't wanna work hard—they want a real easy job, making big bucks—see? And, and when you start talking about hard labor and earning your money with sweat or just once in a while you gotta put out a little bit—you know, that extra effort, I don't, I don't think the guys really wanna do that. And sometimes it comes from, really, not having a, a steady job, or, really, not being out in the work field and just been sittin' back, being comfortable all the time and hanging out.

A 35-year-old welfare mother of eight children from the Englewood neighborhood on the South Side agreed:

> Well, I mean see you got all these dudes around here, they don't even work, they don't even try, they don't wanna work. You know what I mean, I wanna work, but I can't work. Then you got people here that, in this neighborhood, can get up and do somethin', they just don't wanna do nothin'—they really don't.

The deterioration of the socioeconomic status of black men may have led to the negative perceptions of both the employers and the inner-city residents. Are these perceptions merely stereotypical or do they have any basis in fact? Data from the UPFLS survey show that variables measuring differences in social context (neighborhoods, social networks, and households) accounted for substantially more of the gap in the employment rates of black and Mexican men than did variables measuring individual attitudes. Also, data from the survey reveal that jobless black men have a lower "reservation wage" than the jobless men in other ethnic groups. They were willing to work for less than $6.00 per hour, whereas Mexican and Puerto Rican jobless men expected $6.20 and $7.20, respectively, as a condition for working; white men, on the other hand, expected over $9.00 per hour. This would appear to cast some doubt on the characterization of black inner-city men as wanting "something for nothing," of holding out for high pay.

But surveys are not the best way to get at underlying attitudes and values. Accordingly, to gain a better grasp of the cultural issues, I examined the UPFLS ethnographic research that involved establishing long-term contracts and conducting interviews with residents from several neighborhoods. Richard Taub points out:

> Anybody who studies subgroups within the American population knows that there are cultural patterns which are distinctive to the subgroups and which have consequences for social outcomes. The challenge for those concerned about poverty and cultural variation is to link cultural arrangements to larger structural realities and to understand the interaction between the consequences of one's structural position on the one hand and pattern group behavior on the other. It is important to understand that the process works both ways. Cultures are forged in part on the basis of adaptation to both structural and material environments.

Analysis of the ethnographic data reveals identifiable and consistent patterns of attitudes and beliefs among inner-city ethnic groups. The data, systematically analyzed by Taub, reveal that the black men are more hostile than the Mexican men with respect to the low-paying jobs they hold, less willing to be flexible in taking assignments or tasks not considered part of their job, and less willing to work as hard for the same low wages. These contrasts in the behavior of the two groups of men are sharp because many of the Mexicans interviewed were recent immigrants.

"Immigrants, particularly Third World immigrants," will often "tolerate harsher conditions, lower pay, fewer upward trajectories, and other job-related characteristics that deter native workers, and thereby exhibit a better 'work ethic' than others." The ethnographic data from the UPFLS suggest that the Mexican immigrants are harder workers because they "come from areas of intense poverty and that even boring, hard, dead-end jobs look, by contrast, good to them." They also fear being deported if they fail to find employment. . . .

In contrast to the Mexican men, the inner-city black men complained that they get assigned the heaviest or dirtiest work on the job, are overworked, and are paid less than nonblacks. They strongly feel that they are victims of discrimination. "The Mexican-American men also report that they feel exploited," states Taub, "but somehow that comes with the territory." Taub argues that the inner-city black men have a greater sense of "honor" and often see the work, pay, and treatment from bosses as insulting and degrading. Accordingly, a heightened sensitivity to exploitation fuels their anger and gives rise to a tendency to "just walk off the job."

One has to look at the growing exclusion of black men from higher-paying blue-collar jobs in manufacturing and other industries and their increasing confinement to low-paying service laboring jobs to understand these attitudes and how they developed. Many low-paying jobs have predictably low retention rates. For example, one of the respondents in the UPFLS employer survey reported turnover rates at his firm that exceeded 50 percent. When asked if he had considered doing anything about this problem, the employer acknowledged that the company had made a rational decision to tolerate a high turnover rather than increasing the starting salary and improving working conditions to attract higher-caliber workers: "Our practice has been that we'll keep hiring and, hopefully, one or two of them are going to wind up being good."

As Kathryn Neckerman points out, "This employer, and others like him, can afford such high turnover because the work is simple and can be taught in a couple of days. On average, jobs paying under $5.00 or $6.00 an hour were characterized by high quit rates. In higher-paying jobs, by contrast, the proportion of employees resigning fell to less than 20 percent per year." Yet UPFLS data show that the proportion of inner-city black males in the higher-paying blue-collar positions has declined far more sharply than that of Latinos and whites. . . . Increasingly displaced from manufacturing industries, inner-city black males are more confined to low-paying service work. Annual turnover rates of 50 to 100 percent are common in low-skill service jobs in Chicago, regardless of the race or ethnicity of the employees.

Thus, the attitudes that many inner-city black males express about their jobs and job prospects reflect their plummeting position in a changing labor market. The more they complain and manifest their dissatisfaction, the less desirable they seem to employers. They therefore experience greater rejection when they seek employment and clash more often with supervisors when they secure employment.

Residence in highly concentrated poverty neighborhoods aggravates the weak labor-force attachment of black males. The absence of effective informal job networks and the frequency of many illegal activities increases nonmainstream behavior such as hustling. As Sharon Hicks-Bartlett, another member of the UPFLS research team, points out, "Hustling is making money by doing whatever is necessary to survive or simply make ends meet. It can be legal or extra-legal work and may transpire in the formal or informal economy. While both men and women hustle, men are more conspicuous in the illegal arena of hustling."

In a review of the research literature on the experiences of black men in the labor market, Philip Moss and Christopher Tilly point out that criminal activity in urban areas has become more attractive because of the disappearance of legitimate jobs. They refer to a recent study in Boston that showed that while "black youth in Boston were evenly split on whether they could make more money in a straight job or on the street, by 1989 a three-to-one majority of young black people expressed the opinion that they could make more on the street."

The restructuring of the economy will continue to compound the negative effects of the prevailing perceptions of inner-city black males. Because of the increasing shift away from manufacturing and toward service industries, employers have a greater need for workers who can effectively serve and relate to the consumer. Inner-city black men are not perceived as having these qualities.

. . . In summary, the issue of race in the labor market cannot simply be reduced to the presence of discrimination. Although our data suggest that inner-city blacks, especially African-American males, are experiencing increasing problems in the labor market, the reasons for those problems are seen in a complex web of interrelated factors, including those that are race-neutral.

The loss of traditional manufacturing and other blue-collar jobs in Chicago resulted in increased joblessness among inner-city black males and a concentration in low-wage, high-turnover laborer and service-sector jobs. Embedded in ghetto neighborhoods, social networks, and households that are not conducive to employment, inner-city black males fall further behind their white and Hispanic counterparts, especially when the labor market is slack. Hispanics "continue to funnel into manufacturing because employers prefer Hispanics over blacks and they like to hire by referrals from current employees, which Hispanics can readily furnish, being already embedded in migration networks." Inner-city black men grow bitter and resentful in the face of their employment prospects and often manifest or express these feelings in their harsh, often dehumanizing, low-wage work settings.

Their attitudes and actions, combined with erratic work histories in high-turnover jobs, create the widely shared perception that they are undesirable workers. The perception in turn becomes the basis for employers' negative hiring decisions, which sharply increase when the economy is weak. The rejection of inner-city black male workers gradually grows over the long term not only because employers are turning more to the expanding immigrant and female labor force, but also because the number of jobs that require contact with the public continues to climb.

30

Dimensions of the Closet

JAMES WOODS

A gay man faces a decision every time he sets foot in a work environment, whenever he is in the presence of a boss, client, secretary, customer, or some other professional peer. At all times he must balance a whole host of considerations—worries that are largely unknown to his heterosexual peers: How should he handle information about his sexuality? What will be the consequences of coming out? Of remaining in the closet? Of trying to avoid the issue altogether? The decision is not one he can afford to take lightly.

For many employers, however, the scenario is new and unfamiliar. American business is only now waking up to the presence of its lesbian and gay employees, and the awakening has brought the resistance, misunderstanding, and social clumsiness we would expect to accompany so fundamental a change. Lesbians and gay men represent a challenge to some of our most entrenched ideas about the separation of work and sexuality; their mere presence seems to upset our conventional beliefs about privacy, professionalism, and office etiquette. Because they can often disappear, they differ from "traditional" minorities, whose problems are better understood. And because they have historically remained invisible, American business has little idea how numerous they really are.

A curious paradox results. Working under cover, lesbians and gay men are both widely integrated and almost entirely ignored. With the right disguise, they have already been admitted to a club that still formally excludes them. They view the nation's banks, law firms, hospitals, and charities not as outsiders clamoring to be let in, but as insiders afraid to make their presence known. Their collective contributions, productivity, and satisfaction have serious implications for the bottom line. Moreover, their significance to American business is way out of proportion to their actual numbers. As individuals lesbian and gay professionals influence and interact with a much larger circle of peers: the men and women, both gay and straight, who share their offices and secrets, who become their friends and adversaries, who sometimes hire and fire them. As a group they symbolize some of our most basic beliefs about privacy, professionalism, and the role of sexuality in the workplace.

Yet until recently, almost nothing has been said about their role in American business, nor about the changes that must come if they are to participate as full and equal partners. Sexual diversity is a new concept to most companies and as such it requires a certain amount of explaining. Employers too rarely recognize the destructive trade-offs they impose on lesbian and gay workers, and only now are they beginning to see the personal and economic consequences of heterosexism. Even lesbian and gay workers who inhabit the corporate closet are often unaware of its full size and depth. Like their heterosexual peers, they accept certain conventional ways of doing business, blind to the inequities embedded within them. A few preliminary definitions are thus in order.

HETEROSEXISM AT WORK

Though lesbians and gay men are assumed to make up some 10 percent of the population (a figure that has been subject to much recent debate), there is no reliable way to estimate their numbers in the professional workforce. Neither do we know very much about the particular kinds of jobs they hold. Professions like design, travel, and arts management seem to employ a large number of gay men, and popular lore also has them clustered in the world's flower shops, hair salons, and gourmet emporiums. Gay waiters and hairdressers are stock sitcom characters. And who hasn't, on occasion, made assumptions about a male flight attendant, nurse, or ballet dancer?

Indeed, the visibility of gay men in certain professions prompts the frequent claim, usually accompanied by anecdotal evidence, that there are in fact "gay industries." In his 1974 essay "The Homosexual Executive," Richard Zoglin remarks that "certain more 'creative' fields such as advertising and publishing have traditionally had a higher incidence of homosexuality," especially when compared to "the most conservative segments of the business community," like insurance, banking, and utilities. A recent article on the Condé Nast publishing empire, published in the *Advocate,* suggests that the fashion and design industries would virtually grind to a halt without gay men. According to Raul Martinez, art director at *Vogue,* the magazine "could be put together without gay men, but it would be a nightmare. The majority of the fashion magazine world is gay—hair and makeup people, stylists, top photographers like Steven Meisel, Herb Ritts, and Matthew Rolston, and, of course, the people who make the clothes." Having said all this, Martinez reflects for a moment. "Wait a minute," he adds. "Maybe all of this *can't* be done without gays."

In other industries, for whatever reason, gay men seem virtually absent. Writing in *Commentary* in 1980, Midge Decter expressed doubt that homosexuals "have established much of a presence in basic industry or government service or in such classic professions as doctoring and lawyering, but then for anyone acquainted with them as a group, the thought suggests itself that few of them have ever made much effort in these directions." Scott, a gay insurance agent, says that "there are absolutely no other gay people in my industry," while Matt, an executive with Ford Motor Company, assures me that "there aren't too many gays in my business." Darren, a New Jersey dentist, feels that homosexuality is "real uncommon" in his line of work. "It's just one of those professions in which you don't see it that often. You see a lot more gay doctors."

Given the ease with which appearances can (and are) manipulated, however, they are almost certainly deceptive. Like most myths, the notion of "gay industries" sprang from a seed of truth: Gay men are indeed invisible in many professions. The problem is that the myth mistakes invisibility for rarity. Because lesbians and gay men must identify themselves to be counted, their visibility is an unreliable measure of their presence. A quick scan of a particular company or industry may tell us something about the willingness of gay employees to identify themselves, but reveals comparatively little about their actual numbers. Indeed, it is quite plausible that the "most conservative segments" of the business community are fully saturated with gay professionals who simply keep a lower profile than do men in the arts, advertising, and travel.

None of which is to say that gay professionals are distributed evenly across all occupations, companies, or geographic areas—this, too, seems unlikely. Yet there is ample evidence that the true distribution of gay professionals, if known, would be surprising. According to a 1991 survey of 6,075 lesbians and gay men conducted by Overlooked Opinions, a Chicago market research firm, more lesbians and gays are in science than in food service, and more work in finance than in the arts. Among gay men the three most common job categories were management, health care, and education. Whatever the actual distribution and numbers, we can also rest assured that *no* field is entirely devoid of lesbian and gay workers. Certainly anyone who claims that they cannot be found in medicine, banking, or insurance too readily accepts conventional wisdom. "I used to think there were no gay stockbrokers," says a Wall Street veteran, now in his fifties. "Then I got my first job in a brokerage, and little by little they came out of the woodwork. Now I wonder why I didn't see them in the first place."

Under the circumstances, it is hardly surprising that lesbian and gay professionals often elect to remain invisible. While they have always worked in the nation's offices, classrooms, and laboratories, they've never been at home in them. Since 1980 some two dozen studies have documented hiring, promotion, and compensation practices that discriminate against lesbian and gay workers. Of the thousands of gay men surveyed in these reports, roughly one in three believed he had experienced some form of job discrimination. The most recent of these reports also suggest that as lesbians and gay men have become more visible in recent years, the rates are climbing. In a typical survey, conducted in Philadelphia in 1992, 30 percent of gay men (and 24 percent of lesbian women) reported that they had experienced employment discrimination at some point in their careers. A sizable number (12 percent of the men and 9 percent of the women) said that it had occurred within just the past twelve months. The lowest rate reported by any study, for men and women, was 16 percent; the highest was 44 percent. . . .

The numbers, however, are conservative. Because gays often disguise themselves precisely to *avoid* discrimination, we can only guess how its incidence would soar if they made themselves easier targets. The same 1992 survey of Philadelphians found that 76 percent of gay men (and 81 percent of lesbians) remained in the closet at work. Seventy-eight percent of the men (and 87 percent of the women) feared they *would* be the victims of job discrimination if their sexual orientation were known to others. A 1981 study of lesbians came to much the same conclusion. In that report it was found that "nearly two out of three respondents felt that their jobs might be jeopardized if their sociosexual orientation were known. Clearly disclosure in one's work environment is perceived by lesbian women as a high-risk event."

Moreover, actual discrimination is not always recognized as such. With some notable exceptions, most employers avoid the awkward (and increasingly illegal) practice of explicitly denying jobs to homosexuals on the basis of their sexual orientation. Gay men don't always know why they were fired or denied a promotion and can't say for sure what prejudices lurk beneath the generic refusals they are often given: "lack of experience," "overqualified," "job already filled," "wouldn't fit in." While in most cases they can only second-guess an employer's motives, an important study of employers in Anchorage, Alaska, suggests that their fears are well founded. Of the 191 employees who responded to the 1987–88 survey, many said that they would fire (18 percent), not hire (27 percent), or not promote (26 percent) someone they thought to be homosexual.

Even the definitions of discrimination are conservative and narrow to the point that they ignore what may be the most common ways in which lesbian and gay workers are penalized. Surveys of gay professionals typically ask respondents if their sexuality has at any point cost them a job or promotion. Others put the question more broadly, asking respondents if they've ever encountered "any problems at work." The answers reflect the particular way in which the question has traditionally been framed. When he reports discrimination, a gay man usually has in mind an unjust termination, a negative evaluation, a job offer that never materialized, or a promotion that was denied. The question encourages him to think in terms of specific homophobic incidents or individuals—those that can be clearly identified and labeled as such.

Yet discriminatory incidents are just the tip of the iceberg. Even when the most blatant, episodic forms of bias are eliminated, one can identify a process that excludes lesbian and gay professionals in other, more subtle ways. In prejudicial compensation practices, the forced invisibility of gay employees, the social validation of heterosexual mating rituals, the antigay commentary and imagery that circulate through company channels, even the masculine nature of bureaucratic organization itself, a certain kind of heterosexuality is routinely displayed and rewarded. The traditional white-collar workplace is "heterosexist" in the sense that it structurally and ideologically promotes a particular model of heterosexuality while penalizing, hiding, or otherwise "symbolically annihilating" its alternatives. Like racism, sexism, and other isms, heterosexism encompasses not only blatant, isolated displays of prejudice (a bigoted remark, a hate crime) but also the more subtle, unseen ways in which lesbians and gay men are stigmatized, excluded, and denied the support given their heterosexual peers.

When the definition of discrimination is expanded to include these *systemic* forms of heterosexism, the rate soars. When *Out/Look* asked readers if their sexual orientation ever creates "stressful situations at work," 62 percent said that it was "always" or "often" a source of stress, 46 percent said it had influenced their choice of career, and 27 percent thought it had influenced their choice of a particular company. Among the seventy men in my own sample, five (or 7 percent) were convinced that they had lost a prior job due to the prejudice of a boss or client. Another ten (14 percent) feared that they *would* be fired or encouraged to resign if their secret were known. And virtually all (97 percent) thought their sexuality had, at some point, cost them a promotion, a raise, or a relationship with a potential mentor.

Unjust compensation practices are also ignored by traditional measures of discrimination. According to the U.S. Chamber of Commerce, employee benefits like health insurance, relocation expenses, and other perquisites now account for roughly 37 percent of payroll costs. Although married employees routinely receive these benefits for children, husbands, and wives, gay workers are almost always denied them. Neither are they usually granted bereavement, medical, or maternity leave to take care of an unmarried partner. In a 1991 survey by *Partners* magazine, only 8 percent of lesbians and gay men said that employers provided their partners with an employee benefit of some kind. Only 5 percent had company health insurance that covered a partner. As insurance premiums soar and compensation through benefits displaces real wages—having already climbed from 33 percent to 37 percent in just a few years—lesbian and gay workers are being compensated less and less.

A comprehensive definition of discrimination should also take into account the many subtle ways in which heterosexism can distort a career. Even when he is spared an unjust termination or negative evaluation, a gay man faces obstacles that are largely unknown to heterosexual peers. Heterosexism takes its toll across the span of an entire lifetime, and may be visible only in retrospect. Consider Martin, an account executive at Ogilvy & Mather. "After college I wanted to be in New York because I thought it would be a tolerant city," he says, "and I guess I assumed the same thing about advertising. I thought that advertising meant 'creative people, liberal atmosphere, lots of gay men.' To some extent that's proven true." Martin feels that most of his coworkers are "terrific, smart, fun people," and thinks that he's treated reasonable well. Still, he is not untouched by heterosexism. Shortly after joining the company, Martin carefully avoided an assignment that would have required extensive after-hours socializing with a particular client. A few years later he turned down a promotion that would have taken him to a smaller city, one in which he feared it would be "harder to be gay." Then there is the matter of his lover, a physician, who is denied the benefits and social invitations a wife would receive automatically.

Martin also worries about his long-run prospects with the company. He has the reputation of being somewhat aloof and enigmatic at work, someone who doesn't attend all the company parties or outings. He is formal with his superiors and has carefully segregated his personal and professional activities, friends, and identities. "I have a limited relationship with my boss," he explains, "which means I don't always get the mentoring I need." Nor does Martin find himself eager to go the extra mile, to devote long hours to a company that doesn't seem to value him. "My career is less fulfilling than it might be," he says. "I turn to other activities for sustenance."

Even so Martin thinks that by staying in the closet, he has largely avoided discrimination. He recognizes that his sexuality has influenced many of his choices, stunted certain professional relationships, even drawn him to a particular city and line of work. Eventually it may lead him out of advertising and into his own, private consulting practice. Yet he considers himself lucky. "I can't really say I've been the victim of discrimination," he says. "At least not that I'm aware of." . . .

. . . While a man may not recognize the full scope of cultural and institutional heterosexism, he can scarcely ignore the way his boss makes fun of a particular client, the one they say is "just a bit odd." He may not view questions about marital status as a degrading ritual but knows that "sissy" jokes are. Because a man's response to homophobia depends on his perception of it, he must be attentive to the various signals and cues given off by those around him. By necessity he becomes skilled at detecting homophobia. He does so in a number of different ways.

When a man is openly gay, coworkers may express their homophobia directly. Darren, a dentist in New Jersey, remembers the one and only time he revealed himself to someone at work. While Darren was attending dental school, he and his lover Bob became friendly with Rick and Renee, two of Darren's classmates. "They were just wonderful, wonderful people," Darren recalls. "We mixed socially a couple of times a week." As the friendship warmed, Darren thought it was safe to come out. "One night we were out at a club, dancing. Somehow it came up that I lived with Bob, and maybe we were gay. Renee was laughing like it was crazy, a joke. I said, 'Why is that a joke? What's so outrageous about that?' And she says, 'Well, what do you mean?' So I said, 'Well, we *are* gay. We've been lovers for four years.' "

The relationship soured immediately. "The next day Renee called me up and said, 'I need to talk to you.' So she came over, and she says, 'Would you please come out to my car and speak with me?' I didn't know what it was about, so I walked out to the car. She left the motor running, and I remember that she didn't look me in the face. She says, 'Well, I just wanted you to know that Rick and I have talked it over and we've decided—and this isn't a hundred percent because of what you told me last night—that we can't see you or Bob anymore. We can't be your friends anymore.' " Ten years later Darren is still troubled by the episode. "It was like being hit in the face with a sledgehammer," he says. "This was the first person I had ever come out to." Though Renee tried to apologize a few months later, Darren felt it was too late. Today he is guarded and secretive at work. "When you stick your finger in the fire," he says, "you don't put it back in again."

Such episodes leave little doubt that a particular work setting is hostile. But not all gay men have had experiences like these; quite often they base their assessments on more oblique or secondhand evidence of homophobia. Rather than expose himself directly to the climate of opinion, for example, a man may instead observe the way *other* gay men and lesbians are treated at work. Tom, an elementary school teacher in New Jersey, hears comments about the school librarian. "He's real obvious," Tom says. "The kids mimic him, and everybody knows who he is." In Dave's office it is the copier repairman who gets the attention. "He's a real flamer," says Dave, the credit manager for a Philadelphia fuel supply company. "He comes in and everybody talks about how 'flowing and diaphanous' he is. When he asks people for something, they say, 'Honey, it's over there, honey,' as a joke." For Todd, a human resources manager at Bell Atlantic, making personal judgments like these is part of the job. Todd has participated in hundreds of hiring and firing decisions and knows that his staff routinely discriminates against gay candidates; he's see them do it. "In some departments," Todd says, "a candidate will simply be ruled out if he's known to be gay." Not surprisingly these men consider their work environments hostile. All are in the closet.

George has been on both sides of the hiring process, first as a flight attendant and now as a training executive with a large European airline. "In-flight is *so gay,*" he says. "I mean, male flight attendants are like hairdressers. Everyone knows that, and so there's a fear in management. I know that we actively discriminate against gays in the recruiting process. I know it for a fact; they'll terminate them in training for that. I don't think they care if someone's gay or not. They care about how effeminate you are. They really don't care what you *do,* but if you're a flamer—I mean I'm not butch, but if you really were a flamer, they'd have you out of there in a heartbeat. It varies by airline, too. I'm friendly with a recruiter at another airline, and she told me that the head guy in Chicago called her and said, 'Stop sending me all these queers.'" George took this as a warning. Everyone at work knows that he is gay, but he rarely talks about it. He tries not to seem too obvious. . . .

It may be no more than a joke, comment, or anecdote that sets the tone. In its 1990 survey of lesbian and gay journalists, the American Society of News Editors found that in just the twelve months preceding the survey, 81 percent of respondents had heard derogatory comments about gays or lesbians in general, and nearly half had heard comments directed at a particular gay or lesbian employee. Twenty percent of respondents didn't think their newsrooms were a good environment for lesbians or gay men. Among my own seventy respondents, more than half could recall a specific homophobic remark or incident at work; one in five said that such comments were frequent. . . .

In other work environments, explicit antigay remarks are largely unknown. Whatever people think about homosexuality, no one actually *talks* about it. Perhaps there are no other (visibly) gay employees at work. Perhaps the subject has just never come up. If a man is new on the job, he may not yet be privy to conversations at the water cooler. Even if he has been with a company for many years, a man who assiduously avoids personal conversations may have established a climate that encourages others to do likewise. He ensures that conversations about personal and sexual matters will take place out of his earshot.

In these settings a man's assessment of his work environment may be based entirely on indirect evidence: the type of industry he is in, the ages or personal backgrounds of peers, perhaps the way they handle other kinds of unconventional behavior. "I know five or six men who are cheating on their wives or wives who are cheating on their husbands," says Derek, a senior vice president who sees this as evidence of a progressive sexual atmosphere at his firm. "I don't want to say we're *loose,* but people just do their own thing and feel free to discuss it." . . .

Others cite the ages, religious beliefs, or personal backgrounds of coworkers. Grey says that his company is "a very conservative, navy suit, white shirt, real estate developer." To him that portrait includes conservative sexual values. Dave says that the average age at his company is around forty-five. "It's a very old company, with a lot of southerners. They didn't even want to hire black people." Although homosexuality is never mentioned at work, Dave assumes that attitudes are negative. "I can't imagine any of the executives appreciating the fact that they have a gay employee. I'm sure AIDS would probably come up. I'm sure people would probably be worried that, 'Oh my god, he's a gay man.' Who knows what people would do?"

Mark saw evidence of homophobia in the type of work his company, a consulting firm headquartered in New York, was doing for its clients. The firm specializes in the design of compensation and benefits packages, and many of its clients had expressed concern about rising health care costs, especially those associated with AIDS. Several years ago a memo appeared in the firm's electronic mail system. "There was a flyer saying that the firm had put together an addendum of benefits-plan language to minimize costs for prescription drugs or disability or alternative treatment modes or home health care," Mark recalls. "There was a whole laundry list of ways to exclude AIDS. I mean, how much more blatant could you be?"

The type of work or the general reputation of an industry may also be a consideration. Tom, a high school teacher, feels that his profession is conservative by nature. As evidence he recites the familiar myths about gay people: that they "recruit" from the ranks of the young, that they are all pedophiles. He points to the periodic campaigns to remove gay teachers from the classroom—California's "Briggs Initiative" and Anita Bryant's "Save Our Children" campaign, to name two famous examples. He recognizes that there is widespread anxiety about homosexuals who work with children. "In the position I'm in, with kids, you have to worry about it," Tom says. "We've had male teachers accused of molesting the little girls, so you hesitate even putting your arm around a little girl and giving her a hug. And, being gay, you have to be twice as careful because you have to watch out for little boys, too. In that respect being gay is a double-edged sword. So if a kid does something good, you think twice about giving him a hug and saying, 'You did a good job.' It could be misinterpreted." Tom doesn't think he's been the victim of discrimination and hears only an occasional homophobic comment at work (usually directed at the school librarian). Yet he is certain that explicit acknowledgement of his sexuality would be unwelcome.

In situations like these, when there is little concrete evidence of their coworkers' attitudes, gay men tend to play it safe. They hold on to beliefs formed in childhood, attitudes exhibited by family and friends. They import what are usually negative assumptions, formed earlier in life, into the workplace. Perhaps this is why Derek, who describes his own family background as "somewhat tense," remains so nervous around the other men and women at work. He claims to have experienced no job-related discrimination and considers his coworkers a reasonably tolerant group. His boss already knows that Derek is gay and at one point even told him that he's perfectly happy to have a gay man working for him. Even so, Derek is reluctant to lower his guard. "I'm starting to relax," he says, "and I'm going to get happy for the first time I can remember, and I'm not comfortable with that. You know, you're just too vulnerable. When you're most vulnerable, you don't *think* you are. I'm so comfortable now, that I keep thinking, 'You're on borrowed time, darling.'"

Derek's paranoia speaks to the pervasiveness of homophobia both in and out of the workplace. In a homophobic society there is a tendency simply to *presume* that a particular coworker, too, will be intolerant; without compelling evidence to the contrary, one treads lightly. "If I were going for a job interview, I probably wouldn't say, 'By the way, I'm gay,' because of the perception that it could result in my not getting the job," says Jim, a Philadelphia software engineer. "Remember, that's after *years* of having it beat into me that people will react negatively. I've never had a negative experience personally. Every single one has been positive. You could say that I should have learned by now. But I still have the perception that it's negative."

SEX AT WORK?

An obvious solution presents itself: Why not simply keep sexuality out of the workplace? If homosexuality offends or antagonizes one's coworkers, then why bring it to work? . . .

Gay men are often eager to accept this line of thinking, to believe that sexual and professional roles can be disentangled. A familiar refrain, heard from many, is the question: "What does being gay have to do with my career?" Michael, who runs a Philadelphia consulting firm, posed the question early in our interview. He spoke candidly about his job, his plans for the future, and how he felt about his boss and coworkers. But when asked how he managed his sexuality at work, Michael looked apologetic. "I guess I'm not the right person to answer that question. I keep them totally separate. I keep my private life private, and I don't let sexuality interfere with my work."

Not only is sexuality unrelated to work, the men say, but it also reveals little about them *as people.* Dan, the head of a psychiatric clinic in Houston, says, "Everyone on the staff knows who's gay and who's not. Some are more verbal about it than others. But it doesn't matter to anybody. They're people first, and sexual orientation comes second." Scott, a marketing representative with Blue Cross, says that, "My sexuality is secondary to who I am as a person and as a professional. It's really a separate issue." And Keith, a records clerk in a large Houston company, says, "If someone at work knows I'm gay, I'd like to think they can overlook that, that it's secondary to the person I am." . . .

Yet for all their eagerness to detach sexuality from work and to distinguish the sexual (and superficial) from the personal (and essential), gay men find the task a difficult one. Personal and professional roles are firmly entwined, and the reason for this is largely beyond their control: Work is a social activity. In the first issue of the *Harvard Business Review,* Daniel Starch made the commonsense observation that "business consists of human reactions and relations because business is done by human beings, and in that broad sense business is psychological in nature." To this it might be added that humans are also sexual creatures, that our social interactions are always colored by sexual possibilities, expectations, and constraints. We can't help but bring these capacities to work.

Indeed, offices are profoundly sexual places. Inside them sexuality is continuously on display, explicitly or implicitly part of the innumerable interpersonal exchanges that together constitute "work." Sexuality is alluded to in dress and self-presentation, in jokes and gossip, in looks and flirtations. It can also be found in secret affairs and dalliances, in fantasies, and in the range of coercive behaviors that we now call sexual harassment. Even when these more blatant manifestations are excluded, sexuality is part of every professional relationship. No one seriously believes that secretaries spend much time on their bosses' knees or that many executives have really slept their way to the top. Actual sexual contact is the exception rather than the norm, and jokes aside, most work relationships are centered on the tasks at hand. Yet sexual possibilities affect the way professionals perceive themselves and others. Sexual assumptions and attractions can be found in the jokes they tell, the clothes they wear, and the courtesies they extend or deny one another. A sexual subtext is often the basis for such intangibles as rapport, familiarity, and "chemistry." It can generate intense feelings of loyalty and personal commitment. Professional relationships often have the contours of, and are experienced as being like, seductions.

Most jobs involve some degree of socializing, which leads to countless situations that involve sexuality in one way or another. Wives or girlfriends are routinely invited to company outings, to dinners with clients, or to informal stops at the local bar. Even more often, they turn up as subjects of conversation. Dan recalls the excitement a nurse stirred up at the clinic when she announced her plans to have a baby. Her efforts to become pregnant, even her ovulation status, became a regular feature of the office small talk. Another man recalls a formal lunch at which a female coworker described, to an enthusiastic audience, the experience of "breaking water" before she gave birth. . . .

Because our society reads such significance into our sexual behavior, trust inevitably has a sexual dimension. "One of the things that comes up in any relationship that lasts is personal life," says Randy. "People want to talk to someone they have something in common with. They want to talk to someone who's forthcoming in many ways, including about his personal life. They introduce you to their wives, bring you into their lives, and they want to be brought into yours. It's a reciprocal thing. And if they feel they're giving of themselves, letting down their professional defenses, then you should too. It's only fair." Yet Randy has never discussed his sexuality with clients and doesn't always respond to their social overtures. "When you don't," he says, "you risk a little bit of their confidence in you. Outside of work, they may feel that you don't have that much in common, and they may prefer chatting and having conversations with someone else. Or, at worst, they'll just decide, 'I don't trust him.'"

A man's sexuality is thus unavoidably part of his "work." Because it plays a role in all work relationships, sexual identity becomes an informal part of the job description. In a man's sexual biography, potential employers look for clues to his worldview. Judging it to be unusual or unacceptable, they may doubt that he's the right person for the job. When selecting successors or peers, they reflexively seek candidates who share their social characteristics, and this sort of "homosocial reproduction" tends to keep a certain type of person in power. Some 86 percent of management jobs are never advertised through public channels but are filled by word of mouth, encouraging the evolution of restrictive hiring practices and exclusive information loops. Family status becomes symbolic of other shared life experiences, encouraging male executives to hire "family men" who, like them, understand the pressures of feeding a wife and kids. Those who break the mold seem unfamiliar, enigmatic, and less likely to perceive situations in the same way. They cannot be trusted. . . .

LEARNING TO MANAGE

Entering the workplace, a gay man faces a host of decisions. The head of personnel wants to know what kind of guy he is: How does he spend his time? What sort of skills and interests will he bring to the company? Will he get along with the others who work there? His officemates will invite him to the local tavern, to the baseball game, or to the Monday morning chat about their weekend conquests. As he climbs the corporate ladder, those above will want to know if he shares their values, if he faces the same demands from home and family, if he can be trusted. And from the guy next door, he'll hear about a favorite niece or neighbor, a truly remarkable woman who just happens to be single. . . .

Given the heterosexism that pervades most workplaces, it's little wonder that a gay man often assumes a disguise—sheltering himself, as it were, in a vast and crowded corporate closet. He satisfies their curiosity with fabrications. He laughs at their jokes and listens attentively to their stories, trying to act like one of them. He demurs on social invitations. He cultivates a reputation as an eccentric, a liberal, or as an exceedingly private man. He permits them to assume, incorrectly, that he is a heterosexual. However he hides, we say that he is "in the closet."

. . . [D]espite its apparent suitability, the "closet" actually misrepresents the ways gay men reveal and conceal themselves in social settings. A sexual identity encompasses far more than a man's sexual orientation, yet the closet implies that it is a simple, discrete bit of information (like eye color or height) that one could, with a single word or deed, *reveal*. It further depicts the construction of an identity as a choice between possible end states (in the closet, out of the closet), while ignoring the means by which a man arrives at either. Consequently, when we speak of someone who has "come out," we've said almost nothing about what actually took place. What did he reveal? An affinity for other men? For a particular sexual act? For a lover of many years? Closet language collapses these various disclosures into a single, undifferentiated act—an exit from the closet. As Marny Hall points out in her study of lesbian professionals, "'Coming out' is not an end point in the strategy of adjustment. Rather, it is a conceptual short cut, an abbreviated way of thinking which fails to encompass the extremely complex process of managing discrediting information about oneself."

At the same time the closet casts secrecy in passive terms (to disguise himself, a man simply "stays" in the closet), which fails to encompass the countless lies, conversational dodges, and petty deceptions such stasis actually requires. To stay in the closet, a man might fabricate girlfriends or sexual exploits. He might avoid all discussions of sexuality, insisting that others respect his privacy. Or he might complain about his status as a confirmed bachelor, as someone unlucky in love, as someone with an old war wound. Staying in the closet can be a bit like staying on a treadmill.

Closet language makes it impossible to talk about the way a man's sexuality is actually disguised or revealed in any of these situations. When asked if he had "come out" at work, Martin shrugged at the question. "I'm not sure how to respond," he began. "You'll have to tell me what you mean by 'come out.' I worry not only about who knows I'm gay, but how they know, what else they know, and how I have to behave as a result. Do I tell them I'm gay and then drop the subject, or do we make this part of our daily conversation? And what about my love life or about sex? Do we talk about that? The gay pride march I went to? My favorite gay author? Coming out is the beginning of the story." When asked the same question, a Boston executive was more blunt. "There are so many ways to be 'out.' Tell me what the term means, and I'll tell you if I fit the bill."

What is needed is a rhetoric that hovers nearer the lived experiences of gay professionals, a vocabulary that allows us to describe their specific interpersonal maneuvers and countermaneuvers. By necessity gay men become adept at thinking about self-disclosure and learn, sometimes at great cost, that different sexual identities bring different social consequences. They become self-conscious. They learn to control and monitor outward appearances, to distort them when necessary. They learn to dodge. For many the result is a calculating, deliberate way of approaching social encounters. One can say, as I intend to, that they *manage* their sexual identities at work. . . .

Instead sexual orientation is usually inferred from behaviors that do not themselves constitute sex. In employment situations, people communicate *about* sexual behaviors more often then they display them. We symbolize, represent, and talk about sexual orientation. We hear secondhand accounts of actual, intended, or desired sexual contact. We are introduced to girlfriends, boyfriends, and spouses and accept the sexual implications of these labels. We hear that a male coworker saw a new movie—the one about two straight people falling in love, the one featuring the sexy starlet, the one he attended with the single woman in accounting—and discern the multiple, entwined heterosexual scripts that were performed that evening. We also notice that a man has a firm handshake and an impressive knowledge of baseball trivia, and interpret these as signs that he is a masculine man, and thus a heterosexual as well. We make judgments about the sexual behavior of others, in other words, without having observed it.

For gay men, the resulting gap between psychic and social reality poses countless decisions—to display or not to display, to tell or not to tell, and in each case, to whom, how, when, and where—that become central to their navigation of the world. Gay lives and careers are characterized by a preoccupation with self-disclosure and skill in the management of sexual identity. "Coming out" stories figure prominently in gay folklore, and there is an elaborate argot within gay communities for talking about self-disclosure and its consequences. Few other decisions are a source of such intense, recurrent concern. . . .

31

The Indignities of Unemployment

KENNETH W. BROWN

I am a number, a statistic, a peak on a national bar graph. I am not physically or mentally disabled. But I am financially disabled; I am unemployed.

I am unemployed after nearly 14 years of being a loyal, hardworking employee. I bought into the company philosophy, followed the company credo and worked hard to help accomplish the corporate goal. As I was coddled, supported, and groomed, I grew to believe that I was an important member of the corporate family, only to be unceremoniously dumped after I had served my purpose and was no longer needed or welcomed, like a bastard son at a family reunion.

I am making every effort to find work. I have made countless phone inquiries, answered many classified ads and contacted people I have not spoken to in years, looking for job leads. I have attended job fairs in search of another chance to rejoin corporate America, struggling to maintain my sense of dignity as I walked about the crowded rooms—looking along with so many others for the next opportunity to reenter the economic mainstream.

As unemployment refuses to loosen its grasp, I find it increasingly difficult to remain hopeful, positive, and confident. Prolonged unemployment can do a number on your psyche, your soul, and your spirit. Unemployment can make you question your skills and abilities. And it can damage your self-esteem. I don't care how much confidence you have; unemployment can shake that confidence.

As I search for my next position, I do so with growing reluctance and trepidation. When I prepare for the next job interview, I practice my responses to the questions about myself, my strengths, my weaknesses, and my future goals.

I make sure I speak the King's English with perfect enunciation, ever mindful not to let street slang or my ethnicity slip into my conversation.

I often wonder whether my white counterparts are as conscious of their diction as I am of mine. I also wonder whether I am selling out or compromising myself too much. Do white men wonder if they are being too forceful, too eager to please, too assertive? Do they agonize as much as I do over how to appear skilled and capable without intimidating the people who have the power to hire? The thought of always keeping these ideas in mind is exhausting and sad. Is this the cross that African-American men and women must bear in order to be accepted into corporate America?

Unemployment can make you question your faith and spiritual foundation. I pray for my unemployment to end. I pray for direction and guidance. I also pray for the day when I do not have to wonder whether my blackness is a barrier to reentry into the labor force. I am still waiting for a response. Unemployment makes me wonder if anyone is listening.

Being unemployed also makes me question my manhood. I was raised to be a responsible household member who would carry my share of the weight. But now that I am no longer holding up my end of the financial bargain, my wife and I exist on a single income: hers. While I am fortunate that she has been supportive throughout my long

From *Essence,* Sept. 1995, V. 26, No. 5, p. 56.

period of unemployment, I must fight hard against seeing myself as a failure or loser. I have never before considered myself a loser, and I refuse to be one. Every day I must remind myself that being without a job does not make me less of a man.

Some good things have resulted from my unemployment. I have gained a renewed awareness of life's blessings: the value of true friends and the love, support, encouragement and prayers of family. In many ways I have become a stronger person. I have resolved to survive unemployment. I resolve to become a statistic for working Americans, and not just another African-American male without a job.

32

Retirement Dinner

BOB GREENE

The event was a retirement dinner for a man who had spent forty years with the same company. A private dining room had been rented for the evening, and the man's colleagues from his office were in attendance. Speeches and toasts were planned, and a gift was to be presented. It was probably like a thousand other retirement dinners that were being held around the country that night, but this one felt a little different because the man who was retiring was my father.

I flew in for the dinner, but I was really not a part of it; a man's work is quite separate from his family, and the people in the room were as foreign to me as I was to them. To me most of them were names, overheard at the dinner table all my life as my father sat down to his meal after a day at the office; to them I was the kid in the framed photograph on my father's desk.

Names from a lifetime at the same job; it occurred to me, looking at the men and women in the room, that my father had worked for that same company since the time that Franklin Delano Roosevelt was president. Now my father was sixty-five, and the rules said that he must retire; looking at the faces of the men and women, I tried to recall the images of each of them that I had built up over the years at our family dinners.

My father was seated at a different table from me on this night; he appeared to be vaguely uncomfortable, and I could understand why. My mother was in the room, and my sister and brother; it was virtually the first time in my father's life that there had been any mix at all between his family and his work. The people at his office knew he had a family, and we at home knew he had a job, but that is as close as it ever came. And now we were all together.

A man's work, if he is any good at it, is as important to him as his family. That is a fact that the family must, of necessity, ignore, and if the man were ever confronted with it he would have to deny it. Such a delicate balance: the attention that must be paid to each detail of the job, and then the attention that must be paid to each detail of the family, with never the luxury of an overlap.

The speeches began, and, as I had expected, much of their content meant nothing to me. They were filled with references and in-jokes about things with which I was not familiar; I saw my father laughing and nodding his head in recognition as every speaker took his turn, and often the people in the room would roar with glee at something that drew a complete blank with me. And again it occurred to me: a man spends a life with you, but it is really only half a life; the other half belongs to a world you know nothing about.

The speeches were specific and not general; the men and women spoke of little matters that had happened over the course of the years, and each remembrance was like a small gift to my father, sitting and listening. None of us really changes the world in our lifetimes, but we touch the people around us in ways that may last, and that is the real purpose of a retirement dinner like this one—to tell a man that those memories will remain, even though the rules say that he has to go away.

I found myself thinking about that—about how my father was going to feel the next morning, knowing that for the first time in his adult life he would not be driving to the building where the rest of these people would be reporting for work. The separation pains have to be just as strong as to the loss of a family member, and yet in the world of the American work force, a man is supposed to accept it and even embrace it. I tried not to think about it too hard.

When it was my father's turn to speak, his tone of voice had a different sound to me than the one I knew from around the house of my growing up; at first I thought that it came from the emotion of the evening, but then it struck me that this probably was not true; the voice I was hearing probably was the one he always used at the office, the one I had never heard.

During my father's speech a waiter came into the room with a message for another man from the company; the man went to a phone just outside the room. From my table, I could hear him talking. There was a problem at the plant, something about a malfunction in some water pipes. The man gave some hurried instructions into the phone, saying which levers to shut off and which plumbing company to call for night emergency service.

It was a call my father might have had to deal with on other nights, but on this night the unspoken rule was that he was no longer part of all that. The man put down the phone and came back into the dining room, and my father was still standing up, talking about things unfamiliar to me.

I thought about how little I really know about him. And I realized that it was not just me; we are a whole nation of sons who think they know their fathers, but who come to understand on a night like this that they are really only half of their fathers' lives. Work is a mysterious thing; many of us claim to hate it, but it takes a grip on us that is so fierce that it captures emotions and loyalties we never knew were there. The gift was presented, and then, his forty years of work at an end, my father went back to his home, and I went back to mine.

6

Men in Trouble

Students are often puzzled about why men would critique hegemonic masculinity and challenge the social structures within which men and women work, love, and live. After all, in economic and political terms, men are undeniably the privileged gender. On average, they earn or own more than women do, monopolize positions of power, and dominate positions of prestige. These indicators of gender stratification might lead one to conclude that it must be great to be a man. That makes it especially important to look at some of the downside of men's lives.

Without asserting or implying, as some do (e.g., Farrell 1993), that ultimately men are worse off than women or bigger victims of the social system, like women, men experience serious negative consequences as a result of their efforts to conform to expectations attached to their gender. This chapter ties some of the more notable costs that men pay in their health and well-being, and in their propensity to commit and be victimized by crime, to their efforts to live up to expectations attached to masculinity. These issues help us explain why some men ultimately question the ideology of gender and the structure of relationships between women and men. Collectively, they serve as a useful precursor to the final chapter, where we will look at some contemporary challenges to our social and cultural constructions of masculinity.

IN SICKNESS AND IN HEALTH

When we look at issues like disability, disease, and death, we see that *both* men and women suffer hazards to their health and well-being by virtue of their sex and gender. Recall the distinction made earlier between sex and gender. *Sex* refers to the anatomical, chromosomal, and hormonal differences between men and women. The cultural expectations that define what it means to be a woman or a man, feminine or masculine, along with the different status positions men and women occupy, are what we referred to as *gender.* Both sex and gender contribute to the critical differences between men's and women's mortality rates, life expectancies, health, and illness. The relative contributions of sex and gender are more fully assessed in the selection by Christopher Kilmartin (Reading 33).

At every stage of life, men die more often and earlier than women. As James Doyle notes, "throughout most of this century a man's chances of dying have been greater than a woman's at *every* age—from birth to over 80" (Doyle 1995). Male life expectancy trails female life expectancy among both whites and blacks, with black males especially at risk of early death. Male vulnerability is evident even prior to birth. Although more males are conceived than females, males are much more likely than females to be miscarried or stillborn, and they more often die before their first birthdays (Basow 1992; Harrison, Chin, and Ficarrotto 1995). Differences in prenatal and infant mortality are relatively free of cultural influences in that they surface prior to any direct access to and control over the environments in which males and females later live. Thus, they strongly suggest a more sex-based or *biogenic* explanation for the observed differences (Kilmartin 1994 and Reading 33).

As we age, this gender pattern continues, but for more mixed reasons. Young men are much more vulnerable than young women; at age 15–24 the ratio of male to female deaths is nearly 3:1. As described by Kilmartin, at the opposite end of the life span, women are twice as likely to survive past age 80 as are men, and 11 of 12 wives outlive their husbands. Both of these differences, in youth and old age, reflect an interplay between biological and role-related differences that define being male and female.

Whether we are addressing disease-related or lifestyle-related deaths, women appear to be heartier than men. For most of the fifteen leading causes of death, men's death rates exceed women's. This includes a slight male "advantage" in death due to heart disease, a bigger male "advantage" in most cancers and in pneumonia and influenza, and a large male "advantage" in death due to chronic obstructive pulmonary disease, liver disease, HIV, and accidental, suicidal, or homicidal death. Men are also more likely than women to die in auto accidents, be murdered, or commit suicide (although women attempt suicide more often than do men). In the United States, men also die from HIV infection at as much as eight times the rate that women do (Doyle and Paludi 1998; Lindsey 1997). Women, on the other hand, die more than men do from cerebrovascular diseases (such as strokes) and from atherosclerosis, both of which are associated with aging, and likely result from women living longer than men.

Kilmartin identifies a variety of biological factors that affect men's health and curtail their lives. Interacting with these factors are the various cultural and social structural characteristics of gender that exist even within the disease-related deaths noted above. Men abuse their bodies with illegal drugs, alcohol, and tobacco more often and more heavily than do women (Renzetti and Curran 1995). Additionally, some diseases like cancer and certain pulmonary diseases can be linked to occupational risks accompanying certain male-dominated occupations (e.g., construction, mining, and so on), which expose one to various lethal substances (e.g., asbestos, coal dust).

Therefore, *masculinity,* as defined in the inexpressive, competitive, aggressive, risk-taking ways that it is in this culture, helps explain why men live less long, if not less well, than women. The strong, early proscription on anything feminine covers a lot of areas of life, such as displays of affection, disclosure of feelings, expressions of vulnerability. The determination to avoid the stigmatizing epithets "sissy" or "fag" can lead males to many risk-taking or risk-ignoring behaviors. The high expectations for success leads to further health-depleting problems. Men must achieve; men must evoke awe and admiration; men must win. Implied in this broad theme is the competitiveness that drives so many men to conspicuous displays of accomplishment and to exaggeration of the importance of coming out on top. In adolescence that may mean showing athletic prowess or a willingness to fight; later it may shift to sexual conquest, and—where possible—still later, to being a person with significant economic clout.

The toughness, stoicism, and dependability within dominant constructions of masculinity may lead men to subject themselves to excessive stresses, take on more than they can "safely handle," and shoulder loads that begin to wear them down. Definitions of what it means to be female or male may allow women greater ease in reporting illnesses and may discourage men from acknowledging ill health (Reading 33). Finally, masculine aggressiveness is associated with male suicide, homicide, and accidental death rates.

Norms associated with masculinity are also displayed by men suffering emotionally. Terrence Real (Reading 34) shows how the perception that vulnerability is unmanly may explain why men suffer silently, a "covert depression." Because men may hide or deny their sense of despair and fail to seek help, men's experiences with depression are more likely to be missed by the conventional estimates and assessments of the disease. Bill Minutaglio (Reading 36) and Susan Faludi (Reading 37) show how some instances of suicidal and/or homicidal violence can be understood by examining men's inabilities to manage the stresses or understand and control the symptoms they face.

On top of the cultural script defining masculinity, men play the role of breadwinner/provider. The male work role is typically the major source of income for the family. Although most family households are dual-earner households, because of the gendered wage gap and women leaving the labor force to bear and care for young children, many men spend some time as major (or sole) providers. As we saw, they do this in an economic environment that is inflexible and greedy in its expectations of employees (Hochschild 1997). Men bring to these environments the attributes that comprise masculinity. Cumulatively, men's responsibilities, performed in the fashion in which men have been socialized to act, can become "hazardous to one's health" (Harrison, Chin, and Ficarrotto 1995). They may also provoke emotional or behavioral distress when things go awry.

Mark Barton, the Atlanta area day trader who bludgeoned his wife and two children to death before shooting and killing nine others, wounding thirteen more, and finally killing himself, was overwhelmed by mounting, inescapable debt that compounded his deep and growing sense of failure as a provider (see Reading 37). The suicide of former University of Georgia football coach Wayne McDuffie was similarly connected to his economic despair. His loss of work and his difficulties finding another coaching position simply became more than he could bear.

MEN AS VICTIMS AND OFFENDERS

As men move from boyhood to manhood, their lives often detour in troubled directions, including crime, violence, prison, or death. All available data on crime show that criminal behavior is gendered. The F.B.I.'s Uniform Crime Reports of arrest data suggest that crime, especially serious crime, is overwhelmingly a male activity. Males comprise nearly 90 percent of those arrested for violent crimes and three-quarters of those arrested for property crimes. Around 90 percent or more of the arrests for murder and non-negligent homicide, rape, aggravated assault, robbery, burglary, arson, and auto theft are arrests of men. The Bureau of Justice Statistics' National Crime Victimization Survey reflects the same pattern. More than 80 percent of the victims of crime report their assailants as males, further validating the gendered reality of crime.

Patterns of crime victimization add an additional dimension to the understanding of troubled aspects of men's lives. Compared to women, men are far more often the victims of violent crimes like murder, assault, and robbery. In fact, with the exceptions of rape and domestic violence, men are more frequent victims of crime, and their victimizers are other men.

It would be dangerously simplistic to conclude from such statistics that "crime results from masculinity." After all, crime is more complex than such an argument would reveal. Women commit crimes, and their rate of criminal involvement has increased in recent years at a faster rate than men's criminal involvement (McCaghy and Capron 1997; Siegel 1998). The fact that most men are not criminals (and that some women are) indicates that more is going on in the lives of men who commit crimes than just the fact that they are men. Borrowing author Mark Gerzon's comment about connections between masculinity and war: "Only an ideologue would claim that the sole cause of *crime* is masculinity. But only a fool would claim there is no connection at all" (Gerzon 1992). Being neither an ideologue nor a fool, I recognize the deeper complexity of crime causation, but I believe, too, that masculinity plays a part in the processes that produce many criminals and victims.

Characteristics associated with male socialization and dominant constructions of masculinity have been linked by criminologists to a variety of crimes ranging from robbery, through assault (including domestic assault), to rape and homicide (McCaghy and Capron 1997; Daly 1994; Scully 1990). Since men are socialized to be aggressive, that capacity for aggression can spill over into criminal violence. The need for men to be economically successful can lead men of any class to act to acquire more and to use any means necessary to do so. The sense that men should be the "kings of their castles" and heads of their households leads some men to exert excessive and abusive authority over their wives and children in order to achieve compliance in family matters.

Connections can be drawn between elements of male upbringing, hegemonic masculinity, and the perpetration of crimes of sexual assault (Lisak 1997). Research has identified a set of rationalizations and justifications (e.g., "She asked for it," "She said 'no,' but she meant 'yes,'" "She enjoyed it," and the like) used by both convicted rapists and by college men charged with acquaintance rape (Scully 1990; Sanday 1990). Such rape neutralizations are not far from the more general cultural scripts regarding sex, gender, and relationships (McCaghy and Capron 1997). Beliefs that men will initiate and pursue sexual encounters, or that once aroused men cannot control their sexual desires, are two more general cultural beliefs that can be stretched to a point where rapes are tragic outcomes. In Reading 35, these attitudes are evident in but not fully explanatory of campus date or acquaintance rapes. Since some fraternity houses are "safer places" than others, characteristics of the settings are also important elements in understanding such phenomena as fraternity rape.

Acknowledging the connection between masculinity and crime is a beginning. In considering the mechanism that best explains that connection, researchers have gone in a number of different directions. For example, David Lisak considers the social and cultural constructions of masculinity learned during socialization as causal factors in crimes of sexual assault (Lisak 1997). Economist Richard Freeman emphasizes the economic opportunities, pressures, and incentives that drive low-income men toward crime. He asserts that the shortage of jobs for less educated men, especially young black males, is the single most important reason so many young men commit crimes (Freeman 1996). Sociologist Elijah Anderson (Anderson 1996) suggests a more cultural argument, wherein the "meaning of manhood" and the need to be respected as a man take particular turns in inner-city neighborhoods. Displaying a kind of ruthless "nerve" and a willingness to commit acts of violence are parts of an oppositional culture of manhood that emerges when avenues for more conventional measures of respect are lacking (Anderson 1996).

Rather than seeing internalized masculinity as pushing men toward crime, some reverse this connection and argue that crimes are acts men do to express or "accomplish" masculinity. As Robert Connell argues, "[M]asculinity is not something that is performed or settled. It is something that has to be made, and criminal behavior is one of the means for its making. A great deal of crime makes sense only when it is seen as a resource for the making of gender, and in most cases, that means it is a strategy of masculinity" (Connell 1995). In *Masculinities and Crime,* James Messerschmidt declares, "When men enter a setting, they undertake social practices that demonstrate they are 'manly.' The only way others can judge their 'essential nature' as men is through their behavior and appearance. . . . For many men, crime may serve as a suitable *resource* for 'doing gender'—for separating them from all that is feminine" (Messerschmidt 1993, 84). When other resources for accomplishing masculinity are unavailable, men may resort to social practices that are criminal. In so doing, they are not just performing according to mandates of their male role, they are actively constructing a situationally suitable masculinity.

Men versus Men

Thus far we have connected health, crime, and victimization problems to men's attempts to meet the mandates of masculinity, but there are important variations among men. Factors such as class and race, especially, once again merit our attention.

Longevity and good health are among the many life chances that are affected by one's economic status. Exposure to occupational risks or hazards, poor nutrition, substandard housing, and inadequate or unavailable medical care make lower- and working-class men and women more vulnerable to sickness and death than middle- and upper-class men and women. The lesser job security, lower incomes, unpaid bills, higher divorce rates, and greater risk of crime victimization faced by poorer men and women also negatively affect their mental health. In fact, at every stage of life, the lower one's income, the more often she or he is sick and the more likely she or he is to die before the expected age (Henslin 1998). This pattern holds for both men and women, but it means that the lower a man's economic standing, the less well and long he lives compared to his more affluent male counterparts.

Race further differentiates among men. African American men and women both have shorter life expectancies than white men and women. This fits a broader picture in which health and well-being are affected by race. African Americans have twice the rate of infant mortality, four times the rate of maternal mortality (deaths in childbirth), and six to eight years' shorter life spans. The shortest life expectancy is found among African American men, whose life expectancy trails both African American women and white men by more than eight years.

Through the past two decades we have seen the emergence of acquired immune deficiency syndrome, a disease that is the leading killer of 25–44-year-old men, but which surfaces unevenly among men infected by it. The primary sources of AIDS infection continue to be unsafe sexual practices, especially anal intercourse or unprotected vaginal sex, and the sharing of contaminated needles by intravenous drug users. Since its first diagnoses in 1981, AIDS has taken its biggest toll on the gay male population, leading to illness, death, heightened homophobia, and altered lifestyles. Although gay men represent the largest group of men infected with HIV, AIDS also occurs disproportionately among minority males. African American men have a rate of infection nearly five times that of white males (Sabo 1998). Furthermore, though still a mostly male disease in the United States, worldwide AIDS infections are increasing most rapidly among women (Lindsey 1997), and the number of yearly cases of AIDS among women will soon exceed the number of diagnosed cases among men.

Variations can also be found in men's experiences of crime and victimization. Available data on offenders show that the men who commit the majority of the most serious crimes are disproportionately young, lower-income, minority males. One may question whether police reports and arrest data are accurate representations of the reality of crime, but the Uniform Crime Reports consistently show that crimes such as assault, homicide, robbery, and burglary are committed and experienced disproportionately by young, low-income, nonwhite males (McCaghy and Capron 1997; Siegel 1998).

Of course, lower-income, especially nonwhite men are also more vulnerable to arrest and incarceration than white males of higher economic standing. Researchers have shown, for example, that rates of delinquency among white middle-class males and lower-income white males are more similar than different. The types of acts they commit and the chronic nature of such commission, nonetheless, reveal class differences. Men of all classes commit crimes, but serious crimes like assault, robbery, rape, and homicide happen more frequently among lower-income men and disproportionately among nonwhite men (McCaghy and Capron 1997). Vincent Sacco and Leslie Kennedy state, "[O]ffending is associated with various measures of social and economic disadvantage" (Sacco and Kennedy 1996).

Victimization data show that though they are more frequent victims of serious violent and property crimes, all men are not victimized equally. The rate of male victimization drops as family income goes up. Males with family incomes under $10,000 suffer more than twice the rate of violent crime victimization as men whose family incomes exceed $50,000. African American men face higher rates of violent crime victimization than white men, and Hispanic men face higher rates than non-Hispanic men (Sacco and Kennedy 1996). A stunning illustration of racial diversity in relative victimization risk is provided by the F.B.I. A white male, age 21, faced a 1 in 240 chance of being murdered; among black males the odds were a disturbing *1 out of 32* (McCaghy and Capron 1997).

Gay men face their own particular threats, illustrated by the brutal murders of University of Wyoming student Matthew Shepard and Alabama textile mill worker Billy Jack Gaither. Both men were killed because they were gay. Gay men are frequent targets of hate crimes. According to the Uniform Crime Reports for 1997, hate crimes based on sexual orientation represent nearly 14 percent of all reported hate crimes, trailing only racially motivated (59 percent) and religiously motivated (17 percent) offenses. In fact, across the range of hate crimes, gay men and lesbians are the most sought targets of "thrill-seeking" young men. One perpetrator of a brutal stabbing of a gay man in New York City recounted how he and two friends went out searching for "a drug addict, a homo, or homeless" person to attack. They found their victim, twenty-nine-year-old Julio Rivera, and then "stabbed and killed him . . . because he was gay" (Levin and McDevitt 1993).

In surveys on victimization, over 90 percent of gay men and lesbians report having received threats or verbal harassment because of their sexual orientation (Renzetti and Curran 1995). One study of antigay violence in eight U.S. cities said that 19 percent of the gays and lesbians surveyed reported having been punched, kicked, hit, or beaten at least once because of their sexual orientation. Almost half (44 percent) had been threatened with such violence. Almost all (94 percent) had suffered some victimization (i.e., verbal abuse, being spat upon, being chased or pelted with objects, being physically assaulted). McCaghy and Capron report that lifetime assault victimization rates for homosexuals range from 11 to 20 percent (McCaghy and Capron 1997).

REFERENCES

Anderson, E. "The Code of the Streets." In *Crime,* Vol. 1 of *Crime & Society,* edited by R. Crutchfield, G. Bridges, and J. Weis. Thousand Oaks, CA: Pine Forge, 1996.

Basow, S. *Gender Stereotypes & Roles.* Pacific Grove, CA: Brooks/Cole, 1992.

Connell, R. *Masculinities.* Berkeley: University of California Press, 1995.

Daly, K. *Gender, Crime, and Punishment.* New Haven: Yale University Press, 1994.

Doyle, J. *The Male Experience.* 3rd ed. Dubuque, IA: William C. Brown, 1995.

Doyle, J., and M. Paludi. *Sex and Gender: The Human Experience.* 4th ed. Boston: McGraw-Hill, 1998.

Farrell, W. *The Myth of Male Power.* New York: Simon & Schuster, 1993.

Freeman, R. "Why Do So Many Young American Men Commit Crimes and What Might We Do About It?" *Journal of Economic Perspectives* 10, no. 1 (Winter 1996): 25–42.

Gerzon, M. *A Choice of Heroes: The Changing Faces of American Manhood.* 2nd ed. Boston: Houghton Mifflin, 1992.

Harrison, J., J. Chin, and T. Ficarrotto. "Warning: Masculinity May Be Dangerous to Your Health." In *Men's Lives,* 3rd ed., edited by M. Kimmel and M. Messner, 237–249. Needham Heights, MA: Allyn & Bacon, 1995.

Henslin, J. *Essentials of Sociology: A Down-to-Earth Approach.* Needham Heights, MA: Allyn & Bacon, 1998.

Hochschild, A. *The Time Bind: When Work Becomes Home and Home Becomes Work.* New York: Henry Holt, 1997.

Kilmartin, C. *The Masculine Self.* New York: Macmillan, 1994.

Levin, J., and J. McDevitt. *Hate Crimes: The Rising Tide of Bigotry and Bloodshed.* New York: Plenum, 1993.

Lindsey, L. *Gender Roles: A Sociological Perspective.* 3rd ed. Upper Saddle River, NJ: Prentice Hall, 1997.

Lisak, D. "Male Gender Socialization and the Perpetration of Sexual Abuse." In *Men and Sex: New Psychological Perspectives,* edited by R. Levant and G. Brooks, 154–175. New York: John Wiley and Sons, 1997.

McCaghy, C., and T. Capron. *Deviant Behavior: Crime, Conflict, and Interest Groups.* Needham Heights, MA: Allyn & Bacon, 1997.

Messerschmidt, J. *Masculinities and Crime.* Lanham, MD: Rowman and Littlefield, 1993.

Renzetti, C., and D. Curran. *Women, Men, and Society.* 3rd ed. Boston: Allyn & Bacon, 1995.

Sabo, D. "Masculinities and Men's Health: Moving Toward Post-Superman Era Prevention." In *Men's Lives,* 4th ed., edited by M. Kimmel and M. Messner, 347–361. New York: Macmillan, 1998.

Sacco, V., and L. Kennedy. *The Criminal Event: An Introduction to Criminology.* Belmont, CA: Wadsworth, 1996.

Sanday, P. *Fraternity Gang Rape: Sex, Brotherhood, and Privilege on Campus.* New York: New York University Press, 1990.

Scully, D. *Understanding Sexual Violence: A Study of Convicted Rapists.* Boston: Unwin Hyman, 1990.

Siegel, L. *Criminology.* 6th ed. Belmont, CA: West/Wadsworth, 1998.

FOR ADDITIONAL READING

To see how the health status of men is affected by their socialization and lifestyles, read Will Courtenay's study of college men's vulnerabilities ("Men's Health: An Overview and a Call to Action," *Journal of American College Health,* May 1998, vol. 46, no. 6). Courtenay shows how college men live unhealthy lifestyles, engage in more risk-taking behavior, and have difficulty expressing their health problems. In tying these to the dominant construction of masculinity, he suggests that colleges may need to consider and develop gender-specific programming about health. One can also read articles about older men's health ("Update on Men's Health," by John Morley, in *Generations,* Winter 1996, vol. 20, no. 4), Canadian men's health ("Men's Health" by Mark Nichols and Susan Oh, in *Maclean's,* February 22, 1999 issue), and specific health concerns of men (e.g., men's sexual health).

For follow-up reading on connections between masculinity and crime, read Christine Alder and Kenneth Polk's article, "Masculinity and Child Homicide" (*British Journal of Criminology,* Summer 1996, vol. 36, no. 3), or Jo Goodey's "Boys Don't Cry: Fear of Crime and Fearlessness" (also from the *British Journal of Criminology,* Summer 1997, vol. 37, no. 3). Both articles offer disturbing qualitative material that shows how hegemonic masculinity can induce criminal behavior, either by imposing an expectation of fearlessness onto men, or by presenting men with situations that test their masculinity and offer them occasions to display their male suitability.

WADSWORTH VIRTUAL SOCIETY RESOURCE CENTER

http://sociology.wadsworth.com

Visit *Virtual Society* to obtain current updates in the field, surfing tips, career information, and more.

INFOTRAC COLLEGE EDITION: EXERCISE

Go to the *InfoTrac College Edition* web site to find further readings on the topics discussed in this chapter.

33

Surviving and Thriving: Men and Physical Health

CHRISTOPHER KILMARTIN

There are sex differences in the statistical incidences of many physical problems. For example, males are more likely than females to contract heart disease, emphysema, and most forms of cancer. On the average, men die at a significantly earlier age than women. Many researchers believe that sex differences in disease and longevity cannot be explained solely by biological differences between males and females.

A good deal of evidence has led theorists to suggest that certain aspects of traditional masculinity are at least partially responsible for men's problems with disease and longevity. In this chapter, we describe some of these problems and review the relevant psychological literature on physical health as it relates to gender.

SEX DIFFERENCES IN LONGEVITY

In 1991, the United States Bureau of the Census (USBC) reported that the average life expectancy for African-American males born in the United States in 1988 was 64.9 years, compared with 73.4 years for African-American females. For whites, the expectancies were 72.3 years and 78.9 years for males and females, respectively. As you can see from these data, minority men are especially at risk for early death. They are more likely than majority men to live in hazardous and stressful environments, as well as to lack access to health care.

Table 33.1 details sex ratios for United States deaths in 1988. As you can see, males are more likely to die than females at every stage of life until age 75 (the ratios change direction at that point because so many more women than men have survived to that age). At ages 15 to 24, the ratio of male to female deaths is a staggering 297:100! *At best,* U.S. males die "before their time" at about a 5 to 4 ratio to females. At worst, the ratio is almost 3 to 1. Contrary to the social belief that males are heartier and more resistant to disease, the evidence is indisputable that males are more vulnerable than females at every age.

At birth, there is a slight imbalance in the ratio of males to females. Conception favors males because Y-chromosome-bearing sperm (androsperm) are more motile than X-chromosome-bearing sperm (gynosperm), and thus are more likely to fertilize the ovum. Despite the fact that male fetuses are more likely to have problems in utero, leading to spontaneous abortion (miscarriage), there are between 103 and 106 male births for every 100 female births. . . . But, because males die at higher rates than females, parity (an equal number of males and females in the age-group population) is reached somewhere between the ages of 25 and 34. . . . From this age range and up, women outnumber men. Twice as many women as men survive beyond age 80, and 11 out of 12 U.S. wives outlive husbands. . . .

From *The Masculine Self, Macmillan,* 1994, pp. 151–164. Reprinted by permission of the author.

Table 33.1 Ratio of Male to Female Deaths (1988 U.S. Data)

Age in Years	Male: Female
under 15	134:100
15–24	297:100
25–34	265:100
35–44	210:100
45–54	169:100
55–64	158:100
65–74	139:100
75–84	96:100
over 85	49:100

Note: Calculated from data in United States Bureau of the Census, *Statistical Abstract of the United States: 1991* (111th ed.). Washington, DC: U.S. Government Printing Office, 1991.

SEX DIFFERENCES IN DISEASE

There is some evidence to suggest that women get ill more often than men do. Women report feeling sicker than men on a day-to-day basis. For instance, women are more likely than men to report being bothered by headaches, bladder infections, arthritis, corns and calluses, constipation, hemorrhoids, and varicose veins. We cannot be sure to what extent these differences are real and to what extent they reflect the social permission that women have to report illness. When it comes to serious (life-threatening) diseases, which are, of course, hard to ignore, men outnumber women in almost every category. . . .

The two diseases that most often cause death are heart disease and cancer. About 35% of all people eventually die of heart disease. The male: female ratio for cause of death by heart disease is about 101:100 . . . , a very small sex difference. When we look at these data *by age group,* however, a very different picture emerges. Between the ages of 25 and 44, the ratio of male to female deaths from heart disease in 1988 was 283:100. . . . Although many people die of heart disease, males tend to die much earlier than females.

With regard to cancer, sex ratios differ depending on the location of the cancerous tumor in the body. As a cause of death, men lead women in every category except breast cancer. The greatest differences are in mouth and throat cancer (250:100), lung and respiratory cancer (263:100), and cancer of the urinary organs (201:100). . . .

Other diseases (e.g., influenza, liver disease, and diabetes) kill men in greater numbers than women in almost every category. Some of these differences are relatively large (e.g., respiratory diseases, 191:100), others are smaller (e.g., pneumonia and influenza, 113:100), and women lead men in a few categories, such as cerebrovascular accident (stroke), in which the male to female ratio is 66:100. . . .

In the United States in 1989, Acquired Immunodeficiency Syndrome (AIDS) killed 19,499 males and 2,176 females, a ratio of almost 9:1. In 1986 this ratio was 12:1. . . . Sex ratios from AIDS deaths will probably continue to change. As with many other causes of death, AIDS strikes minority men disproportionately. . . .

OTHER CAUSES OF DEATH

There are sex differences in the incidence of accidents, suicide, and homicide, and they are very large ones. Males are much more likely than females to die in motor vehicle or other accidents (210:100), to commit suicide (382:100), or to be victims of homicide (315:100). . . . Thus, a staggering number of physically healthy men die "before their time" from causes that are somewhat preventable. In fact, homicide is the leading cause of death for African-American men between the ages of 15 and 24. . . .

WHY DO MEN LIVE LESS LONG THAN WOMEN?

The preceding might strike you as an awkwardly worded title. It would be smoother to ask "Why do women live longer than men?" However, it appears that the difference in life span may be due more to men's lives being shortened rather than women's lives being lengthened.

Around the beginning of the 20th century, men's and women's lives, on the average, were about the same length. In modern industrial nations such as the United States, there has been a dramatic reduction in the risk of death from pregnancy and childbirth, which were relatively dangerous at the beginning of the 20th century. The decrease of this risk resulted in the life-span sex differential. It could be said that both women's and men's lives were shortened 100 years ago, and that we have found ways to stop shortening women's lives. Hopefully, the same can be done for men. However, it may become apparent to you that this is a complicated process.

There are two basic types of explanations for the sex difference in average life span. The first is a *biogenic* explanation. From this viewpoint, men die earlier because of genetic, hormonal, or other biological differences between the sexes. The second type of explanation is a *psychogenic* one in which sex differences in life span are attributed to gender differences in psychological and social areas such as behaviors, socialization, and ways of dealing with the self.

It should be noted that these two types of explanations are not necessarily competitive with each other. It is possible for there to be both biogenic and psychogenic reasons for sex differences. In fact, there is good evidence to suggest that both factors are operating. The question is one of the relative contribution of each factor. There has been a trend among researchers in recent years to speak of *biopsychosocial* models that take biology, individual psychology, and the effects of other people and social systems into account in constructing comprehensive pictures of phenomena.

It is also possible for biogenic and psychogenic factors to interact with one another. For instance, a man who is at high risk for heart disease because of his physiology (biogenic factor) might be less likely than a woman to see a physician for regular checkups because to do so would be an admission of weakness and vulnerability, which he views as being unmasculine (psychogenic factor). The man in this example might have a shorter life than would be the case if either factor were operating in isolation.

In the preceding example, the biogenic factor might be the major contribution. It would also be possible for a psychogenic factor to make a major contribution while still interacting with a biogenic factor. For instance, a man might drink alcohol heavily in response to the pressure of meeting masculine gender-role demands, thus damaging his body and shortening his life, or he might use tobacco to enhance his masculine image, with a similar result.

BIOGENIC EXPLANATIONS

As mentioned earlier, male fetuses are more vulnerable in utero than female fetuses, and the death rate for males aged 1 to 4 exceeds that of females. These data are ample evidence that biological factors operate in the life-span sex differential, as it would be difficult to argue that male socialization could have a profound effect on mortality at such an early age. Explanations of biological factors include genetic and hormonal sex differences.

Genetic Differences

The differences in males' and females' genetic makeup is that females have two X chromosomes and males have one X and one Y chromosome. When recessive disease genes are found on the X chromosome, having a second X chromosome turns out to be quite a genetic advantage. The second X chromosome often contains a dominant corresponding gene that protects the female from contracting the genetic disease. For example, if there is a gene for hemophilia on one X chromosome, the female will not contract the disease unless there is also a hemophilia gene on the other X chromosome.

Because the form of the Y chromosome does not correspond exactly to that of the X chromosome, the male is not always afforded such protection. Genetic abnormalities on the X chromosome are much more likely to appear in the male because of the absence of a second (corrective) X chromosome. Some "X-linked" abnormalities such as color blindness or baldness are relatively innocuous. Some are more serious. For instance, there is some speculation that dyslexia (a learning disability) and hyperactivity might be X-linked. A few genetic abnormalities, such as hemophilia, can be life threatening.

In the search for explanations of the life-span sex differential, genetic differences make a very small contribution because of the rarity of life-threatening, X-linked diseases. As Waldron . . . stated, "Most of the common X-linked diseases aren't fatal, and most of the fatal X-linked diseases aren't common." . . .

Hormonal Differences

The major sex differences in hormones is in males' higher levels of testosterone and females' higher levels of estrogen. These two hormones account for physiological sex differences such as muscle size, body fat percentage, and metabolic speed. There is evidence to suggest that testosterone may render men somewhat more physiologically vulnerable to certain diseases, and that estrogen may have some protective effect.

The most demonstrable effect of these two hormones is in the area of heart disease. In recent years, the effect of cholesterol on heart disease has been the subject of much research and discussion. There are two kinds of cholesterol: high-density lipoprotein (HDL), called "good cholesterol" because it protects against heart disease, and low-density lipoprotein (LDL), called "bad cholesterol" because of its damaging effects.

In prepubescent males and females, HDL levels are about equal. At puberty, HDL levels drop rapidly in boys, but they hold steady in girls. This change coincides with the large surge of testosterone in boys and estrogen in girls. It is assumed that adolescent testosterone production is responsible for the reduction of HDL cholesterol, while estrogen has little or no effect on HDL levels. . . .

LDL ("bad") cholesterol begins to rise in both males and females after puberty. Males show a more rapid rise, however, leaving them more susceptible to heart disease. After menopause, when women's estrogen level is greatly reduced, LDL levels show this same kind of sharp increase. It is assumed, therefore, that estrogen has a protective effect against LDL cholesterol, while testosterone probably has little effect. . . .

These hormonal effects are important because heart disease is the leading cause of death. However, there is also some evidence that testosterone may shorten men's lives in other ways that are not fully understood.

Male cats who are neutered (which drastically lowers testosterone) live a good deal longer than those who are not (there is no corresponding effect for spayed female cats). Part of this difference is due to the fact that unneutered cats are more likely to fight, as testosterone is related to aggression. Thus, these cats are more likely to die in fights. When cats who died in fights are eliminated from the data, however, a large lifespan difference between neutered and unneutered cats remains. . . . Still, we cannot be sure of the possible life-shortening effects of fighting, even if the cat does not actually die in a fight. Human boxers and football players do not live as long as most men, but we can't necessarily attribute this difference solely to participation in the sport.

If we wanted to make a true experimental study of the effects of testosterone on human longevity, we could take a group of males, castrate them, and see if they live as long as their counterparts. This kind of experimentation is obviously not possible for ethical reasons. However, there have been times in history when castrations were performed on humans for various reasons, as recently as 1950! In China and the Ottoman Empire, castrated males (eunuchs) were employed as palace guards. They could guard harems of women without the possibility of sexual liaisons. In Europe, boy singers were castrated in order to keep their singing voices in prepubescent ranges. . . . In Kansas, mentally ill and retarded men were sometimes castrated in order to reduce their aggressiveness. . . .

Historical anecdotes tell us that Chinese, Turkish, and Italian eunuchs seemed to live longer than other men, but no data on life span were collected. However, a research team did study 297 men who had been castrated at a Kansas institution for the mentally retarded. When compared with a matched group of inmates, the eunuchs' lives were an average of 14 years longer. . . .

What is the relative contribution of biogenic factors to sex differences in longevity? Estimates range from two-thirds of the approximate 7-year difference . . . down to one-quarter, . . . and, because of the aforementioned interactions with psychogenic factors, apportioning fractions to either type of cause may be difficult and misleading. Almost all researchers would agree that biogenic factors are in operation. However, almost no researchers would say that these are the *only* factors.

PSYCHOGENIC EXPLANATIONS

There are at least four ways in which psychological processes can contribute to illness, injury, and/or the shortening of a man's life. First, behaviors can be directly self-destructive. Suicide is obviously the best example of this type of psychogenic factor, but we might also consider the use of tobacco products or the excessive use of alcohol and other drugs in this category. These behaviors involve a man's active harm of his own body. It is also possible for a man to passively harm his health by neglecting to perform behaviors that maintain health. For example, a man with high blood pressure who refuses to take the medication to control it or someone who does not see a physician even though he has detected a symptom of cancer (when he has medical resources available to him) adversely affects his health through his behavior.

Third, some behaviors involve physical risk of illness, injury, or death. These behaviors include sharing needles in intravenous drug use, drunk driving, and contact sports. Fourth, some psychological processes seem to have adverse effects on the body. For instance, the effects of stress on physical health are well documented, and certain personality characteristics are also predictive of some physical conditions. These four categories of psychogenic factors will be separated for purposes of discussion.

Self-Destructive Behaviors

Suicide Suicide is, of course, the ultimate self-destructive behavior. Although females in the United States are more likely than males to make suicidal gestures or attempts, males make "successful" (a strange use of the word) attempts over three times more often than females. . . . Among older people, the ratio of male to female suicides approaches a staggering 10:1, and elderly suicide is on the rise. . . .

There is some conjecture that more women than men use suicide attempts to "cry for help" rather than as determined efforts to die, which are more common in men. . . . Women are also more likely to use suicide methods that have relatively low potential for death, such as overdose or wrist slashing, whereas men are more likely to use violent and highly lethal methods such as firearms or motor vehicles. . . . In 1988, males committed suicides with firearms six times more often than females. Firearm suicides accounted for 65% of all male suicides, compared to 39% of all female suicides. . . .

What connections does traditional masculinity have to suicide? There are several. Foremost among these is a differential gender-role socialization for dealing with psychological pain. Whereas women have been socialized to think about and express feelings, gain social support, and take care of themselves, men are socialized to act on problems, be hyperindependent, and disdain emotional self-care. The hypermasculine man in severe emotional distress is often alone with his pain. He cannot express it, and he cannot ask for help with it. If the pain becomes great enough, he may feel that suicide is his only option.

The sex differential in suicide for older people is a powerful clue to the effects of traditional masculinity on suicide. Men are culturally defined by physical abilities and the work role. The older man must face the facts that his body is declining and, after he retires, that he is no longer a valued contributor in the working world. If his sense of self is overly consumed by this narrow and distorted standard of masculinity, body and work-role decline can seriously undermine his sense of self-worth. Additionally, the man may also be faced with a loss of independence at some point during his physical decline, which is also antithetical to the traditional male role.

Use of Tobacco Tobacco products are the only commodities legally sold in the United States that, when used as intended, will kill the user. The most common results of extended tobacco use are bronchitis, emphysema, asthma, and cancers of the respiratory system, mouth, and throat. In 1982, men outnumbered women in deaths by a 3½ to 1 ratio for bronchitis, emphysema, and asthma, and a 2½ to 1 ratio for lung, mouth, and throat cancer. . . .

In 1976, Waldron estimated that one-third of the sex difference in longevity was attributable to the sex difference in smoking. Deaths from bronchitis, emphysema, and asthma are proportional to the number of cigarettes smoked. . . . The use of "smokeless" tobacco (chewing tobacco and snuff) has also been linked to mouth and throat cancer, and these kinds of tobacco are used almost exclusively by males. Because the sex difference in smoking has been shrinking steadily since 1960, we might expect some of the disease proportions to also change. . . .

Socialization of destructive male behaviors can certainly be implicated in tobacco use. Advertisers have long used masculine mystique approaches (e.g., the Marlboro Man) to sell their products by associating tobacco with desirable images of masculinity (self-assuredness, independence, and adventurousness). Advertisers know that they can sell a great deal by playing on people's insecurities. When he is asked to live up to vague and impossible standards of masculinity, what man would not feel insecure? This advertising is both reflective and encouraging of certain cultural values for men: Do whatever you want; don't worry about dying.

Neglectful Behaviors

Men sometimes shorten their lives or become ill because they fail to perform the behaviors necessary to maintain their health. For example, 30 to 50% of hypertension (high blood pressure) patients stop taking their medication, leaving them at increased risk for heart attack. . . . Men are disproportionately represented in this group.

Men can also create problems by failing to seek help or take time off from work when it is indicated, such as when they are injured, sick, emotionally distraught, or when they have not had a physical examination for a long time. Taking necessary medication and seeking help are admissions of weakness, vulnerability, and dependence, which go against the masculine cultural prescriptions to handle problems on one's own, focus outside of the self, be strong and invulnerable, and "take it like a man." Negative or extreme masculinity is related to poor health practices. . . .

In a recent national survey conducted by the American Medical Association and the Gallup Poll, 40% of doctors endorsed the belief that over 50% of men aged 50 or older undermine potential lifesaving treatment for prostate or colorectal cancer (which kill an estimated 200,000 men per year in the United States) by ignoring symptoms, delaying treatment, or refusing to discuss symptoms. Embarrassment was cited as the major reason for failing to discuss medical problems. . . . Fifty percent of the male population does not know the symptoms of these cancers or those of prostate enlargement, which affects 75% of men over 50. . . .

Risk Behaviors

Men sometimes choose to engage in behaviors that involve the risk of injury, death, or legal sanction. For instance, habitual drinking to excess puts a man at increased risk for liver disease and accidents. Drunk driving, high-speed driving, drug dealing, sharing hypodermic needles, using firearms, engaging in gang violence, and working in dangerous jobs are other risky behaviors. Again, these are all engaged in by many more men than women. For instance, more than 90% of those arrested for alcohol and drug-abuse violations are men. . . . Our discussion will focus on three areas of risk: dangerous sports, war, and unsafe sexual practices.

Dangerous Sports To say that sports and masculinity are strongly connected in U.S. culture is an understatement. The almost religious fervor with which many men approach athletics is evidence that sports have more importance to men than mere physical fitness. An unfortunate minority of men and boys has suffered debilitating injuries and even death as a result of overexertion, physical contact, or accidents in sporting events. The most dangerous sports would seem to be auto racing and professional boxing. The object of the latter is to pummel one's opponent into unconsciousness. Some (like Korean boxer Du Koo Kim) have died from blows to the head. (Kim was killed by U.S. fighter Ray Mancini.) Others—such as former heavyweight champion Muhammad Ali, who now suffers from Parkinsonian symptoms and general intellectual decline—have suffered irreparable brain damage.

Few men participate in auto racing and boxing relative to other violent sports such as ice hockey and football. In one research study . . . it was reported that an average of 13 U.S. high school football players die every year. Approximately half die from injury and half from overexertion. Other estimates of the number of fatalities range as high as 40. . . . There are approximately 30 catastrophic injuries (such as permanent brain injuries or paralysis) per year . . . and an estimated 600,000 other injuries. . . . At least one professional or college football player has suffered paralysis in each of the last several years.

Football is one of the ultimate expressions of hypermasculinity in the United States. It involves sacrifice of the body for a task and denial of basic instincts for self-preservation and safety. Television commentary of football injuries (and sports injuries in general) almost always involves reference to the task: "Will he return in the second half?"; "How will they contain the pass rush without their best blocker?"; "Will the other team exploit the substitute?" Imagine a television commentator reacting to an injury with, "I wonder how he feels about that?"

The U.S. Army does a great deal of advertising during televised professional football games. One Army commercial depicted a football player who acted as a "wedge cracker" on kickoff plays. (This player's job is to throw his body into as many opposing players as possible.) The commentary on the commercial said the following about this man: "He's always willing to sacrifice his body for the greater cause of victory"! Is there any better statement that epitomizes self-destructive masculinity?

War Generation after generation, we have marched young men off to be killed in wars. In U.S. society, dying or being maimed in war is considered an act of heroism rather than the victimization of a young man. While participating in the defense of their country might be considered to be a loving thing to do, only recently have men been given a choice as to whether or not they would serve in the armed forces.

Goldberg . . . compared the men's issue of the military draft to the women's issue of reproductive rights. In both cases, the individual's control over his or her own body has been restricted by the law. Although there is currently no mandatory draft, and the legal right to abortion still exists, mechanisms are in place that can remove what many consider to be basic human freedoms.

The wars of the 20th century have been described as "holocausts of young men" in which millions of men were killed and over 100 million men were injured. The average age of World War I and II casualties was 18.5 years. . . . Thus, the victims of war tend to be the youngest men, who feel (and often are) less powerful and who feel the strongest need to establish a sense of masculinity. Men of color, who are marginalized by mainstream U.S. culture, have also been disproportionately represented among the war dead. Many men who have survived the modern, technological war in Vietnam returned with physical and emotional scars of profound proportions. It is perhaps no coincidence that the modern men's movement began at about this time, when it was becoming obvious that soldiers were finding it difficult to "take it like a man."

The recent war in the Persian Gulf raised the question of whether women should be able to serve in combat. There seemed to be little awareness that perhaps *nobody* should be in combat. A letter to advice columnist Ann Landers . . . argued against women participating in combat because of its hardships. In this letter, the author states that "combat means sleeping in a muddy foxhole, eating a can of beans for your main meal of the day, urinating and defecating wherever you happen to be, and keeping your toilet paper on your head under your helmet because it's the only dry spot on your entire body when it rains. Combat means watching your buddy step on a land mine and get blown to pieces." It is implicit in this person's letter that it is acceptable to subject men to these hardships.

Unsafe Sexual Practices Kimmel and Levine . . . reported that 93% of U.S. adults with AIDS are men and that AIDS is the leading cause of death for 30- to 44-year-old men who live in New York City. It is only possible to contract AIDS through introduction of the human immunodeficiency virus (HIV) into the bloodstream. This may happen due to unfortunate accidents such as receiving an HIV-tainted blood transfusion, but the overwhelming majority of HIV infections occur through sharing hypodermic needles by intravenous drug users and through unsafe sexual practices.

It is not difficult to find connections between sexual-risk behaviors and cultural prescriptions for masculinity. Foremost among these are the expectations that males will be sexually promiscuous and adventurous. Condom use reduces the risk of infection, but there is a good deal of resistance to using condoms, perhaps because their use involves an acknowledgement of vulnerability, as well as a caring for the self and the sexual partner. These go against masculine norms. As Kimmel and Levine . . . put it, "Abstinence, safer sex, and safer drug use compromise manhood. The behaviors required for the confirmation of masculinity and those required to reduce risk are antithetical." . . .

Adverse Effects of Psychological Processes

Only recently has science begun to understand in some detail the connections between psychological and physiological processes. The relatively young field of behavioral medicine is focused on understanding and treating physical disorders that are thought to be strongly influenced by the person's psychological functioning. . . . Masculine socialization may well contribute to some physical disorders that are disproportionately experienced by men. . . . Chief among these are cardiovascular disorders and peptic ulcers.

Cardiovascular Disorders It has long been suspected that coronary artery disease and hypertension have strong relationships to stressful work environments and certain behavioral patterns of response to those environments. Hypertension is common among workers in highly stressful occupations (e.g., air traffic controllers, policemen). It is also common among those described as projecting an image of being easygoing but at the same time suppressing a good deal of anger. . . . It is not surprising that many African-American men have problems with hypertension, given the emphasis in many African-American cultures on "cool pose"—an outward appearance of calmness. . . . At the same time, many of these men experience the anger that is created by membership in an oppressed group and by harsh masculine socialization. These men may experience the powerful internal conflict between these feelings and the demand to be "cool," and this conflict may be partially manifested in high blood pressure.

Several decades ago, cardiologists Meyer Friedman and Ray Rosenhan coined the term *Type A behavior* to describe a personality pattern commonly found in people who had suffered myocardial infarction (heart attack). This pattern described the classic compulsive, hostile, competitive, emotionally inexpressive "workaholic." Friedman . . . defined *Type A* as "a characteristic action-emotion complex which is exhibited by those individuals who are engaged in a relatively chronic struggle to obtain an unlimited number of poorly defined things from their environments in the shortest period of time and, if possible, against the opposing efforts of other things or persons in this same environment." . . .

Again we see vestiges of the destructive aspects of traditional masculinity in this pattern: the aggressive attempt to measure up to vague standards of achievement and competition. Type A individuals tend to be hyperindependent; they seize authority and dislike sharing responsibility. Contrary to popular belief, they tend to be less successful than those who are more relaxed and less aggressive. . . . Type A behaviors are significantly related to masculine sex typing. . . . Negative or extreme masculinity is also related to heart-attack severity. . . .

Peptic Ulcer Peptic ulcers are cuts in the lining of the stomach. They are a common condition, affecting about 12% of men at some time in their lives. This is approximately double the rate for women. . . . Ulcers may well be the result of hydrochloric acid in the stomach, which is part of the stress response. . . . There are significant positive correlations between ulcer and chronic feelings of anger, frustration, and resentment. Among middle-class men, there is a subgroup of ulcer sufferers who are described as hyperindependent, striving, ambitious, competitive, having poor diets, and overusing alcohol. . . .

CONCLUSION

It is clear that men are biologically predisposed toward certain physical health problems. This predisposition interacts with certain aspects of masculine socialization, as well as with the negative effects of dangerous, historically male environments. Women who expose themselves to these environments and exhibit traditionally masculine behaviors increase their health risks. . . . Thus, it is hazardous to be male, and it is also hazardous to be negatively masculine.

Sidney Jourard said it best in his landmark 1971 article, "Some Lethal Aspects of the Male Role." Jourard described a set of human needs: to know and be known, to depend and be depended upon, to love and be loved, and to find some purpose and meaning in one's life. The masculine role is poorly designed to fill these human needs. It requires men to "be noncommunicative, competitive and nongiving, and to evaluate life successes in terms of external achievement rather than personal and interpersonal fulfillment." . . . Jourard's belief was that this not only limited the quality of men's emotional lives, but that it also had the effect of slowly destroying them physically. . . .

34

Men's Hidden Depression

TERRENCE REAL

In the middle of the journey of our lives,
I found myself upon a dark path.

DANTE

There is a terrible collusion in our society, a cultural cover-up about depression in men.

One of the ironies about men's depression is that the very forces that help create it keep us from seeing it. Men are not supposed to be vulnerable. Pain is something we are to rise above. He who has been brought down by it will most likely see himself as shameful, and so, too, may his family and friends, even the mental health profession. Yet I believe it is this secret pain that lies at the heart of many of the difficulties in men's lives. Hidden depression drives several of the problems we think of as typically male: physical illness, alcohol and drug abuse, domestic violence, failures in intimacy, self-sabotage in careers.

We tend not to recognize depression in men because the disorder itself is seen as unmanly. Depression carries, to many, a double stain—the stigma of mental illness and also the stigma of "feminine" emotionality. Those in a relationship with a depressed man are themselves often faced with a painful dilemma. They can either confront his condition—which may further shame him—or else collude with him in minimizing it, a course that offers no hope for relief. Depression in men—a condition experienced as both shame-filled and shameful—goes largely unacknowledged and unrecognized both by the men who suffer and by those who surround them. And yet, the impact of this hidden condition is enormous.

Eleven million people are estimated as struggling with depression each year. The combined effect of lost productivity and medical expense due to depression costs the United States over 47 billion dollars per year—a toll on a par with heart disease.[1] And yet the condition goes mostly undiagnosed. Somewhere between 60 and 80 percent of people with depression never get help.[2] The silence about depression is all the more heartbreaking since its treatment has a high success rate.[3] Current estimates are that, with a combination of psychotherapy and medication, between 80 and 90 percent of depressed patients can get relief—if they ask for it.[4] My work with men and their families has taught me that, along with a reluctance to acknowledge depression, we also often fail to identify this disorder because *men tend to manifest depression differently than women.*[5]

Few things about men and women seem more dissimilar than the way we tend to handle our feelings. Why should depression, a disorder of feeling—in psychiatric language, an *affective disorder*—be handled in the same way by both sexes when most other emotional issues are not?[6] While many men are depressed in ways that are similar to women, there are even more men who express depression in less well-recognized ways, ways that are most often overlooked and misunderstood but nevertheless do great harm. What are these particular male forms of depression? What are their causes? Is the etiology of the disorder the same for both sexes? I think not. Just as

From *I Don't Want to Talk About It: Overcoming the Secret Legacy of Male Depression,* by Terrence Real, New York: Scribner, 1997.

men and women often express depression differently, their pathways toward depression seem distinct as well.

Traditional gender socialization in our culture asks both boys and girls to "halve themselves."[7] Girls are allowed to maintain emotional expressiveness and cultivate connection.[8] But they are systematically discouraged from fully developing and exercising their public, assertive selves—their "voice," as it is often called. Boys, by contrast, are greatly encouraged to develop their public, assertive selves, but they are systematically pushed away from the full exercise of emotional expressiveness and the skills for making and appreciating deep connection. For decades, feminist researchers and scholars have detailed the degree of coercion brought to bear against girls' full development, and the sometimes devastating effects of the loss of their most complete, authentic selves. It is time to understand the reciprocal process as it occurs in the lives of boys and men.

Current research makes it clear that a vulnerability to depression is most probably an inherited biological condition. Any boy or girl, given the right mix of chromosomes, will have a susceptibility to this disease.[9] But in the majority of cases, biological vulnerability alone is not enough to bring about the disorder. It is the collision of inherited vulnerability with psychological injury that produces depression.[10] And it is here that issues of gender come into play. The traditional socialization of boys and girls hurts them both, each in particular, complementary ways. Girls, and later women, tend to internalize pain. They blame themselves and draw distress into themselves. Boys, and later men, tend to externalize pain; they are more likely to feel victimized by others and to discharge distress through action. Hospitalized male psychiatric patients far outnumber female patients in their rate of violent incidents; women outnumber men in self-mutilation. In mild and severe forms, externalizing in men and internalizing in women represent troubling tendencies in both sexes, inhibiting the capacity of each for true relatedness. A depressed woman's internalization of pain weakens her and hampers her capacity for direct communication. A depressed man's tendency to extrude pain often does more than simply impede his capacity for intimacy. It may render him psychologically dangerous. Too often, the wounded boy grows up to become a wounding man, inflicting upon those closest to him the very distress he refuses to acknowledge within himself. Depression in men, unless it is dealt with, tends to be passed along. That was the case with my father and me. And that was the situation facing David Ingles and his family when we first met. . . .

David did not know it, but he was depressed. Along with whatever biological vulnerabilities he may have carried, David's depression was born from the pain of [his boyhood]—not just from this one incident with his father, but from hundreds, perhaps even thousands of similar moments, small instances of betrayal or abandonment, perhaps more subtle than this one but just as damaging. For those with a biological vulnerability to the disorder, such moments can become the building blocks of depression, a condition which, conceived in the boy, erupts later on in the man. David's unrecognized pain ticked inside like a bomb, waiting for its appointed time. The force of that ticking pushed him from his family. It sped him toward mood buffers and self-esteem enhancers like work, alcohol, and occasional violence. By the time I first met him, his son was on the edge of school failure and his wife was on the verge of filing for divorce. The bomb inside was due to release itself and his life was about to explode. And neither he nor anyone close to him would have understood why. But I knew why.

I knew what it felt like to have the breath knocked out of you by your own father, what it meant to be thrown against a wall and dared to fight back. Intimacy with the sticky threads of loving violence that bind parents to sons across generations helped me recognize David's secret. Deep inside his bullying and drinking, his preoccupations and flight, lay that little boy. The depressed part of David, his unacknowledged child, waited in darkness, resentfully, for its moment in the light, wreaking havoc upon anyone near. Showing great courage, David allowed, on that afternoon in my office, the pain he had carried within for decades to break through to the surface. His vulnerability drew the people he loved back toward him. The appearance of his hidden depression permitted him to touch and to be touched after a long, bristling time behind armor. In his struggle, David Ingles is not alone.

In order to treat a man like David, I must first "get at" him, "crack him open." The patient needs help bringing his depression up to the surface. Depressed women have obvious pain; depressed men often have "troubles." It is frequently not they who are in conscious distress so much as the people who live with them.

If you had asked David what was bothering him, before that session, it is uncertain what he would have answered, or even if he would have given an answer at all. Like a lot of the successful men I treat, David was unpracticed in, even wary of, introspection.

What David might have told you was that he was unhappy at work, where he had a new senior partner to deal with, whom he neither liked as well as his old mentor nor felt particularly favored by. He might have told you that over the last several years he had grown increasingly restless—to the point where it had become difficult for him to sleep at night without pills and hard to get through a dinner at a friend's house without a few cocktails. David knew—though he would not have bored anyone other than Elaine with the details—that he was bothered more and more by stomachaches and by backaches, which his internist chalked up to "stress," a medical opinion that David dismissed as "the great twentieth-century catchall."

David's physician was right, however, although his diagnosis did not go nearly far enough. David was "stressed." At forty-seven, he had begun to feel old. He did not like the spare tire that no amount of racquetball seemed to touch. He did not like the receding hairline. And he did not like looking at the kind of women he had always admired only to have them now look away with disinterest or sometimes with outright disdain. If asked, David would gladly have unloaded his feelings of disappointment about his difficult son, Chad. He might even have voiced his sense of betrayal by Chad's "overprotective mother," who, from the day of Chad's birth, had undercut his attempts to be firm with the boy. Toward the end of an evening, after a sufficient number of drinks, David might have confessed his unhappiness in his marriage—how unsupported he felt, how much like a stranger in his own home. Not once would it have occurred to him that he might be suffering from a clinical condition. But the depression David neither felt nor recognized was close to fracturing his family. It was eating away at his relationship with his son and eroding his marriage. In his efforts to escape his own depression, David had let himself sink into behaviors—like irritability, dominance, drinking, and emotional unavailability—that pushed away the very people whom he most loved and needed. As Elaine described it, he was no longer himself. Like Shakespeare's Lear, David, without realizing it, had lost his estate. "What do you see when you look at me?" demanded the broken king of his fool. And the fool replied, "Lear's shadow."[11] Depression was whittling David, fading him to a shadow state as surely and inexorably as a physical disease like cancer or AIDS. As one of my clients put it, depression was "disappearing" him.

We do not generally think of driven men like David as being depressed. We tend to reserve the concept of *depression* for a state of profound impairment, utter despair, thorough debilitation. A truly depressed man would lie in bed in the morning, staring up at the ceiling, too apathetic to drag himself off to another meaningless day. By comparison, what David faced seemed barely to qualify as midlife malaise. As Thoreau once wrote: "The mass of men lead lives of quiet desperation."[12] Others, not so quiet. When we think of depression, it is to those "others, not so quiet" that our thoughts usually turn.

For close to twenty years, I have treated those others—men with the kind of depression most of us easily discern. I call this state *overt depression*. Acute and dramatic, the pain inflicted by overt depression is writ large. In contrast, David's type of depression was mild, elusive, and chronic. The kind of depression from which David suffers is not even referenced in most of the literature about the disorder. The guidebook for diagnosis used by most clinicians throughout the country is the American

Psychiatric Association's *Diagnostic and Statistical Manual of Mental Disorders* (DSM IV) which labels a person as having a clinical depression only if he or she shows, for a duration of at least two weeks, signs either of feeling sad, "down," and "blue," or having a decreased interest in pleasurable activities, including sex.[13] In addition, the person must exhibit at least four of any of the following symptoms: weight loss or gain, too little or too much sleep, fatigue, feelings of worthlessness or guilt, difficulty making decisions or forgetfulness, and preoccupation with death or suicide.

The condition described in the DSM IV is the classic form of depression most of us think of. Although many men may be reluctant to admit that they are suffering from overt depression, the disorder itself has been recognized since ancient times. As early as the fourth century, B.C., Hippocrates, "the father of medicine," reported a condition whose symptoms included "sleeplessness, irritability, despondency, restlessness, and aversion to food"—a description of overt depression easily recognizable today. Hippocrates saw the malady as caused by an imbalance of black bile, one of the four humors, and he therefore named the disease simply "black bile," which in Greek reads *melanae chole,* or *melancholia.*[14]

Overt depression preys upon men, women, and sometimes children from all walks of life, all classes, all cultures. Epidemiologists have found descriptions resembling overt depression throughout the world—both in developed and in developing societies.[15] And the number of overtly depressed people seems to be on the rise. Researcher Myrna Weissman and her colleagues checked medical records going back to the beginning of the century. They calculated that, even allowing for increased reporting, each successive generation has doubled its susceptibility to depression.[16] Such trends were corroborated worldwide in a random sampling of 39,000 subjects from such diverse countries as New Zealand, Lebanon, Italy, Germany, Canada, and France. Researchers have found depression in greater numbers and at earlier ages than ever before throughout the world.[17]

The National Institute for Mental Health reports that in the United States somewhere between 6 and 10 percent of our population—close to one out of every ten people—are battling some form of this disease.[18] And yet, as sobering as these figures may be, I believe they greatly underestimate the full impact of depression in men's lives. A man like David Ingles, whose condition manifests itself in ways more subtle than those described by the DSM IV, would not have been included in these figures, even while the effects of his less obvious disorder are powerful enough to threaten his health and break up his family. Why is it that not only the general public but even the medical and psychiatric community give credence to depression in only its most obvious and most severe form? In a recent national survey, over half of the people questioned did not see depression as a major health issue. In another survey, in which 25 percent of the respondents had themselves experienced depression, and another 26 percent had observed it in family members, close to half of the respondents still viewed the disorder not as a disease or a psychological problem deserving of help, but rather as a sign of personal weakness.[19]

Our current patterns of judgment and denial about depression are reminiscent of the older moralistic attitudes toward the disease of alcoholism, and the source of our minimization is much the same now as it was then. The issue is shame. While depression may carry some sense of stigma for all people, the disapprobation attached to this disease is particularly acute for men. The very definition of manhood lies in "standing up" to discomfort and pain. It is sadly predictable that David would be more likely to react to depression by redoubling his efforts at work than by sitting still long enough to feel his own feelings. Until therapy, "giving in" to his pain would have been experienced by David not as a path toward relief, but as a humiliating defeat. In the calculus of male pride, stoicism prevails. All too often, denial is equated with tenacity—"Under the bludgeonings of chance/My head is bloody, but unbowed."[20]

When David Ingles runs from his own internal distress, he plays out our culture's values about masculinity. As a society, we have more respect for the walking wounded—those who deny their difficulties—than we have for those who "let" their conditions "get to them." Traditionally, we have not liked men to be very emotional or very vulnerable. An overtly depressed man is both—someone who not only has feelings but who has allowed those feelings to swamp his competence. A man brought down in life is bad enough. But a man brought down by his own unmanageable feelings—for many, that is unseemly.

This attitude often compounds a depressed man's condition, so that he gets depressed about being depressed, ashamed about feeling ashamed. Because of the stigma attached to depression, men often allow their pain to burrow deeper and further from view. Physician John Rush spoke in an interview about the stain of "unmanliness" attached to the condition and its possible consequences.

> [Depression] doesn't mean I'm weak, it doesn't mean I'm incurable, it doesn't mean I'm insane. It means I've got a disease and somebody better treat it. One of my friends says, "Depression? Hell, boy, that's wimp disease." Wimp disease? Oh, yeah, it's wimp disease. And I guess the ultimate wimp kills himself.[21]

What John Rush implies is correct. For many men—ashamed of their feelings and refusing help "wimp disease" can kill. Men are four times more likely than women to take their own lives.[22]

Over the last twenty years, researchers have investigated the relationship between traditional masculinity and physical illness, alcohol abuse, and risk-taking behaviors—and have demonstrated what most of us already know from common experience: many men would rather place themselves at risk than acknowledge distress, either physical or emotional.[23] In *The Things They Carried,* Tim O'Brien gives a clear example of the force of men's shame, when he remembers his fellow "grunts" in Vietnam:

> They carried their reputations. They carried the soldier's greatest fear, which was the fear of blushing. Men killed, and died, because they were embarrassed not to. It was what brought them to the war in the first place, nothing positive, no dreams of glory or honor, just to avoid the blush of dishonor. They crawled into tunnels and walked point and advanced under fire. They were too frightened to be cowards.[24]

Preferring death to the threat of embarrassment, the men O'Brien describes remind me of Harry, the old-fashioned Irish father of one of my clients, who was too ashamed to see a doctor until cancer had eaten away half of a testicle. . . .

Men's willingness to downplay weakness and pain is so great that it has been named as a factor in their shorter life span. The ten years of difference in longevity between men and women turns out to have little to do with genes. Men die early because they do not take care of themselves. Men wait longer to acknowledge that they are sick, take longer to get help, and once they get treatment do not comply with it as well as women do.[25] . . .

One factor mitigating against the recognition of David's condition is that mental health professionals, no less than anyone else, tend to look for what they expect to find. The conventional wisdom that women are depressed while men are not leads some therapists away from an accurate diagnostic assessment.

A number of studies looking at who gets labeled as being depressed have been carried out nationwide. Some, like the Potts study involving no less than 23,000 volunteer subjects, have been conducted on a massive scale. The results of most of them show a tendency for mental health professionals to overdiagnose women's depression and underdiagnose the disorder in men.[26]

In a study of a different nature, psychologists were given hypothetical psychiatric "case histories" of patients with a variety of complaints. Only one variable was changed, the sex of the client. Consistently, psychologists diagnosed the depressed "male" clients as more severely disturbed than depressed "female" clients. On the other hand, women alcoholics were viewed as being more severely disturbed than their male counterparts. These conflicting results show that an overlay of gender expectations complicates the judgment of clinicians. It seems that they are punishing clients of both sexes with a more severe diagnosis for crossing gender lines. If it is unmanly to be depressed and unwomanly to drink, then a depressed man must be *really* disturbed, just like an alcoholic woman.[27]

While a great many men conceal their condition from the outside world, and while those close to them—loved ones, doctors, even psychotherapists—may miss a diagnosis of overt depression, a man like David Ingles goes even further with the deception. David not only managed to camouflage his condition from those around him; he managed to hide it even from himself. A great many men never make it into the official roll call of the depressed because their overt depression remains undiagnosed. But other men, like David, fail to get help because their expression of the disease does not fit the classic model as described in the DSM IV. David suffers from what I call *covert depression.* It is hidden from those around him, and it is largely hidden from his own conscious awareness. Yet it nevertheless drives many of his actions. David Ingles buries himself in work; he wraps his disquiet in anger and numbs his discontent with alcohol. Everywhere in his life, the prohibition against bringing his vulnerable feelings into the open fosters behaviors that leave him and the people around him ever more disconnected. An unrecognized swell of abandonment washes over David when Elaine does not respond to him and causes him to wall her off in subtle retaliation—throwing the couple into an escalating cycle of alienation. David's unacknowledged desperation to be involved with Chad—to be the kind of father his father was not—leads him, paradoxically, to bully his son, to reenact the very drama he wishes to avoid. My work with families like the Ingleses has convinced me that many of the difficult behaviors one sees in men's relationships are depression driven.

Under the names of "masked depression," "underlying characterological depression," and "depression equivalents," the kind of disguised condition David suffers from has been written about sporadically for years. But it has rarely been systematically studied. Researcher Martin Opler observed as far back as 1974: "Masked depression is one of the most prevalent disorders in modern American society, yet it is perhaps the most neglected category in psychiatric literature."[28] That neglect continues. If *overt depression* in men tends to be overlooked, *covert depression* has been rendered all but invisible.

NOTES

1. P. F. Greenberg, L. F. Stiglin, L. Finkelstein, S. Ernst, R. Berndt, "The Economic Burden of Depression in 1990," special report in *The Journal of Clinical Psychiatry,* vol. 54 (11) 408–418 1990: "The Economics of Depression," 1990. See also P. Greenberg, L. Stiglin, S. Finkelstein, "Depression: A Neglected Major Illness," *Journal of Clinical Psychiatry,* Nov. 1993, pp. 419–24. For reactions to and discussions of Greenberg's report, see D. Goleman, "Depression Costs Put at $43 Billion," *The New York Times,* Dec. 3, 1993.
2. D. A. Reiger, R. M. A. Hirschfeld, F. K. Goodwin, et al., "The NIMH Depression Awareness Recognition and Treatment Program: Structure, Aims, and Scientific Basis," *American Journal of Psychiatry,* 1988, vol. 145 pp. 1351–57.
3. L. Eisenburg, "Treating Depression and Anxiety in Primary Care: Closing the Gap Between Knowledge and Practice," *New England Journal of Medicine,* 1992–93, vol. 326 pp. 1080–84.
4. Reiger, Hirschfeld, Goodwin, et al., "The NIMH Depression Awareness Recognition and Treatment Program." See also M. Weissman, "The Psychological Treatment of Depression: Evidence for the Efficacy of Psychotherapy alone, in Comparison with, and in Combination with Pharmacotherapy," *Archives of General Psychiatry,* vol. 36 Oct. 1979, pp. 1261–69.
5. I am neither the first nor the only person reviewing the research on depression and gender to posit this explanation. See, for example, epidemiologists Hammen and Padesky, who remark: "It is unclear whether women are in fact more depressed than men, or whether male and female experiences with depression differ in ways that lead women to express symptoms. . . . seek help, or receive labels of depression in ways [that are] different from men." C. I. Hammen and C. A. Padesky, "Sex Differences in the Expression of Depressive Responses on the Beck Depression Inventory," *Journal of Abnormal Psychology,* 1977, vol. 86 (b) pp. 609–14. See also S. J. Oliver and B. B. Toner, "The Influence of Gender Role Typing on the Expression of Depressive Symptoms," *Sex Roles,* vol. 22 (11–12) pp. 175–262, 1990.
6. E. S. Chevron, D. M. Quinlan, and S. J. Blatt, "Sex Roles and Gender Differences in the Experience

of Depression," *Journal of Abnormal Psychology,* 1978, vol. 87 pp. 680–83. See also A. F. Chino and D. Funabiki, "A Cross-Validation of Sex-Differences in the Expression of Depression," *Sex Roles,* 1984, vol. 11 pp. 175–87, and C. A. Padesky, "Sex Differences in Depressive Symptoms Expression and Help Seeking Among College Students," *Sex Roles,* vol. 7 1981, pp. 309–20.

7. The phrase "halve themselves" and many of my ideas on boys' socialization I owe to Olga Silverstein. See O. Silverstein and B. Roshbaum, *The Courage to Raise Good Men* (New York: Viking, 1994).

8. C. Gilligan, *In a Different Voice: Psychological Theory and Women's Development* (Cambridge: Harvard University Press, 1982).

9. D. L. Duner, "Recent Gender Studies of Bipolar and Unipolar Depression," in J. C. Coyne, ed., *Essential Papers on Depression* (New York: New York University Press, 1985). See also E. S. Gershon, W. E. Bunney, Jr., J. E. Lehman, "The Inheritance of Affective Disorders: A Review of Data and Hypothesis," *Behavioral Genetics,* 1976, vol. 6 pp. 227–61.

10. P. Gilbert, *Depression: The Evolution of Powerlessness* (New York: Guilford, 1982).

11. [Lear] "Who is it that can tell me who I am?" [Fool] "Lear's shadow." *King Lear,* I, 4. *Shakespeare's Complete Works* (Baltimore: Penguin, 1969).

12. H. D. Thoreau, cited in *The Columbia Dictionary of Quotations* (New York: Columbia University Press, 1993).

13. American Psychiatric Association, *Diagnostic and Statistical Manual of Mental Disorders,* 4th ed. (Washington, D.C.: American Psychiatric Association, 1994).

14. Stanley Jackson has written a remarkably comprehensive history of the idea of depression, from early Greek writers to the present. See S. Jackson, *Melancholia & Depression* (New Haven: Yale University Press, 1986). See also P. Gilbert, *Depression.*

15. See D. Goleman, "A Rising Cost of Modernity: Depression," *The New York Times,* Dec. 8, 1992, science section.

16. Reiger, Hirschfeld, Goodwin, et al., "The NIMH Depression Awareness Recognition and Treatment Program," pp. 1351–57. See also N. M. Weissman, M. L. Bruce, P. J. Leaf, et al., "Affective Disorders," in L. N. Robbins, D. A. Reiger, eds., *Psychiatric Disorders in America: The Epidemiologic Catchment Area Study* (New York: The Free Press, 1991).

17. Goleman, "A Rising Cost of Modernity: Depression."

18. National Institute of Mental Health, *Number of U.S. Adults (in millions) with Mental Disorder, 1990,* March 25, 1992. See also E. S. Schneidman, "Overview: A Multidimensional Approach to Suicide," in D. Jacobs and H. N. Brown, eds., *Suicide: Understanding and Responding* Harvard Medical Perspectives (Connecticut: International Universities Press, 1989).

19. See J. Brody, "Myriad Masks Hide an Epidemic of Depression," *The New York Times,* Sept. 30, 1992, for both surveys.

20. W. F. Henley, "Perseverance," cited in *The Columbia Dictionary of Quotations.*

21. John Rush spearheaded the recent committee assigned with devising federal guidelines for the recognition of depression to be used by medical practitioners. He is interviewed in K. Cronkite, *On the Edge of Darkness: Conversations about Conquering Depression* (New York: Doubleday, 1994), p. 79.

22. National Center for Health Statistics, U.S. Department of Health and Human Services, 1993. USDH & HS (NCHS, Center for Disease Control, Division of Vital Statistics, Office of Mortality Statistics, Monthly Vital Statistics Report), vol. 40, no. (8), supp. 2, Jan. 7, 1992.

23. A. P. Douglas, D. Hull, J. Hull, "Sex Role Orientation and Type A Behavior Pattern," *Personality and Social Psychology,* 1981, vol. 9 (2) pp. 600–604; F. Conrad, "Sex Roles as a Factor in Longevity," *Sociology and Social Research,* 1962, vol. 46 pp. 195–202; D. McCelland, N. D. William, B. Kalen, K. Rudolph, *The Drinking Man* (Riverside, N.J.: Free Press, 1972).

24. T. O'Brien, *The Things They Carried* (New York: Penguin, 1990), pp. 20–21.

25. I. Waldron, "Why Do Women Live Longer than Men?" *Journal of Human Stress,* 1976, vol. 2 pp. 1–13; G. E. Good, D. M. Dell, and L. B. Mintz, "Male Role and Gender Role Conflict: Relations to Help Seeking in Men," *Journal of Counseling Psychology,* 1989, pp. 295–300; J. Harrison, "Warning: The Male Sex Role May Be Dangerous to Your Health," *Social Issues,* vol. 34, 1978, pp. 65–86.

26. M. K. Potts, M. A. Burnam, and K. B. Wells, "Gender Differences in Depression Detection: A Comparison of Clinician Diagnosis and Standardized Assessment," *Psychological Assessment,* 1991, vol. 3 (4) pp. 609–15. See also I. M. Werbrugge and R. P. Steiner, "Physician Treatment of Men and Women Patients. Sex Bias or Appropriate Care?" *Medical Care,* 1981, vol. 19 pp. 609–32.

27. See J. Waisberg and S. Page, "Gender Role Nonconformity and Perception of Mental Illness," *Women and Health,* 1988, p. 3. See also J. Wallen and J. D. Waitzkin, "Physician Stereotypes about Female Health and Illness: A Study of Patient's Sex and the Informative Process During Medical Interviews," *Women and Health,* vol. 4, 1979, pp. 135–46.
28. M. K. Opler, "Cultural Variations in Depression: Past and Present," in S. Lesse, ed., *Masked Depression* (New York: J. Aronson, 1974).

35

Fraternities and Collegiate Rape Culture: Why Are Some Fraternities More Dangerous Places for Women?

A. AYRES BOSWELL
JOAN Z. SPADE

Date rape and acquaintance rape on college campuses are topics of concern to both researchers and college administrators. Some estimate that 60 to 80 percent of rapes are date or acquaintance rape (Koss, Dinero, Seibel, and Cox 1988). Further, 1 out of 4 college women say they were raped or experienced an attempted rape, and 1 out of 12 college men say they forced a woman to have sexual intercourse against her will (Koss, Gidycz, and Wisniewski 1985).

Although considerable attention focuses on the incidence of rape, we know relatively little about the context or the *rape culture* surrounding date and acquaintance rape. Rape culture is a set of values and beliefs that provide an environment conducive to rape (Buchwald, Fletcher, & Roth 1993; Herman 1984). The term applies to a generic culture surrounding and promoting rape, not the specific settings in which rape is likely to occur. We believe that the specific settings also are important in defining relationships between men and women.

Some have argued that fraternities are places where rape is likely to occur on college campuses (Martin and Hummer 1989; O'Sullivan 1993; Sanday 1990) and that the students most likely to accept rape myths and be more sexually aggressive are more likely to live in fraternities and sororities, consume higher doses of alcohol and drugs, and place a higher value on social life at

From *Gender & Society,* Vol. 10, No. 2, April 1996.

college (Gwartney-Gibbs and Stockard 1989; Kalof and Cargill 1991). Others suggest that sexual aggression is learned in settings such as fraternities and is not part of predispositions or preexisting attitudes (Boeringer, Shehan, and Akers 1991). To prevent further incidences of rape on college campuses, we need to understand what it is about fraternities in particular and college life in general that may contribute to the maintenance of a rape culture on college campuses.

Our approach is to identify the social contexts that link fraternities to campus rape and promote a rape culture. Instead of assuming that all fraternities provide an environment conducive to rape, we compare the interactions of men and women at fraternities identified on campus as being especially *dangerous* places for women, where the likelihood of rape is high, to those seen as *safer* places, where the perceived probability of rape occurring is lower. Prior to collecting data for our study, we found that most women students identified some fraternities as having more sexually aggressive members and a higher probability of rape. These women also considered other fraternities as relatively safe houses, where a woman could go and get drunk if she wanted to and feel secure that the fraternity men would not take advantage of her. We compared parties at houses identified as high-risk and low-risk houses as well as at two local bars frequented by college students. Our analysis provides an opportunity to examine situations and contexts that hinder or facilitate positive social relations between undergraduate men and women.

The abusive attitudes toward women that some fraternities perpetuate exist within a general culture where rape is intertwined in traditional gender scripts. Men are viewed as initiators of sex and women as either passive partners or active resisters, preventing men from touching their bodies (LaPlante, McCormick, and Brannigan 1980). Rape culture is based on the assumptions that men are aggressive and dominant whereas women are passive and acquiescent (Buchwald et al. 1993; Herman 1984). What occurs on college campuses is an extension of the portrayal of domination and aggression of men over women that exemplifies the double standard of sexual behavior in U.S. society (Barthel 1988; Kimmel 1993).

Sexually active men are positively reinforced by being referred to as "studs," whereas women who are sexually active or report enjoying sex are derogatorily labeled as "sluts" (Herman 1984; O'Sullivan 1993). These gender scripts are embodied in rape myths and stereotypes such as "She really wanted it; she just said no because she didn't want me to think she was a bad girl" (Burke, Stets, and Pirog-Good 1989; Jenkins and Dambrot 1987; Lisak and Roth 1988; Malamuth 1986; Muehlenhard and Linton 1987; Peterson and Franzese 1987). Because men's sexuality is seen as more natural, acceptable, and uncontrollable than women's sexuality, many men and women excuse acquaintance rape by affirming that men cannot control their natural urges (Miller and Marshall 1987).

Whereas some researchers explain these attitudes toward sexuality and rape using an individual or a psychological interpretation, we argue that rape has a social basis, one in which both men and women create and recreate masculine and feminine identities and relations. Based on the assumption that rape is part of the social construction of gender, we examine how men and women "do gender" on a college campus (West and Zimmerman 1987). We focus on fraternities because they have been identified as settings that encourage rape (Sanday 1990). By comparing fraternities that are viewed by women as places where there is a high risk of rape to those where women believe there is a low risk of rape as well as two local commercial bars, we seek to identify characteristics that make some social settings more likely places for the occurrence of rape.

METHOD

We observed social interactions between men and women at a private coeducational school in which a high percentage (49.4 percent) of students affiliate with Greek organizations. The university has an undergraduate population of approximately 4,500 students, just more than one third of whom are women; the students are primarily from upper-middle-class families. The school, which admitted only men until 1971, is highly competitive academically.

We used a variety of data collection approaches: observations of interactions between men and women at fraternity parties and bars, formal interviews, and informal conversations. The first author, a former undergraduate at this school and a graduate student at the time of the study, collected the data. She knew about the social life at the school and had established rapport and trust between herself and undergraduate students as a teaching assistant in a human sexuality course.

The process of identifying high- and low-risk fraternity houses followed Hunter's (1953) reputational approach. In our study, 40 women students identified fraternities that they considered to be high risk, or to have more sexually aggressive members and higher incidence of rape, as well as fraternities that they considered to be safe houses. The women represented all four years of undergraduate college and different living groups (sororities, residence halls, and off-campus housing). Observations focused on the four fraternities named most often by these women as high-risk houses and the four identified as low-risk houses.

Throughout the spring semester, the first author observed at two fraternity parties each weekend at two different houses (fraternities could have parties only on weekends at this campus). She also observed students' interactions in two popular university bars on weeknights to provide a comparison of students' behavior in non-Greek settings. The first local bar at which she observed was popular with seniors and older students; the second bar was popular with first-, second-, and third-year undergraduates because the management did not strictly enforce drinking age laws in this bar.

The observer focused on the social context as well as interaction among participants at each setting. In terms of social context, she observed the following: ratio of men to women, physical setting such as the party decor and theme, use and control of alcohol and level of intoxication, and explicit and implicit norms. She noted interactions between men and women (i.e., physical contact, conversational style, use of jokes) and the relations among men (i.e., their treatment of pledges and other men at fraternity parties). Other than the observer, no one knew the identity of the high- or low-risk fraternities. Although this may have introduced bias into the data collection, students on this campus who read this article before it was submitted for publication commented on how accurately the social scene is described.

In addition, 50 individuals were interviewed including men from the selected fraternities, women who attended those parties, men not affiliated with fraternities, and self-identified rape victims known to the first author. The first author approached men and women by telephone or on campus and asked them to participate in interviews. The interviews included open-ended questions about gender relations on campus, attitudes about date rape, and their own experiences on campus.

To assess whether self-selection was a factor in determining the classification of the fraternity, we compared high-risk houses to low-risk houses on several characteristics. In terms of status on campus, the high- and low-risk houses we studied attracted about the same number of pledges; however, many of the high-risk houses had more members. There was no difference in grade point averages for the two types of houses. In fact, the highest and lowest grade point averages were found in the high-risk category. Although both high- and low-risk fraternities participated in sports, brothers in the low-risk houses tended to play intramural sports whereas brothers in the high-risk houses were more likely to be varsity athletes. The high-risk houses may be more aggressive, as they had a slightly larger number of disciplinary incidents and their reports were more severe, often with physical harm to others and damage to property. Further, in year-end reports, there was more property damage in the high-risk houses. Last, more of the low-risk houses participated in a campus rape-prevention program. In summary, both high- and low-risk fraternities seem to be equally attractive to freshmen men on this campus, and differences between the eight fraternities we studied were not great; however, the high-risk houses had a slightly larger number of reports of aggression and physical destruction in the houses and the low-risk houses were more likely to participate in a rape prevention program.

RESULTS

The Settings

Fraternity Parties We observed several differences in the quality of the interaction of men and women at parties at high-risk fraternities compared to those at low-risk houses. A typical party at a low-risk house included an equal number of women and men. The social atmosphere was friendly, with considerable interaction between women and men. Men and women danced in groups and in couples, with many of the couples kissing and displaying affection toward each other. Brothers explained that, because many of the men in these houses had girlfriends, it was normal to see couples kissing on the dance floor. Coed groups engaged in conversations at many of these houses, with women and men engaging in friendly exchanges, giving the impression that they knew each other well. Almost no cursing and yelling was observed at parties in low-risk houses; when pushing occurred, the participants apologized. Respect for women extended to the women's bathrooms, which were clean and well supplied.

At high-risk houses, parties typically had skewed gender ratios, sometimes involving more men and other times involving more women. Gender segregation also was evident at these parties, with the men on one side of a room or in the bar drinking while women gathered in another area. Men treated women differently in the high-risk houses. The women's bathrooms in the high-risk houses were filthy, including clogged toilets and vomit in the sinks. When a brother was told of the mess in the bathroom at a high-risk house, he replied, "Good, maybe some of these bar wenches will leave so there will be more beer for us."

Men attending parties at high-risk houses treated women less respectfully, engaging in jokes, conversations, and behaviors that degraded women. Men made a display of assessing women's bodies and rated them with thumbs up or thumbs down for the other men in the sight of the women. One man attending a party at a high-risk fraternity said to another, "Did you know that this week is Women's Awareness Week? I guess that means we get to abuse them more this week." Men behaved more crudely at parties at high-risk houses. At one party, a brother dropped his pants, including his underwear, while dancing in front of several women. Another brother slid across the dance floor completely naked.

The atmosphere at parties in high-risk fraternities was less friendly overall. With the exception of greetings, men and women rarely smiled or laughed and spoke to each other less often than was the case at parties in low-risk houses. The few one-on-one conversations between women and men appeared to be strictly flirtatious (lots of eye contact, touching, and very close talking). It was rare to see a group of men and women together talking. Men were openly hostile, which made the high-risk parties seem almost threatening at times. For example, there was a lot of touching, pushing, profanity, and name calling, some done by women.

Students at parties at the high-risk houses seemed self-conscious and aware of the presence of members of the opposite sex, an awareness that was sexually charged. Dancing early in the evening was usually between women. Close to midnight, the sex ratio began to balance out with the arrival of more men or more women. Couples began to dance together but in a sexual way (close dancing with lots of pelvic thrusts). Men tried to pick up women using lines such as "Want to see my fish tank?" and "Let's go upstairs so that we can talk; I can't hear what you're saying in here."

Although many of the same people who attended high-risk parties also attended low-risk parties, their behavior changed as they moved from setting to setting. Group norms differed across contexts as well. At a party that was held jointly at a low-risk house with a high-risk fraternity, the ambience was that of a party at a high-risk fraternity with heavier drinking, less dancing, and fewer conversations between women and men. The men from both high- and low-risk fraternities were very aggressive; a fight broke out, and there was pushing and shoving on the dance floor and in general.

As others have found, fraternity brothers at high-risk houses on this campus told about routinely discussing their sexual exploits at breakfast the morning after parties and sometimes at house meetings (cf. Martin and Hummer 1989; O'Sullivan 1993; Sanday 1990). During these sessions, the brothers we interviewed said that men bragged about what they did the night before with stories of sexual conquests often told by the same men, usually sophomores. The women involved in these exploits were women they did not know or knew but did not respect, or *faceless victims.* Men usually treated girlfriends with respect and did not talk about them in these storytelling sessions. Men from low-risk houses, however, did not describe similar sessions in their houses.

The Bar Scene The bar atmosphere and social context differed from those of fraternity parties. The music was not as loud, and both bars had places to sit and have conversations. At all fraternity parties, it was difficult to maintain conversations with loud music playing and no place to sit. The volume of music at parties at high-risk fraternities was even louder than it was at low-risk houses, making it virtually impossible to have conversations. In general, students in the local bars behaved in the same way that students did at parties in low-risk houses with conversations typical, most occurring between men and women.

The first bar, frequented by older students, had live entertainment every night of the week. Some nights were more crowded than others, and the atmosphere was friendly, relaxed, and conducive to conversation. People laughed and smiled and behaved politely toward each other. The ratio of men to women was fairly equal, with students congregating in mostly coed groups. Conversation flowed freely and people listened to each other.

Although the women and men at the first bar also were at parties at low- and high-risk fraternities, their behavior at the bar included none of the blatant sexual or intoxicated behaviors observed at some of these parties. As the evenings wore on, the number of one-on-one conversations between men and women increased and conversations shifted from small talk to topics such as war and AIDS. Conversations did not revolve around picking up another person, and most people left the bar with same-sex friends or in coed groups.

The second bar was less popular with older students. Younger students, often under the legal drinking age, went there to drink, sometimes after leaving campus parties. This bar was much smaller and usually not as crowded as the first bar. The atmosphere was more mellow and relaxed than it was at the fraternity parties. People went there to hang out and talk to each other.

On a couple of occasions, however, the atmosphere at the second bar became similar to that of a party at a high-risk fraternity. As the number of people in the bar increased, they removed chairs and tables, leaving no place to sit and talk. The music also was turned up louder, drowning out conversation. With no place to dance or sit, most people stood around but could not maintain conversations because of the noise and crowds. Interactions between women and men consisted mostly of flirting. Alcohol consumption also was greater than it was on the less crowded nights, and the number of visibly drunk people increased. The more people drank, the more conversation and socializing broke down. The only differences between this setting and that of a party at a high-risk house were that brothers no longer controlled the territory and bedrooms were not available upstairs.

Gender Relations

Relations between women and men are shaped by the contexts in which they meet and interact. As is the case on other college campuses, *hooking up* has replaced dating on this campus, and fraternities are places where many students hook up. Hooking up is a loosely applied term on college campuses that had different meanings for men and women on this campus.

Most men defined hooking up similarly. One man said it was something that happens

> when you are really drunk and meet up with a woman you sort of know, or possibly don't know at all and don't care about. You go home with her with the intention of getting as much sexual, physical pleasure as she'll give you, which can range

> anywhere from kissing to intercourse, without any strings attached.

The exception to this rule is when men hook up with women they admire. Men said they are less likely to press for sexual activity with someone they know and like because they want the relationship to continue and be based on respect.

Women's version of hooking up is different. Women said they hook up only with men they cared about and described hooking up as kissing and petting but not sexual intercourse. Many women said that hooking up was disappointing because they wanted longer-term relationships. First-year women students realized quickly that hook-ups were usually one-night stands with no strings attached, but many continued to hook up because they had few opportunities to develop relationships with men on campus. One first-year woman said that "70 percent of hook-ups never talk again and try to avoid one another; 26 percent may actually hear from them or talk to them again, and 4 percent may actually go on a date, which can lead to a relationship." Another first-year woman said, "It was fun in the beginning. You get a lot of attention and kiss a lot of boys and think this is what college is about, but it gets tiresome fast."

Whereas first-year women get tired of the hook-up scene early on, many men do not become bored with it until their junior or senior year. As one upperclassman said, "The whole game of hooking up became really meaningless and tiresome for me during my second semester of my sophomore year, but most of my friends didn't get bored with it until the following year."

In contrast to hooking up, students also described monogamous relationships with steady partners. Some type of commitment was expected, but most people did not anticipate marriage. The term *seeing each other* was applied when people were sexually involved but free to date other people. This type of relationship involved less commitment than did one of boyfriend/girlfriend but was not considered to be a hook-up.

The general consensus of women and men interviewed on this campus was that the Greek system, called "the hill," set the scene for gender relations. The predominance of Greek membership and subsequent living arrangements segregated men and women. During the week, little interaction occurred between women and men after their first year in college because students in fraternities or sororities live and dine in separate quarters. In addition, many non-Greek upperclass students move off campus into apartments. Therefore, students see each other in classes or in the library, but there is no place where students can just hang out together.

Both men and women said that fraternities dominate campus social life, a situation that everyone felt limited opportunities for meaningful interactions. One senior Greek man said,

> This environment is horrible and so unhealthy for good male and female relationships and interactions to occur. It is so segregated and male dominated. . . . It is our party, with our rules and our beer. We are allowing these women and other men to come to our party. Men can feel superior in their domain.

Comments from a senior woman reinforced his views: "Men are dominant; they are the kings of the campus. It is their environment that they allow us to enter; therefore, we have to abide by their rules." A junior woman described fraternity parties as

> good for meeting acquaintances but almost impossible to really get to know anyone. The environment is so superficial, probably because there are so many social cliques due to the Greek system. Also, the music is too loud and the people are too drunk to attempt to have a real conversation, anyway.

Some students claim that fraternities even control the dating relationships of their members. One senior woman said, "Guys dictate how dating occurs on this campus, whether it's cool, who it's with, how much time can be spent with the girlfriend and with the brothers." Couples either left campus for an evening or hung out separately with their own same-gender friends at fraternity parties, finally getting together with each other at about 2 a.m. Couples rarely went together to fraternity parties. Some men felt that a girlfriend was just a replacement for a hook-up. According to one junior man, "Basically a girlfriend is someone you go to at 2 a.m. after you've hung out with the guys. She is the sexual outlet that the guys can't provide you with."

Some fraternity brothers pressure each other to limit their time with and commitment to their girlfriends. One senior man said, "The hill [fraternities] and girlfriends don't mix." A brother described a constant battle between girlfriends and brothers over who the guy is going out with for the night, with the brothers usually winning. Brothers teased men with girlfriends with remarks such as "whipped" or "where's the ball and chain?" A brother from a high-risk house said that few brothers at his house had girlfriends; some did, but it was uncommon. One man said that from the minute he was a pledge he knew he would probably never have a girlfriend on this campus because "it was just not the norm in my house. No one has girlfriends; the guys have too much fun with [each other]."

The pressure on men to limit their commitment to girlfriends, however, was not true of all fraternities or of all men on campus. Couples attended low-risk fraternity parties together, and men in the low-risk houses went out on dates more often. A man in one low-risk house said that about 70 percent of the members of his house were involved in relationships with women, including the pledges (who were sophomores).

Treatment of Women

Not all men held negative attitudes toward women that are typical of a rape culture, and not all social contexts promoted the negative treatment of women. When men were asked whether they treated the women on campus with respect, the most common response was "On an individual basis, yes, but when you have a group of men together, no." Men said that, when together in groups with other men, they sensed a pressure to be disrespectful toward women. A first-year man's perception of the treatment of women was that "they are treated with more respect to their faces, but behind closed doors, with a group of men present, respect for women is not an issue." One senior man stated, "In general, college-aged men don't treat women their age with respect because 90 percent of them think of women as merely a means to sex." Women reinforced this perception. A first-year woman stated, "Men here are more interested in hooking up and drinking beer than they are in getting to know women as real people." Another woman said, "Men here use and abuse women."

Characteristic of rape culture, a double standard of sexual behavior for men versus women was prevalent on this campus. As one Greek senior man stated, "Women who sleep around are sluts and get bad reputations; men who do are champions and get a pat on the back from their brothers." Women also supported a double standard for sexual behavior by criticizing sexually active women. A first-year woman spoke out against women who are sexually active: "I think some girls here make it difficult for the men to respect women as a whole."

One concrete example of demeaning sexually active women on this campus is the "walk of shame." Fraternity brothers come out on the porches of their houses the night after parties and heckle women walking by. It is assumed that these women spent the night at fraternity houses and that the men they were with did not care enough about them to drive them home. Although sororities now reside in former fraternity houses, this practice continues and sometimes the victims of hecklings are sorority women on their way to study in the library.

A junior man in a high-risk fraternity described another ritual of disrespect toward women called "chatter." When an unknown woman sleeps over at the house, the brothers yell degrading remarks out the window as she leaves the next morning such as "Fuck that bitch" and "Who is that slut?" He said that sometimes brothers harass the brothers whose girlfriends stay over instead of heckling those women.

Fraternity men most often mistreated women they did not know personally. Men and women alike reported incidents in which brothers observed other brothers having sex with unknown women or women they knew only casually. A sophomore woman's experience exemplifies this anonymous state: "I don't mind if 10 guys were watching or it was videotaped. That's expected on this campus. It's the fact that he didn't apologize or even offer to drive me home that really upset me." Descriptions of sexual encounters involved the satisfaction of men by nameless women. A brother in a high-risk fraternity described a similar occurrence:

> A brother of mine was hooking up upstairs with an unattractive woman who had been pursuing him all night. He told some brothers to go outside the window and watch. Well, one thing led to another and they were almost completely naked when the woman noticed the brothers outside. She was then unwilling to go any further, so the brother went outside and yelled at the other brothers and then closed the shades. I don't know if he scored or not, because the woman was pretty upset. But he did win the award for hooking up with the ugliest chick that weekend.

Attitudes Toward Rape

> The sexually charged environment of college campuses raises many questions about cultures that facilitate the rape of women. How women and men define their sexual behavior is important legally as well as interpersonally. We asked students how they defined rape and had them compare it to the following legal definition: the perpetration of an act of sexual intercourse with a female against her will and consent, whether her will is overcome by force or fear resulting from the threat of force, or by drugs or intoxicants; or when, because of mental deficiency, she is incapable of exercising rational judgment. (Brownmiller 1975, 368)

When presented with this legal definition, most women interviewed recognized it as well as the complexities involved in applying it. A first-year woman said, "If a girl is drunk and the guy knows it and the girl says, 'Yes, I want to have sex,' and they do, that is still rape because the girl can't make a conscious, rational decision under the influence of alcohol." Some women disagreed. Another first-year woman stated, "I don't think it is fair that the guy gets blamed when both people involved are drunk."

The typical definition men gave for rape was "when a guy jumps out of the bushes and forces himself sexually onto a girl." When asked what date rape was, the most common answer was "when one person has sex with another person who did not consent." Many men said, however, that "date rape is when a woman wakes up the next morning and regrets having sex." Some men said that date rape was too gray an area to define. "Consent is a fine line," said a Greek senior man student. For the most part, the men we spoke with argued that rape did not occur on this campus. One Greek sophomore man said, "I think it is ridiculous that someone here would rape someone." A first-year man stated, "I have a problem with the word rape. It sounds so criminal, and we are not criminals; we are sane people."

Whether aware of the legal definitions of rape, most men resisted the idea that a woman who is intoxicated is unable to consent to sex. A Greek junior man said, "Men should not be responsible for women's drunkenness." One first-year man said, "If that is the legal definition of rape, then it happens all the time on this campus." A senior man said, "I don't care whether alcohol is involved or not; that is not rape. Rapists are people that have something seriously wrong with them." A first-year man even claimed that when women get drunk, they invite sex. He said, "Girls get so drunk here and then come on to us. What are we supposed to do? We are only human."

DISCUSSION AND CONCLUSION

These findings describe the physical and normative aspects of one college campus as they relate to attitudes about and relations between men and women. Our findings suggest that an explanation emphasizing rape culture also must focus on those characteristics of the social setting that play a role in defining heterosexual relationships on college campuses (Kalof and Cargill 1991). The degradation of women as portrayed in rape culture was not found in all fraternities on this campus. Both group norms and individual behavior changed as students went from one place to another. Although individual men are the ones who rape, we found that some settings are more likely places for rape than are others. Our findings suggest that rape cannot be seen only as an isolated act and blamed on individual behavior and proclivities, whether it be alcohol consumption or attitudes. We also must consider characteristics of the settings that promote the behaviors that reinforce a rape culture.

Relations between women and men at parties in low-risk fraternities varied considerably from those in high-risk houses. Peer pressure and situational norms influenced women as well as men. Although many men in high- and low-risk houses shared similar views and attitudes about the Greek system, women on this campus, and date rape, their behaviors at fraternity parties were quite different.

Women who are at highest risk of rape are women whom fraternity brothers did not know. These women are faceless victims, nameless acquaintances—not friends. Men said their responsibility to such persons and the level of guilt they feel later if the hook-ups end in sexual intercourse are much lower if they hook up with women they do not know. In high-risk houses, brothers treated women as subordinates and kept them at a distance. Men in high-risk houses actively discouraged ongoing heterosexual relationships, routinely degraded women, and participated more fully in the hook-up scene; thus, the probability that women would become faceless victims was higher in these houses. The flirtatious nature of the parties indicated that women go to these parties looking for available men, but finding boyfriends or relationships was difficult at parties in high-risk houses. However, in the low-risk houses, where more men had long-term relationships, the women were not strangers and were less likely to become faceless victims.

The social scene on this campus, and on most others, offers women and men few other options to socialize. Although there may be no such thing as a completely safe fraternity party for women, parties at low-risk houses and commercial bars encouraged men and women to get to know each other better and decreased the probability that women would become faceless victims. Although both men and women found the social scene on this campus demeaning, neither demanded different settings for socializing, and attendance at fraternity parties is a common form of entertainment.

These findings suggest that a more conducive environment for conversation can promote more positive interactions between men and women. Simple changes would provide the opportunity for men and women to interact in meaningful ways such as adding places to sit and lowering the volume of music at fraternity parties or having parties in neutral locations, where men are not in control. The typical party room in fraternity houses includes a place to dance but not to sit and talk. The music often is loud, making it difficult, if not impossible, to carry on conversations; however, there were more conversations at the low-risk parties, where there also was more respect shown toward women. Although the number of brothers who had steady girlfriends in the low-risk houses as compared to those in the high-risk houses may explain the differences, we found that commercial bars also provided a context for interaction between men and women. At the bars, students sat and talked and conversations between men and women flowed freely, resulting in deep discussions and fewer hook-ups.

Alcohol consumption was a major focus of social events here and intensified attitudes and orientations of a rape culture. Although pressure to drink was evident at all fraternity parties and at both bars, drinking dominated high-risk fraternity parties, at which nonalcoholic beverages usually were not available and people chugged beers and became visibly drunk. A rape culture is strengthened by rules that permit alcohol only at fraternity parties. Under this system, men control the parties and dominate the men as well as the women who attend. As college administrators crack down on fraternities and alcohol on campus, however, the same behaviors and norms may transfer to other places such as parties in apartments or private homes where administrators have much less control. At commercial bars, interaction and socialization with others were as important as drinking, with the exception of the nights when the bar frequented by under-class students became crowded. Although one solution is to offer nonalcoholic social activities, such events receive little support on this campus. Either these alternative events lacked the prestige of the fraternity parties or the alcohol was seen as necessary to unwind, or both.

In many ways, the fraternities on this campus determined the settings in which men and women interacted. As others before us have found, pressures for conformity to the norms and values exist at both high-risk and low-risk houses (Kalof and Cargill 1991; Martin and Hummer 1989; Sanday 1990). The desire to be accepted is not unique to this campus or the Greek system (Holland and Eisenhart 1990; Horowitz 1988; Moffat 1989). The degree of conformity required by Greeks may be greater than that required in most social groups, with considerable pressure to adopt and maintain the image of their houses. The fraternity system intensifies the "groupthink syndrome" (Janis 1972) by solidifying the identity of the in-group and creating an us/them atmosphere. Within the fraternity culture, brothers are highly regarded and women are viewed as outsiders. For men in high-risk fraternities, women threatened their brotherhood; therefore, brothers discouraged relationships and harassed those who treated women as equals or with respect. The pressure to be one of the guys and hang out with the guys strengthens a rape culture on college campus by demeaning women and encouraging the segregation of men and women.

Students on this campus were aware of the contexts in which they operated and the choices available to them. They recognized that, in their interactions, they created differences between men and women that are not natural, essential, or biological (West and Zimmerman 1987). Not all men and women accepted the demeaning treatment of women, but they continued to participate in behaviors that supported aspects of a rape culture. Many women participated in the hookup scene even after they had been humiliated and hurt because they had few other means of initiating contact with men on campus. Men and women alike played out this scene, recognizing its injustices in many cases but being unable to change the course of their behaviors.

Although this research provides some clues to gender relations on college campuses, it raises many questions. Why do men and women participate in activities that support a rape culture when they see its injustices? What would happen if alcohol were not controlled by groups of men who admit that they disrespect women when they get together? What can be done to give men and women on college campuses more opportunities to interact responsibly and get to know each other better? These questions should be studied on other campuses with a focus on the social settings in which the incidence of rape and the attitudes that support a rape culture exist. Fraternities are social contexts that may or may not foster a rape culture.

Our findings indicate that a rape culture exists in some fraternities, especially those we identified as high-risk houses. College administrators are responding to this situation by providing counseling and educational programs that increase awareness of date rape including campaigns such as "No means no." These strategies are important in changing attitudes, values, and behaviors; however, changing individuals is not enough. The structure of campus life and the impact of that structure on gender relations on campus are highly determinative. To eliminate campus rape culture, student leaders and administrators must examine the situations in which women and men meet and restructure these settings to provide opportunities for respectful interaction. Change may not require abolishing fraternities; rather, it may require promoting settings that facilitate positive gender relations.

REFERENCES

Barthel, D. 1988. *Putting on appearances: Gender and advertising.* Philadelphia: Temple University Press.

Boeringer, S. B., C. L. Shehan, and R. L. Akers. 1991. Social contexts and social learning in sexual coercion and aggression: Assessing the contribution of fraternity membership. *Family Relations* 40:58–64.

Brownmiller, S. 1975. *Against our will: Men, women and rape.* New York: Simon & Schuster.

Buchwald, E., P. R. Fletcher, and M. Roth, eds. 1993. *Transforming a rape culture.* Minneapolis, MN: Milkweed Editions.

Burke, P., J. E. Stets, and M. A. Pirog-Good. 1989. Gender identity, self-esteem, physical abuse and sexual abuse in dating relationships. In *Violence in dating relationships: Emerging social issues,* edited by M. A. Pirog-Good and J. E. Stets. New York: Praeger.

Gwartney-Gibbs, P., and J. Stockard. 1989. Courtship aggression and mixed-sex peer groups. In *Violence in dating relationships: Emerging social issues,* edited by M. A. Pirog-Good and J. E. Stets. New York: Praeger.

Herman, D. 1984. The rape culture. In *Women: A feminist perspective,* edited by J. Freeman, Mountain View, CA: Mayfield.

Holland, D. C., and M. A. Eisenhart. 1990. *Educated in romance: Women, achievement, and college culture.* Chicago: University of Chicago Press.

Horowitz, H. L. 1988. *Campus life: Undergraduate cultures from the end of the 18th century to the present.* Chicago: University of Chicago Press.

Hunter, F. 1953. *Community power structure.* Chapel Hill: University of North Carolina Press.

Jenkins, M. J., and F. H. Dambrot. 1987. The attribution of date rape: Observer's attitudes and sexual experiences and the dating situation. *Journal of Applied Social Psychology* 17:875–95.

Janis, I. L. 1972. *Victims of groupthink.* Boston: Houghton Mifflin.

Kalof, L., and T. Cargill. 1991. Fraternity and sorority membership and gender dominance attitudes. *Sex Roles* 25:417–23.

Kimmel, M. S. 1993. Clarence, William, Iron Mike, Tailhook, Senator Packwood, Spur Posse, Magic . . . and us. In *Transforming a rape culture,* edited by E. Buchwald, P. R. Fletcher, and M. Roth. Minneapolis, MN: Milkweed Editions.

Koss, M. P., T. E. Dinero, C. A. Seibel, and S. L. Cox. 1988. Stranger and acquaintance rape: Are there differences in the victim's experience? *Psychology of Women Quarterly* 12:1–24.

Koss, M. P., C. A. Gidycz, and N. Wisniewski. 1985. The scope of rape: Incidence and prevalence of sexual aggression and victimization in a national sample of higher education students. *Journal of Consulting and Clinical Psychology* 55:162–70.

LaPlante, M. N., N. McCormick, and G. G. Brannigan. 1980. Living the sexual script: College students' views of influence in sexual encounters. *Journal of Sex Research* 16:338–55.

Lisak, D., and S. Roth. 1988. Motivational factors in nonincarcerated sexually aggressive men. *Journal of Personality and Social Psychology* 55:795–802.

Malamuth, N. 1986. Predictors of naturalistic sexual aggression. *Journal of Personality and Social Psychology* 50:953–62.

Martin, P. Y., and R. Hummer. 1989. Fraternities and rape on campus. *Gender & Society* 3:457–73.

Miller, B., and J. C. Marshall. 1987. Coercive sex on the university campus. *Journal of College Student Personnel* 28:38–47.

Moffat, M. 1989. *Coming of age in New Jersey: College life in American culture.* New Brunswick, NJ: Rutgers University Press.

Muehlenhard, C. L., and M. A. Linton, 1987. Date rape and sexual aggression in dating situations: Incidence and risk factors. *Journal of Counseling Psychology* 34:186–96.

O'Sullivan, C. 1993. Fraternities and the rape culture. In *Transforming a rape culture,* edited by E. Buchwald, P. R. Fletcher, and M. Roth. Minneapolis, MN: Milkweed Editions.

Peterson, S. A., and B. Franzese. 1987. Correlates of college men's sexual abuse of women. *Journal of College Student Personnel* 28:223–28.

Sanday, P. R. 1990. *Fraternity gang rape: Sex, brotherhood, and privilege on campus.* New York: New York University Press.

West, C., and D. Zimmerman. 1987. Doing gender. *Gender & Society* 1:125–51.

36

The Coach, the Player . . . and Their Demons

BILL MINUTAGLIO

There was the perseverance, the brutal sacrifice. The kind of work that usually leaves little time for melancholy or acidy drops of self-doubt.

The player did it because he loved the sweep and flow of the game, the loping triumphs and the easy smiles—and because he knew he would use the sport to shepherd his family to a better place. The stern-faced assistant coach did it because he believed he could squeeze perfection from an imperfect world. He mercilessly drove himself and everyone in his orbit—and he told the few people he confided in that it was all for a greater good.

Once, years ago, the player had been locked in an unexpected slow dance with death. Somehow, the graceful young man recovered and walked away. Those who knew him said his laughter was richer, that he exuded the air of someone whose life had been touched by heaven's fingerprint.

Once, years ago, the highly intelligent coach felt that he, too, had been invited onto the floor with death. He made a rare, serious admission to his wife. Maybe he had finally driven himself too hard—had been loathed too often—and come too close to dying.

Finally, this year, the coach and the player came to share one last thing.

This year they became achingly bound when each man raised a .380 handgun and pressed it against his body.

Reprinted from *Sporting News,* Aug. 5, 1995, Vol. 220, No. 32, p. 38.

In separate explosions, in distant thunderclaps, the coach and the player abandoned this world the very same, horrible way.

"It's just like a mind game. If you talk to yourself, they say you're not supposed to answer, but I feel if you answer yourself you can probably solve some things," Willie Smith, then 22, was carefully telling a staffer at Louisiana Tech University two years ago. The lean, smiling Smith was talking football, about how he liked to converse with himself in the middle of a game. He didn't say it, but those closest to him said he frequently went back-and-forth with his inner voices off the field.

It was still early in his career but the walk-on had built a reputation in the locker room. Sturdy Willie Smith was a leader. The almond-eyed cornerback was quick with the humor, his hands, the interceptions. He had his father's sly sixth sense, the uncanny ability to picture something, feel something, before it happened—like a receiver's move, a quarterback's pump fake. By late last year, there wasn't any question that Willie Smith could play—the junior tied for the Division I-A lead in interceptions. The real question was whether he would continue to play in college or try the pros.

"Bug" is what his three sisters called him, and it had a lot to do with those fast hands. The ones that would catch flies and cause the Nintendo games to stutter and blink because no one had ever hit such a high score. Bug was 5-10 and 175 pounds and a survivor of the meanest streets in south Dallas—where he played dirt-lot football while homeboys on Grand Avenue were shooting each other over new sneakers and Pennsylvania Avenue gangsters were gunning down a cop.

Bug's cancer-riddled mother, Sarah, made sure he got away, and she took him to a supposedly safer Dallas suburb called Garland. He went to a new school, and it was as if he had gone to another planet. The move uncorked his humor, and his teammates and coaches were enthralled by this bull-necked prankster, a silky defensive back who would wink and cackle at his own ferocious abilities.

Bug liked math, he liked history and, if he had a choice, he'd eat chicken three times a day. He'd grab the remote control and stab at it until there was a nature or wildlife show on TV. Bug could dunk a basketball and he read the Bible at night. On the tape player, he leaned toward Tupac Shakur's gritty inner-city visions or R Kelly's seductive verses. He called his sisters Keathley, Anita and Andrea almost every day—and offered lectures on how Keathley should be raising her son, his little nephew, Tevin. There was someone else he called every day—the slender homegirl named Anchesionique Drake. He and "Nique" had dated and talked marriage for five years, even after she had gone to work driving a city bus. And Bug revered his father, the one who always left him with the words: "It's gonna be all right." But, most of all, above all, there was his mother—the person whom Bug knew loved him the most.

In the summer of 1991, on a Saturday at 11:30 P.M., Bug was standing with some friends on a Garland street corner. It was the year after he had played his first season of college football at Tyler (Tex.) Junior College. A car rolled through the lingering heat, a gun emerged and a bullet pierced Bug. The shooters were aiming for someone else, maybe someone who had messed with a girl, but one bullet found Bug. It passed through his stomach, somehow missing vital organs, and exited his side.

For the next 10 days in the hospital, Bug had a lot to think about. He dwelled on how his mother had insisted the family move—but how they hadn't been able to afford to move far enough. He thought, not about dying, but about saving his mother and the rest of the family.

"It, well, it changed him. He started going to church," says his father, Willie George Smith Sr., late one evening. A wiry normally energized man, the elder Smith has slowed down for a few seconds. He stares at the ground and idly spins his son's old football helmet in his hands. "He wanted to take care of us."

And maybe it was that time in the hospital that caused one more thought to form in Bug's head. It was something that he would recite, chant, to his sisters, coaches and teammates: "I'm taking my family to a big, fine, fat home far away in the country. They're getting away, and I'm the one taking them."

"I really thought I wouldn't survive this year. I'm so exhausted from trying to put pieces together that don't fit," Wayne McDuffie was saying in 1994 to his wife of more than two decades. It was at the end of another grueling season as offensive coordinator at the University of Georgia. And it was one of the most serious discussions they had ever had. Those who knew him would never imagine Wayne McDuffie admitting anything resembling weakness. "I'm trying to make something from nothing. I really thought I would die. I thought I would have a heart attack and die because I worked so hard. I worried so much and tried so desperately to hold this thing together." The team had what was—for Wayne McDuffie, at least—a disastrous season. The Bulldogs went 6-4-1, but endured a homecoming loss to Vanderbilt, a last-quarter defeat against Alabama and a 38-point drubbing at Florida in which McDuffie's offensive unit contributed only 14 points.

Toni McDuffie, an artist, teacher and horse trainer, lived for those moments when her tough, chiseled husband would open himself up to her. When he would express some vulnerability. When she could help Wayne carry the weight accumulated through his carefully regimented climb from playing at Florida State and coaching for more than two decades, including two stints each at his alma mater, at the University of Georgia and even for the Atlanta Falcons.

Of course, Wayne would rise above his pain. She knew that was what he always did—that he had wrestled with manic depression for years and that with her help and the help of his medication, he could cope. He would always, always pick himself up—and never admit weakness to anyone but her. He would be back to being the Wayne McDuffie that she alone knew—someone far removed from the grim, oppressive, aloof, abrasive perfectionist so many others encountered. Even today, in 1996, she wishes that people knew the real Wayne McDuffie—though she remains hesitant to speak publicly about him.

Over the years, Wayne McDuffie would always return to being the stoic, steady, loving father and husband. The man who—before they were married—found out that Toni had always wanted a horse and promptly got her one. The husband who never sat in a chair without a book, a biography, a history, something about business—or one of the travel magazines he subscribed to. The one who loved golf and the fact that he and his wife continued to live in a home overlooking the fourth hole of a Tallahassee-area country club. The one who sometimes went hunting with buddies, mainly to sip cool beverages and tell lies. The Wayne McDuffie who took the family hiking in northern Georgia; whose daughter had him wrapped around her little finger; who once asked his wife to create a painted memento for a player whose mother had died—a player he had hounded without compassion.

She knew that, as Wayne turned 50, he allowed most people to see him only as cold. Mean. Egotistical. Someone to be feared, someone who could be brutally sarcastic and humiliating.

But she alone knew that his mind was forever racing, analyzing, reviewing. That there was never a moment when he wasn't sorting out some problem inside his head. His distant demeanor wasn't egotistical or intentional, she thought—it was just a byproduct of an extraordinarily active, preoccupied mind.

Wayne's mind, thought his wife, was always going in a thousand directions.

Willie Smith was a relatively fast healer. And as soon as he was out of the hospital, he began to chart it all out.

Play pro ball for five years, probably cornerback, maybe for the New York Giants. He wouldn't be like Deion, wearing his wallet on his sleeve. He'd take care of business, then retire. Be a social worker. Provide for his mother and maybe rescue the old man from his cycle of jobs at the post office, Sears and the clothing factory—and he'd personally screen any young men who came calling on his sisters. "I'm gonna take care of the family. My way," he'd say with a laugh.

Bug's father participated in masters track meets. Now, after Bug got out of the hospital, the father would nurse the son back into shape. They went into some open fields in the South Oak Cliff part of Dallas and ran together, doing 200s, 300s and eventually some miles. And they always made sure to listen for that telltale crack of a weapon, the ubiquitous and ominous sound of guns being fired. When they heard it, it was time to head back to Bug's '91 white Buick Skylark. "It's gonna be all right," his father would simply say, just like he always did.

Bug and his father drove the Skylark to Louisiana Tech in 1992. Bug liked being away from the big city and the crime. His father talked to the coaches, and they said, at the very least, they would take care of his son. After not playing football for two years, he walked on in 1993 and was redshirted. "The first time he was on the field, I'm watching him and I was shocked. He was with one of our fastest receivers step by step. I said, 'Man, where did that guy come from?'" remembers Joe Raymond Peace, former Louisiana Tech coach. "He was one of those unusual kids who had a burning desire to play. He wasn't noticed coming out of high school. It's very special for someone to do that."

In 1994, Smith earned a scholarship and a starting role on Tech's defense. He had five interceptions in 10 games, including one he returned 75 yards for a touchdown against West Virginia. There were circus catches, and there were spectacular interception returns. Scouts began noticing No. 33. He told coaches he was going to get even better. And he told them he was doing it all for his family.

"He had a real desire to take care of his family. He felt like he had a real obligation to get them into a good environment," Louisiana Tech assistant coach Pete Fredenburg says. "He was a fierce competitor. And he had the knack for the game. I've been around a few guys who are that way. I wish you could bottle it and sell it. There are so many guys who are up and down with their successes and failures. But Willie wasn't that way. He always was just so steady."

Wayne McDuffie also wanted to chart it all out. The game, the finances, his career. Everything, it seemed, since he grew up in Hawkinsville, Ga.

Sometimes, the assistant coaches at Florida State, where he coached for most of the 1980s, would hear a strange flapping sound echoing from one of the football offices. It could be 6 A.M. or even 5 A.M.

As they held their cups of coffee and looked inside, there would be Wayne McDuffie asleep on a conference table, his Clint Eastwood face and body oddly illuminated by the flickering light coming from the movie projector. The film he had been studying, rethreading and rerunning all night long was still spinning wildly in the reels. But, when someone woke him up, he would simply, wordlessly, move to the football field—where he had ordered his offensive lineman to show up before sunrise.

"When Wayne was here, when it came to the fourth quarter, we knew his people were going to be Rambos. They weren't going to cheese up. They were going to be tough," says Florida State assistant coach Jim Gladden, a friend for 22 years. "Wayne was a great coach, but he didn't have a lot of tolerance. If a guy showed any cowardice or any inkling of not being as tough as he ought to be, Wayne would get him in a drill and not let up on him. Wayne probably ran off as many good players as he made. Wayne never knew when to let up. It was full-bore every minute."

Back to his days as a Florida State and Continental Football League player, Wayne McDuffie had always gone full-bore. He married the homecoming queen and clawed toward those prestigious assistant positions—including the 1990 stint with the Falcons and a reported $100,000-plus package starting in 1991 at Georgia. By 1995, he had a sprawling home, with a swimming pool, on a Tallahassee golf course. There were three bright children, two boys and a girl. Ruggedly handsome, he was 50 with the body of a 30-year-old man in perfect shape.

He had something else: The grudging admiration from that tight big-time coaching fraternity. Wayne McDuffie was the ultimate hard-ass, the common sentiment went, but he got results. During seven years as offensive coordinator at Florida State, he had six players earn All-American first-team designation—and plenty more who earned second- and third-team honors and honorable-mention recognition. In his last three years at Florida State, the team went 32-4. Back at Georgia in 1991 some said he single-handedly retooled the team. In his first season, the offense increased by almost 1,200 yards and the school went 9-3—after going 10-13 in the previous two seasons.

Confident in his abilities, Wayne told people he also wouldn't suffer any fools. That he had no time for small talk with boosters and hangers-on. He had that chart, that plan, the one that he was constantly refining in his mind.

"Wayne was one of those guys who had dedicated a long part of his career to map out a path, to always attain the goal of being a head coach at a major Division I school," says David Kelly, the former running backs coach at Georgia. Sometimes, he and Kelly would sit in McDuffie's office and talk late into the night. McDuffie admitted his wife constantly reminded him of his shortcomings. "And I'm very well aware of them," McDuffie added. "But I can't help the way I am. I know about my lack of personal skills, but I'm not going to compromise my principles."

Kelly personally experienced McDuffie's withering verbal attacks. He tested Kelly, who joined Georgia from the high school ranks, all the time. And Kelly saw McDuffie regularly rattle players and other coaches—pushing buttons to the bursting point as McDuffie went around the table at Georgia staff meetings. Later, McDuffie would simply tell Kelly: "In order for me to finish on top, I attack the people I know and love the best. From a personal standpoint, I try to destroy them. But then I come back later and say, 'Ah, Kelly, you know I'm crazy.' "

Kelly, like many others, told people that McDuffie was brilliant with the X's and O's. And, Kelly said to himself, McDuffie never received the credit he was due because so many people couldn't see beyond his personality.

"He was absolutely the most unique character I've ever met. I'd see him in the weight room late at night, killing himself," says Matt Braswell, a former All-Southeastern Conference offensive lineman at Georgia. And when Braswell and other players would drive by the jogging McDuffie, they would lower their car windows and listen as the coach violently cursed himself for not running harder and faster. "He was a son-of-a-bitch. The closest analogy I can draw would be a drill instructor. But Wayne taught me more football than any other coach. I'm not sure it was his mantra 'to never give up.' I think it was, 'If you're going to do it, then be the best you can be . . . and if you can't do it, then you quit.'"

"He was a tough, hard-nosed football coach. You won't run across any harder," says Ray Goff, a close friend and the former Georgia coach who worked hard to lure McDuffie to his staff. "People would recruit against you because of Wayne. They'd say, 'You don't want to go there (to Georgia), the guy is too tough, he's too hard.' He wanted to be the best at everything. He could not stand anything not being the best. Maybe he tried to keep that same persona off the field that he had on the field—and he had a hard time distinguishing where to cut it loose."

When Goff and McDuffie first met, it was 1977 and the two were in a car on a recruiting trip into deep Georgia. McDuffie, beginning his first stint on the Georgia staff, was virtually silent—as usual. He spoke only once to Goff, then a Bulldogs graduate assistant coach, during the entire eight-hour trip: "Ray, what church do you go to?" After Goff told him, McDuffie added, "Ray, I like you, you enjoy life."

Goff realized McDuffie was unlike anyone else he had ever met. "He was truly the most intense guy I've ever been around in my life. I've never seen the likes of Wayne McDuffie."

Willie's father was running, again, with his son. But this time it was up in the stands.

When he was able to travel from Dallas to Ruston, La., for a game, he would jump and ramble through the stadium, paralleling his son's runs after yet another interception. After a game, Willie's father would carefully look at the pages and pages of handwritten notes he had scrawled outlining his son's life on and off the field. He could remember plays, the time of day and the field conditions from every game—even Willie's high school games. Every day, he would carefully, gingerly, place his notes, his game-day programs, every newspaper account, into a cardboard box.

Sometimes Willie would give his father the team, state and conference trophies that were streaming in. Defensive Player of the Week. Defensive Player of the Year. Big Hit of the Week. Big West Conference Defensive Player of the Year. Second-team UPI All-America. Third-team Associated Press All-America. They, too, would be carefully wrapped in plastic and taken to the elder Smith's home in Dallas.

In the summer of 1995, the cancer spread too far through Willie's 47-year-old mother. He had grown increasingly close to his mother, and her death pressed down on him like a cold, heavy hand. Willie began carrying her key chain, the one that had "God Loves Me" stamped on it. And, though his father told him to dedicate his season to her, Willie often was inconsolable.

"Sometimes I feel the pressure. Sometimes I feel up and down. Why did she die? She was a good mom. Why would this happen to her? She was a Christian, she raised good kids," Willie quietly told his father.

His sister Keathley watched him closely. Sometimes, she thought, it seemed as though Willie didn't want to be here. She knew that he listened, over and over again, to Tupac Shakur's ghostly incantation: "Only God Can Judge Me Now." Meanwhile, there was another death—his girlfriend Nique's brother had been killed. Willie had been close to him, and Willie's buoyant personality seemed to stumble a little more.

"Well, y'all are gonna miss me when I'm gone," Willie would suddenly say to Nique. "I'm gonna leave first, before anybody else leaves."

His girlfriend and anyone else listening chalked it up to crazy talk.

One day, Willie's dad told him: "You can't take your life. You won't get to heaven and see your mother if you take your life." For the millionth time, he added, "It's gonna be all right."

Willie listened to his father, nodded and seemed to agree.

The contract for a $40,000 lakefront lot in Tallahassee was on Wayne's desk at home. He and Toni could envision their future. Erin, Kyle and Scott were getting older. The lot, with a new home built on it, would be a place to grow old in as the kids moved out. It was big, about 1.5 acres, and near a golf course. It even had plenty of room for Toni's two horses, including the one Wayne had given her 27 years ago. But Wayne left the contract on his desk. Unsigned. "We'll have to wait and see. We can't afford to buy it . . . in case anything happens," he told his wife.

Something did happen. At the end of 1995, he was fired for the first time as a coach. Fired after five years, mostly successful, at Georgia—including a year when he thought he had almost given his life to the school. It was, truth be told, something Wayne saw coming. Something, said some, he had invited. Last October, he spoke to the Athens Touchdown Club and publicly suggested that Goff's staff had already been fired by athletic director Vince Dooley. McDuffie's animosity toward Dooley was thinly veiled.

"I had to address him professionally on a couple of issues that I thought he was wrong on. I called him down. But when I did it, it was over. It was just a professional thing," Dooley says. "But he may have carried it with him . . . he could have." After the Touchdown Club speech, Dooley waited three or four days, thinking that Wayne would come in to apologize or explain. "I thought he had not conducted himself the way he should have. What he did at the Touchdown Club was shocking to everybody. . . . I had a responsibility to talk to him about it."

A month later, Wayne and the rest of Ray Goff's staff were fired. Some old colleagues, admirers, heard the news and wondered what it would mean to Wayne McDuffie.

Jim Gladden at Florida State remembered how Wayne would act when coach Bobby Bowden called a different play than the one Wayne had recommended from the press box. Sometimes Wayne would storm away from the press box, maybe in the third or fourth quarter. And Gladden remembered how Wayne would get when he wasn't on the medication he had been taking to combat his bouts of manic depression. Gladden could tell, like others, if he wasn't on his Prozac: "He was very different, cantankerous, hard to get along with."

Through it all, there was Wayne McDuffie's insistent belief that he had earned a rightful opportunity at the head-coaching table. That the people he had worked for, from Bowden to Dooley, should have found him a head-coaching position.

"Wayne was a great coach, one of the finest coaches I've ever been around. We all loved him in spite of his faults," Gladden says late one day after finishing work in his north Florida yard. "But Wayne's personality was never going to allow anyone to hire him as head coach. He wanted to be a head coach desperately. He was so obsessed and possessed with his profession that he failed to see anything else in the world that went by him. He was the guy who marched to a different drummer. He was out of sync."

And Gladden, for one, was truly concerned. "I always thought he was highly subject to . . . I thought he could easily snap and have to be institutionalized. I thought that might occur or that he might have a nervous breakdown. But I never thought he would kill himself. It never crossed my mind."

In the first game of the 1995 season, Willie Smith skyjacked three interceptions against Bowling Green. Against Arkansas State, he had eight tackles and another interception. Against Tulsa, he picked off two more passes—including a stunning, game-saving interception at the 3-yard line in the final two minutes. Louisiana Tech's promotion department had begun touting him as the school's finest defensive back since Doug Evans, who plays for the Packers. NFL scouts began paying more attention. At the end of the season, he had eight interceptions and some All-American acclaim.

Through it all, Willie admitted to a few friends that he still wondered if he was really as good as other cornerbacks. William Amos, his teammate and roommate when the team traveled, saw a deep but sometimes troubled friend. They would talk late into the night, staring up at the ceiling and letting their minds wander.

"On the outside, he was always laughing and joking. On the inside, he had a lot of pressures. He always had so much on his mind. He was debating whether or not to come back for his senior year. He would jump up in the middle of the night to check on his family—he would just leave and drive in the night to Dallas. He always, always, thought about his mother. He was worried about his grades, about them hurting his chances to play ball and graduate."

And sometimes, Willie would drink. "Sometimes all the talk in the world wouldn't help," Amos says. "It was a way to forget, for a short time. It felt good not to have to worry . . . at least for a short time."

The last time his roomie saw Willie was before Christmas. Willie had celebrated his birthday December 20. He was 24 and still thinking about pro ball, about how it would be a miracle to make $300,000 and finally save his family. "Stay out of trouble," he said to his roomie, as he left to go to Dallas.

Back in the city, he and his father watched the Blue-Gray college all-star game on TV on Christmas Day. In midgame, Willie abruptly said: "Dad, I have nine lives."

His father studied his son. "Well, Junior you need to be careful. Everybody has to die sometime." He added his favorite words: "It's gonna be all right."

Later that week, Willie, his sister and two of her girlfriends made a late-night trip to the edgy south Dallas neighborhood that their mother had once insisted they all abandon. It was December 30 and there was a club there, and, maybe, a chance to see some of the survivors in the old neighborhood. Before they left on that Saturday night, they carefully took a handgun, in a small pouch, for protection.

The quartet spent the night dancing, sipping beer and laughing. At 2 A.M., they headed home. Willie, as usual, drove. He seemed lost, locked in a conversation with himself—something Keathley had sometimes seen before. "Hey, yeah, what's up?" Willie was saying. Sometimes he would laugh. He would mumble more words. To everyone in the car Willie Smith was going back-and-forth with those inner voices.

Outside his sister's apartment, Willie found a parking space facing an empty lot bathed in moonlight. Keathley jumped from the Skylark and reached back to grab the gun hidden under the seat. As usual, Willie's cornerback hands were quicker. He reached it first.

"Bug, give me the gun," she demanded. Behind them they could hear the roar of the traffic on the nearby LBJ Freeway and sometimes feel a chill wind blowing across that empty lot. Willie smiled and stared at her. "What if I kill myself? What are you going to do?"

Time slowed down. Willie was in the car by himself. The car was still running. Keathley went to open her apartment so her friends could get their purses. Every step she took toward her door, she had a gnawing feeling. In the weak light, she turned and watched Willie step out of the car. He reached back and shut the motor off. Willie walked halfway toward Keathley.

She heard her brother mutter something about his mother's precious key chain, the one that read "God Loves Me."

As she watched, Willie Smith raised the Lorcin .380 handgun to his head and squeezed the trigger. He fell on his back and the blood began flowing toward his sister. It was 3:20 A.M.

For the next several hours, Keathley, Nique and Willie Sr. tried to talk to Bug in the hospital. His girlfriend, Nique, couldn't shake the thought that Willie had called her six times in the hours before he shot himself. He had begged her to come over. She had meant to drop by but just never did. Now, nothing "made sense; Willie had the world ahead of him. It was right there for him, right there in his hands," she kept saying to herself.

Meanwhile, Willie's father, who passed his sixth sense to his son, remembered feeling that something was wrong that night—even before his daughter told him Willie had shot himself. At the time, Willie's father didn't know it was death. But he sensed it was something awful. And now, in the hospital, he cradled his dying son. He patted Willie on the face.

At 10:52 A.M. on the last day of 1995, his son died. The short, bittersweet life of Willie Smith had ended—and so had his plans to rescue his family.

"It's gonna be all right," Willie George Smith Sr. whispered to his son for the last time.

Wayne McDuffie, everyone said would be coaching somewhere soon. But, only those handful of friends knew that was small comfort for someone who wanted to be a head coach and who studied and remapped that career path—until the reels were mindlessly flapping over and over again. "When the situation at the University of Georgia disintegrated, he saw that dream die. And that was a big part of his life that was dying right there in front of him," close friend David Kelly tells people.

But for the three months after he was told he was less than perfect for the coaching position at Georgia, Wayne continued his mapping. He watched members of the old staff move on to other jobs. He even knew that Goff was spending more time at the little farm he had in Georgia. Wayne jogged in his golf-course neighborhood, pushing himself hard. He lifted weights. And, with his wife, he wrestled with plans for the future. He hoped a professional team would come calling. He had feelers in with the Dolphins.

But his birthday (December 1) and the holidays passed—as did the big bowl games, the pro playoffs and the Super Bowl—and Wayne McDuffie still was unemployed. The contract for that lakeside retirement paradise still sat unsigned on his desk. He was 51 years old. The chart, the map, had led nowhere.

By mid-February, there were times when he and Toni didn't talk. She knew, without asking, that he was struggling. But, like he had in the past, she also knew he would rise above it. On February 16, she left for her job at a Tallahassee middle school. It was 7:30 A.M. Wayne told her that she might not see him when she got home, because he was going hunting. Toni took it as a good sign that Wayne was communicating with her. As she left the house, she saw her husband watching her from the kitchen window.

Toni returned home at 4 P.M. Wayne's red car was still in the driveway, and she assumed someone else had driven on the hunting trip. There was an unerased message on the answering machine from the Dolphins: Wayne hadn't gotten the job. She left for a while to feed her horses and run errands. As Toni cleaned up the house close to 7 P.M., she saw Wayne's hunting boots on a back-porch table. She looked on the porch and stared at the blood.

Sometime that day, Wayne McDuffie had taken two shotguns, two handguns, guncleaning supplies and several rounds of ammunition and placed them on his patio table. He partially disassembled one shotgun and took the cylinder from one handgun. Dressed in blue jeans, white socks, brown leather shoes and a white Atlanta Falcons shirt, Wayne McDuffie raised the other handgun and shot himself in the chest.

The police found one wound on his left side. It was, maybe, a wound meant to appear accidental. He was still able to move. From the blood trail, it appears he began walking slowly around his pool and patio overlooking the golf course. Then, Wayne McDuffie took aim at his heart.

Months later, Toni McDuffie is blanketed by a deep, profound, lingering sadness. She wonders what she could have said. Why she wasn't there for her husband. Why her husband wasn't admired more for his talents, his good side, his abilities as a husband, father and coach. There is guilt, and there are the inescapable stacks of letters from sympathizers—even from the other women whose husbands took their lives. She knows now that Wayne was being eaten alive, that this enormous nightmare had surrounded him and convinced him that life was not worth living.

And always, she asks Wayne why he left. Always, she asks Wayne to come back. To please come back.

At his funeral in Tallahassee, some were surprised to hear them play the independent anthem "My Way." A family member said, for Wayne McDuffie, it was a natural death. He was not going to live to be 60 or 70. The coaching fraternity brothers, the short-haired men who looked uncomfortable in their suits, dutifully assembled and told Wayne McDuffie stories.

Ray Goff had received a letter from Wayne the day he died. Goff's wife had brought it to Goff's farm and it was stamped "McDuffie" on the outside. Goff thought it was from Toni. Wayne wasn't the kind of guy to write him a letter. But Wayne had written and mailed it a few days before he killed himself—and he talked about losing his insurance, his retirement, his job. Goff called

Wayne, but it was too late. He learned his old friend had died that same day.

"He will go down as a friend of mine and someone I will always cherish and love. . . . I think the saddest thing of all is that I don't think he realized how many lives he touched," Goff says.

David Kelly, who had found a job at LSU, heard the news when his wife phoned. She said, "You won't believe what I just heard about Wayne." Instantly, Kelly sensed what was coming next. He was shocked, he said, but not surprised. Kelly weighed the paradox of the hard-nosed coach—"Wayne always taught never give up; persevere and fight to the end"—who seemed to betray his own philosophy.

Well, Wayne McDuffie wasn't going to do anything in the ordinary. He was going to do it his way. Bottom line, Wayne McDuffie was a man's man," Kelly would say. "That's what made Wayne McDuffie so special. We will continue to speculate, and maybe that's the way he wanted it."

Willie Smith is buried about five tombstones from his mother. His father sleeps with all those painstakingly written notes, the trophies, the helmet, the yellowed newspaper articles in a box at the foot of his bed. He drives his son's car, and he carries the key chain with "God Loves Me" on it. When he made that long, lonely trip to Louisiana to clean out Willie's college room, he was surprised to find some flyers from a church Willie had been attending. Willie died on a Sunday, but it didn't really hit his father until Monday. Now, and probably for a long time, Mondays are bad days for Willie's father.

Sometimes, he listens to his son's cassette tapes as he drives. And sometimes, he remembers that on Willie's birthday, less than two weeks before he died, Willie had suddenly blurted: "I'm going away, Daddy. I'm going a long way away."

His father didn't know what Willie meant.

"Well, just don't leave me," he responded. He thought, maybe, that Willie just needed to be by himself. Just needed some peace.

Now, Willie's father spreads the shards of his son's life on the living room floor. He smooths out the pictures, programs and articles. He cradles Willie's helmet like it is a precious, ancient heirloom weighted with memories.

"I knew you were coming. I have a sixth sense," he says, in a weary voice, to the writer who has come to ask him about his son. It is late and Willie's father seems as small as a child. "I knew people would want to know about Willie. Tell them he didn't kill himself on purpose. It was an accident. Tell them that he loved his mother, that he loved his family. Tell them Willie had everything to live for."

The coach and the player never knew each other, but they shared some things:

The player had unearthed a buried fortitude and transformed himself into an unlikely All-American. The coach had commanded, extracted, perfection.

The player could look at "God Loves Me" on his key chain and know, unlike many people, that it was probably true. The coach could measure his career and know, unlike many people, that he had often won and rarely compromised.

Both had survived the threat of death. Both labored heavily, quietly and internally. Both had stringent, unyielding destinations. Both the coach and the player worked late.

They spent endless hours working and watching as the sun set and the dark outlines began to slither over the field.

And in a wicked instant, the coach and the player vanished into those shadows.

37

Rage of the American Male

SUSAN FALUDI

A man who responds to financial reversal by whacking his wife and kids to death and gunning down a score of people is not someone you'd want to confuse with an average American. Yet, in the aftermath of Mark O. Barton's murderous rampage through two brokerage firms in Atlanta, we pore over the features of his life like tea leaves—just as we do the lives of the many schoolyard shooters, employees who "go postal" and, most recently, the 34-year-old Alabama man charged with killing three people at his current and previous job sites last week.

We can't quell the suspicion that their crimes reveal something meaningful about our society, about us. Particularly, since the shooter is always the same sex, we wonder: what does it mean about the struggles of American men?

We receive little insight from the rampagers, who so often save the final bullet for themselves. But Mark Barton did us the dubious favor of providing documentation. He wrote us a letter, a suicide note addressed not to family or friends but, more globally, "To Whom It May Concern." Its words hold some clues, if we knew how to decode them.

"I wake up at night so afraid, so terrified that I couldn't be that afraid while awake," Barton wrote. "I have come to hate this life and this system of things. I have come to have no hope." It is unhappily a comment that could have been written by many ordinary men in America, who sense that some vague and shifting "system" has let them down. And hasn't it? More and more, the American community fails to offer its postwar sons and grandsons what it used to offer all men: a chance to ground their manhood on utility, dedication and loyalty, whether as a GI serving a nation and caring for his fellow grunts or as a civilian plying a craft essential to his society. For all the grim aspects of industrial labor and World War II-era sacrifice, men could at least feel they belonged to a meaningful brotherhood and provided a utility beyond mere earning power.

But the heirs of the GI generation increasingly find themselves stranded in a different world: computerized, consumerized, celebritized. In an ornamental culture where worth is measured by bicep and SUV size, by image and celebrity, men feel severed from fellowship and a tangible craft, valued only for their stock-market portfolios. In that way, Mark Barton was the garish distillation of the modern male predicament—a Dockers-and-polo-shirted figure seated alone in his suburban home, wired to the Internet so many hours a day that no one else could make a phone call. Meanwhile, his ignored children roamed the streets.

Like so many men in this telemarketed, outsourced economy, Barton's earning power came from enterprises far removed from him. As a day trader, he gambled on abstractions, riding, pilot-fish style, the tiny gyrations of companies whose products he had no hand in producing. But he gave it his every waking moment; something told him his utility at home went little further than his winnings in this jackpot game. Even as men have been freed (thanks largely to the women's movement) to be more involved fathers, their progress is undermined by a sweepstakes culture where only the biggest winner is valued.

That Barton's demons were masculine demons was evident in the words he left behind. "The fears of the father are transferred to the son," he wrote. "It was from my father to me and from me to my son . . . I had to take him with me." To be a man has always been to receive and pass on a patrimony of skills and a place within a system. But many men suspect that all they have inherited are their fathers' fears—of being found wanting, incapable, not needed. They haven't inherited the tools to deal with those fears, because so many tools of the fathers are obsolete. Barton's father was a lifelong Air Force man, a quaint calling in a world where even the military discourages lifers. But he might as well have been an engineer or a steelworker.

When the sources of our agonies are not visible, we invent enemies—typically, the people closest at hand. Barton killed colleagues, and like so many men going under, he saw another enemy: a woman whose only sin was witnessing his humiliating descent. In his note, Barton wrote of his wife, "I killed Leigh Ann because she was one of the main reasons for my demise . . ." Barton went on a "Mortal Kombat"-style search for his persecutors, ignoring the culprit in front of him, a culture that feeds the fears of many American men. That culture holds up a frightening mirror. Reflected there is an image of a man in a room alone—isolated from his fellows, unneeded by his family, staring into a computer screen on which he seeks a disembodied fortune or, if that fortune fails, types a suicide note.

7

Movements of Men and the Politics of Masculinity

A major objective of this book has been to illustrate how broadly and deeply men are affected by gender. As we have seen, the effects of gender in men's lives make many of their experiences quite different from women's experiences. In earnings, opportunities for advancement, freedom from domestic labor, power in marital relationships, men are the more privileged gender.

Looking at this same coin from the other side, these realities represent burdens that women disproportionately bear. They work a larger and longer "second shift," suffer the economic disadvantages of wage and mobility gaps, experience sexual harassment at work (and school, and in the armed forces), and are victims of subtle and extreme forms of male power (from control of remote control devices to domestic and sexual assault). Thus, on a daily basis, women are victimized in ways that men are not; men benefit in ways that women do not. The fact that women are more critical of these unfair gendered arrangements or of the construction of masculinity and femininity is therefore not surprising.

But we have more than one coin in our pockets. We have seen how men's lives are constrained by gender. Their socialization makes them less equipped for sharing intimacy (at least of the kind valued in our culture). Their lives come to be heavily and narrowly focused around their economic success, and ultimately shortened, because they're men. Identifying these gendered privileges and problems shows men to be both victors and victims of the social structure. In other words, alongside the benefits men reap (compared to women), they suffer major costs. Although they enjoy the daily advantages of higher wages, they suffer long-term costs such as earlier deaths. Do those benefits and costs "even out"? Some of you might trade longer lives for larger salaries, or deeper personal relationships for power and dominance, but some might prefer to think about that a little bit.

As we have already seen, there is wide variation in men's lives. Those men who are other than white, middle-class, and heterosexual have their experiences further shaped by their race, class, sexual orientation, and age. Thus, they are not as richly rewarded by the gender inequalities that privilege men as a group over women as a group. In fact, like women, they may be systematically deprived of those opportunities and benefits that go to the males who dominate this society.

In this last chapter, we will look at how this duality in men's lives is differently experienced and understood by different factions of men. Depending on one's perspective on the nature of men's lives (i.e., more privilege than problem? unfair exploitation of women and advantage over minority men? filled with serious, uniquely male problems?), one will advocate solutions that follow from that perspective. Different analytical interpretations inspire different political ideologies regarding gender and men's lives.

THE POLITICS OF MASCULINITY

Twenty years ago, sociologist John Scanzoni observed that a theme of "fairness" ran through much of the early analyses of men's lives. Proponents of this "prosocial" position advocated that men should support and make change because it is what is right, fair, and best for everyone, especially women and children, particularly in improved family relations (Scanzoni 1979). This sort of appeal (men ought to change in fairness to women) can still be heard among some in organized movements, but it is joined by other appeals for fairness from across a broader range of perspectives. Where some seek more equity between women and men, others seek to correct some of the inequalities among men across lines of race, class, and/or sexual orientation. They advocate sharing more of the opportunities for economic success and mobility and opening up the previously segregated or stratified arenas of the workplace, the military, and political office to women and to various previously excluded categories of men.

Scanzoni identified a second early appeal for change: the idea that men ought to change because they would be better off if they abandoned traditional masculinity (1979). Extending this to the present discussion, they might have better relationships with other men and within their families, face less pressure to provide and thus demonstrate less competitiveness and aggressiveness, and feel less need to be in control (and occasionally to exercise that control malevolently). As a consequence, they might live longer, suffer fewer of the social and behavioral problems that men monopolize, and be "healthier" overall. Such an emphasis on self-interest can still be heard today, though it leads some to resist calls for change and advocate, instead, that things ought to stay just as they are or return to an earlier model. Ultimately, the fairness and self-interest emphases might converge at the same endpoint—supporting change in the meaning and manifestations of masculinity—but they rely on different appeals.

The various positions from which scholars and activists address or advocate change in men's lives still include appeals for fairness and self-interest, but cannot be divided into a simply dichotomy. Recent books by Kenneth Clatterbaugh (1997) and Michael Messner (1997) assess the broad range of organized responses to the status of men in American society, ranging across the political spectrum from right to left (i.e., conservative to socialist) and including both men's rights and profeminist positions.

Clatterbaugh (1997) compares and contrasts these perspectives (conservative, men's rights, spiritual, group-specific, profeminist, socialist) in terms of the responses they offer to the following questions about men and masculinity: (1) What is the nature of social reality for men in American society? That is, what is it like to be a man in American society? What do people think men are like? What do people believe men should do? (2) How is this reality maintained? That is, is it biologically produced and maintained, or is it a product of learning? (3) What would be a better social reality for men, and for women or society in general? (4) How can we achieve the more desirable reality? Messner (1997) looks across the same broad spectrum of organized responses to the status of men and the consequences of hegemonic masculinity in men's lives. He differentiates between them in terms of the relative weight they place on the importance of men's institutionalized privileges, the costs men pay for adhering to narrow notions of masculinity, and the differences between men.

In keeping with these two fine analyses of the politics of masculinity, the range of perspectives on men's lives can be better understood by identifying the differing objectives each emphasizes. Those men most concerned about issues of fairness and justice tend to emphasize men's *responsibilities,* the need for *recognition* (and respect), and the distribution of *resources.* Those more concerned with taking steps to improve men's lives emphasize the *restoration* of earlier versions of masculinity, the extension and protection of *men's rights,* or the maintenance of the status quo via *resistance to change.* Although not exhaustive, these emphases represent the range of viewpoints about the most important issue to which men should respond.

"Life's Not Fair"

Clatterbaugh suggests that within the range of perspectives men have about masculinity, one can find some (e.g., the profeminist) that are both aware of and opposed to the extent of gender inequality and the oppression of women in our society (Clatterbaugh 1990). Profeminist men advocate working with women to oppose gender injustice in both personal and public domains and to create a more equitable and just society. In advancing this argument about fairness they are joined by others (e.g., gay men, black men, Socialist men) who question and critique the inequities among men that result from the overlapping of gender with class, race, sexual orientation, or age. They call upon men to share more of the responsibilities for home and children, to acknowledge and respect women's contributions and concerns, and to share more of the opportunities for achieving economic success.

Responsibility In emphasizing fairness, some call for greater equality between women and men and across categories of men. The division of housework and childcare is not only unequal but also one source of women's and men's different economic opportunities. They argue for men sharing more of the responsibility for their families, especially for domestic work and child care. Earlier (Reading 20), Risman and Johnson-Sumerford depicted how some ideologically committed couples "did away with gender" as a basis for organizing their household responsibilities, acting instead on an ethic of "fairness" as the determination of "who does what." They rejected the "hegemonic notions of gender" and opted for more egalitarian relationships (Risman and Johnson-Sumerford 1998). These kinds of couples are not only useful illustrations of a profeminist perspective on gender, but demonstrate people's ability to "co-construct" roles and relationships that are far removed from gendered constraints. In a similar fashion, Jon Morgan (Reading 41) questions why gender must determine who initiates relationships or pays for the things couples do.

While both Risman and Johnson-Sumerford's article and Morgan's essay fit with a feminist critique of the division of labor in families, an emphasis on responsibilities is not limited to domestic life or to progressive ideologies. The "rightward" slant on emphasizing men's responsibilities is well illustrated by the recent movement known as the Promise Keepers. Citing how men have abdicated responsibility as husbands and fathers, the Promise Keepers promise to "build strong marriages (and) families" (Kimmel 1996, 1997). They come at this from a fundamentalist Christian perspective on men's and women's "scripturally ordained" obligations to each other. Their promises are cast in terms of "becoming godly influences in their world" and being committed to their duty to God and family. Notably oriented toward making men accountable for familial responsibilities, the Promise Keepers also recognize men as the "head of the wife" and family (Messner, Reading 38). Thus, despite relying upon an appeal similar to the profeminist perspective (i.e., men need to be more responsible), and pointing to a similar sphere (the home), the issue of responsibility leads them in very divergent directions.

In the more politically mixed message that surrounded the Million Man March, one found both more and less traditional messages about fairness and responsibility. Organizers and participants strongly rejected and opposed the racial oppression that disadvantages African American families and communities. They also argued, in ways that are not that far from those of the Promise Keepers, for more of a commitment by men to their roles as husbands, fathers, and providers. In their rhetoric, male accountability was, again, a key issue around which men were urged to rally.

Some look both inside and outside of the domestic arena in arguing that men have to accept responsibility for such things as domestic violence, rape, sexual assault, and sexual harassment. Given the responsibility men have for these kinds of victimizations of women, they also bear a responsibility to actively fight against them. Rather than seeing these issues as they are often referred to, as "women's issues," there is recognition that deeply and ultimately they are important "men's issues."

Recognition, Respect, and Resources Profeminist men join feminist women's appeals for greater economic opportunity and political equality for women and men, and for an end to sexism. Although the gender stratification of American society benefits men as a group over women as a group, as we know, it does not equally benefit all men. Thus, the appeal for more even distribution and more equal access to resources and opportunities can be heard in the more specialized "movements" of men, among those more marginal men. Kenneth Clatterbaugh (Reading 39) addresses the "group-specific" viewpoint of black men and the ways in which their experiences shape their responses to the profeminist men's movement. It drives home the point that for some men, "life is not fair" even though they are men and live in a male-dominated society. Clatterbaugh's analysis illustrates the difficulty in analyzing and/or mobilizing men when different groups of men possess such distinctive experiences as men.

Michelle Fine, Lois Weis, Judi Addelston, and Julia Marusza (Reading 40) add another "group-specific" viewpoint by articulating the perspectives of white working-class boys and men who perceive their place in society to be threatened. Noteworthy are their scapegoating of blacks and their refusal to see how their economic circumstances put them on common ground both with some women and with nonwhite men who should be their allies, not rivals or enemies.

What Men Need

The other issues are more often the emphases of those who highlight men's suffering and identify problematic consequences for men of the current state of gender relations. This is often a reaction against or resistance to feminism, though it too can be heard across the political spectrum from left to right. For example, the more feminist version of the "men's movement" recognizes that by challenging and dismantling a system that disadvantages women, men, too, will gain (NOMAS 1991). More extreme expressions of the need to recognize men's needs, though, come from those who see men as disadvantaged, their lives as lacking, and their positions under attack.

Men's Rights The "men's rights" perspective argues that men are discriminated against through a variety of legal inequities (e.g., custody, visitation, and child-support policies). Thus, they lobby for more "fairness" to men in those areas in which they perceive women to have unfair advantages (Farrell 1993). First articulated in the 1970s as a reaction against feminism, this viewpoint suggests that men are a "victimized and oppressed segment of the population" (Kilmartin 1994, 312).

There are a number of areas in which men's rights advocates see men as disadvantaged (besides the family-related areas noted above). These include the social acceptability of "male bashing" and the pressures imposed on men to initiate heterosexual relationships (Morgan, Reading 41), and a variety of domains in which men are "victimized" (including the threat of false accusations of rape or sexual harassment, the toll men face in war, the existence of a male-only draft, and men's less recognized victimizations in spouse abuse or prison rape) (Kilmartin 1994; Kimmel 1996). Besides occasional assertions that "make the rational mind reel" (Kimmel 1996), men's rights groups tend to downplay and disregard the breadth and depth of male "advantage" over women that the more profeminist men's groups recognize and oppose.

Restoration One of the most visible perspectives on men's lives has come from what has been called the "mythopoetic" men's movement. As Kilmartin (1994) points out, tens of thousands of men have been moved by the messages of Robert Bly (1990), Sam Keen (1991), and others who seek to reconnect men with their "deep masculinity." This movement is built around reclaiming the "transhistorical, transcultural, and transsituational" (Kilmartin 1994) essential masculinity that is currently lacking in men's lives. Advocates suggest that through a series of historical and cultural shifts, men have lost their masculine nature and their lives suffer as a consequence. Asserting that men are "disconnected" from themselves and from other men, the mythopoetic movement is organized around retreats and ritual through which masculinity can be restored and the "wild warrior within" can be reclaimed (Kimmel 1996).

The mythopoetic men's movement creates opportunities outside of everyday life situations in which men can bond with each other and enjoy a sense of community in the absence of women. Such connections are seen as necessary but missing in men's lives, leaving them soft, diminished, and lacking healthy nurturing by other males. The mythopoetic movement, along with the Promise Keepers, is the focus of the chapter by Michael Messner (Reading 38) from his book *The Politics of Masculinities* (Messner 1997).

Resistance There are those who find suggestions that men abandon elements of masculinity to be in direct opposition to "basic" or "essential" differences between the genders. They resist claims that men ought to change, and advocate instead a maintenance of the status quo. Within this viewpoint there is variation. Clatterbaugh, for example, differentiates between "moral" and "biological" conservatives. The former emphasize the societal benefits that accrue when men play their roles as fathers, protectors, and providers, and the damage that would result if men vacated the traditional male role (Clatterbaugh 1997). Biological conservatives see traditional masculinity as the product of an evolutionary process, eventuating in certain biologically driven male tendencies. Because they perceive these tendencies as "natural," they reject the notion that men could or should try to change.

Merging Messages Of course, we should not treat those themes that center around fairness and those that emphasize self-interest as mutually exclusive. Some groups act with explicit recognition of both issues. For example, National Organization for Men Against Sexism is representative of the profeminist, gay-affirmative men's movement. NOMAS opposes the oppression of women and of gay men, as well as the oppression of people because of "race, class, age, religion, and physical condition," which are "vitally connected to sexism, with its fundamental premise of unequal distribution of power" (National Organization for Men Against Sexism 1991). NOMAS also advocates "positive changes for men," noting that "Men can live as happier and more fulfilled human beings by challenging the old-fashioned rules of masculinity that embody the assumption of male superiority . . . (which) contains qualities that have limited and harmed us" (National Organization for Men Against Sexism 1991).

Men versus Women

Divisions of opinion and multiple perspectives on men and masculinity constitute a basic similarity between women and men. Just as there is no one perspective on how women should be or what they should do, neither is there unanimity about men's lives. Just as there are multiple feminisms, each with its own agenda, so, too, are there different viewpoints on whether, in what direction, and how men ought to change (Clatterbaugh 1997; Renzetti and Curran 1995).

Of course, a major distinction between women's and men's reactions to gender issues is that the system that both might critique is not one that each had an equal part in making. If, like women, men suffer as a consequence of the structure of gender relations, they suffer from a system that was "man-made" and is still male-dominated. When feminists raise this point, men often react defensively. After all, as William Goode pointed out in his classic essay, "Why Men Resist," individual men do not see what role they played in constructing this system, and most do not see this system as providing them "only benefits" (Goode 1992). Strategically, what women and profeminist men must do is convince more men that their interests are in common, and that although some male privilege will be lost, so will much of what otherwise ails and afflicts men's lives also be lost.

Men versus Men

Just as men's experiences are diverse so, too, are there a variety of issues and strategies that organize and energize men. Class, race, sexuality, age, religion, all give a special cast to men's experiences as men, and these unique experiences then lead to different agendas and varying reactions to any one platform for change. In recent years, at one time or another, we have witnessed the crowning of each of the following as "the men's movement": the mythopoetic men's movement, the men's rights/fathers' rights advocates, the Christian Men's Movement (e.g., Promise Keepers), and the profeminist, gay-affirmative men's organization, NOMAS (National Organization for Men Against Sexism). Each represents just one of a number of movements.

The readings that follow reflect some of the diversity within the politics of masculinity. Messner examines both the mythopoetic men's movement and the Promise Keepers (Reading 38). In Reading 39, Clatterbaugh describes how some black men have responded to the profeminist men's movement. Fine, Weis, Addelston, and Marusza (Reading 40) report some of the antagonism white working-class men feel as they see the foundations of race, class, and gender shift beneath their feet. Finally, Jon Morgan (Reading 41) personalizes the issue by talking about his own wants and needs as a man at the beginning of a new millennium.

MAKING CHANGE

Assuming one wants to make change in men's lives, how does one do it? Certainly we've seen that there are troubling dimensions to male socialization and that both men and women suffer some of the consequences of that socialization. Those who stress "gender-neutral socialization" as the way to make change are often discouraged by how deeply and often subtly gendered socialization is, and how hard it would be for any set of parents to fully eliminate gender socialization in the lives of their children.

Socialization is important, but by itself is neither necessary nor sufficient to make change. Gerson (1993), Risman (1989), Risman and Johnson-Sumerford (Reading 20), and Cohen and Durst (Reading 28) all illustrate how change can and does occur independent of the early socialization that would predict it. Optimistically, this suggests that we can make change more immediately than a socialization emphasis implies. We don't have to wait a generation, writing off all but the very young, whom we then expose to gender-free or more flexible socialization (Risman and Schwartz 1989).

There is a discouraging side to the limits of socialization, too. Assuming that starting tomorrow we socialized all newborn baby girls and boys similarly, to believe that anything is within their grasp and that many qualities are acceptable for them to display, would this be enough? Here I remind you that institutions have their own influence. Earlier, Arlie Hochschild (Reading 26) showed that even well-intentioned men, who might have welcomed the opportunity to spend more time with their children, didn't because of the institutional expectations within which they worked. Institutionalized sexism is something that socialization alone could not end. Our social institutions would almost have to be fully dismantled to accomplish a gender-free society (Lorber 1994). As it is, they will have to be significantly altered to create more equality between the genders. It is not impossible, but understand this: It is more difficult than just deciding that you will neither live your life nor raise your children (should you have them) in traditional ways.

REFERENCES

Bly, R. *Iron John: A Book About Men.* Reading, MA: Addison Wesley, 1990.

Clatterbaugh, K. *Contemporary Perspectives on Masculinity: Men, Women, and Politics in Modern Society.* 2nd ed. Boulder: Westview Press, 1997. 1st ed. Boulder: Westview Press, 1990.

Farrell, W. *The Myth of Male Power.* New York: Simon & Schuster, 1993.

Gerson, K. *No Man's Land: Men's Changing Commitment to Family and Work.* New York: Basic Books, 1993.

Goode, W. "Why Men Resist." In *Rethinking the Family,* edited by B. Thorne and M. Valon. Boston: Northeastern University Press, 1992.

Keen, S. *Fire in the Belly.* New York: Bantam Books, 1991.

Kilmartin, C. *The Masculine Self.* New York: Macmillan, 1994.

Kimmel, M. *Manhood in America: A Cultural History.* New York: The Free Press, 1996.

Kimmel, M. "Promise Keepers: Patriarchy's Second Coming." *Tikkun* 12, no. 2 (March/April 1997): 46–50.

Lorber, J. *Paradoxes of Gender.* New Haven: Yale University Press, 1994.

Messner, M. *The Politics of Masculinities: Men in Movements.* Beverly Hills, CA: Sage, 1997.

National Organization for Men Against Sexism. "Statement of Principles." 1991.

Renzetti, C., and D. Curran. *Women, Men, and Society.* 3rd ed. Boston: Allyn & Bacon, 1995.

Risman, B. "Can Men Mother? Life as a Single Father." In *Gender in Intimate Relationships: A Microstructural Approach,* edited by B. Risman and P. Schwartz. Belmont, CA: Wadsworth, 1989.

Risman, B., and D. Johnson-Sumerford. "Doing It Fairly: A Study of Postgender Marriages." *Journal of Marriage and the Family* 60(1) (1998): 23–40.

Risman, B., and P. Schwartz. "Being Gendered: A Microstructural View of Intimate Relationships." In *Gender in Intimate Relationships: A Microstructural Approach,* edited by B. Risman and P. Schwartz, 1–9. Belmont, CA: Wadsworth, 1989.

Scanzoni, J. "Strategies for Changing Male Family Roles: Research and Practice Implications." *The Family Coordinator* 28, no. 4 (October 1979): 435–444.

FOR ADDITIONAL READING

As we have seen, men today face potentially difficult issues, especially in the context of the personal relationships they have and the work that they do. Ronald Levant characterizes these difficulties as a *masculinity crisis* which he attributes to some major social changes over the past few decades ("The Masculinity Crisis," in *The Journal of Men's Studies,* February 1997, vol. 5, no. 3). After briefly assessing the separate elements that have led to his sense of crisis, Levant looks at some of the different ways men have responded to these issues confronting them.

A major theme of this volume is that men vary both in how they experience and perceive masculinity issues. One source of such variation is ideological. In the article, "Do Promise Keepers Dream of Feminist Sheep?" (*Sex Roles: A Journal of Research,* May 1999), Louise B. Silverstein, Carl F. Auerbach, Loretta Grieco, and Faith Dunkkel explore the appeal of the Promise Keepers Movement. Although they are especially interested in the appeal the movement has for some men concerned about their roles as fathers, they frame their article around the bigger question of the appeal of the Promise Keeper ideology versus a feminist (or profeminist) ideology.

A second source of variation grows out of one's cultural background or socioeconomic circumstances. Peter Chua and Diane C. Fujino, in "Negotiating New Asian-American Masculinities: Attitudes and Gender Expectations" (*The Journal of Men's Studies,* Spring 1999, vol. 7, no. 3), look at college-age Asian American men's attempts to redefine their masculinity and construct new forms of nonhegemonic masculinities. They connect this broader, sometimes contradictory, and developing masculinity to the social and economic position of Asian American men as economically privileged but racially subordinated.

Finally, Kevin Chappell's "Don't Wanna Be a Player No More" (*Ebony,* vol. 54, April 1999) illustrates how aspects of masculinity are experienced on deeply personal levels. His articulation of what it means to be a "2000 man" is largely an expression of seeking freedom from the life-constraining expectations that comprise masculinity.

WADSWORTH VIRTUAL SOCIETY RESOURCE CENTER

http://sociology.wadsworth.com

Visit *Virtual Society* to obtain current updates in the field, surfing tips, career information, and more.

INFOTRAC COLLEGE EDITION: EXERCISE

Go to the *InfoTrac College Edition* web site to find further readings on the topics discussed in this chapter.

38

Essentialist Retreats: The Mythopoetic Men's Movement and the Christian Promise Keepers

MICHAEL MESSNER

There's a general assumption now that every man in a position of power is or soon will be corrupt and oppressive. Yet the Greeks understood and praised a positive male energy that has accepted authority. They called it Zeus energy, which encompasses intelligence, robust health, compassionate decisiveness, good will, generous leadership. Zeus energy is male authority accepted for the sake of the community.

ROBERT BLY . . .

Christian men all over our nation and around the world are suffering because they feel they are on a losing streak and they can't break the pattern. . . . God's eyes are moving to and fro for men with a full passion for the gospel message. The Lord is calling men from across our nation to lead a new uprising of men filled with God's Spirit. Now is the time for that uprising.

BILL McCARTNEY . . .

Because feminism has had such a profound effect on U.S. society over the past three decades, it is virtually impossible for men to entirely avoid confronting "gender issues" in daily life, whether in personal relations, schools, workplaces, or the media. But some men clearly have made collective attempts to disengage themselves from feminism. The irony is that these organized attempts to grapple with the meanings of masculinity by retreating from women would not exist if feminism had not raised "the man question" in the first place. The two groups I will focus on . . . —the mythopoetic men's movement and the Christian Promise Keepers—are movements of men that have grown with astonishing rapidity in recent years. Although these movements differ in some important ways, they are strikingly similar in other ways. Leaders of both share an aversion to what they see as a recent "feminization" of men. The mythopoetic movement, though, is more apt to blame modernization for this feminization of men, whereas Promise Keepers is more apt to blame feminism, gay liberation, sexual liberation, and the "breakdown of the family" for men's problems. Both groups see a need for men to retreat from women to create spiritually based homosocial rituals through which they can collectively recapture a lost or strayed "true manhood." And these movements are asserting men's responsibility to retake their natural positions of leadership in their communities.

THE MYTHOPOETIC MEN'S MOVEMENT

The mythopoetic men's movement began quietly in the 1980s with a few men attending lectures and weekend retreats. By 1990, thousands of

From *Politics of Masculinities: Men in Movements,* pp. 16–35, by Michael Messner. Reprinted by permission of Sage Publications, Inc.

men—most of them white, middle-aged, heterosexual, and of the professional class—had attended mythopoetic events, and Robert Bly's book, *Iron John,* was a national best-seller. Through the use of old fairy tales and poetry, Bly and other mythopoetic leaders, such as Michael Meade and James Hillman, attempted to guide men on spiritual journeys aimed at rediscovering and reclaiming "the deep masculine" parts of themselves that they believed had been lost. In a nutshell, Bly argues that tribal societies had masculinity rituals, through which adult men initiated boys into a deeply essential (natural) manhood. Furthermore, urban industrial society, by severing the ritual ties between the generations of men and replacing them with alienating, competitive, and bureaucratic bonds, obliterated masculinity rituals, thus cutting men off from each other and ultimately from their own deep masculine natures. In place of these healthy masculinity rituals, as Bly's . . . words suggest, modern men revert either to destructive hypermasculinity or to a "femininity" that softens and deadens their masculine, life-affirming potential:

> We have to accept the possibility that the true radiant energy in the male does not hide in, reside in, or wait for us in the feminine realm, nor in the macho/John Wayne realm, but in the magnetic field of the deep masculine. It is protected by the *instinctive* one who's underwater and who has been there we don't know how long. . . .

Bly's curious interpretations of mythology, his highly selective use of history, psychology, and anthropology, and his essentialist ("instinctive") assumptions about masculinity, which run counter to the past 30 years of social scientific research on the social construction of gender, have been soundly criticized as shoddy scholarship. . . . But more important than a critique of Bly's ideas is a sociological interpretation of why the mythopoetic men's movement has attracted so many predominantly white, college-educated, middle-class, middle-aged men in the United States over the past decade. I suspect that Bly's movement attracts these men *not* because it represents any sort of radical break from "traditional masculinity" but precisely because it is so congruent with shifts that are already taking place within current constructions of hegemonic masculinity.

First, mythopoetic discourse appears congruent with the contemporary resurgence of belief in essential differences between women and men. Michael Schwalbe . . . , who has conducted the only systematic sociological study of mythopoetic men, has noted that the Jungian basis of the mythopoetic belief system has laid the groundwork for a complex and contradictory "loose essentialism." On the one hand, mythopoetic men tend to treat "gender, masculinity, and the category 'men' as if they were primitive constituent elements of the universe" rather than social constructions. . . . This belief in essential natures of women and men appealed to mythopoetic men because

> it provided an ideological defense against feminist criticisms of men. Such a defense was necessary precisely because the men saw it as natural. The men were aware of generic feminist criticisms of men as brutish, insensitive, power hungry, and so on. However, the men did not see these criticisms as aimed at social arrangements that produced a lot of genuinely bad men. Rather, they interpreted these . . . as criticisms of the essential nature of men. Feminist criticism of men was thus experienced as indicating the morality of all men. A defense had to respond in kind; it had to somehow redeem the category. . . .

This essentialist redemption of the category "men" in effect allows mythopoetic men to assert, "This is what I am as a man—take it or leave it. I won't feel guilty about it. I won't apologize for my gender." . . . On the other hand, Schwalbe argues, the "loose essentialism" that mythopoetic men subscribe to allows for agency and flexibility in individual men's constructions of their own masculinity. In short, mythopoetic men's work offers a collective ritual structure within which individual men can explore, discover, and reconstruct their inner selves. As Schwalbe . . . summed it up,

> The reason *loose* essentialism was so appealing is that it left room for change. A stricter form of essentialism would have implied that a man's way of being was immutable. . . . Loose essentialism, however, allowed the mythopoetic men to have it both ways. They got the moral license for possessing the feminine and masculine traits they already had, and they got the theoretical possibility of changing what they wanted to change. . . .

There do appear to be several things that mythopoetic men would like to change. Many of the men who attend mythopoetic gatherings are acutely aware of some of the problems, limits, and costs that are attached to narrow conceptions of masculinity. A major preoccupation of men at mythopoetic gatherings is the poverty of men's relationships with fathers and with other men in workplaces. These concerns are based on real and often painful experiences. Indeed, industrial capitalism undermined much of the structural basis of middle-class men's emotional bonds with each other, as wage labor, market competition, and instrumental rationality largely supplanted primogeniture, craft brotherhood, and intergenerational mentorship. . . . Mythopoetic "male initiation" rituals are intended to heal and reconstruct these masculine bonds, and they are thus probably experienced as largely irrelevant to men's relationships with women. . . .

But in focusing on how myth and ritual can reconnect men with each other, and ultimately with their own deep masculine essences as a means of dealing with men's pain, mythopoetic discourse and practice manage to sidestep the central point of the feminist critique—that men, as a group, benefit from a structure of power that oppresses women, as a group. The "loose essentialism" that underlies mythopoetic thought does allow these men to "have it both ways," as Schwalbe . . . has pointed out. They can ignore—or even become defensive about—feminist criticisms of men's institutional power and privileges. At the same time, they can construct practices that confront their own major preoccupation with the "costs of masculinity" or, in mythopoetic terms, with "men's wounds." . . .

In ignoring the social structure of power, Bly and other mythopoetic leaders conveyed a false symmetry between the feminist women's movement and the mythopoetic men's movement. Bly . . . assumes a natural dichotomization of "male values" and "female values" and states that feminism has been good for women, in allowing them to reassert "the feminine voice" that had been suppressed. But Bly states (and he carefully avoids directly blaming feminism for this), "the masculine voice" has now been muted—men have become "passive . . . tamed . . . domesticated." Men thus need a movement to reconnect with the "Zeus energy" that they have lost. And "Zeus energy is male authority accepted for the good of the community." . . .

At least on its surface, this mythopoetic discourse appears to be part of a contemporary antifeminist backlash. . . . But Michael Schwalbe . . . observes that most mythopoetic men are not overtly concerned with creating a backlash against feminist women—in fact, they are not consciously attempting to articulate a rational political vision or practice at all. They believe that industrial society has trapped men into straitjackets of rationality, thus blunting the powerful emotional communion and collective spiritual transcendence that they believe men in tribal societies typically enjoyed. Schwalbe notes that there is a "widening recognition" in society today that social problems are not technical ones that can be solved by science or rationalism. "Into this breach of faith have stepped fundamentalist religion and various New Age philosophies. Jungian psychology [the major basis of mythopoetic thought] has also found a niche in the gap between science and religion. . . . Jungian psychology sends us back to the unconscious." . . . Thus, mythopoetic discourse is at its core antimodernist and antirationalist, and the aims of mythopoetic practice are primarily therapeutic and spiritual. Schwalbe . . . describes mythopoetic men as

> selectively apolitical. They did not want to see that it was other *men* who were

> responsible for many of the social problems they witnessed and were sometimes affected by. To do so, and to talk about it, would have shattered the illusion of brotherhood among the men. . . . The mythopoetic men believed that engaging in political or sociological analysis would have led them away from their goals of self-acceptance, self-knowledge, emotional authenticity, and *communitas*. . . . They wanted untroubled brotherhood in which their feelings were validated by other men, and in which their identities as men could be infused with new value. . . .

The implications of this anti-intellectual and apolitical core of mythopoetic discourse were brought home to me when I attended a mythopoetic event a few years ago. In a speech to these men, I presented what I hoped was a carefully reasoned argument, grounded in social scientific research; in short, when we look at nonindustrial societies, we see that the more rape-prone societies tend to be those that have high levels of male-dominated sex segregation in public spaces and those that celebrate war. . . . Thus, I concluded, it seemed dangerous that the mythopoetic men were celebrating the image of "the male warrior," within the context of their expressed need to create homosocial rituals that emphasized and reinforced men's separation from women. To this, the mythopoetic men responded that the form of my analysis and critique was an example of "traditional masculine" discourse. They had no interest in engaging in "rational, sociological analysis" of what they were doing. In fact, this kind of rational analytical thought ran counter to the largely spiritual and emotional men's work that they were engaged in. I (and the rest of the guests) were invited from this point on to share only thoughts that came "from the heart, not from the head."

This anti-intellectual stance, in effect, insulated the group from dealing with criticisms or troubling questions about how their "men's work" might fit in with concrete social relations, power, and inequality. But despite its apolitical intent, the political implications of mythopoetic discourse and practice can still be critically analyzed. First and foremost, the mythopoetic notion that men need to be empowered *as men* echoes the views of some of the early men's liberation activists . . . , who saw men and women as "equally oppressed" by sexism. But the view that "everyone is oppressed by sexism" strips the concept of "oppression" of its political meaning and obscures the social relations of domination and subordination. *Oppression* is a concept that describes a relationship between social groups; for one group to be oppressed, there must be an oppressor group. . . . This is not to imply that an oppressive relationship between groups is absolute or static. To the contrary, oppression is characterized by a constant and complex state of play. Oppressed groups actively participate in their own domination and actively resist that domination. The state of play of the contemporary gender order is characterized by men's individual and collective oppression of women. . . . Men continue to benefit from this oppression of women but, importantly, in the past 30 years, women's compliance with masculine hegemony has been counterbalanced by active feminist resistance.

Men do tend to pay a price for their power. They are often emotionally limited and commonly suffer poor health and a shorter life expectancy than women. These problems, however, are best viewed not as gender oppression but as the "costs of being on top" . . . —or at least of *trying* to be on top. This is well illustrated by recent shifts in masculine styles that we see among privileged men. These shifts indicate that these men would like to stop paying these costs but not that they desire to cease being "on top." In recent years, it has become commonplace to see powerful and successful men weeping in public: Ronald Reagan shedding a tear at the funeral of slain U.S. soldiers, basketball player Michael Jordan openly crying after winning the NBA championship. In this shifting context, I would argue that the easy manner in which the media lauded U.S. General Schwartzkopf as a "new man" for shedding a public tear for the U.S. casualties in the Persian Gulf is indicative of the importance placed on *styles of masculine gender display* rather than the institutional *position of power* that men such as Schwartzkopf still enjoy.

What this emphasis on the significance of public displays of crying indicates, in part, is a native belief that if boys and men can learn to "express their feelings," they will no longer feel a need to dominate others. This assumption is based partly on the fact that much of the earlier research on men and masculinity was conducted by psychologists who tended to focus on the "tragedy" of "male inexpressivity" but ignored the social structures of power within which male personalities are constructed. . . . In a highly influential critique of the psychological literature on male inexpressivity, Jack Sattel . . . argued,

> A boy must become inexpressive not simply because our culture expects little boys to be inexpressive but because our culture expects little boys to grow up to become decision-makers and wielders of power. . . . To effectively wield power, one must be able both to convince others of the rightness of the decisions one makes and to guard against one's own emotional involvement in the consequences of that decision; that is, one has to show that decisions are reached rationally and efficiently. One must also be able to close one's eyes to the potential pain one's decisions have for others and one's self. . . .

Sattel's perspective would imply that, although "male inexpressivity" is socially constructed, mere therapeutic interventions to get individual men to "open up" will have a limited appeal unless we also change the structural positions of power that men have been expected to hold. But for men of the professional classes, these positions—and thus the types of personalities that it takes to operate within these positions—may have been shifting in recent decades. Clearly, the kind of masculine personality that was ascendant (hegemonic) during the rise of entrepreneurial capitalism was extremely instrumental, stoic, and emotionally inexpressive. . . . But there is growing evidence (e.g., Schwartzkopf) that today there is no longer a neat link between class-privileged men's emotional inexpressivity and their willingness and ability to dominate others. . . . Perhaps a situationally appropriate public display of sensitivity such as crying, rather than signaling weakness, has become a legitimizing sign of the "new man's" class status. . . .

Thus, relatively privileged men may be attracted to the mythopoetic men's movement because, on the one hand, it acknowledges and validates their painful "wounds," while guiding them to connect with other men in ways that are nurturing and mutually empowering. On the other hand, and unlike feminism, it does not confront men with the reality of how their own privileges are based on the continued subordination of women and other men. In short, the mythopoetic men's movement may be seen as facilitating the reconstruction of a new form of hegemonic masculinity—a masculinity that is less self-destructive, that has revalued and reconstructed men's emotional bonds with each other, and that has learned to feel good about its own "Zeus power."

PROMISE KEEPERS

In 1995, 600,000 Christian men, led by Bill McCartney, packed several football stadiums throughout the United States to listen to sermons, sing, and pray about their roles as men. The rapid growth of McCartney's Christian men's organization, Promise Keepers, has been dramatic. Launched with its first meeting in 1990 with only 72 men in attendance, the organization's yearly meeting swelled to 4,200 in 1991, 22,000 in 1992, and 50,000 in 1993. In 1994, 278,600 men attended Promise Keepers rallies in seven cities, and in 1995, over 600,000 men reportedly attended Promise Keepers meetings in 13 cities in the United States. The Promise Keepers organization raises about 3 million dollars per event, and the money is spent on organizing future rallies and on developing a nationwide network of men. Their phenomenal growth rate led organizers to plan a mass rally in 1996, preceding the U.S. presidential elections, where they hoped "that 1 million men will descend on Washington to kneel in prayer and ask God for forgiveness as men—and to restore America." . . . But Promise Keepers decided to put off its own rally in Washington, D.C. until perhaps 1997 after hearing of the plan by African Americans for their own 1996 Million Man March on the Capitol. . . .

Promise Keepers may seem to have suddenly appeared from virtually nowhere, but in fact, there is a historical ebb and flow of overt masculinity politics within fundamentalist Christianity in the United States. And the "flow" tends to follow in the wake of feminist challenges to taken-for-granted assumptions about men's positions of authority in families and in communities. Shortly after the turn of the 20th century, during what is now commonly known as the "first wave of feminism," a popular wave of "Muscular Christianity" swept the United States. The most famous Muscular Christian leader was an evangelist named Billy Sunday. Like Promise Keepers' most visible leader Bill McCartney a century later, Sunday entered his ministry after a successful career in sports (he had played professional baseball), and according to one journalist of the time, he "brought bleacher-crazy, frenzied aggression to religion." . . . Muscular Christianity was part of a larger turn-of-the-century masculine response to a crisis of masculinity brought on by feminism, modernization, and widespread fears that boys and men were becoming "feminized." According to Kimmel . . . ,

> The goal of the Muscular Christians was to revitalize the image of Jesus and thus remasculinize the Church. Jesus was "no dough-faced, lickspittle proposition," proclaimed evangelist Billy Sunday, but "the greatest scrapper who ever lived." Look to Jesus, counseled Luther Gulick of the YMCA, for an example of "magnificent manliness." Books such as *The Harvest Within* (1909), *Building the Young Man* (1912), *The Call of the Carpenter* (1913), *The Manhood of the Master* (1913), *The Manliness of Christ* (1900), *The Manly Christ* (1904), and *The Masculine Power of Christ* (1912) portrayed Jesus as a brawny carpenter, whose manly resolve challenged the idolators, kicked the money changers out of the temple, and confronted the most powerful imperium ever assembled. He was no "Prince of peace at any price." . . .

Muscular Christian organizations, such as the Men and Religion Forward Movement of 1911–1912, "swept the country like a spiritual storm," increasing the number of men coming to church by up to 800% in some communities. . . . More recently, in the wake of the second wave of feminism, a similarly virile version of fundamentalist Christianity began to emerge in the United States. Its first stirrings appeared in the 1970s as a right-wing Christian movement flexing its muscles in opposition to feminism, gay and lesbian liberation, and sexual liberation. . . . By the early to mid-1980s, an overt politics of masculinity had begun to coalesce within this conservative Christian coalition, especially in the teachings of Edwin Louis Cole. In his early 1980s workshops, television appearances, and cassette tapes, Cole began to articulate a late 20th-century version of Muscular Christianity. For instance, in his 1982 book, *Maximized Manhood,* Cole assures men that "women just want [men] to be the leaders in the home in every way." . . . Echoing the views of the Muscular Christians, Cole passionately argues that the model for masculine toughness and leadership is Jesus himself:

> Some "sissified" paintings of Jesus come nowhere near showing the real character of Him who was both Son of Man and Son of God. Jesus was a fearless leader, defeating Satan, casting out demons, commanding nature, rebuking hypocrites. . . . God wants to reproduce this manhood in all men. . . . since to be like Jesus—Christlike—requires a certain ruthlessness, manhood does also. . . .

By the early 1990s, this call for a remasculinized image of Jesus, and its concomitant call for men to retake leadership roles in families, had become the central message of Promise Keepers. For example, Promise Keepers leader Dr. Tony Evans . . . stated,

> I am convinced that the primary cause of this national crisis is the feminization of the American male. When I say *feminization,* I am not talking about sexual preference. I'm trying to describe a misunderstanding of manhood that has produced a nation of "sissified" men who abdicate their role as spiritually pure leaders, thus forcing women to fill the vacuum. . . .

This fear of male feminization and a call for a remasculinization of men has clearly caught on among certain groups of Christian men. Sociologist Jay Coakley . . . , who attended the 1992 meeting of Promise Keepers in Colorado, observed that most of the men appeared to be "between 35 and 50 years old . . . and at least 98% of them were white, and most were middle or upper middle class." . . . The discourse, Coakley observed, was also exclusively heterosexual. Observers of 1995 rallies agree that the organization remains a predominantly white, male, middle-class Protestant and heterosexual affair. . . . But in 1996, Promise Keepers organized its conferences around the theme of "Break Down the Walls," with the stated intention of directly confronting racial barriers between men. The 1996 rallies all included prominent religious leaders from African American, Latino, Native American, and Asian American communities. And Promise Keepers' official magazine, *New Man,* focused many of its articles and photo spreads on the lives of Latino and Asian immigrant men and African American men. Despite this impressive aim to "break down" racial "walls" between men, Promise Keepers' demographic composition as still a mostly white organization . . . and its stated aim to help men "reclaim the spiritual leadership in their families and communities" echo the white, middle-aged heterosexual mythopoetic movement's aim to reassert men's "Zeus power." But there are two significant differences between the mythopoetics and the Promise Keepers. The first is grounded in different ways that the two movements assert an essentialist view of masculinity; the second relates to their different views of what constitutes a desirable family structure.

Biblical Essentialism Reasserted

Whereas the mythopoetics' "loose" gender essentialism contains a belief in individual agency and flexibility in the shaping of gender, Promise Keepers relies entirely on a fundamentalist biblical interpretation of essentially fixed and categorically different natures of women and men. This categorical essentialism underlies Promise Keepers' rejection of feminist critiques of men's institutional power, and it encourages a blurring or ignoring of differences among men. Promise Keepers is, however, aware of and concerned with a number of problems that contemporary men share. . . . Promise Keepers argues that men's problems today result largely from *departures* from men's natural roles.

As a movement, Promise Keepers is on a leading edge of an antifeminist reassertion of essentialist views of male and female differences, a reassertion that expresses itself in religion, popular culture, and mainstream political life. For instance, in early 1995, Speaker of the U.S. House of Representatives Newt Gingrich explained why he thought that military women are not suited for combat duty: "If combat means living in a ditch," the Republican Congressman asserted, "females have a biological problem staying in a ditch for 30 days because they get infections. . . . Males are biologically driven to go out and hunt giraffes." Though many people were likely puzzled by the shortage of giraffes in their neighborhoods, Gingrich's statement reflected a widely held view that women and men are suited for different activities due to natural, inborn differences between the sexes.

This biological essentialism—the belief that women and men are essentially and categorically different—is grounded historically in conservative readings of Judeo-Christian texts, in Victorianism, in degenerate Freudianism, and in 19th- and early 20th-century medical science. . . . A major implication of this kind of essentialism, as Gingrich's words suggest, is the assumption that there are natural, biological reasons for women and men to occupy different *social* positions, and not just in the military. The belief that women are biologically best suited for the care and nurturance of children but that men's tendency toward rationality and aggressive competition best suits them to be captians of industry or workers in the paid labor force has provided the foundation for the development of a gendered public-domestic split (that, until recent years, was most pronounced among the white, middle and upper classes). Essentialism also provided the justification (mostly in the past) for the belief that women should not be allowed to vote or hold political office and for the passing of protective legislation that barred women from doing certain jobs, for fear that these jobs would endanger their health (and especially their childbearing abilities).

By the beginning of the 1970s, a body of social scientific research had begun to emerge that demonstrated that nearly all of the things that we assumed were natural differences between women and men were in fact socially constructed. This kind of research led social scientists to insist on the use of two distinct concepts: *sex* and *gender.* Sex, social scientists argued, is a biological category that describes whether a person is male or female. Gender was proposed as a social category that describes the masculine or feminine traits that are socially assigned to biological males and females. Many of the differences that we considered to be "natural sex differences," this research revealed, were actually socially constructed gender differences. As Maccoby and Jacklin . . . concluded from their survey of the sex difference research, though we assume women and men to be categorically different, in actuality, the research reveals that, despite some average differences, there is more overlap than difference between women's and men's physical traits, emotional states, and attitudes. This analytic distinction between biological sex and cultural gender provided a challenge to social inequalities that rest on essentialist thought.

But still, isn't it true that women's and men's bodies are fundamentally different? And doesn't it make sense that these differences, grounded in our biology, should have implications for the ways in which we organize and live our lives? For instance, it is common knowledge that because men have the hormone testosterone, they are naturally more physically powerful, aggressive, and violent than women. According to advocates of this view, patriarchy is "inevitable," and feminism is an affront to nature. . . . These beliefs can be summarized as a "common sense" causal proposition about the behavioral and social implications of supposed biological differences between men and women:

> men have testosterone ⟶ men are more aggressive ⟶ men dominate women

First, this proposition begins with a *categorical* premise: "Men have testosterone." This premise is correct, except to the extent that it implies that women do *not* have testosterone. In fact, the common practice of calling testosterone a "male hormone" and estrogen a "female hormone" tends to mask the fact that women and men have *both.* On average, men do have about 10 times the amount of testosterone that women have. But the range of testosterone levels among men varies greatly, and some women actually have higher levels than do some men. . . .

Second, the above proposition implies that the existence of testosterone in males *causes* them to be naturally more physically powerful, aggressive, and violent than women. In fact, taken as a whole, the numerous studies that have aimed to establish a clear correlation between testosterone and aggression levels have been inconclusive. . . . Theodore Kemper . . . , in his fascinating book *Social Structure and Testosterone,* has argued that if there is a relationship between testosterone levels and aggression, it is clearly not a one-way causal relationship, as the above proposition implies. Kemper notes several studies that measured the links between men's changing testosterone levels and their experiences within specific social settings. These studies with tennis players, medical students, wrestlers, nautical competitors, officer candidates, and parachutists all came to remarkably similar conclusions. Put simply, "winners" experienced dramatic surges in testosterone levels; testosterone levels of the "losers" stayed the same or dropped. The key factors leading to men's testosterone surges are the experience of *dominance,* which "refers to elevated social rank that is achieved by overcoming others in a competitive confrontation," and/or *eminence,* where "elevated social rank is earned through socially valued and approved accomplishment." . . . Significantly, these studies indicated that a man's "before" testosterone level could not be used to predict whether he would be a "winner" or a "loser." Instead, it was the *experience* of rising status due to one's successful efforts in a competitive endeavor that led to rising testosterone levels. Studies also suggest that the experience of dominance and eminence also leads to elevations in *women's* testosterone levels, thus leading Kemper to hypothesize a widespread increase in women's testosterone levels as they increase their positions of status and power in society. In short, our bodies respond and change as a result of our experiences and our changing social positions within society.

Clearly, the third part of the above proposition, that biological sex differences make a specific form of *social organization* inevitable, is based on questionable assumptions. In fact, rather than assuming that "biology is (social) destiny," Kemper's and others' research suggests that our biology is shaped, even at times dramatically transformed, by our social experiences. This is not to say that the errors of biological determinism should be replaced by a simplistic social determinism that ignores biology. Instead, R. W. Connell . . . argues that the old "nature vs. nurture" debates (and, indeed, the social-scientific dichotomy between biological "sex" and cultural "gender") tended to oversimplify and overdichotomize what is really a continuous circuit of "body-reflexive practices." . . .

> Through body-reflexive practices, bodies are addressed by social process and drawn into history, without ceasing to be bodies. . . . Their materiality (including material capacities to engender, to give birth, to give milk, to menstruate, to open, to penetrate, to ejaculate) is not erased, it continues to matter. The *social* process of gender includes childbirth and child care, youth and aging, the pleasures of sport and sex, labor, injury, death from AIDS. . . .

In short, a *social constructionist* perspective parts from biological essentialism by asking the questions, How does our social structure shape our experiences in ways that limit, enable, injure, stimulate, or repress our bodies, thus eliciting a range of possible bodily responses and meanings? And in turn, How do our bodily responses to these socially constructed experiences then enter history and serve to reinforce or challenge the current social order? In the 1990s, biological essentialists, who are attempting to reassert a *scientifically based* view of natural and categorical male-female difference, face three decades of empirical research that reveals a much more complex relationship between social structure and gender. Simply put, it cannot be scientifically demonstrated that biology determines women's and men's social destinies. But Promise Keepers is claiming a higher authority than the scientific method for its essentialist beliefs. Promise Keepers' discourse relies on little or no scientific justification or basis for its essentialist beliefs. Rather than expressing a biological essentialism, Promise Keepers holds to a *biblical essentialism.* Based on faith, rather than on scientific argument, this essentialism allows Promise Keepers' discourse about women to be couched in terms of "respect" for women (in their proper places as mothers, wives, and emotional caretakers of house and home). For example, when John Stoltenberg . . . attended a 1995 Promise Keepers rally in Texas Stadium, he observed,

> No overtly misogynist slurs—neither from the Promise Keepers stage nor in the milling corridors. Rarely does a presenter even use the word "woman," much less pronounce a loopy opinion about women's ostensible nature. There is no tit-for-tat accusing of women; nor is there any pop-psych defense of the double standard, as in the secular mega-seller, *Men Are from Mars, Women Are from Venus.* The Promise Keepers' message is much simpler: Men are from God. This obviates the sorts of gender-defender dramas that would-be real men are prone to—contests and put-downs

> to prove who's got "manhood" and whose is "greater." *God's* manhood is greater—'nuff said.

Thus, although Promise Keepers sometimes draws on modern scientific and social scientific discourses, its biblical essentialism—as opposed to a purely biological essentialism—is largely impervious to empirical refutation.

Taming Men with Patriarchal Bargains

Promise Keepers differs from the mythopoetics in another fundamental way. Whereas we might say that the mythopoetics are antimodern and are seeking to rediscover a premodern, preindustrial essence of manhood, Promise Keepers is attempting to reassert what it sees as a "traditional family" that is based on a God-given division of labor between women (as mothers and domestic caretakers) and men (as providers, protectors, and leaders). In fact, according to Stacey . . . , what is commonly believed to be the traditional family—a married heterosexual couple, with the male performing the paid labor and bringing home a family wage and the female doing the unpaid housework and child care in the home—is actually the modern nuclear family that very briefly achieved prominence as the white middle-class ideal in the post-World War II era. Starting in the 1960s, social conditions (including economic shifts, feminism, and gay and lesbian liberationism) have led to a proliferation of varying family forms and a breakdown in the ideological (and demographic) hegemony of this 1950s middle-class nuclear family. Nevertheless, Promise Keepers seems to be claiming that there is a divine basis for this *Leave It to Beaver* family form.

And whereas mythopoetic discourse on gender and families is, on its surface, apolitical, Promise Keepers has clear political aims. It intends to be a force in confronting and turning back the "downward spiral of morality" that has undermined traditional family values. A Promise Keepers' bumper sticker says it all: "If You Want to Go to Heaven, Take a Right and Go Straight." And it is clearly men who are intended to be at the wheel of this return drive to a mythical pre-feminist and pre-gay liberation era. According to Coakley . . . , McCartney ended the 1992 meeting by proclaiming that "the strongest voice in America is the Christian male" and implored these males "to take the nation for Jesus Christ." . . .

The antifeminist and antigay backlash potential of Promise Keepers is obvious. After all, its most visible leader Bill McCartney "stumped for Colorado's anti-gay Amendment 2 and has been featured at [the militant antiabortion group] Operation Rescue events." . . . But it would be a mistake to conclude that political backlash is the only—or even the major—motivating factor for men who join Promise Keepers. Coakley . . . observes that a dominant theme in the meetings is men's feeling of having lost control:

> . . . control over their children, wives, family lives, themselves, or over the "morally decaying" social worlds in which they lived. The content of the program appealed to men who were searching, fearful, confused, or guilty for their own "sins of insensitivity and ignorance" about how to be responsible fathers, husbands, and "manly men." . . .

For example . . . , one of the 65,000 men who attended the 1995 Promise Keepers rally in Denver's Mile High Stadium said,

> I came to pick up the seed of belief that can make life better for my family. Right now, my family is like an airplane going down. My wife is the right wing, my children are the left wing. I'm the pilot and we're about to crash. . . .

This assertion of men's responsibility to seize the reins of family leadership is a key to understanding the appeal of Promise Keepers to men. Beal and Gray's . . . content analysis of the writings of 60 Promise Keepers leaders revealed that a major theme of Promise Keepers' discourse is the "degradation of egalitarianism" and the reassertion of a natural hierarchy of authority that stretches from God the Father, to His Son, to the father of a family, and finally, down to his wife and children. For instance, Promise Keeper leader

Dr. Tony Evans . . . wrote that "I'm not suggesting that you *ask* for your role back, I'm urging you to *take it back*. . . . There can be no compromise here. If you're going to lead, you must lead. Be sensitive. Listen. Treat the lady gently and lovingly. But *lead!*" . . .

This reassertion of responsible male leadership has not led to a defensive backlash among Christian women. To the contrary, it might very well be the basis for an apparent groundswell of support for Promise Keepers among women in fundamentalist Christian communities. Why is this? Starting in the 1970s and accelerating in the 1980s and 1990s, middle- and lower middle-class families and communities have been profoundly affected by current crisis tendencies in the gender order. Economic shifts and pressures—higher levels of structural unemployment, fewer jobs paying a family wage, rapidly rising costs of housing, and so on—have undermined the likelihood of achieving the male breadwinner and female homemaker middle-class ideal. Moreover, feminism and gay liberationism have undercut the ideological basis for the unquestioned authority of heterosexual males.

Feminism, often viewed as a purely secular movement, has actually had a profound impact on various traditional Judeo-Christian religions. Far from remaining insulated from the feminist revolution of the past 30 years, Catholic nuns . . . , charismatic Pentecostals . . . , African American women church activists . . . , and Orthodox Jewish women . . . have creatively (and selectively) incorporated aspects of feminism into their religions. Similarly, feminist ideals of egalitarian marriages filtered into fundamentalist Christian communities in the 1970s and 1980s. . . . But many married fundamentalist Christian women eventually concluded that feminist activism (especially that aimed at reforming individual husbands) had gotten them nowhere. Moreover, the lack of improvement in the economic opportunities and realities for most women left these particular women (especially those who were mothers) economically dependent on men. Thus, the major issue often became not whether the man would agree to clean the toilet but whether he would consistently be a responsible breadwinner, husband, and father. This concern was codified in the 1970s and early 1980s in Phyllis Schlafly's Eagle Forum organization, which viewed the Equal Rights Amendment as a threat to "traditional families" and to the support and protection that these families gave to women. According to Schlafly . . . , if the ERA passed, it would "take away the traditional rights of wives (such as financial support) and give new rights to homosexuals (such as marriage licenses). . . . Laws that say 'husband must support wife' . . . would not be permitted under the ERA." . . .

The combination of men's intransigence in the face of women's pleas for a fair family division of labor, along with women's continued economic dependence, led many women to see the fundamentalist Christian view of "the traditional family" as a secure refuge for themselves and their children, especially if, as an integrated part of a Christian community, men could be tamed into being sober, monogamous, responsible breadwinners and fathers. What has resulted is an attempt to strike what Kandiyoti . . . calls a new "patriarchal bargain": In return for a domesticated, responsible masculinity, Christian women concede to a clear gendered division of labor (usually with their taking on the vast majority of the housework and child care, even if they are also in the paid labor force) and they concede formal leadership of the family to the man. Thus, there is a great deal of evidence that fundamentalist Christian women are actively supportive of Promise Keepers. Stoltenberg observed at the Texas Promise Keepers rally in 1995 that "several hundred young female volunteers" helped staff the cash registers and merchandising tents and served soft drinks to the press and that "two twentyish women employees of Promise Keepers tell me that they themselves would sure like to meet and marry a man of Promise Keepers caliber some day." . . . Recently, in fact, these women who support Promise Keepers have formed their own organization, "The Promise Reapers."

Potentially, men who join Promise Keepers receive far more than a secure position of authority in their families. Many of these middle-aged men are weary of paying the costs that are attached to trying to live up to the confusing and fragmented modern conceptions of masculinity. Real men, they have learned, have to constantly fight it out among themselves—either literally through violent encounters or through daily battles in workplaces—to prove and then re-prove that they are men. Real men deal with their own hurt and doubts not by talking with ministers, friends, or family members but, instead, through destructive and self-destructive practices, such as drinking too much alcohol, taking other drugs, and/or multiple sexual conquests of women. What a relief it is for many of these men to learn that a real man is simply one who faithfully keeps his promises to be a responsible husband, father, and family breadwinner. In short, like many therapeutic human potential or 12-step programs, Promise Keepers offers these men a clear set of principles and practices through which they may deal with their pain, doubts, and anxieties in ways that are less self-destructive. A man who is secure in his position as a man has no need for alcohol, has no need to destroy his own body or other men's bodies through violence, has no need to resort to sexual promiscuity to prove himself. The sense of relief at being given permission—by thousands of other men, in the masculine environs of a football stadium—to relax one's masculine posturing with one's self and with other men appears to be a great draw for men who attend the Promise Keepers events. Jesus, after all, has already paid the "costs" of masculinity for these and all men.

So far, no outcome research has been done on these men to determine how much participants' lives and relationships change after they leave Promise Keepers events. As with many therapies and 12-step programs that aim to eliminate destructive behaviors, such as alcohol addiction, it may be that the effects are short-lived, especially if they are not continually followed up with more meetings or groups. It does appear, though, that although the most public face of Promise Keepers is its mass public meetings, a top priority of theirs is to organize and maintain small, local groups of men who continue to meet regularly. In this way, it may come to resemble the kinds of therapeutic movements that have modeled themselves on organizations such as Alcoholics Anonymous.

In sum, Promise Keepers can be viewed, simultaneously, as Christian masculinity therapy in a confusing and anxiety-producing era, as organized and highly politicized antifeminist and antigay backlash, and as a religious intervention through which Christian women hope to secure men's agreement to change their behaviors in ways that are far more dramatic than any amount of "feminist browbeating" was ever able to accomplish. This bargain involves some concrete gains for women; if men keep their "promises," they are likely to become more responsible breadwinners, fathers, and husbands. As Stoltenberg . . . concludes, "From the point of view of any married woman in their social universe, neither men's talk, nor men's walk ever gets much better than this: Keep your promises to your wife and kids and be a man of your word." . . . But despite these obvious benefits to women, this is still clearly a bargain struck within the rules of a gendered hierarchy that has now been reaffirmed and renaturalized. In short, in the context of the current crisis tendencies in the gender order, Promise Keepers constitutes fundamentalist Christian men's organized agreement to settle into the driver's seat of a new patriarchal bargain with their wives. Indeed, as the wife of one Promise Keeper told him, "If you follow God, I'll follow you anywhere." . . .

39

African American Men: The Challenge of Racism

KENNETH CLATTERBAUGH

As a starting point, I see the black male as being in conflict with the normative definition of masculinity. This is a status which few, if any, black males have been able to achieve. Masculinity, as defined in this culture, has always implied a certain autonomy and mastery of one's environment.

ROBERT STAPLES

"Black Male Sexuality"

Black males start life with serious disadvantages. They are, for instance, more likely to be born to unwed teenage mothers who are poorly educated and more likely to neglect or abuse their children; the children of these mothers are also more likely to be born underweight and to experience injuries or neurological defects that require long-term care. Moreover, such children are more likely to be labeled "slow learners" or "educable mentally retarded," to have learning difficulties in school, to lag behind their peers in basic educational competencies or skills, and to drop out of school at an early age. Black boys are also more likely to be institutionalized or placed in foster care.

RONALD L. TAYLOR

"Black Males and Social Policy: Breaking the Cycle of Disadvantage"

HISTORICAL SKETCH AND PRIMARY SOURCES

We have already noted that being masculine does not require that a man be white, middle class, Anglo-Saxon, Protestant, or heterosexual; in fact, the men who possess all these qualities constitute a minority. But because of the emphasis on dominant or hegemonic masculinities, men who are outside of these boundaries, when they read the literature on men and masculinity, feel left out. . . . They see that others who do not share their experiences are speaking for them. Indeed, men who are black, Jewish, gay, or physically unique may question whether they, like gay men, are being defined as "not measuring up" to the "masculine norm." As Michael Kimmel has noted, "Historically, the Jewish man has been seen as less than masculine, often as a direct outgrowth of his emotional 'respond-ability.' The historical consequences of centuries of laws against Jews, of anti-Semitic oppression, are a cultural identity and even a self-perception as 'less than men,' who are too weak, too fragile, too frightened to care for our own." . . . In a similar vein, Franklin speaks of the black male experience: "Understanding Black men means recognizing that in America adult Black males have been Black 'men' for only about twenty years. In addition, even during this time Black males have not been recognized as 'societally approved' men." . . .

No group of men has received more attention in the late 1980s and 1990s than black men (. . . "black man," "African American man," and "Black man" are all conventions adopted by different authors or even the same author at different times; I shall use these terms interchangeably). There are many reasons for this public scrutiny.

First, an increasing number of scholars, many of whom are black males, study and write on the social reality for black males. For example, the National Council of African American Men (NCAAM) was founded in 1990 by Richard Majors and Jacob Gordon. . . . This umbrella organization also publishes the new *Journal of African American Male Studies* and the *Annual State of Black Male America.* Programs such as the Albany State Center for the Study of the Black Male and the Morehouse Institute for Research add to the growing scholarship about the status of Black males in American society. Even prior to the Million Man March in Washington, D.C., on October 16, 1995, there was ongoing concern in the mass media about the situation of black males in America. . . . Much of the recent interest derives from books like Jewelle Taylor Gibbs's *Young, Black, and Male in America: An Endangered Species* and popular films like *Hoop Dreams* (1994), which seek to capture the dreams and difficulties of young black males.

Second, men's studies and the men's movements, which have always been dominated by white males, have continually struggled publicly with how to address issues of black masculinities. How much of the middle-class white male experience is applicable to the black experience? And how can black men become more involved in both academic programs and political movements? Black scholars, in return, have argued that in spite of efforts to counter academic racism and exclusion, men's studies and the men's movements remain tarnished by antiblack racism. . . .

Third, conservatives and neoconservatives have launched a series of ideological and political attacks on social policies and programs that were intended to help groups, such as black Americans, who are subject to discrimination. These attacks, although they do not focus exclusively on the black population, often draw attention to the lives of African Americans. And books like *The Bell Curve,* for example, do target black Americans as intellectually inferior. The argument in *The Bell Curve* is that the lower average IQ among black Americans is mostly determined by genes and that IQ differences in a population determine whether that population is disadvantaged relative to other populations or whether programs that enhance opportunities are cost-effective. . . .

Fourth, closely connected to the conservative attacks on black Americans has been a national preoccupation with crime and the black criminal. . . . Although black males remain a minority in the criminal class, they are a highly visible minority. Almost one in four black males between the ages of twenty and twenty-nine is either in prison, in jail, on probation, or on parole. . . . Forty-four percent of the prisoners in the United States are black males. . . . Much of black male involvement with the prison system is due to differential sentencing for crack offenses (more likely to affect black men) and powder cocaine offenses (more likely to affect white men). Thus, "the U.S. Sentencing Commission reported in 1995 that 88 percent of offenders sentenced for crack offenses are African-Americans and 4.1 percent, white. Sentences for selling a certain amount of crack are equal in length to sentences for selling 100 times that amount of powder cocaine. If crack and powder cocaine were treated similarly, the average sentence for convicted crack traffickers would be 47 months, as opposed to 141 months. . . . More black men go to prison for crack use in spite of studies like the *Oakland Tribune* study, which concluded that the typical crack addict is "a middle-class, white male in his forties." . . . In short, arrest, prosecution, and sentencing practices are based in favor of putting the black male in prison, which only adds fuel to the national preoccupation with the black man as criminal.

A fifth reason for the sociological and political attention paid to black men is that being a African

American male in the United States has always been a sociopolitical issue. Since being brought to the United States as slaves, black men and women have been resistant to the social reality in which they find themselves. . . . When black men struggle, whether it is through the Congressional Black Caucus, the NAACP, or the Million Man March, their struggle becomes the focus of attention in the media, among racist vigilante groups, and among the general population. . . .

There are many voices among black writers who discuss the African American male's reality. Much of the black liberation movement is tied to a Christian Protestant tradition represented by the Reverends Martin Luther King Jr., Ralph Abernathy, and Jesse Jackson. But there are evangelical Christians as well, such as Tony Evans and Crawford Loritts, spokespersons for the Promise Keepers. These movements are generally moralistic and integrationist in nature. In contrast, the Black Panther Party for Self Defense, founded in 1967 by Huey P. Newton and Bobby Seale; the Congress for Racial Equality (CORE); and the Nation of Islam, founded in the 1930s, generally embrace separatist strategies calling for a "nation within a nation." . . . There are also black socialists such as W. E. B. DuBois, Richard Wright, and Manning Marable, who are a crucial part of the history of U.S. radical politics. . . . Thomas Sowell, Clarence Thomas, and J. C. Watts are prominent conservative thinkers who generally reject explanations in terms of slavery, racism, or poverty in favor of explanations that blame liberal government programs and the decline of family values. . . . Clyde W. Franklin II, who coedited *Changing Men*'s special issue on black men (Winter 1986), represents a profeminist men's studies perspective. Michael Meade, working with Malidoma Some, frequently does mythopoetic work on black-white relations in his gatherings under the Mosaic of Multicultural Foundation. And Earl Ofari Hutchinson offers a men's rights perspective of black masculinity in *The Assassination of the Black Male* (1994).

In spite of the variety of viewpoints on black masculinities, there is remarkable agreement about the nature of black male reality. Richard G. Majors and Jacob U. Gordon's *The American Black Male: His Present Status and His Future* (1994) serves as our first primary source. This comprehensive collection offers history as well as descriptions and explanations of the present social reality for black males. Our second primary source is Manning Marable's *How Capitalism Underdeveloped Black America* (1983). Marable, a widely read social theorist, writes from a standpoint that is best described as black socialist feminism. His work emphasizes the impact of black oppression and its ongoing political struggle on black masculinities. The writings of Clyde W. Franklin II in the Majors and Gordon book as well as in *The Making of Masculinities* constitute a third primary source. . . . Franklin is a profeminist concerned with the messages being given to young black men. In order to bring out the diversity of voices, we shall use a variety of materials, such as Louis Farrakhan's speech at the Million Man March, articles from the *Journal of African American Male Studies* and *emerge,* and essays by Sowell and Hutchinson.

DESCRIPTION AND EXPLANATION OF MALE REALITY

Masculinity

Almost any approach to black masculinity begins with a dismal set of figures descriptive of black male reality. Richard Majors summarizes some of the most dramatic statistics:

- Over 30 percent of black males (but only about 15 percent of white males) were not in the work force because they were unemployed or unaccounted for.
- Although 54 percent of the black men are married, for every 100 unmarried black women, there are only 63 marriageable black men.
- Homicide is the leading cause of death of black males between the ages of fifteen and thirty-four.

- A young Black male has a 1 in 21 chance of being a victim of homicide. . . .
- African-American males are the only segment of the U.S. population with a decreasing life expectancy. . . .
- One-third of all Blacks live in poverty.
- One-half of all Black children live in poverty. . . .

For black men there is an acute shortage of work opportunities. . . . "In 1978 only 1.8 million out of 18.1 million Black persons over 14 years of age could find employment." . . . Sociologist Robert Staples estimates that 46 percent of black men between the ages of sixteen and sixty-two are not in the labor force. . . . And "literally millions of Black Americans . . . have absolutely no meaningful prospects for future work." . . . There are many reasons for the exclusion of black Americans from the work force. . . . Technological replacement is certainly the cause of a decline of work opportunities in fields such as agriculture. Forty years ago, two out of five black men worked on farms; in 1960 only 5 percent of black men worked in agriculture. By the early 1980s the number declined to less than 130,000 men. . . . However, according to the U.S. Commission on Civil Rights, the exclusion of black Americans from the work force is due primarily to the racist structuring of work rather than to lack of skill or education. . . .

As a result of their exclusion from employment, "black men do not expect to assume a traditional male sex role." . . . Young black men often turn to illegal work and gang activity to make a living. Marable states that "in Chicago, over one-fourth of all Black youth between the ages of 14 to 25 belong to gangs, which often deal in small robberies, drugs and prostitution." . . . And Michael Dyson notes that gang membership in Los Angeles "has risen from 15,000 to almost 60,000 . . . as gang warfare claims one life per day." . . . Of course, black men do not have a monopoly on illegal activities. As noted, white males commit the majority of crimes in the United States, are convicted less frequently, and use most of the drugs, while black men are scapegoated. . . .

Throughout U.S. history, antiblack racism has been particularly violent: "Between 1884 and 1900 more than 2,500 black men were lynched, the majority of whom were [falsely] accused of sexual interest in white women." . . . Some black men were lynched simply for being "too successful" as farmers or tradespeople. . . . Such figures do not include the thousands of black men who have been harassed daily by police and other institutions and convicted by a racist judicial system that too often cannot even distinguish one black man from another. . . .

Violence against black men is often thought of as a thing of the past or something that happens to low-income uneducated black men. But racism is contemporary and pervasive; daily harassment and denials of opportunity continue in the 1990s, affecting all African American men, including middle-class and upper-class black males. Michael Dyson tells of trying to get a cash advance on his credit card while traveling. Because he was black, the teller (who was white) immediately confiscated his card, and the bank manager cut it in half, assuming it was stolen. . . . The *Atlanta Journal/Constitution* tracked home-loan applications in the nation's banks from 1983 to 1988. Black Americans were rejected for loans at a much higher rate than were white Americans. Coontz notes:

> In many areas, rejection rates for high-income blacks were higher than for low-income whites. An Asian or Hispanic who finished only the third grade or who earns less than $2,500 a year has a higher chance of living in an integrated neighborhood than does a black person who has a Ph.D. or earns more than $50,000. College educated black men now make 75 percent as much as their white counterparts when employed, but their unemployment rate is four times higher. . . .

This pervasive antiblack racism carries over into many masculinities that are embraced by white males. White men simply behave differently, and many of these behaviors indicate a level of distrust or fear. . . . Just as deeply ingrained patterns of sexism are included in most masculinities, so, too, do deeply entrenched patterns built around race infect these masculinities, especially North American hegemonic masculinities, which invariably include being white as a component. Kimmel cites Goffman's observation that "in America, there is only one complete, unblushing male" who is white, heterosexual, Protestant, a married father, attractive, and with a recent record in sports. . . . This racism is not simply a racism that sees white males as superior and as the normative men, but it is a racism that thrives on negative stereotypes of black males as violent, hypersexual, unintelligent, and insensitive. These attitudes since the time of slavery have resulted in a continuous series of violent and hateful acts aimed at black men. . . .

Confronted with a seemingly endless series of oppressive structures that strip black men of much control over their lives, those men respond in a variety of ways, each of which tends to create a particular form of black masculinity. One form of masculinity is achieved by being "cool." Majors and Billson identify "cool pose" as a "ritualized form of masculinity that entails behaviors, scripts, physical posturing, impression management, and carefully crafted performances that deliver a single, critical message: pride, strength, and control." . . . The purpose of this stance is clear: "As a performance, cool poise is designed to render the black male visible and to empower him; it eases the worry and pain of blocked opportunities. Being cool is an ego booster for black males comparable to the kind white males more easily find through attending good schools, landing prestigious jobs, and bringing home decent wages. . . .

Cool pose is hardly the only kind of black masculinity to which black men turn. Franklin identifies at least five distinctive masculinities: conforming black masculinity, ritualistic black masculinity, innovative black masculinity, retreatist black masculinity, and rebellious black masculinity. . . . *Conforming* masculinity is familiar to anyone who knows hegemonic masculinity: the acceptance "of mainstream society's prescriptions and proscriptions for heterosexual males." . . . Franklin notes, "They [black males] do so despite the fact that, when society teaches men to work hard, set high goals, and strive for success, it does not teach Black men simultaneously that their probability of failure is high because blocked opportunities for Black males are endemic to American society." . . . *Ritualistic* black masculinity resembles conforming black masculinity except that its practitioners do not believe in the rules and the institutions they obey—it is "'playing the game' without purpose or commitment." . . . *Innovative* black masculinity is a form of masculinity that has abandoned conformity or even ritualistic conformity. Often in this masculinity traits of hegemonic masculinity are exaggerated, as in 2 Live Crew's "As Nasty as They Wanna Be," which "debases women" while making the group wealthy. . . . In other forms of innovative masculinity the pursuit of material success leads to black-on-black homicide, drug dealing, and theft. *Retreatist* black masculinity gives up any hope of success, and members of this group slip into drug addictions, alcoholism, and homelessness. These men give up looking for work or meaningful existence. . . . *Rebellious* black masculinity was symbolized by the Black Panther movement of the 1960s, but it can be found among black activists who work in many organizations committed to black liberation.

There are many other forms of black masculinity that combine many features of Franklin's categories; for example, Louis Farrakhan's vision of a black man is someone who is rebellious and angry—yet cool—a black separatist who espouses the values of traditional male-headed families, Christianized Islamic faith, and free enterprise. . . . Thus, in spite of the many forms of oppression, or perhaps because of them, African American masculinities are at least as diverse as the masculinities exhibited by any other group of men in modern society.

In light of its ongoing struggle for liberation and equality, it is not surprising that the black liberation movement itself identifies black liberation with the assertion of black manhood. . . . This conception of black liberation is prominent in the writings of such leaders as Malcolm X, Stokely

Carmichael, and Louis Farrakhan. . . . "Even Frederick Douglass, the leading male proponent of women's rights in the nineteenth century, asserted in 1855 that the struggle for racial liberation meant that blacks 'must develop their manhood.'" . . . The black movement recognizes the fact that black male power was severely restricted under slavery—not just in terms of personal freedom but also in terms of exclusion from the power relations of white patriarchy. . . . By the same token, if black men are to achieve equality with white men, they need to adopt a patriarchal structure—which, in the United States, would entail acceptance of the norms of white masculinity. . . . This identification of black liberation with male liberation has been reinforced by the fact that the churches, the press, the business establishment, and the black liberation movement itself are male dominated.

How black masculinities are explained varies greatly among social theorists. Socialists tend to emphasize the material conditions of work that keep black men in poverty. . . . Coontz writes:

> The experience of black families has been qualitatively different from that of white, or even other minorities, all along the line. . . .
>
> More than other minorities, blacks encountered periodic increases in discrimination and segregation, first as democratic politicians tried to justify the continuation of slavery, then as blacks were pushed not up but off the job ladder by successive waves of immigrants.
>
> No other minority got so few payoffs for sending its children to school, and no other immigrants ran into such a low job ceiling that college graduates had to become Pullman porters. No other minority was saddled with such unfavorable demographics during early migration . . . or was so completely excluded from industrial work during the main heyday of its expansion. . . .

Black profeminists tend to stress that the patriarchal norms of masculinity set standards that are impossible for black men to achieve, thereby generating alternative but marginalized forms of manhood. . . . Many black sociologists point to the abandonment of the war on poverty—a war that was being waged successfully in the 1960s—as a cause of the 1970s decline in the well-being of black males. . . . During the 1960s, black family income doubled as federal spending on antipoverty programs went from $1 billion to $10 billion from 1964–1968. Poverty dropped among black Americans from 58 percent to 30 percent during these years, a decrease of 28 percent. . . .

Conservatives such as Thomas Sowell, George Gilder, and David Blankenhorn paint a very different picture: They argue that these same programs were a failure, creating the very conditions that now plague young black males—fatherlessness, crime, and poverty. . . . Among conservatives it is fashionable to blame not racism or poverty but a failure of values, a lack of responsibility: "Robert Rector of the Heritage Foundation, writing in the *Wall Street Journal,* asserts that 'the primary cause of black poverty' is neither economic nor racial inequality but 'disintegration of the family.'" . . . The disintegration of the family is laid squarely on the shoulders of antipoverty programs that "destroy the father's key role and authority." . . .

Still other theorists look to the history of slavery, Jim Crow segregation, and their accompanying racism. . . . Black authors such as Earl Ofari Hutchinson focus on negative images in the media and in popular literature. . . . Of course, there are multiple causes for the fact that black males are considerably worse off today then they were twenty years ago. The preceding causal factors are not all mutually exclusive, but different thinkers tend to give weight to different explanations.

However, for all the differences, most writings about black men clearly implicate the role of racism in the development of white masculinities, especially hegemonic masculinities. This racism is further causally implicated in the decimation and degradation of the black male. White men learn a stereotype of blacks; it is a stereotype that attributes to blacks the traits that whites do not want to attribute to themselves. For example, in 1933 there was a zero percent overlap between white and black traits as determined by whites; by 1982 that overlap had increased to only 22 percent. . . . If antiblack racism is so important to the way in which whites think about themselves, it is safe to assume that other forms of racism and prejudice also help to form white behaviors and attitudes. In short, the earlier perspectives simply do not capture the role of prejudice in the formation of the dominant white masculinities.

Feminism

In many ways, the black liberation movements in North America contributed to the rise of various feminist movements. Within many civil rights organizations and many leftist organizations women were relegated to such mental tasks as mimeographing and making coffee; they were also confronted with a "gross sexism" that denied them opportunities to express their views in public. . . . Thus, a women's movement demanding that women be treated equally under the law, that women be heard, that violence and objectification be stopped, was bound to arise. Many leftist and civil rights organizations simply did not live up to their own rhetoric. At the same time, the emerging women's movement adopted much of the demands and language of the black liberation movements. Gayle Rubin's "Woman as Nigger" is a prime example. . . . Rubin's essay is clearly an effort to merge the notions of sexism and racism, thus:

> The basic premise of women's liberation is that women are an exploited class, like black people, but that unlike blacks, they are not marginal to our technocratic society. So that one might expect that social control of women is less slipshod and more subtle than that of black people. In other words, women suffer from some form of racism, as that word is currently used. Racism has come to refer . . . to something enormously more complex than what it meant in the days of the first sit-ins. It refers to any dynamic system of social, political, economic, and psychological pressures that tend to suppress a group. . . .

Yet for all feminism's indebtedness to the language and ideas of the black liberation movements, feminism was met with considerable skepticism by many black Americans. "At the outset, the majority of Blacks who wrote on feminism were decidedly hostile. In one widely read 1971 essay published in *Ebony* magazine, Helen King denounced 'women's lib' as a white petty bourgeois fad that had little to nothing to do with the interests of Black women." . . . Other arguments from the black community included the charge that the women's movement in the past had been racist and white dominated . . . , and that it projected the domination by white males over white women onto the black community. This projection was seen by some black feminists as totally inappropriate. As Gloria Joseph put it, "Historically, Black men were definitely not afforded supremacy over any females. . . . During slavery the Black male was disallowed a superior position in relation to the Black female and there is really no question about Black men having control over white women." . . . In other words, given the authoritarian structure of slavery and the continued impoverishment of blacks, black men have never dominated women (including black women) in the same way that white men have. This is not to suggest that there is no violence between black men and black women or between black men and white women. . . . Joseph's claim is simply that black men have never been able to *control* women, either sexually or in terms of labor power, because such control has been preempted by white males. . . . Some black feminists reject this claim, however. They argue that black women have always been subject to sexism and control by black men. Even under slavery, black men "regarded tasks like cooking, sewing, nursing, and even minor farm labor as woman's work." . . . Moreover, "women in black communities have been reluctant to publicly discuss sexist oppression but they have always known it exists. We too have been socialized to accept sexist ideology and many black women feel that black male abuse of women is a reflection of frustrated masculinity." . . .

Whatever the final arbitration of this dispute, it is clear that black and white men differ in terms of their power relations with women. In the first place, because of racism black men lack many of the privileges and powers over white women that white men have. In the second place, because of their tenuous place in the relations of production, black men wield less power over black women than white men wield over white women. Under these conditions, what becomes of the profeminist claim that masculinity manifests itself as power over women or the socialist claim that men control women's labor power? Indeed, these two perspectives fail to account for the differences among men based on racial privilege or disadvantage.

Black feminists such as Bell Hooks remain critical of the white feminist movement. "Despite the current focus on eliminating racism in the feminist movement, there has been little change in the direction of theory and praxis. . . . Women will know that feminist activities have begun to confront racism in a serious and revolutionary manner when they are not simply acknowledging racism in the feminist movement or calling attention to personal prejudice, but are actively struggling to resist racist oppression in our society." . . .

Hooks's skepticism toward the white feminist movement is mirrored in the observations of black men who have worked in the profeminist men's movement. Tony Bell, for instance, indicates that, "while I have not regretted any of my involvement over the past three and a half years, I have had my share of worries and fears about the 'men's movement.' I continue to be discouraged that we have not attracted more men of color. I fear that we will become just another well-meaning liberal group that 'talks a good game.'" . . .

As of the 1990s black feminism is an established fact. . . . The contributions of black feminists to feminist theory and practice are undeniable and indispensable, yet still considered marginal. As black feminism moves more to the center of feminist theory, black men may yet respond by adopting a profeminist perspective or by moving into socialist organizations. If this is to happen, however, the marginalization of racial issues, the suspicion of white-led organizations, and the tradition of alternative black-focused movements that *do* put antiracism at the center of their agendas must be honestly confronted. Only then will the men's movement earn the trust of black men.

ASSESSMENT OF MALE REALITY AND AGENDA FOR CHANGE

Of all the assessments of male reality, the assessments of black male reality are the most discouraging and the most pessimistic. Because there is such an obvious need to address the problems facing black males of every age, there are probably more "solutions" for these problems than apply to any other group. For many writers, black men in America are facing nothing less than a literal, or at least a social, extinction. According to these writers, black men face "genocide" and "decimation"; they are an "endangered species." . . .

> Staples characterizes the crisis as "genocidal" because he believes it is being fueled by economic necessities. . . . Staples cited these statistics to bolster his grim assessment: while black men are only 6 percent of the U.S. population they compose half its male prisoners in local, state and federal jails; more than 18 percent of black males drop out of high school; more than 50 percent of black men under age of 21 are unemployed; . . . approximately 32 percent of black men have incomes below the poverty level. . . . By the year 2000 it is estimated that 70 percent of all black men will be in jail, dead or on drugs or in the throes of alcoholism. . . .

Some black writers consider *genocide* too strong a word, but decimation is occurring. It is also a practice that serves racist society in distinctive ways. . . . The pattern of decimation is such that, by the age of 45, there are 10 percent fewer black males than white (adjusted to their relative proportions); by the age of 65, 20 percent fewer; and by the age of 75, 30 percent fewer. . . . As noted earlier, black males are the only group in North America for whom life expectancy is declining. . . .

The agendas to end this "decimation" vary according to the viewpoint. The black men who write about themselves and their communities tend to focus on the issues they can actually influence. Profeminists such as Clyde W. Franklin want to see a change in the "socialization message" for black youth as well as a black male community effort toward "a return to more peaceful and productive relations between black men, and between black men and black women." . . . Robert Staples calls for educational programs that sensitize black males to the need for responsibility in family planning, sex education for both sexes, opportunities for meaningful employment, and better communication between men and women. . . .

Ronald Taylor's solution is liberal although it shares much with Marable's. Taylor argues that black males (and females) experience an "ecology of deprivation," which includes malnutrition, ill health, poverty, poor schools, family discord, and prejudicial treatment, throughout their lives. But one does not have to eradicate each of these negative factors in order to give black boys a chance to compete. Their natural resilience will compensate for many factors if a few disadvantages can be eliminated or small amounts of help can be given at the right times. Thus, programs like Head Start can help children enormously throughout their lives, even though the program helps only modestly with basic academic skills. . . . Taylor also calls for a renewal of many of the training programs launched in the 1960s that do improve "the long-term employment prospects and life chances of disadvantaged Black males." . . . Sowell and other conservatives of course disagree. Sowell would dismantle all the 1960s programs and let the market and personal responsibility decide who survives and who does not, although Sowell's "facts" that prove the failure of anti-poverty programs have been seriously challenged by Boston's *Race, Class and Conservatism*. . . .

Probably no single event has so focused the nation's attention on the struggle of black men than the October 16, 1995 Million Man March in Washington, D.C. This march, conceived by Louis Farrakhan and organized by Faye Williams, drew approximately a million black men from various communities in North America. Politically, the marchers were diverse: 31 percent claimed liberal ideology, 21 percent moderate, 13 percent conservative, 11 percent nationalist, 4 percent socialist, and 21 percent something else. . . . About one-third came from families with yearly incomes below $35,000. . . .

Although the audience was addressed by a long list of black leaders, male and female, the featured speaker was Louis Farrakhan, head of the Nation of Islam. Farrakhan's speech was a blend of religious conservative capitalism. His focus was on the moral crisis facing the black community, which he blamed on the degradation of slavery, the attitude of white supremacy, and the fractured status of the black community. His solutions were essentially separatist and psychological. Black communities must come together and change the way they think, which means they must recognize individual and collective wrongs and take action to change these wrongs ("atonement"). . . . The undeserved "furnace of affliction" of the black community is allowed by God because suffering is redemptive. Men are to be the leaders in this renewal because "in the beginning God made man" and a new beginning requires a new man. . . . The new man is captured by a pledge recited by the marchers:

> I pledge that from this day forward I will strive to love my brother as I love myself. I, from this day forward, will strive to improve myself spiritually, morally, mentally, socially, politically and economically for the benefit of myself, my family and my people. I pledge that I will strive to build business, build houses, build hospitals, build factories and enter into international trade for the good of myself, my family and my people.
>
> I pledge that from this day forward I will never raise my hand with a knife or a gun to beat, cut, or shoot any member of my family or any human being, except in self-defense.
>
> I pledge from this day forward I will never abuse my wife by striking her, disrespecting

> her, for she is the mother of my children and the producer of my future. I pledge that from this day forward I will never engage in the abuse of children, little boys or little girls, for sexual gratification. For I will let them grow in peace to be strong men and women for the future of our people.
>
> I will never again use the "B word" to describe any female, but particularly my own Black sister. I pledge from this day forward that I will not poison my body with drugs or that which is destructive to my health and my well-being.
>
> I pledge from this day forward I will support Black newspapers, Black radio, Black television. I will support Black artists who clean up their acts to show respect for themselves and respect for their people and respect for the ears of the human family. I will do all this, so help me God. . . .

In short, Farrakhan's message stresses moral responsibility and black capitalism; in this he agrees with conservatives. But he also identifies social structures that are harmful to the black community, and he urges black voters to support programs that help the poor and vulnerable. Other speakers identified these programs; they include liberal programs such as affirmative action and Aid to Families with Dependent Children (AFDC). And in what was clearly a profeminist message, Farrakhan denounced lack of respect for women. Of note to religious conservatives is that Farrakhan believes that it is *man* and not *woman* who is made in God's image and on whom falls the task of righting what is wrong with the black community. It is small wonder that the march had widespread appeal to African American men of diverse political persuasions.

Manning Marable's socialist solution challenges both the liberalism of Taylor and the religious conservatism of Farrakhan. Marable connects contemporary racism with the current needs of capitalism and rejects the idea that any strategy "limited to Black Americans and their conditions" can succeed. . . . In the language of the socialist perspective covered in Chapter 6, the decimation of black men and the alienations of black masculinity can be addressed conclusively only within a system of meaningful work in which black men have greater control over their own labor. Teaching responsibility when there is little opportunity to exercise that responsibility and creating meaningful work when capitalism by design maintains black men primarily as surplus labor are liberal strategies doomed to fail. Marable concludes that "the road to Black liberation must also be a road to socialist revolution." . . . Staples, too, seems to support this conclusion when he attributes the crisis facing black men to "the ravages of an unbridled and dying American capitalism." . . . A chilling corollary of this socialist assessment is that the decimation of black men and black male alienation will continue so long as capitalism refuses to use fairly black male labor.

Criticism 1

The danger of a focus on antiblack racism is always a risk. First, there is the risk of oversimplification. For example, many black men have lost jobs due to technology rather than to racism. And although racism is certainly common among white males who adopt hegemonic masculinities, it is not common in all such masculinities and therefore is hardly a necessary component. Hence, to build an analysis of dominant masculinities out of its common but unnecessary component—antiblack racism—distorts what creates and maintains these masculinities. Second, there is the risk of victimization. It is true that black men and women have been victims of racism, but so have Native Americans, gay men, Latinos, Asian groups, and many others. To see only one's own victimization may itself lead to racist attitudes toward other groups. Generally, black perspectives put too much emphasis upon racism. . . . By itself, racism explains very little.

Response

The black American experience is unique in many ways. Blacks in the United States are the only group deprived of its national and local heritage, the only group whose name and religion was taken away, the only group to be enslaved, and the group that suffers the greatest economic privation and denial of opportunity in modern society. . . . And black men have carried the brunt of these afflictions. Thus, black Americans must focus on the antiblack racism, institutional and personal, that they face. Their oppression is unique in the history of the United States.

Criticism 2

Regarding the Million Man March: "There are friends of mine who would have absolutely nothing to do with the conservative fundamentalism of this platform under any other circumstance who find this call for a coming together irresistible—and they will not argue about it; they do not care to see beyond that Great Coming of togetherness. . . . I also worry about a 'personal responsibility' march, as some of the organizers have called it, on the site of the civil rights marches of the past." . . .

Response

It is obvious from the breakdown of political opinions that most of the marchers did not share the conservative political opinions of Farrakhan that are referred to above; most of the march speakers did not share his views. But sometimes a gathering can transcend the political rhetoric of a particular speaker. Black men who were interviewed came together for the greater purpose of a show of strength and support for one another. . . . Commentators like Williams underestimate the independence of black men and overestimate the impact of the words of religious conservatives like Farrakhan.

Criticism 3

Adolph Reed, a professor of political science at Northwestern University, noted that the march reflected the desperation of the black community, which fears a return to the days of even fewer opportunities. "People are feeling pinched and threatened, and a lot of political leaders don't seem to have any answers. . . . People like Farrakhan feed on that sort of desperation. The substance of his message and agenda meshes well with right-wing Republicans." . . .

Response

"Professor Reed's concerns seemed far removed from the black men who had listened to Mr. Farrakhan for more than two hours. After the minister had finished, they hugged and shook hands with friends and strangers. Then they headed home past dozens of vendors selling T-shirts, videos, books and candy bars. Most of the items had a picture of a smiling Louis Farrakhan." . . .

Criticism 4

The view that liberal programs were working in the 1960s and would so continue if only we had kept at it is simply wrong.

> For example, the employment situation of young Black males, eighteen to nineteen, continued to deteriorate in the 1970s, even as federal employment and training programs and related efforts became better organized and more extensive. . . . Moreover, the labor force participation rate of Black males continued its downward trend in the early 1970s, particularly among teenagers, despite a tight labor market. Similarly, the percentage of Black husband-wife households declined between 1968 and 1973 from 72 to 63 percent of all Black families, a precipitous decline in just five years. . . .

Response

To single out government programs as the cause of such troubles in the black community is irresponsible. The 1970s were a period of double-digit inflation, and a severe economic recession that set in in 1974 undid most of the post–World War II gains by black men. . . . What the destruction of social programs in the late 1970s and 1980s did was to amplify the devastating effects of the economic downturn of the 1970s.

SUMMARY AND CONCLUSIONS

The issues raised by black men are instructive for all the perspectives already discussed. Let us begin with the perspective of gay men. Black men who are gay often find that gay perspectives are racist. And gay black men may feel that they have no place in either community because of homophobia in the black community.

Larry Icard, who studies how gay black men seek their identities, notes:

> The black community . . . holds negative attitudes about homosexuality and the gay community. This negative attitude is influenced by the black community's emphasis on group survival, survival against social pressures from a white-oriented society. For many blacks, homosexuality is a culture phenomenon of whites—a white problem inimical to the interest of blacks.
>
> As a member of the gay community, black gays are viewed as inferior members. This negative image limits their receiving the same kinds of positive psychological benefits that mixing in the gay community offers their white counterpart. Black gay men are frequently greeted with subjugation in bars, clubs, and other gay social gatherings. Often there are signs in the gay community which state, "No Blacks, Fems, or Faggots." These signs convey the restriction that blacks and other ethnic minority groups experience along with effeminate gays. . . .

What these attitudes reveal is that sensitivity to one kind of oppression does not translate into sensitivity to another kind. When one happens to fall into two oppressed groups—as gay black men do—there is a good chance that one will suffer from the prejudices of both groups rather than benefit from dual membership.

Profeminists—liberal, radical, or socialist—must also face questions that are raised by the black male experience. How has male domination been affected by the history of slavery and the domination of black men and women by white men? Do black men dominate black women in the same way that white men dominate white women? Is the socialist claim that men control women's labor power true for black men? Or is it largely white women and men who control black men's labor? If feminism marginalizes racism, why should black men join the profeminist men's movement? Indeed, writers like Hutchinson argue that feminist black women such as Bell Hooks or Alice Walker continue to contribute to the assassination of the black male image. . . . Is there a case, given the very different experiences of white men and black men, for a separate black men's movement? Given the different behaviors of white men around black men, is there even a chance that a combined movement will work?

Conservatives also must continually struggle with the experience of black men. Black men have not done well in the unbridled marketplace. White men who are already advantaged seem simply to increase their advantages at the expense of black men. The best years for black men were in the 1960s, when the government programs that most conservatives now would like to dismantle took effect. It seems to be an undeniable fact that as antipoverty programs are discontinued, black men have greater difficulty in achieving the American dream.

Andrew Kimbrell, in *The Masculine Mystique,* suggests that the plight of black men should be moved to the forefront of men's issues. Such a strategy would "roll back the Social Darwinist 'competition man' dogma of winner and losers." . . . But Kimbrell hints at more: that the social reality of black men may well be the indicator of the future of white men. The economic enclosure of black men has been the most severe of all groups of men, but if such an enclosure can be allowed to succeed for one group of men, surely it can succeed for other groups. Unless all men support these causes, no group of men will be secure. Kimbrell's view probably underestimates the differences between groups—how the enclosure of some groups of men by other men means a lessening of their own enclosure. And Kimbrell tends to see the black male experience as just a special case, though perhaps the most dramatic, of how men are victimized in North American society.

Certainly there is no "quick fix" for the serious social problems that beset a racist society. But for all the identified ugliness in the lives of many black men, one must never forget the millions of black men who do overcome obstacles to live with dignity. Reflecting on the brutal murder of his best friend, Brian Gilmore writes:

> But regardless of any political or sociological analysis, the struggle for any black man is simply not to be next. This is the legacy that my friend Chico has left me. In D.C. and everywhere in America where black men feel hunted by their own and hunted by the system, thousands, even millions, of young black men are trying to avoid the destructiveness of other black men and not be next. The black man may be marked wherever he goes and whatever he does, but he must overcome these incredible obstacles placed in his path and live a life of dignity and beauty. It is not simply the violence alone that is an obstacle, but also this society's growing distaste for his continued presence. This might sound like an exaggeration. . . . But the daily body count that is currently a central part of that oppressive truth is too real to suggest even for a second that what is going on is not really happening. . . .

SUGGESTED READING

Excellent sources for historical accounts of antiblack racism in the United States are Herbert Aptheker's *The Negro in the Civil War* (1938) and W. E. B. DuBois's *Black Reconstruction in America* (1935). And the powerful novel by Ralph Ellison, *Invisible Man* (1952), is probably the best account of antiblack racism in the early and mid-1900s. A more recent source that echoes many of the black viewpoints in this chapter is *Young, Black and Male in America: An Endangered Species,* edited by Jewelle T. Gibbs (1988). Richard Majors and Janet Mancini Billson's *Cool Pose: The Dilemmas of Black Manhood in America* (1992) and Jay MacLeod's *Ain't No Makin' It* (1987) both give a vivid picture of the difficulties facing young black males in America.

A look at the lives of men in other racial groups is important, too. David Suzuki's autobiography *Metamorphosis* (1988) is a rich personal account of a Japanese Canadian's struggle with racism, feminism, and cultural heritage. A more general history that discusses the role of republicanism in encouraging anti-Indian, anti-Asian, and antiblack racism is Ronald T. Takaki's *Iron Cages: Race and Culture in 19th Century America* (1979). In addition, Carlos Munoz, Jr., has written a powerful account of the liberation struggles of Chicanos in *Youth, Identity and Power* (1989), which begins with the 1960s and covers a dimension of the civil rights movements that is seldom presented. And Alfredo Mirandé's "Qué gacho es ser macho: It's a Drag to Be a Macho Man" (1988) not only offers a good discussion of *machismo* in Latino culture but also provides a good bibliography for further work.

The National Council of African American Men, Incorporated is located at 1028 Dole Development Center, Kansas University, Lawrence, KS 66045. The same address will serve as a subscription address for *The Journal of African American Male Studies.*

40

(In) Secure Times: Constructing White Working-Class Masculinities in the Late 20th Century

MICHELLE FINE
LOIS WEIS
JUDI ADDELSTON
JULIA MARUSZA

This article documents a moment in history when poor and working-class white boys and men are struggling in their schools, communities, and workplaces against the "Other" as a means of framing identities. Drawing on two independent qualitative studies, the authors investigate distinct locations where poor and working-class boys and men invent, relate to, and distance from marginalized groups in an effort to create self. First the authors look at an ethnography of "the Freeway boys," a community of urban white working-class high school boys who must deal with the economic ravagement of their neighborhood and their insecure place in a world different than that of their fathers. Next, the authors draw from a large-scale survey of young adults to hear how a combined sample of urban poor and working-class white men narrate identities that are carved explicitly out of territory bordered by white women and African American men. Across sites, the authors theorize how high schools and workplaces both, in part, create white masculinities and interrupt them, at a time when white working-class boys and men feel under siege.

In the late 1980s and early 1990s, the poor and working-class white boys and men whom we interviewed have narrated "personal identities" as if they were wholly independent of corroding economic and social relations. Drenched in a kind of postindustrial, late twentieth-century individualism, the discourse of "identity work" appears to be draped in Teflon. The more profoundly that economic and social conditions invade their personal well-being, the more the damage and disruption is denied. Hegemony works in funny ways, especially for white working-class men who wish to think they have a continued edge on "Others"—people of color and white women.

Amid the pain and anger evident in the United States in the 1990s, we hear a desperate desire to target, to pin the tail of blame on these "Others" who have presumably taken away economic and social guarantees once secure in a nostalgic yesteryear. Our work in this article follows this pain and anger, as it is narrated by two groups of poor and working-class white boys in the Northeast, in high school and at their public sector jobs. Through pooled analyses of two independent qualitative studies, we look at the interiors and fragilities of white working-class male culture, focusing on the ways in which both whiteness and maleness are constructed through the setting up of "Others." Specifically, the two

From *(In) Secure Times,* pp. 52–68, by M. Fine, et al. Reprinted by permission of Sage Publications, Inc.

populations in this study include white working-class boys in high school and poor and working-class white men in their communities and workplaces—including a group of firefighters—between the ages of 24 and 35. These two groups were purposefully selected to demonstrate how white working-class men construct identities at different stages of adulthood. Although some of the men in this study are poor, the analytic focus remains on the identity formation of white working-class men, as the poor men come from working-class backgrounds and, as their articulations indicate, they routinely fluctuate between poor and working-class status.

Through these narratives we cut three analytic slices, trying to hear how personal and collective identities are formed today by poor and working-class white men. The first slice alerts us to their wholesale *refusal to see themselves inside history,* drowning in economic and social relations, corroding the ever-fragile "privilege" of white working-class men. The second slice takes us to the *search for scapegoats* and the ways in which these men scour their "local worlds" for those who have robbed them of their presumed privilege—finding answers in historically likely suspects, Blacks and white women. The third slice, taken up in the conclusion, distressingly reveals the erosion of union culture in the lives of these boys and men and the *refusal to organize along lines of class or economic location,* with women and men across racial/ethnic groups, in a powerful voice of protest or resistance. These themes document the power of prevailing ideologies of individualism and meritocracy—as narrated by men who have, indeed, lost their edge but refuse to look up and fetishistically only look "down" to discover who stole their edge. These are men who belong to a tradition of men who think they "did it right," worked hard and deserve a wife, a house, a union job, a safe community, and public schools. These are men who confront the troubled pastiche of the 1990s, their "unsettled times," and lash out at pathetically available "Others." By so doing, they aspire toward the beliefs, policies, and practices of a white elite for whom their troubles are as trivial as those of people of color. Yet, these boys and men hold on, desperate and vigilant, to identities of white race and male gender as though these could gain them credit in increasingly class-segregated worlds.

The poor and working-class white boys and men in this [study] belong to a continua of white working-class men who, up until recently in U.S. history, have been relatively privileged. These men, however, do not articulate a sense of themselves inside that history. In current economic and social relations that felt sense of privilege is tenuous at best. Since the 1970s, the U.S. steel industry has been in rapid decline as have other areas of manufacturing and production, followed by the downward spiral of businesses that sprung up and around larger industry (Bluestone and Harrison 1982). In the span of a few decades, foreign investment, corporate flight, downsizing, and automation have suddenly left members of the working class without a steady family wage, which, compounded with the dissipation of labor unions, has left many white working-class men feeling emasculated and angry (Weis, 1990; Weis, Proweller, and Centrie, 1996). It seems that overnight, the ability to work hard and provide disappeared. White working-class men, of course, are not more racist or sexist than middle-class and upper-class white men. In this analysis, however, we offer data that demonstrate how white working-class male anger takes on virulent forms as it is displaced in a climate of reaction against global economic change.

As they search for someone who has stolen their presumed privilege, we begin to understand ways in which white poor and working-class men in the 1980s and 1990s manage to maintain a sense of self in the midst of rising feminism, affirmative action, and gay/lesbian rights. We are given further insight into ways in which they sustain a belief in a system that has, at least for working- and middle-class white men, begun to crumble, "e-racing" their once relatively secure advantage over white women and women and men of color (Newman 1993). As scholars of the dominant culture begin to recognize that "white is a color" (Roman 1993; Wong 1994), our work makes visible the borders, strategies, and fragilities of white working-class male culture, in insecure times, at a moment in history when many feel that this identity is under siege.

Many, of course, have theorized broadly about the production of white working-class masculinity. Willis (1977), for example, focuses on how white working-class "lads" in the industrial English Midlands reject school and script their futures on the same shopfloor on which their fathers and older brothers labor. Because of the often tense and contradictory power dynamics inherent in any single cultural context, Connell (1995) draws attention to the multiplicity of masculinities among men. In the absence of concrete labor jobs in which poor and working-class white men partially construct a sense of manhood, Connell also explores how the realm of compulsory heterosexuality becomes a formidable context for the production of white working-class male subjectivities. Various strands within the literature on masculine identity formation consider the construction of the "Other." For instance, researchers who explore all-male spaces in schools for white working-class and middle-class boys indicate that they often become potent breeding grounds for negative attitudes toward white women and gay men, whether in college fraternities (Sanday 1990), high school and college sports teams (Messner and Sabo 1994), or on an all-male college campus (Addelston and Stirratt forthcoming). The look at the formation of white working-class masculinity in this study draws on this significant literature, while bringing to the forefront of analysis the current effects that the deindustrializing economy has on the meaning-making processes among poor and working-class white boys and men, particularly as it translates to the construction of a racial "Other."

ON WHITENESS

In the United States, the hierarchies of race, gender, and class are embodied in the contemporary "struggle" of working-class white men. As their stories reveal, these boys and men are trying to sustain a *place* within this hierarchy and secure the very *hierarchies* that assure their place. Among the varied demographic categories that spill out of this race/gender hierarchy, white men are the only ones who have a vested interest in maintaining both their position and their hierarchy—even, ironically, working-class boys and men who enjoy little of the privilege accrued to their gender/race status.

Scholars of colonial thought have highlighted the ways in which notions about non-Western "Others" are produced simultaneously with the production of discourse about the Western white "self," and these works become relevant to our analyses of race/gender domination. Analysts of West European expansion document the cultural disruptions that took place alongside economic appropriation, as well as the importance of the production of knowledge about groups of people that rendered colonization successful. As Frankenberg states,

> The notion of "epistemic violence" captures the idea that associated with West European colonial expansion is the production of modes of knowing that enabled and rationalized colonial domination from the standpoint of the West, and produced ways of conceiving other societies and cultures whose legacies endure into the present. (1993, 16)

Central, then, to the colonial discourse is the idea of the colonized "Other" being wholly and hierarchically different from the "white self." In inventing discursively the colonial "Other," whites were parasitically producing an apparently stable Western white self out of a previously nonexistent self. Thus the Western (white) self and the colonial "Other" both were products of discursive construction. The work of Chakravorty Spivak (1985), which explores how Europe positioned itself as sovereign in defining racial "Others" for the purposes of administration and expanding markets, is useful on this point.

One continuing effect of colonial discourse is the production of an unnamed, unmarked white/Western self against which all others can be named and judged. It is the unmarked self that must be deconstructed, named, and marked

(Frankenberg 1993). This article takes up this challenge. As we will argue here, white working-class male identity is parasitically coproduced as these men name and mark others, largely African Americans and white women. Their identity would not exist in its present form (and perhaps not at all) if these simultaneous productions were not taking place. At a moment of economic crisis in which white working-class men are being squeezed, the disparaging constructions of others proliferate.

RACISM AND THE CONSTRUCTION OF THE "OTHER"

The first study we focus on involves an ethnographic investigation conducted by Lois Weis in the mid-1980s. This is an exploration of white working-class high school students in a deindustrializing urban area called "Freeway." Data were collected in the classrooms, study halls, during extracurricular activities, and through in-depth interviews with over 60 juniors, most of their teachers, the vice-principal, social workers, guidance counselors, and others over the course of an academic year. Data collection centered on the junior class since this is the year when some students begin to plan for further schooling, and in the state where Freeway is located, college entrance exams are administered.

While there are several facets to the production of the boys' identity, we focus on the ways in which young white boys coproduce African American male identities and their own identities. For the most part, these young white boys narrate a sense of self grounded in the sphere of sexuality, in which they script themselves as the protectors of white women whom they feel are in danger of what they regard as a deviant African American male sexuality. Not only are these young working-class boys unable to see themselves as belonging to a tradition of privilege in their being white and male, their felt loss of that historic status in a restructuring economy leaves them searching in their school, their neighborhood, and surrounding communities for those responsible. Perhaps due, in part, to student peer culture contextualized within the lived culture of the school in which these interviews took place, this examination of white male working-class youths of high school age reveals meaning-making processes that are strikingly uniform, at least in relation to the construction of a racial "Other."

Freeway is a divided city and a small number of Arabs and Hispanics live among African Americans largely on one side of the "tracks," and whites on the other, although there are whites living in one section of Freeway just adjacent to the steel mill, which is in the area populated by people of color. Virtually no people of color live in the white area, unlike many large cities in the United States, where there are pockets of considerable mix. Most African Americans came up from the South during and after World War II, drawn by the lure of jobs in the steel industry. Having been relegated to the dirtiest and lowest paid jobs, most are now living in large public housing projects, never having been able to amass the necessary capital to live elsewhere. Although we have no evidence to this effect, we also assume that even had they been able to accumulate capital, mortgages would have been turned down if African Americans had wished to move into the white area. Also, there are no doubt informal agreements among those who rent, not to rent to African Americans in the white areas, further contributing to the segregated nature of the town. Today, most of project residents receive welfare and have done so for a number of years.

Among these white adolescent men, people of color are used consistently as a foil against which acceptable moral, and particularly sexual, standards are established. The goodness of white is always contrasted with the badness of Black—Blacks are involved with drugs, Blacks are unacceptable sexually, Black men attempt to "invade" white sexual space by talking with white women, Black women are simply filthy. The binary translates in ways that complement white boys. As described by Jim, there is a virtual denial of anything at all good being identified with Blackness and of anything bad identified with whiteness:[1]

> The minorities are really bad into drugs. You're talking everything. Anything you want, you get from them. A prime example, the ______ ward of Freeway; about 20 years ago, the ______ ward was predominately white, my grandfather used to live there. Then Italians, Polish, the Irish people, everything was fine. The houses were maintained; there was a good standard of living. . . . The Blacks brought drugs. I'm not saying white people didn't have drugs; they had drugs, but to a certain extent. But drugs were like a social thing. But now you go down to the ______ ward; it's amazing; it's a ghetto. Some of the houses are okay. They try to keep them up. Most of the homes are really, really terrible. They throw garbage on the front lawn; it's sickening. You talk to people from [surrounding suburbs]. Anywhere you talk to people, they tend to think the majority of our school is Black. They think you hang with Black people, listen to Black music. . . . A few of them [Blacks] are starting to go into the ______ ward now [the white side], so they're moving around. My parents will be around there when that happens, but I'd like to be out of there.

Much expressed racism centers on white men's entitled access to white women, thus serving the dual purpose of fixing Blacks and white women on a ladder of social relations. Clint expresses these sentiments as he relays that the fighting between Blacks and Whites in the community is a result of white men protecting white women:

> [The Blacks] live on the other side of town. . . . A lot of it [fights] starts with Blacks messing with white girls. That's how a lot of them start. Even if they [white guys] don't know the white girl, they don't like to see [it] . . . I don't like it. If I catch them [Blacks] near my sister, they'll get it. I don't like to see it like that. Most of them [my friends] see it that way [the same way he does] . . . I don't know many white kids that date Black girls.

This felt need to protect white girls also translates as a code of behavior for white male students inside school. Within school walls, white working-class male anger toward African American men is magnified. As Bill bitterly accounts, white male students are not seen as doing the right thing:

> Like my brother, he's in ninth grade. He's in trouble all the time. Last year he got jumped in school . . . about his girlfriend. He don't like Blacks. They come up to her and go, "Nice ass," and all that shit. My brother don't like that when they call her "nice ass" and stuff like that. He got suspended for saying "fucking nigger"; but it's all right for a Black guy to go up to whites and say stuff like that ["nice ass"]. . . . Sometimes the principals aren't doing their job. Like when my brother told [the assistant principal] that something is going to happen, Mr. ______ just said, "Leave it alone, just turn your head." . . . Like they [administrators] don't know when fights start in this school. Like there's this one guy's kid sister, a nigger [correction]—a Black guy—grabbed her ass. He hit him a couple of times. Did the principal know about it? No!

These young white men construct white women as if they were in need of their protection. The young men fight for these young women. Their complaints are communicated through a language of property rights, Black boys intruding onto *white property*. It is the fact that *Black* men are invading *white* women, the property of *white* men, that is at issue here. The discursive construction of Black men as oversexualized enables white men to elaborate their own "appropriate" heterosexuality. At a time of heightened concern with homosexuality, by virtue of their age, the collective nature of their lives, the fear of being labeled homosexual and the violence that often accompanies such labeling in high school, these boys assert virulently and publicly their concern with Black men, while expressing their own heterosexuality and their ability to "take care of their women."

There is a grotesqueness about this particular set of interactions, a grotesqueness that enables white men to write themselves as pure, straight, and superior, while authoring Black men as dirty, oversexualized, and almost animal-like. The white female can be put on a pedestal, in need of protection. The Black female disparaged; the Black male avenged. The elevation of white womanhood, in fact, has been irreducibly linked to the debasement of both Black women and men (Davis 1900). By this Davis asserts that in the historic positioning of Black females as unfeminine and Black males as predators, the notion of what is feminine has become an idealized version of white womanhood. It is most interesting that not one white female in this study ever discussed young Black men as a "problem." This is not to say that white women were not racist, but this discursive rendering of Black men was quite noticeably the terrain of white men.

The word *nigger* flows freely from the lips of white men and they treat Black women far worse than they say Black men treat white women. During a conversation at the lunch table, for example, Mike says that Yolanda [a Black female] should go to "Niggeria" [Nigeria]. In another conversation about Martin Luther King Day, Dave says, "I have a wet dream—about little white boys and little Black girls." On another occasion, when two African American women walk into the cafeteria, Pete comments that "Black people . . . they're yecch. They smell funny and they [got] hair under their arms." The white boys at this table follow up this sentiment by making noises to denote disgust.

Young white men spend a great deal of time expressing and exhibiting disgust for people of color. This is done at the same time they elaborate an uninvited protectionist stance toward white women. If white women are seen as the property of white men, it is all the more acceptable for them to say and do anything they like. This set of discursive renditions legitimates their own "cultural wanderings" since they are, without question, "on the top." For the moment, this symbolic dominance substitutes for the real material dominance won during the days of heavy industry. Most important, for present purposes, is the coproduction of the "white self," white women, and the African American male "Other."

YOUNG ADULTS: WHITE POOR AND WORKING-CLASS MEN IN AN ECONOMIC STRANGLEHOLD

The second set of narratives stems from an ongoing study of poor and working-class young adults who grew up in the Reagan-Bush years, conducted by Michelle Fine, Lois Weis, and a group of graduate students, including Judi Addelston. In broad strokes we are investigating constructions of gender, race, ethnic, and class identities; participation in social and community-based movements for change; participation in self-help groups; participation in religious institutions; experiences within and outside the family; and experiences within and outside the new economy. We have adopted a quasi-life history approach in which a series of in-depth interviews are conducted with young people—poor and working-class—of varying racial backgrounds. Data were gathered in Buffalo, New York and Jersey City, New Jersey. Seventy-five to 80 adults were interviewed in each city. While the larger aspects of the project are as stated above, in this [study] we focus on the bordered constructions of whiteness as articulated by young white men—a combined sample of poor and working-class men, some of whom are firefighters.[2]

As with the Freeway boys, we hear from these somewhat older men a set of identities that are carved explicitly out of territory bordered by African Americans and white women. Similar to the Freeway study, these groups are targeted by young white adult men as they search their communities, work sites, and even the local social service office for those who are responsible for stealing their presumed privilege. While most of these men narrate hostile comparisons with "Others," some offer sympathetic, but still bordered, views. Like cartographers working with different tools on the same geopolitical space, all these men—from western New York and northern New Jersey—sculpt their identities as if they were discernibly framed by, and contrasted through, race, gender, and sexuality.

As with the teens, the critique by young adult white men declares the boundaries of acceptable behavior at themselves. The white male critique is, by and large, a critique of the actions/behaviors taken by African Americans, particularly men. This circles around three interrelated points: "not working," welfare abuse or "cheats," and affirmative action.

Because many, if not most, of the white men interviewed have themselves been out of work and/or received welfare benefits and food stamps, their critique serves to denigrate African Americans. It also draws the limits of what constitutes "deserving" circumstances for not working, receiving welfare, and relying on government-sponsored programs at themselves.

By young adulthood, the target site for this white male critique shifts from sexuality to work but remains grounded *against* men of color. When asked about the tensions in their neighborhood, Larry observes,

> Probably not so much [tension] between them [Blacks and Hispanics]. But like for us. I mean, it gets me angry sometimes. I don't say I'm better than anybody else. But I work for the things that I have, and they [Blacks and Hispanics] figure just because you're ahead, or you know more and you do more, [that's it] because you're white. And that's not really it. We're all equal, and I feel that what I've done, I've worked for myself to get to where I'm at. If they would just really try instead of just kind of hanging out on the street corners. That's something that really aggravates me, to see while I'm rushing to get to work, and everybody is just kind of milling around doing nothing.

In Larry's view, he is a hardworking man, trying to live honestly, while African Americans and Hispanics do nothing all day long. Larry talks about the anger he feels for those who are Black and Hispanic and in so doing sets up a binary opposition between whites and "Others," with whites as morally superior. From this flows an overt racial critique of affirmative action programs, as well as a more racially coded critique of welfare abusers and cheats.

We take up the issue of affirmative action first. Many of these white men focus on what they consider to be unfair hiring practices, which they see as favoring people of color and white women. Pete, for example, has a great deal to say about his experience at work and the Civil Rights movement more generally, and then how such movements have hurt him as a white man:

> For the most part, it hasn't been bad. It's just that right now with these minority quotas, I think more or less, the white male has become the new minority. And that's not to point a finger at the Blacks, Hispanics, or the women. It's just that with all these quotas, instead of hiring the best for the job, you have to hire according to your quota system, which is still wrong . . . Civil rights, as far as I'm concerned, is being way out of proportion . . . granted, um, the Afro-Americans were slaves over 200 years ago. They were given their freedom. We as a country, I guess you could say, has tried to, well, I can't say all of us, but most of us, have tried to, like, make things a little more equal. Try to smooth over some of the rough spots. You have some of these militants who are now claiming that after all these years, we still owe them. I think the owing time is over for everybody. Because if we go into that, then the Poles are still owed [he is Polish]. The Germans are still owed. Jesus, the Jews are definitely still owed. I mean, you're, you're getting cremated, everybody wants to owe somebody. I think it's time to wipe the slate clean . . . it's all that, um, you have to hire a quota

> of minorities. And they don't take the best qualified, they take the quota number first. . . . So that kind of puts you behind the eight ball before you even start. . . . Well, I'm a minority according to some people now, because they consider the white male now a minority.

Larry focuses on what he interprets as a negative effect of the Civil Rights movement—government-sponsored civil service tests. For Larry, these exams favor white women and "minorities" and exclude qualified white men from employment:

> I mean, in theory, a whole lot of it [Civil Rights movement] is good. I feel that is worthwhile, and there has to be some, not some, there has to be equality between people. And just because of . . . I feel that the federal government sometimes makes these laws or thinks that there's laws that are bad, but they themselves break them. I mean, I look at it as where—this is something that has always irked me—taking civil service exams. I feel that, I mean, I should be given a job based on my abilities and my knowledge, my background, my schooling, everything as a whole, rather than sometimes a Black man has to have a job just because he's Black. And really you're saying, you're not basing it on being Black or whether you're a male or female, but that's exactly what they're doing . . . I really, I completely disagree with quotas. I don't feel it's, they're fair. I mean, me as a white male, I'm really being pushed, turned into a minority. I mean, it doesn't matter. We have to have so many Blacks working in the police department or in the fire department, or women. And even though, well, say, I'm not just saying this because I'm a white male, but white males, you know, will be pushed, you know, pushed away from the jobs or not given the jobs even though they might qualify more so for them, and have more of the capabilities to do the job. And they just won't get it because they're white males.

According to Tom, "color" is not an issue—there are lazy people all over and he even has friends who are Black. Tom, however, accuses African Americans and white women of unfairly playing up minority status to get jobs. From politicians to other lazy minorities, in Tom's view, Blacks in particular have a lock hold on all the good jobs:

> I have nothing against Blacks. Whether you're Black, white, you know, yellow, whatever color, whatever race. But I don't like the Black movement where, I have Black friends. I talk to them and they agree. You know, they consider themselves, you know. There's white trash and there's white, and there's Black trash, and there's Blacks. And the same in any, you know, any race. But as soon as they don't get a job, they right away call, you know, they yell discrimination. That's where I think some of our, you know, politicians come in too. You have your [council members in Buffalo], and I think they do that. But I think maybe if you went out there, and educated yourself. And you know, there's a lot of educated Blacks, and you don't hear them yelling discrimination because they've got good jobs. Because they got the know-how behind them. But the ones that are really lazy, don't want it, they, they start yelling discrimination so they can just get the job and they're not even qualified for it. And then they might take it away from, whether it's a, you know, a woman or a guy.

The white male critique of affirmative action is that it is not "fair." It privileges Blacks, Hispanics, and at times white women, above white men. According to these men, white men are today being set up as the "new minority," which contradicts their notions of equal opportunity. Nowhere in these narratives is there any recognition that white men as a group have historically been privileged, irrespective of individual merit. These assertions about affirmative action offer white men a way of "Othering" African Americans, in particular. This theme is further elaborated in discussions of welfare abusers and cheats. Like talk of sexuality among the younger men, as exemplified by Pete, the primary function of discussions about welfare abusers is to draw the boundaries of acceptable welfare at their own feet:

> [The Welfare system] is a joke. . . . They treat you like absolute garbage. They ask you everything except your sexual preference to be quite honest with you. They ask how many people are in the house. What time do you do this? What time do you do that? Where do you live? Do you pay your gas? Do you pay your electric? Um, how come you couldn't move into a cheaper apartment? Regardless of how much you're paying to begin with. If you ask them for a menial item, I mean . . . like your stove and refrigerator. They give me a real hard time. . . . There's definitely some people who abuse the system, I can see that. But then there are people who, when you need it, you know, it's like they have something to fall on to. And they're [the case workers] basically shoving everybody into one category. They're all users. But these [case workers] are the same people that if the county closes them off, they won't have a job and they're going to be there next too.

Ron, a white working-class man who has been in and out of instances of stable employment, makes observations on welfare and social services that are based on his own varied experiences. Ron says that he has never applied for welfare and takes pride in this fact, and he compares himself with those who abuse the system—whom he believes are mostly Black. Later, Ron reveals that he has used social services:

> You know, we [spouse] look at welfare as being something, um, less than admirable . . . I think for the most part, I think most people get out of life what they put into it. You know, because some people have more obstacles than others, there's no doubt about it. But I think a lot of people just expect things to come to them, and when it doesn't, you know, they've got the government to fall back on. . . . You know I think it [falling back on the government] is more common for black people. I mean social services, in general, I think, is certainly necessary, and Kelly and I have taken advantage of them. We've got food stamps several times. Um, one of the things about the home improvement [business he was in], when I first got into that, before I really developed my skills better and, and the first company, like I said, when they were doing some change over. And, just before they left [the city], we were at a point where business was starting to slack off and um, especially in the winter time. So, a lot of times in the winter when my income was quite low, we'd go on food stamps, and I think, I think that's the way it should be used. I mean, it's help there for people. But, you know, as soon as I was able to get off it, I did. And not for any noble reasons, but just, you know, I think I'd rather be able to support myself than have things handed to me.

Since most of the case workers are white, Ron is aligning himself with the hardworking white people who have just fallen on hard times, unlike the abusers, largely Black, who exploit the system. Along these same lines, Pete's criticism of the case workers is that they treat *all* welfare recipients as cheats. Many of the white men who have been out of work, or are now in a precarious economic state, speak with a strong disdain for African American men and, if less so, for white women as well. Others, however, narrate positions relative to white women and people of color within a discourse of concern and connection. This more liberal discourse is typically spoken by working-class men who occupy positions of relative economic security. But even here the borders of their identity nevertheless fall along the same fault lines of race and gender (Roediger, 1994).

The white working-class firefighters interviewed in our study narrate somewhat similar views. Joe, for instance, works in a fire department in Jersey City. He, like so many of our informants, insists that he is "not a racist," but he vehemently feels that "Civil Rights has [*sic*] gone far enough." As we discovered, the fire and police departments in Jersey City have historically garnered a disproportionate share of the city's public sector investment and growth over the last decade, and they employ a disproportionate share of white men. We began to hear these departments as the last public sector spaces in which white working-class men could at once exercise identities as white, working, and men. Joe offers these words to describe his raced and gendered identity:

> No, I'm not racist. I'm not prejudiced. There are definitely lowlifes in this community where we live in. If you see somebody do something stupid, you call them stupid. You don't call them a stupid Black person because there's no need for those extra words. Just stupid. That's how I feel, I look at things. I'm not racist at all. If there is such a thing, racist towards a person. That's how I see it.

Although Joe makes the disclaimer that he is not racist, ironically, he specifically marks the Black person "who does something stupid." Later, in his interview, we hear greater clarity. Joe is tired of hearing about race and has come to some frightening conclusions about how such issues should be put to rest:

> Civil rights, I think they're going overboard with it. Everything is a race issue now. Everything you see on TV, all of the talk shows. You have these Black Muslims talking, preaching hate against Whites, the Whites should be dead. And then you got these Nazi fanatics who say Blacks and Jews should be dead. That's fine, let them [Blacks and Jews] go in their own corner of the world.

In characterizing African Americans as "lowlifes" and "stupid," Joe ostensibly creates a subclass used to buttress what he sees as the higher moral character of whites. Sick of the race "issue," Joe also critiques gains won during progressive movements for social change. On the streets of his community and on television, Joe maintains that he is bombarded with examples of irresponsible African Americans and others whom he feels are taking over and, therefore, should be pushed back into "their own corner of the world." Interestingly, Joe revises his otherwise critical look at affirmative action because it *positively* affected him:

> I would say, what you call affirmative action, I would say that helped me to get this job. Because if it wasn't for minorities pressing the issue two or three years ago about the test being wrong, I would have scored a 368 on the test [and would have failed].

Joe passed the exam only after it had been modified to be more equitable. For Joe, public sector commitments to equity, including affirmative action and welfare, could be helpful if they help whites. But they are racist if they don't. In talking about his sister, Joe points out how she is being discriminated against because she is white. In Joe's logic, because so many Blacks and Asians are using and abusing the system, whites, unfairly, are the ones who are being cheated. Again, Joe places his sister in a position of superiority in relation to people of color. While his sister is a hard worker, "Others," who do not really need the assistance, are simply bilking the system:

> She just had a baby. She works as a waitress. Not too much cash in there because they cut her hours, and she's getting welfare and from what I understand from her, there are people, Black or Asian people, that aren't having as a much problems as she is. It seems that the system is trying to deter her from using it. The impression she gets is you're white, you can get a job. If it's true, and I think that's definitely not right. You could be Black, white, gold, or brown, if you need it, you should have it.

Mark is another white firefighter. Echoing much of what Joe has said, Mark portrays the firehouse as a relatively protected and defended space for whiteness. By extension, the firehouse represents the civic goodness of [white] public institutions. In both Joe's and Mark's interviews, there is a self-consciousness about "not sounding racist," yet both consistently link any mention of people of color and the mention of social problems—be it child neglect, violence, or vandalism. Whiteness preserves the collective good, whereas people of color periodically threaten the collective good:

> I wouldn't say there is tension in the fire department but people are prejudiced. I guess I am to a certain degree. I don't think I'm that bad. I think there's good Blacks and bad Blacks, there's good Whites and bad Whites. I don't know what the percentage of minorities are, but Jersey City is linked with other cities and they have to have a certain percentage of minorities. Where I live right now, it's not too bad. I don't really hang out. . . . I have no problems with anybody. Just the vandalism. You just got to watch for that.

Mark doesn't describe how he got to be a firefighter, and he also does not know how Joe successfully landed the job. Although he is secure in his vocation, Joe is somehow certain that "minorities" have gained access unfairly. When asked what he might like to see changed about the job, Mark responds,

> Have probably testing be more well-rounded. More straightforward and fair. It seems to be a court fight every time to take a test. Everybody takes the same test. I just don't understand why it's so difficult. I understand you have to have certain minorities in the job and that's only fair, but sometimes I think that's not fair. It's not the fire department, it's the people that fight it . . . I think everybody should take the same test and that's it. The way you score is the way you score.

Frank, on the other hand, embodies the white working-class "success" story. He has completed college and graduate school and speaks from an even greater distance about his community's sentiments about race, safety and crime. Frank complicates talk of race/ethnicity by introducing social class as the social border that cannot easily be crossed. His narrative of growing up unravels as follows:

> Well, because, you know, we were white, and these other places were, were much less white, and I think there was kind of that white fear of, minorities, um, particularly Blacks and Hispanics. And you know, I'm not proud of that, but I mean, that's just, that's part of the history of it. But it was also perceiving that things were changing, very radically, very dramatically. And what's happened over the years is that a lot of people who lived there for generations have moved away. But, you know, it was, I think, they just, a fear of, of the changes going on in the 60s and '70s, and seeing, you know, crime increase. . . . And wanting to keep, you know, this neighborhood as intact as possible. . . . I sense that there's a lot of apprehension [among Whites]. You know, I think . . . I mean, a lot of it comes out of people talking about, um, their fear, you know, um, getting mugged, or getting their, you know their car stolen.

Seemingly embarrassed by racialized biases embedded in the community in which he was raised, Frank nevertheless shifts responsibly off of whites and onto "minorities" when he discusses solutions to racial problems:

> Indian women have these . . . marks on their foreheads. And um, you know, they're apparently just racists (referring to white youths who beat on these women). . . . You know, ignorant. Yeah, they're, they're young white, ignorant people who go around beating up Indians, in particular because the Indians tend to be passive. Um, it's something they need to learn to do, which is to be more assertive, I think and to be, um, you know, to stand up for their, their basic human rights.

We hear, from these young white males, a set of identities carved inside, and against, demographic and political territories. The borders of gender, race/ethnicity, and, for Frank, class, mark the borders of self, as well as "Other." While all of our interviewees are fluent in these comparisons, those who sit at the collapsing "bottom" of the economy or in sites of fragile employment rehearse identities splintered with despair, verbal violence, and hostile comparisons of self and "Other." Those more economically secure also speak through these traditional contours of identity but insist that they have detached from the moorings of hostile attitudes and oppositional identities. Even this last group, however, has little social experience from which to invent novel constructions of self, as white, working-class, male, and positively engaged with others.

From men like Frank we hear the most stretch, the greatest desire to connect across borders. But even these men feel the pull of tradition, community, historic, and contemporary fears. They are simply one job away from the narrations of their more desperate and hostile or perhaps more honest white brothers. With few noticing that the economy has produced perverse relations of scarcity, along lines of race, class, and gender, these white men are the mouths that uphold, as if truth, the rhetoric of the ruling class. Elite white men have exploited these men's fears and provided them with the language of hate and the ideology of the "Other." To this end, many of the working-class men in this sample believe that there are still good jobs available for those who work hard, only "minorities" are blocking any chance for access to such employment. Refusing analyses of collapsing urban economies and related race relations, these young adult white men hold Black and Latino men accountable for their white misery and disappointments.

CONCLUSION

The U.S. economy is rapidly changing, moving from an industrial to a postindustrial society. Jobs that once served to secure the lives and identities of many working-class people are swiftly becoming a thing of the past. The corrosion of white working-class male felt privilege—as experienced by the boys and men in this analysis—has also been paralleled by the dissipation of labor unions, which are being washed away as quickly as industry. Even though capital has traditionally used fundamental cleavages such as racism and sexism as tools to fracture a working-class consciousness from forming, labor unions have typically played a strong role in U.S. history in creating a space for some workers to organize against capital for change (Roediger 1994). Historical ties to white working-class union activity is fading fast, particularly among young white working-class men, whose fathers, uncles, and older brothers no longer have a union tradition to pass on to the next generation. With the erosion of union culture and no formal space left to develop and refine meaningful critique, some white working-class men, instead, scramble to reassert their assumed place of privilege on a race-gender hierarchy in an economy that has ironically devalued all workers. Unorganized and angry, our data indicate that white working-class boys and men consistently displace their rage toward historically and locally available groups.

We have offered two scenes in which white men in various stages of adulthood, poor and working-class, are constructing identities on the backs of people of color and white women. Clearly this is not only the case for white working-class men, nor is it generalizable to *all* white working-class men, but these men are among the best narrators of virulent oppositional hostility. It is important that the boys and young adult men in both studies in different geographic locations exhibit similar sentiments. These white men are a race/class/gender group that has been dramatically squeezed relative to their prior positions. Meanwhile, the fantasies and stereotypes of "Others" continue to be promoted, and these delicate, oppositional identities constantly require "steroids" of denigration to be maintained. As the Freeway data suggest, white working-class men also virulently construct notions of identity around another historically available "Other"—gays. Many studies, such as that by Messner and Sabo (1994), evidence how homophobia is used as a profound foil around which to forge aggressive forms of heterosexuality. Heteromasculinity, for the working class in the United States, may indeed be endangered.

As these white boys and men comment on their sense of mistreatment, we reflect, ironically, on their stone-faced fragility. The 1980s and 1990s have marked a time when the women they associate with got independent, their jobs got scarce, their unions got weak, and their privileged access to public institutions was compromised by the success of equal rights and affirmative action. Traditional bases of white male material power—head of the family, productive worker, and exclusive access to "good" public sector and/or unionized jobs—eroded rapidly. Sold out by elites, they are in panic and despair. Their reassertions of status reveal a profound fragility masked by the protection of "their women," their fight for "fairness" in the workplace, and their demand for "diversity" among (but not within) educational institutions. As they narrate a precarious white heteromasculinity, perhaps they speak for a narrow slice of men sitting at the white working-class nexus. More likely, they speak for a gendered and raced group whose privilege has been rattled and whose wrath is boiling over. Their focus, almost fetishistically, is on themselves as victims and "Others" as perpetrators. Research conducted by Janoff-Bulman (1979) documents that an exclusive focus on individual "perpetrators" of injustice [real or imagined] is the *least* likely strategy for transforming inequitable social conditions and the *most* likely strategy for creating poor mental health outcomes. Comforted by Howard Stern and Rush Limbaugh, these men are on a treacherous course for self, "Others," and the possibilities for broad-based social change.

The responsibility of educators, researchers, and citizens committed to democratic practice is not simply to watch passively or interrupt responsively when these boys/men get "out of hand." We must embark on serious social change efforts aimed at both understanding and transforming what we uncover here. Spaces must be located in which men/boys are working together to affirm white masculinity that does not rest on the construction of the viral "other." Such spaces must be imagined and uncovered, given the attention that they deserve. Schools, churches, and work sites all offer enormous potential for such transformative cultural activity. We need to make it our task to locate spaces in which white men and boys are reimagining what it means to be white and male in the 1990s. Activists and researchers can profitably work with such groups to chronicle new images of white masculinity that are not based on the aggressive "othering" that we find to be so prevalent.

NOTES

1. We must point out that although we focus on only the white boys' construction of Blacks, we do not mean to imply that they authored the race script in its entirety nor that they wrote the meaning of Black for the African American students. We are, for present purposes, simply focusing on the ways in which young white men discursively construct the "Other."
2. We include men of different ages and statuses to represent an array of voices that are white, male, and working-class.

REFERENCES

Addelston, J., and M. Stirratt. Forthcoming. The last bastion of masculinity: Gender politics and the construction of hegemonic masculinity at the Citadel. In *Masculinities and organizations,* edited by C. Chang. Thousand Oaks, CA: Sage.

Bluestone, B., and B. Harrison. 1982. *The deindustrialization of America.* New York: Basic Books.

Connell, R. 1995. *Masculinities.* Berkeley: University of California Press.

Davis, A. 1990. *Women, culture, and politics.* New York: Vintage.

Frankenberg, R. 1993. *The social construction of whiteness: White women, race matters.* Minneapolis: University of Minnesota Press.

Janoff-Bulman, R. 1979. Characterological versus behavioral self-blame: Inquiries into depression and rape. *Journal of Social Psychology* 37:1798–1809.

Messner, M., and D. Sabo. 1994. *Sex, violence, and power in sports: Rethinking masculinity.* Freedom, CA: Crossing.

Newman, K. 1993. *Declining fortunes: The withering of the American dream.* New York: Basic Books.

Roediger, D. 1994. *The wages of whiteness: Race and the making of the American working class.* New York: Verso.

Roman, L. 1993. White is a color!: White defensiveness, postmodernism, and anti-racist pedagogy. In *Race, identity and representation in education.* New York: Routledge.

Sanday, P. R. 1990. *Fraternity gang rape: Sex, brotherhood, and privilege on campus.* New York: New York University Press.

Spivak, C. 1985. The Rani of Sirmur. In *Europe and its others,* edited by F. Barker. Colchester, United Kingdom: University of Essex Press.

Weis, L. 1990. *Working class without work: High school students in a de-industrializing economy.* New York: Routledge.

Weis, L., A. Proweller, and C. Centrie. 1996. Re-examining a moment in history: Loss of privilege inside white, working class masculinity in the 1990s. In *Off white,* edited by M. Fine, L. Powell, L. Weis, and M. Wong. New York: Routledge.

Willis, Paul. 1977. *Learning to labor: How working-class kids get working-class jobs.* New York: Columbia University Press.

Wong, L. M. 1994. Di(s)-secting and di(s)-closing whiteness. Two tales from psychology. *Feminism and Psychology* 4:133–53.

41

The Feminist Man's Manifesto

JON MORGAN

An ad recently ran in Ohio Wesleyan's *Daily Bulletin* advertising self-defense classes with the sensationalistic tag, "Did you ever want to kick the crap out of a man?" This jibe was less defensive than offensive, especially to a feminist man opposed to all violence.

Consider how you'd react to a similar ad asking, "Did you ever want to kick the crap out of a woman?"

Neither of these is better than the other. Both are sexist. Both are wrong. I don't question the need for self-defense classes; I applaud them. But the way in which they were advertised recently was extremely irresponsible.

I agree that women remain disadvantaged by a sexist status quo, but I think my feminist credentials are solid. I am a women's studies minor, I spoke at Take Back the Night in 1997, I wrote and collected 75 signatures on the original letter to *The Transcript* on the Hooters issue, and I have been active in NARAL and the Women's Task Force. I absolutely do not tolerate sexist language,

Reprinted by permission of the author.

remarks, jokes, or e-mail, and a Women's Week organizer remarked in March 1998 that I attended more Women's Week events than she did.

But too often people don't realize that sexism works both ways. If you don't accept jokes or comments degrading women, how can you accept those degrading men? We will never end sexism and create fair, new paradigms if we can't stop insulting each other. Let's cut the putdowns on both sides and sit down to real discussions about gender issues—in mixed company.

While I'm discussing anti-male sexism, I think it's past time to air some other gripes held by feminist men. We recognize that we have long held unfairly advantaged positions in society, and try as much as possible to do only what is fair and equal. We reject the old definitions of masculinity and femininity, and behave in more realistic, nonsexist ways than our traditional brethren. We don't, for example, feel any need to act aggressively or hide our emotions to prove that we are men. Likewise, we don't think women should be passive, or paid less than men, or be barred from anything men do, or suffer any disadvantage because of their gender. In short, feminist men are those "nice guys" you often hear of.

We pay a price for our feminist behavior to more "masculine" men. They often call us fags, sissies, or wimps because we dare to be sensitive, advocate that women have more power and men relinquish their unearned advantages, don't compete physically, and sometimes (God forbid) cry. Many "nice guys" I know are intimidated by such men and feel more comfortable in the company of women.

And when we pay daily costs to other men for being who we are, it especially frustrates us to pay further costs to women; we get hit from both sides. Women say we nice guys are the ones they want to date, but all too often we are told, "You're too mature—wait till you're 30." What are we supposed to do until then? We also hear that age-old sexist dichotomy between fun/sex/dating and marriage. "Oh, I can't date him—he's the kind of guy I'd marry." (Women have told me that repeatedly about nice guys.) Or, "Sure, I'd sleep with her, but I wouldn't marry her." (Conversely, a roommate once scoffed at the idea that he'd be interested in sex with his girlfriend, because she was the type to marry.)

This "logic" is nonsense. Why should there be any difference between the kind of person you want to date and the kind you want to marry? Why can't you date a nice guy or marry someone you'd like to have sex with?

In advancing from the sexist status quo to a paradigm based on equality, men are usually the ones who must give something up, and women the ones who gain. In the name of equality, men should surrender unfair advantages, and women should reap the benefits of feminism. But in those cases where equality requires women to shoulder a new responsibility, they should do so. It's not fair for women to enjoy the benefits of sexual equality and sexism simultaneously. It's not fair for us to bear the burdens of the sexist men we aren't as well as the nonsexist men we are.

One such example is relationships. Men should not always have to ask women out on dates. Men should not have to pay for everything, and they should not always have to initiate physical intimacy, including sex. These things should be done equally often by men and women. Besides putting men (mostly the shy ones who are often nicest) at a disadvantage, these sexist social conventions hurt women. Doesn't this comfortable position leave them passive and weak? How surprised should we be that when all men are forced to be initiators, some become aggressors? What is implied if men always pay for dates? Giarrusso et al. (1979) reported that 12 percent of high school girls and 39 percent of high school boys found it acceptable for a boy to force sex on a girl if he spent a lot of money on her. Muehlenhard and Linton (1987) linked date rape with men initiating and paying for dates. They labeled men asking and paying for dates as risk factors in rape, and recommended that rape prevention programs encourage women to take a more active role in dating.

Kilianski and Rudman (1998) asked if women "want it both ways," avoiding "hostile sexism" but accepting "benevolent sexism." They concluded that many women do, underestimating the link between hostile and benevolent sexism, and thus the risks of rewarding the latter. A female friend exemplifies this. She opines that she has asked out all her boyfriends, but admits women are often wedded to remnants of sexism: "I'm not opening the door myself!" And my own April 1999 survey of Ohio Wesleyan students found that the clear majority of those who ask for and pay for dates are still men.

I've never heard a good reason why women shouldn't share these responsibilities equally with men, because there are none. I've heard that asking men out would be difficult and awkward, risking rejection. What do you think it's like to be a man? I've heard that women don't always know what a particular man wants; men have the same problem. None of us can afford to expect all men or all women to be alike. I'm sick of biological arguments because I don't understand why having a penis better suits someone to ask for a date than having a vagina. I've been told that men should pay for everything or always lead physically because "that's the man's role" or "that's what men are supposed to do"—a male friend at UC Berkeley crows, "Let men be men!" The same ludicrous, sexist argument was used to keep women out of the workforce. It still doesn't work.

The credibility of the feminist movement and its adherents is staked on points where sexism disadvantages men. Unless women are committed to eliminating these injustices as much as those to which they fall prey, we can only conclude that they are false feminists, in it for themselves, not for sexual equality. Having worked as I have against anti-female sexism, that thought makes me want to kick the crap out of someone.

REFERENCES

Giarrusso, R., P. Johnson, J. Goodchilds, and G. Zellman. "Adolescents' Cues and Signals: Sex and Assault." In P. Johnson (Chair), "Acquaintance Rape and Adolescent Sexuality." Symposium conducted at the meeting of the Western Psychological Association, San Diego, CA, April 1979.

Kilianski, S., and L. Rudman. "Wanting It Both Ways: Do Women Approve of Benevolent Sexism?" *Sex Roles* 5/6 (1998): 333–352.

Muehlenhard, C., and M. Linton. "Date Rape and Sexual Aggression in Dating Situations: Incidence and Risk Factors." *Journal of Counseling Psychology* 2 (1987): 186–196.